Lise Meitner

A CENTENNIAL BOOK

One hundred books
published between 1990 and 1995
bear this special imprint of
the University of California Press.
We have chosen each Centennial Book
as an example of the Press's finest
publishing and bookmaking traditions
as we celebrate the beginning of
our second century.

UNIVERSITY OF CALIFORNIA PRESS

Founded in 1893

Lise Meitner

A Life in Physics

Ruth Lewin Sime

UNIVERSITY OF CALIFORNIA PRESS
Berkeley   Los Angeles   London

The publisher gratefully acknowledges the contribution provided by the General
Endowment Fund of the Associates of the University of California Press.

California Studies in the History of Science, Volume 13
J. L. Heilbron, editor

University of California Press
Berkeley and Los Angeles, California
University of California Press
London, England

Library of Congress Cataloging-in-Publication Data
Sime, Ruth Lewin, 1939–
    Lise Meitner : a life in physics / Ruth Lewin Sime.
        p. cm.
    Includes bibliographical references and index.
    ISBN 0-520-08906-5 (cloth: alk. paper)
    ISBN 0-520-20860-9 (pbk.: alk. paper)
    1. Meitner, Lise, 1878–1968.   2. Women physicists—Austria—Biography.   I. Title.
QC774.M4S56   1996
530'.092—dc20
[B]   95-35246
                                                                       CIP

Printed in the United States of America
2   3   4   5   6   7   8   9
The paper used in this publication meets the minimum requirements of American
National Standard for Information Sciences—Permanence of Paper for Printed Library
Materials, ANSI Z39.48-1984

CONTENTS

PREFACE AND ACKNOWLEDGMENTS

It seems to me that I have always known of Lise Meitner. As a child I must have seen her picture in *Life*, or in *The New York Times*, or perhaps in the *Aufbau*, the German refugees' newspaper that my parents and grand-mother often read. In America just after World War II, Lise Meitner was a celebrity: the tiny woman who barely escaped the Nazis, the physicist responsible for nuclear fission, "the Jewish mother of the atomic bomb"—although she was a Jew by birth, not affiliation, and she had refused to work on the bomb. When I was six, the details didn't matter. To me, she was a hero, like Eleanor Roosevelt.

I came back to Meitner thirty years later, in the 1970s, by way of a class I taught at California State University, Sacramento. Then, as now, I was on the chemistry faculty at Sacramento City College, a community college. At the university, I was known as the woman the all-male chemistry department did not want to hire; under such circumstances one becomes, and remains, a feminist. When the women's studies board asked me to put together a "Women in Science" course, I accepted right away, although at that moment I could think of only two: Marie Curie (of course) and Lise Meitner. So successful was feminist scholarship, however, that I was sure I would find more women in science and perhaps even begin to answer the question, Why so few?

As it turns out, they were not so few. Throughout history, everywhere, women have been active in science and mathematics and medicine. What these women shared, over the centuries, was the irregularity of their education and a determined undervaluation at the hands of historians. The great exceptions, from Hypatia to Laura Bassi to Sofia Kovalevskaia, were so recognized by their contemporaries that later historians, try as they might, could not make them disappear entirely. Among the less famous but still well known—Caroline Herschel and Marie-Anne Lavoisier, for example—are women who worked with male collaborators, an arrangement that gave them a chance to work but tended to obscure their contributions. There have been other women whose contributions are barely recorded, still others we know only from private correspondence or incidental references, and many more, surely, of whose existence and work we will never know. On the whole it is clear, however, that women have always done science and that the accomplishments of women, like men, have ranged from minor to extraordinary.

Historically, however, women scientists are far less visible than men, a "Matthew effect" in which the already famous attract repeated study and the lesser known are neglected. Although women's studies (like ethnic studies) has brought forth great treasures from neglected history, many historians, even today, are reluctant to bring women in from the cold. (Recently, when one historian of science decided to omit women physicists, including Lise Meitner, from his edited volume, he explained that women and gender questions have always fallen below a certain "historiographic threshold"—a tradition he was evidently content to perpetuate.) The result is a persistent double exclusion: of women from history, of their work from the scientific record. The neglect has been anything but benign. Over the centuries, the apparent paucity of women in science has been used to deny women equitable access to education and the professions. And although many more women are scientists today, sociologists note that as a group they are still, to a significant extent, at the margins.

Lise Meitner almost broke that pattern. She was born in 1878 (eleven years after Marie Curie), her timing just about right to begin cracking open the doors that were still closed to women. Her schooling in Vienna ended when she was fourteen, but a few years later, the university admitted women, and she studied physics under the charismatic Ludwig Boltzmann. As a young woman she went to Berlin without the slightest prospects for a future in physics, but again she was fortunate, finding a mentor and friend in Max Planck and a collaborator in Otto Hahn, a chemist just her age. Together Meitner and Hahn made names for themselves in radioactivity, and then in the 1920s Meitner went on, independent of Hahn, into nuclear physics, an emerging field in which she was a pioneer. In the Berlin physics community she was, as Einstein liked to say, "our Marie Curie"; among physicists everywhere, she was regarded as one of the great experimentalists of her day. Her career was a string of firsts for the inclusion of women in science and academia. The painfully shy young woman had become an assertive professor—"short, dark, and bossy," her nephew would tease—and although at times she was haunted by the insecurity of her youth, she never doubted that physics was worth it. She never married, or even, as far as one can tell, had a serious love affair, but her capacity for friendship was very great. To the end, she was grateful to physics for bringing joy and meaning to her life and for surrounding her with friends and colleagues who were "great and lovable personalities." In the end, these were the only things she did not lose. Everything else—work, position, even, to a great degree, her scientific reputation—was taken from her when she fled Germany in the summer of 1938. Had she stayed longer, she would have lost her life as well.

When I began this study, less than ten years after her death in 1968, Lise Meitner was curiously in and out of view—odd, I thought, for someone who had been so well known. In the autobiographies of Otto Hahn, for thirty years her best friend and closest colleague, there was almost nothing of her personality and very little of her science; in the general literature, her pioneering work in nuclear physics was hardly mentioned. When her name appeared at all, it was for the discovery of nuclear fission, but then only at the margins. Like many of the women scientists I had recently studied, Lise Meitner seemed about to vanish.

Fission seemed to hold clues to her near-disappearance. Seen from Meitner's perspective, the story was fragmented, like torn-up snapshots thrown together. Here are Meitner, Hahn, and the chemist Fritz Strassmann, working as a team in Berlin from 1934 to 1938; there, in December 1938, is the discovery of nuclear fission, published under the names Hahn and Strassmann only; here we have Meitner again, in Sweden with her nephew, Otto Robert Frisch, providing the first theoretical interpretation for the fission process; there, finally, is the Nobel Prize, to Hahn alone.

For the rest of his life, Hahn provided a standard explanation: fission was a discovery that relied on chemistry only and took place after Meitner left Berlin; she and physics had nothing to do with it, except to prevent it from happening sooner. Hahn was believed: he was a Nobel laureate, and a very famous man. Strassmann, very much in his shadow, saw it differently. Lise Meitner had been the intellectual leader of their team, he insisted, and she remained one of them, through her correspondence with Hahn, even after she left. Meitner herself said little, other than to point to the essential interdependence of physics and chemistry throughout the long investigation. Privately, she described Hahn's behavior as "simply suppressing the past." And, she added, "I am part of that suppressed past."

The distortion of reality and the suppression of memory are recurrent themes in any study of Nazi Germany and its aftermath. By any normal standard of scientific attribution, there would have been no doubt about Meitner's role in the discovery of fission. For it is clear from the published record and from private correspondence that this was a discovery to which Meitner contributed from beginning to end—an inherently interdisciplinary discovery that would, without question, have been recognized as such, were it not for the artifact of Meitner's forced emigration. But nothing about this discovery was untouched by the politics of Germany in 1938. The same racial policies that drove Meitner out of Germany made it impossible for her to be part of Hahn and Strassmann's publication, and dangerous for Hahn to acknowledge their continuing ties. A few weeks after the discovery was made, Hahn claimed it for chemistry alone; before long, he suppressed and denied not only his hidden collaboration with a "non-Aryan" in exile but the value of nearly everything she had done before as well. It was self-deception, brought on by fear. Hahn's dishonesty distorted the record of this discovery and almost cost Lise Meitner her place in its history. The unrecognized dishonesty, its careless acceptance, and deliberate perpetuation are among the most disturbing issues I address in this biography.

Given what is known about the systemic repression and "forgetting" of postwar Germany, it is, perhaps, not surprising that Hahn did not look back; he saw no need, and it was not to his advantage to correct the record with respect to Lise Meitner. A chorus of followers eagerly echoed his view, and a generation of journalists, writers, and casual historians of science uncritically propagated it. They may have been dazzled by Hahn's solo Nobel Prize (here, too, is an interesting issue), or motivated by nationalism. They also, apparently, found it entirely natural to suppose that a woman scientist would only be incompetent, or subordinate, or wrong. Or invisible: for thirty-five years Germany's leading science museum displayed the fission apparatus—the physical instruments Meitner used in her laboratory in Berlin—without mentioning her name at all. Were it not for her earlier achievements and her scientific reputation outside Germany, she might well have slipped permanently below the historiographic threshold.

Lise Meitner lived to be ninety years old; she knew what was taking place. Except for a few brief statements, she did not campaign on her own behalf; she did not write an autobiography, nor did she authorize a biography during her lifetime. Only seldom did she speak of her struggle for education and acceptance, although the insecurity and isolation of her formative years affected her deeply later on. And she almost never spoke of her forced emigration, shattered career, or broken friendships. She would have preferred that the essentials of her life be gleaned from her scientific publications, but she knew that in her case that would not suffice. Scientist that she was, she preserved her data. Her rich collection of personal papers, in addition to archival material from other sources, provides the basis for a detailed understanding of her work, her life, and the exceptionally difficult period in which she lived. In expressing my gratitude to those who have made this biography possible, I begin with Lise Meitner.

I have not worked in isolation. It is gratifying to note that in Germany, especially, a new generation of writers and historians of science has taken an interest in Lise Meitner; among the earliest were Fritz Krafft, Charlotte Kerner, Helga Königsdorf, and Renate Feyl. It may be, however, that Lise Meitner's greatest visibility lies just ahead. Recently, the Society for Heavy Ion Research (GSI) in Darmstadt has proposed that element 109, one of the heaviest yet, be named for her; in 1994, IUPAC, the commission that decides such things, approved the name (see Appendix fig. 4). It is not the first such proposal. In 1918, when Meitner and Hahn discovered element 91, her friend Stefan Meyer jokingly suggested the name "lisonium," or possibly "lisottonium"; the discoverers chose protactinium instead. It may take a while before a new periodic table makes its way into my classroom, but when it does, "meitnerium," Mt, will be there. I'm looking forward to it.

Many people have helped me. My first contact was Hahn's former secretary, the late Marie-Luise Rehder, Göttingen, who generously shared with me her knowledge of Otto Hahn and his papers and helped me contact others who were associated with Lise Meitner. I am indebted to the late Otto Robert Frisch for several interviews and for access to the Meitner Collection in the Churchill College Archives Centre, Cambridge, and to Ulla Frisch for interviews, family photographs, and continued access to the Meitner Collection. I am also deeply grateful to the late Fritz Strassmann and to Irmgard Strassmann, Mainz, for documents and photographs. Lilli Eppstein, Danderyd, Sweden, has made Meitner's Swedish experience accessible to me with her astute recollections of Meitner's personality and her friends in Sweden; she has generously permitted me to quote from her private correspondence with Meitner. I also wish to thank Sigvard Eklund, Vienna, for sharing with me his memories of his work and friendship with Meitner in Sweden. I am grateful to Hilde Levi, Copenhagen, for her assistance at the Niels Bohr Archive and for her memories of Lise Meitner's Copenhagen circle. Theodore Von Laue, Worcester, Massachusetts, shared childhood memories, photographs, and copies of his father's automobile guest book. Auguste Dick, Vienna, was an invaluable resource for documents and information about Meitner's education and family in Vienna. Franco Rasetti told me about his stay in Meitner's laboratory in the 1930s, and Emilio Segrè provided insight into the early neutron irradiation experiments in Rome and the discovery of technetium. Stephen Weininger interviewed Tikvah Alper, Sarisbury Green, England, for her memories of her student years with Meitner; Leslie G. Cook, Summit, New Jersey, recalled his experiences in Hahn's laboratory around the time of the fission discovery. I am grateful to Hans P. Coster, Belleaire, Texas, and Ada Klokke-Coster, Epse, Netherlands, for letters, photographs, and memories of their parents, Dirk and Miep Coster, and I owe special thanks to L. K. ter Veld, Groningen, for his interest in this project, for photographs, and for copies of Dirk Coster's correspondence from the University of Groningen.

The Office for History of Science and Technology, Berkeley, has been an invaluable resource; I wish to thank Bruce Wheaton for introducing me to the facility and John Heilbron for his ongoing advice and help. Ulla McDaniel and Eleonore Watrous translated letters for me from Danish, Swedish, and Dutch. Roger Stuewer has been a boundless source of encouragement and information. I am indebted to him and to David Cassidy and Susan Quinn for their extremely helpful reviews of the manuscript; they improved it greatly. Elizabeth Knoll, my editor at the University of California Press, was the driving force in getting me to finish the book.

I have been generously assisted by the following institutions and individuals: Marion Stewart and Alan Kucia of the Churchill College Archives Centre, Cambridge; Marion Kazemi of the Archiv zur Geschichte der Max-Planck-Gesellschaft, Berlin; Finn Aaserud, of the Niels Bohr Archive, Copenhagen; Wolfgang Kerber, Zentralbibliothek für Physik, Vienna; Elisabeth Vaupel and the staff of the Deutsches Museum, Munich; the Museum Boerhaave, Leiden; Urban Wråkberg, of the Royal Academy of Sciences, Stockholm; Spencer Weart of the American Institute of Physics; the Joseph Regenstein Library, University of Chicago; and the Bancroft Library, University of California, Berkeley.

Along the way I have been helped by many other people as well. I wish to thank Helmuth Albrecht, Mitchell Ash, Lawrence Badash, Dan Bar-On, Ingmar Bergström, Alan Beyerchen, Peter Brix, Anna Borelius-Brodd, Joan Bromberg, H. B. G. Casimir, Ute Deichmann, T. De Vries-Kruyt, Steven Dickman, Inga Fischer-Hjalmars, Vincent Frank-Steiner, Robert Marc Friedman, Stanley Goldberg, Dietrich Hahn, Günter Herrmann, Erwin Hiebert, Darleane C. Hoffman, Dieter Hoffmann, Roald Hoffmann, Walter Höflechner, Teri Hopper, Thomas Kaiserfeld, Bettyann Kevles, Daniel Kevles, Christa Kirsten, Kerstin Klein, Lester R. Kleinberg, Fritz Krafft, Arnold Kramish, Svante Lindqvist, Evelies Mayer, Herbert Mehrtens, Anne Meitner, Barbara Orland, Diane Paul, Sir Rudolf Peierls, Max Perutz, Thomas Powers, Hildegard Pusch, Paul Lawrence Rose, Margaret Rossiter, Kurt Sauerwein, Elvira Scheich, Glenn T. Seaborg, M. D. Sturge, Lieselotte Templeton, Sheila Tobias, Pieter Van Assche, Angela Von Laue, Mark Walker, Sallie Watkins, Burghard Weiss, and Carl Friedrich von Weizsäcker. I apologize to anyone I may have inadvertently omitted.

I have received generous grants from the National Endowment for the Humanities, the National Science Foundation, and the Alfred P. Sloan Foundation, for which I am truly grateful.

During the years I worked on this project, my daughters, Karen and Jennifer, have grown from children to young women. I thank them for their love and patience; I think they understand that this book is, in many ways, for them. And finally my love and gratitude goes to my husband, Rod, without whose interest, understanding, kindness, and help I might not have started this book, and certainly would never have finished it.

CHAPTER ONE

Girlhood in Vienna

And even today I am filled with deep gratitude for the unusual goodness of my parents, and the extraordinarily stimulating intellectual atmosphere in which my sisters and brothers and I grew up.

Lise Meitner was born in Vienna in 1878, the third child of Hedwig and Philipp Meitner. She would live in Vienna twenty-nine years, and then she would leave, not realizing how permanently, to make her professional home in Berlin. Part of her remained sentimentally, irreversibly Viennese. She gave in to it, laughing at herself each time she paid the special fee to maintain her Austrian residency. "Na ja," she would shrug. "Foolishness costs money." And later still, after she fled Germany for Stockholm, after every member of her family was gone from Vienna, after the community from which she came was lost forever, even then she clung to her Austrian past, refusing to take Swedish citizenship until she could have both.[1] Had she stayed longer in Vienna, she might not, perhaps, have remained so strongly bound.

Of Lise's childhood we have few details. Even her date of birth is not entirely certain. In the birth register[2] of Vienna's Jewish community it is listed as 17 November 1878, but on all other documents it is 7 November, the day Lise herself observed. It may be that her parents, already ambivalent about their Jewish affiliations,[3] somehow delayed the record, or perhaps the discrepancy was merely a case of *Schlamperei*, that well-known imprecision that contributed to Vienna's charm. Neither explanation is satisfactory. Lise's name also changed slightly, from its original Elise. In Berlin such things might have caused a flurry of paperwork; in Vienna it made no difference.

Like many of their generation, Lise's parents were recent arrivals in the capital, a move whose sense of future may explain their lack of attention to a detailed family history. The Meitners traced themselves back only a few generations,[4] to the village of Meiethein in Moravia, the fertile region north of Vienna that is now part of the Czech Republic. Toward the end of the eighteenth century, not long before the Rights of Man began drifting toward Austria, Kaiser Josef II initiated a series of reforms designed to consolidate power and secure the loyalty of all his subjects: he made German the official language of government, curtailed the Church, gave peasants some relief from serfdom, and granted Jews their first very limited access to civic employment, military service, and education. The Kaiser's tolerance did not extend to his own environs—fewer than two hundred Jewish families were permitted to live in Vienna—but he cracked the ghetto walls, so that Jews flooded the schools, joined the military, and looked to German language and culture for its promise of emancipation, opportunity, and humanism.[5]

Among Kaiser Josef's administrative reforms was the requirement of a family name. Lise's great-great-grandfather took the name Meietheiner, an indication that the family had lived in the village a long time; the name eventually shortened to Meitheiner, Meithner, Meitner. The family lived modestly;[6] if some achieved special distinction, it was for their character and good deeds. Lise's great-grandfather, it was told, crept through the town after dark every Friday night to lay a loaf of challah, the Sabbath bread, at the door of every poor Jew. He did this as secretly as possible and did not permit anyone to thank him, but everyone knew it was the work of Reb Meitner. "Reb" did not mean "rabbi"—there were none in the Meitner family—but was a traditional title of respect.

Reb Meitner's son Moriz, Lise's grandfather, married Charlotte Kohn Lowy, a widow with two small boys who had inherited an inn, some property, and a guest house in the town of Wsechowitz. Her granddaughters would remember her as beautiful, well dressed, and as cheerful as she was self-disciplined. "The house might burn down," it was said, "and grandmother sings; there is cholera in the village, and still grandmother sings!" Moriz and Charlotte's son, Philipp, was blond and blue-eyed like his mother; like his grandfather, Reb Meitner, he would later be known for his integrity and kindness. In 1873 he married petite, dark-eyed Hedwig Skovran, whose grandfather had emigrated from Russia to Slovakia to escape the ongoing persecution of Jews.

Philipp and Hedwig Meitner grew up with Austria's transition from late feudalism to a recognizably modern society. The liberal revolutions of 1848 were crushed in Austria, but the struggle for individual freedoms and national autonomy went on. Industrialization came to Vienna and with it, a great internal migration from throughout the empire. In 1858, the medieval fortifications ringing the old inner city were torn down; in their place came the imposing Ringstrasse, grand new public buildings, and a parliament with little real power whose Liberal majority pressed for a modern secular state and constitutional government. At a time when the old order was failing and the very notion of empire was threatened by nationalist dissensions, the Habsburg monarchy was humiliated by a string of unwise military ventures and diplomatic blunders. By 1867, Kaiser Franz Josef saved what he could by dividing the empire and letting Hungary go. He granted his people a number of constitutional laws: national and religious toleration, a laissez-faire economy, an impartial judiciary, greater individual freedoms of education, belief, speech, and press. For Jews, this meant full civic equality, including access to professions from which they had previously been barred.[7] Philipp Meitner was among the first group of Jewish men who were free to study law and be admitted to its practice.

In the twenty years from revolution to constitution (so the saying went), Austria had been dragged into the nineteenth century. By the time Lise Meitner was born in 1878, imperial Vienna was mostly theater, set with palaces of impossible opulence and a Kaiser, the popular and long-lived Franz Josef. It hardly mattered any more. The new Vienna was bursting with life of its own, sprawling into the countryside, its population doubling and doubling again with an influx so constant that for generations most Viennese would be born somewhere else: overwhelmingly Catholic with some Jews and virtually no Protestants, mostly German-speaking with large contingents of Czechs, Hungarians, Italians, Poles, Croats, Ukrainians, and others who retained their languages and national identities in newspapers and ethnic associations. To many of the new arrivals, Vienna was a place of marginal work and much unemployment, water shortages, and summer cholera, with congestion so severe that even the wealthy lived in apartments and the very poor shared beds and slept in shifts. The most heterogeneous city in Europe, it was among the most crowded and unsanitary; it had the highest rate of suicide. Still people came: conditions in the provinces were not better. Vienna at least promised improvement and pleasure: music of every sort, opera and theater, newspapers by the dozens, a renowned university, famous physicians and scientists, good food, vineyards at the edge of town, and blue hills shimmering hazily in the distance. If the Danube seemed muddy or the waltz overrated, Vienna was beguiling nonetheless, drawing from every stream of European culture, layered with history and beauty every newcomer could aspire to make his own. The intellectual ferment was very great. By the end of the century, Vienna had given birth to Viktor Adler's democratic socialism and Theodor Herzl's Zionism; it was the home Sigmund Freud loved to hate and the political base for Karl Lueger, the city's longtime mayor, whose heady mix of populism and anti-Semitism drew the rapt attention of the young Adolf Hitler. If the nineteenth century came late to this society, the twentieth arrived early.[8]

When Philipp Meitner entered the legal profession in the early 1870s, it was possible not only to practice law but to have a hand in the creation of a new political order. The new constitution called for sweeping reforms of Austria's entire legal system, and in Vienna, after years of neglect by Crown and Church, the Liberal city council built an ample water supply and provided flood control and improved public health, hospitals, and schools.[9] It was a time when progress seemed the natural order of things, each decade a noticeable improvement on the one before. As an attorney, "freethinker," and humanist, Philipp Meitner was committed to the Liberal ideals of reason and civic progress, sympathetic to the Social Democratic goals of justice and individual improvement. He immersed himself in Vienna's political life. Although he never sought elective office, he and Hedwig made their home a gathering place for interesting people—legislators, writers, chess players, lawyers. The children stayed up and listened. Years later, when Lise was asked about her childhood, she remembered most of all "the unusual goodness of my parents, and the extraordinarily stimulating intellectual atmosphere in which my brothers and sisters and I grew up."[10]

During Lise's childhood the family lived in the second district, known as Leopoldstadt, just north across the Danube canal from the old city. Originally a ghetto, the community was named for Leopold I, who expelled Vienna's Jews in the 1600s, then grudgingly permitted them to return. For the next two centuries, the number of Jews in the capital remained small, but in the 1860s, when residence restrictions were abolished and Jews from throughout the empire converged on Vienna, Leopoldstadt grew.[11] Crowded and run-down in some areas, it was pleasant, even somewhat prosperous, in others.

Lise was born in the family apartment at 27 Kaiser Josefstrasse,[12] a tree-lined avenue that traversed Leopoldstadt from a commercial district at one end to the Prater, Vienna's huge park, at the other. There on a Sunday the family could enjoy amusements and cafés, wooded paths and open fields, and even on occasion glimpse the Kaiser riding by. On the whole Leopoldstadt was a comfortable place to raise a family. The first three Meitner children, Gisela, Auguste (Gusti), and Lise, were born only a year apart, followed not quite so rapidly by five more: Moriz (Fritz), Carola (Lola), another boy, Frida, and finally Walter, the baby brother Lise adored, who was born in 1891.[13] The large family could afford few luxuries, but Philipp Meitner's law practice did provide the middle-class essentials: books, a few summer weeks in the mountains, and—virtually a necessity in Vienna—music lessons.[14] Gusti was the family's most talented musician, a child prodigy who became a composer and pianist of concert rank.[15] Lise played the piano too; all her life music would be a passion for her, as necessary as food. But she was especially curious about mathematics and science, an eight-year-old who kept a math book under her pillow and would ask about the colors of an oil slick and remember what she was told about thin films and the interference effects of reflected light.[16] In this family children were seen and heard—and expected to think for themselves. Once, when Lise was still very young, her grandmother warned her never to sew on the Sabbath, or the heavens would come tumbling down. Lise was doing some embroidery at the time and decided to make a test. Placing her needle on the embroidery, she stuck just the tip of it in and glanced anxiously at the sky, took a stitch, waited again, and then, satisfied that there would be no objections from above, contentedly went on with her work.[17] Along with books, summer hikes, and music, a certain rational skepticism was a constant of Lise's childhood years.

Judaism was not one of those constants. In Leopoldstadt the Meitner children lived among Jews, in a neighborhood dotted with synagogues and shuls, fully aware that they too were of Jewish origin. And yet it is clear that the family distanced itself from its Jewish past. One of Lise's nephews,

Gusti's son, Otto Robert Frisch, would later have the firm impression that his mother and all the Meitner children had been baptized and raised as Protestants.[18] In fact, this was not so: the children were all registered with the Jewish community at birth and accepted baptism only as adults—Lola and Gisela as Catholics in 1908, Lise as Protestant the same year.[19] But Frisch's impression was in essence true: the Meitners did leave the old religion for the new.

Their reasons were never explicitly stated. Opportunism was apparently not one of them: Philipp and Hedwig Meitner never baptized their children or themselves and thus derived none of the advantages conversion would have offered, particularly in the legal profession where discrimination remained strong and conversion was still a passport to judgeships and other civil service positions.[20] One can only assume that the Meitner couple lost interest in Judaism, regarding it as a ghetto relic perhaps, or an undesirable ethnic division; they surely felt little kinship with Leopoldstadt's many *Ostjuden*, Jews from Galicia and other Austrian-held Polish provinces whose language, dress, and orthodoxy set them apart.[21] Enlightened and progressive, Hedwig and Philipp Meitner were drawn to German culture; freshly emancipated, with optimism bordering on faith, they embraced the culture that freed them.[22] By the turn of the century, such optimism must have dimmed somewhat, as the most charismatic Viennese mayor of all time, the handsome Karl Lueger (*der schöne Karl*), rallied his voters by appealing to their Catholicism, nationalism, and anti-Semitism. It is worth noting that none of the Meitner children followed their father into politics, or even law. But their parents' idealism influenced them nonetheless. It was part of the "unusual goodness" Lise remembered, the basis for the extraordinary intellectual atmosphere that nurtured Lise and the other children in their parents' home.

In this atmosphere all the Meitner children, including the five daughters, pursued an advanced education. Even today such a family record would be notable, but at the time it was truly extraordinary, for until the end of the nineteenth century women were by law excluded from Austrian universities and, by the same logic, from rigorous secondary schools as well. While a bright boy might attend a *Gymnasium* and take the *Matura*, a leaving examination that was required before entering the university, public school for girls was over at age fourteen, and it was poor. Lise attended the Mädchen-Bürgerschule at Czerninplatz, a crowded intersection not far from home. On 15 July 1892, she received her final *Jahres-Zeugnis*, a report card that was also an *Entlassungs-Zeugnis*, a completion certificate.[23] She had learned bookkeeping arithmetic but not algebra, a smattering of history, geography, and science, the requisite drawing, singing, and "feminine handwork," a little French and gymnastics. Although her grades were all good and her behavior "entirely appropriate" (*vollkommen entsprechend*), her diligence was rated only "satisfactory" (*befriedigend*) rather than "industrious" (*ausdauernd*), an indication that she did not find school very challenging. Inked at the bottom of her Jahres-Zeugnis was the line: "*vom weiteren Schulbesuch befreit*" (released from further schooling). Lise had gone as far in public school as an Austrian girl could go.

Not yet fourteen, her choices were few. Most girls would spend the next few years helping at home, sewing, and daydreaming of marriage. The only way for a girl to go on was to attend a private *höhere Töchterschule* for young ladies of the middle class; the only profession she could seek was teaching a subject that did not require university education. Lise chose French. Nothing in her contemporary records or later memoirs indicates that she ever had a real interest in it. Instead, she lavished her energy and love on her baby brother, Walter; he would always be her closest sibling. She also tutored younger girls to help pay for Gusti's advanced music lessons and volunteered with the poor in relief organizations and schools.[24]

Of these years Lise would remember little but a sense of loss. "Although I had a very marked bent for mathematics and physics from my early years, I did not begin a life of study immediately," she wrote later.[25] "Thinking back to . . . the time of my youth, one realizes with some astonishment how many problems then existed in the lives of ordinary young girls, which now seem almost unimaginable. Among the most difficult of these problems was the possibility of normal intellectual training."[26]

In Austria the issue of higher education for women had been simmering for a generation, certainly since 1867 when universities were first opened to men without regard to economic class, religion, or national origin. Over the years a small number of women had approached the universities, petitioned professors, begged to attend a class or two. At best they were permitted to sit in as unofficial auditors, not expecting and certainly never receiving any credit or documentation. Most of these women were teachers whose prior education did not qualify them for university admission. But even the few who did qualify—occasionally a young woman from Bohemia or Austrian Poland would somehow manage to attend her local Gymnasium and pass the Matura—were also denied admission. Daughters of the wealthy and the aristocracy were routinely educated in Switzerland. The rest were trapped in a cycle willed by the state: since the universities (all public institutions) excluded women, the government did not see fit to establish schools that would prepare women for university admission. In Europe, only Germany and Turkey offered more resistance to women's education.[27]

Toward the end of the nineteenth century, however, the resistance began to falter. Women's groups, often led by headmistresses of girls' schools, regularly petitioned for improved secondary education; a private *Mädchengymnasium* was established in Vienna in 1891 even though its graduates were not permitted to take the Matura; the government itself, urgently needing female physicians for Moslem women in occupied Bosnia and Herzegovina, recruited foreign women for many years, hired the first Austrian (Swiss-trained) in 1892, but still denied medical certification to other Swiss-trained Austrian women physicians who wished to enter private practice, although, at the same time, a highly competent eye surgeon who was born in Russia and trained in Zurich received special permission from the Kaiser to establish a clinic with her husband in Salzburg. Publicity accompanied each case, and opinion gradually softened. It seemed plausible, finally, to suppose that in Austria—as in America, France, and Switzerland—women could be educated without suffering mental illness or infertility or social catastrophe. By the mid-1890s, even conservative university professors regarded women students as a flood that could no longer be held back. In 1897 the government granted women access to the philosophical faculties (letters and sciences) of Austrian universities; a few years later women were admitted to medical schools as well.[28]

With this, the cycle of exclusion was thrown into reverse. Justice, and the need for university educated women teachers, required that universities admit women at once, even without Gymnasium preparation. For the interim women would be required only to pass the Matura, any way they could. This came as good news—late, but not too late—for Lise and her sisters.

Gisela, already twenty-one, came first. After two years of intensive private lessons, Gisela passed the Matura and entered medical school in

1900. Lise meanwhile completed her teacher training (as insurance, her father advised) and in 1899 began her own lessons in a group with two other young women. Together they compressed eight missing school years into two: Greek and Latin, mathematics and physics, botany, zoology, mineralogy, psychology, logic, religion, German literature, history. Lise studied night and day. "You'll fail," her younger brothers and sisters would tease. "You've just walked across the room without picking up a book."[29] A photograph shows a pale young woman with dark circles under her eyes.

For physics and mathematics, Lise's group was tutored by Arthur Szarvassy, a young physicist who had just completed his doctorate at the University of Vienna.[30]

> Dr. Szarvasy [sic] had a real gift for presenting the subject matter of mathematics and physics in an extraordinarily stimulating manner. Sometimes he was able to show us apparatus in the Vienna University [Physics] Institute, a rarity in private coaching—usually all one was given were figures and diagrams of apparatus. I must confess that I did not always get correct ideas from these, and today it amuses me to think of the astonishment with which I saw certain apparatus for the first time.[31]

Lise took the Matura in July 1901 at the Akademisches Gymnasium, a distinguished boys' school on Beethovenplatz in the old city.[32] The course of study had been so intense and the examination conditions so terrifying—as *Externisten* (outside students), Lise and the other women were examined in strange surroundings by teachers they had never met—that Lise never failed to mention it in her later remembrances. Of fourteen who took the exam, only four passed;[33] three were the students of Dr. Szarvassy. The fourth was Henriette Boltzmann,[34] whose father would soon be a formative influence in Lise's life.

Lise would always think of Arthur Szarvassy as her first true teacher. And she was grateful to her parents, who made it possible for her to achieve what few other young women of her generation could.

> Many parents shared the prejudice of the time against [women's] education, so that their daughters either had to forgo the education they desired, or fight for it. . . . [I knew] a young woman who at age 24 wanted to be privately tutored by her cousin to prepare for the Matura; her parents—in other respects very loving, I'm sure—literally kept her prisoner in their apartment to keep her from carrying out her intentions. Only when she disappeared from the apartment one day and let her parents know that she would not return unless she had permission to study, did they give in.[35]

Although Lise herself had no such obstacles, she sensed that for her mother, at least, it was not always easy.

> I had the feeling that in the beginning, when first my older sister, and then I passed the Matura, that my mother was inwardly somewhat depressed by it. But she was much too loving a mother ever to express it in any way.[36]

From her father there was no such ambivalence. On the contrary, he was a steady source of support and advice.

> Even as a child I was strongly interested in mathematics and physics, and as I grew up I also developed a very pronounced inclination for social responsibility. . . . When I was 23 years old and about to enter the university, I entertained the idea of primarily pursuing medicine, for its social usefulness, and studying mathematics and physics only at the side. My father kept me from this incorrect choice by making it clear to me that such a course of study might be possible for a genius like Hermann Helmholtz, but not for another person.[37]

Lise entered the University of Vienna in October 1901. Small and slender, with a faraway expression and serious dark eyes, she looked younger than her twenty-three years. A bluestocking, her nephew would judge later, a young woman who cared for nothing but study. He was probably right. Anxious to make up for lost time, Lise filled her university registration book with physics, calculus, chemistry, and botany—twenty-five hours a week of lectures, laboratories, demonstration and discussion sections.[38]

> No doubt, like many other young students, I began by attending too many lectures. . . . I cannot say I have a very lively recollection of the lectures on experimental physics. These were delivered almost without experiments, between noon and one P.M., when most of the students were already very tired. Sometimes I was really afraid I would slip off my chair.

But for calculus, at eight o'clock in the morning, she was awake.

> My first term I studied differential and integral calculus with Professor Gegenbauer. In my second term he asked me to detect an error in the work of an Italian mathematician. However I needed his considerable assistance before I found the error, and when he kindly suggested to me that I might like to publish this work on my own, I felt it would be wrong to do so, and so unfortunately annoyed him forever.

Here was Lise, a first-year student, refusing to publish as her famous professor asked. Assertive in one way, self-deprecating in another—neither to her academic advantage. "This incident did make it clear to me, however, that I wanted to become a physicist, not a mathematician."[39]

In fact, the physics course Lise attended, her drowsiness nothwithstanding, had the reputation of being exceptionally well taught. It had been designed for pharmacy students, but Professor Franz Exner brought such clarity and perspective to the subject that students from all disciplines thronged to it.[40] The laboratory was directed by Anton Lampa, a promising young physicist and teacher.[41] Lise may have been drowsy in lecture, but she vividly remembered the laboratory: the somewhat aloof instructor, the primitive equipment, the experiments requiring ice that could be done only when there was snow in the courtyard below.[42] For this young woman who had never had science in school, whose only previous encounter with apparatus was to view it with astonishment, the laboratory was of paramount interest. She would study physics after all.

The physics institute was on the Türkenstrasse, a short side street in Vienna's ninth district, on the same block as the institutes for pharmaceutical chemistry and medicinal chemistry, not far from the renowned medical school and its clinics. The university had no central campus; its buildings were interspersed among the residences and shops of the neighborhood. A photographer's studio and a coffee house stood on either side of Türkenstrasse 3; Sigmund Freud lived and worked on the steep Berggasse nearby. Originally the structure had been a small apartment house, already run-down when the university purchased it as a temporary building in 1875 (a permanent physics building opened in 1913). Its entrance reminded Lise Meitner of the door to a hen house. "I often thought, 'If a fire breaks out here, very few of us will get out alive.'"[43] Inside were worn stairs and shaky floors, makeshift laboratories with untold amounts of mercury in the floor cracks, a lecture room with neither podium nor desks, ceiling beams so rotten they looked as though they had been chewed by termites.[44]

> The lecture halls in particular were downright life-threatening. This was so widely known that the Viennese newspaper *Arbeiterzeitung* once carried this notice: "Once again a student has registered at the Physics Institute on the Türkenstrasse; unhappiness in love is said to be the motive for the deed."[45]

But in that shabby building the quality of teaching and research was very high. Exner, the first professor students encountered, was a multifaceted experimental physicist whose research included electrochemistry, atmospheric electricity, crystal physics, spectroscopy, and optics. A friend of Wilhelm Röntgen, Exner had introduced x-ray research and its medical applications to Vienna; one of the first to take an interest in radioactivity, Exner secured uranium ore residues for Marie and Pierre Curie, received an enriched radium sample in return, and made Vienna an early center for radioactivity research. Although Exner lectured only to first-year students, he directed the advanced physics laboratories and supervised a large number of doctoral candidates. One of Lise's fellow students, Karl Przibam, remembered Exner for his contagious enthusiasm and for the community spirit that went far beyond the usual relationship between teacher and students.[46]

This sense of community was essential for Lise in finding her way. She had come to the university on her own, very conscious of how few women there were and how visible she was, how some of the men went out of their way to be pleasant and others, just as conspicuously, did the opposite. Never having gone to a Gymnasium, she could only imagine that she had missed some vital aspect of normal student life, in academics, perhaps, or student friendships, or relationships with teachers. With Professor Gegenbauer she had apparently been awkward and then embarrassed by her awkwardness; not ready to be singled out, she needed first to be convinced that she could be a student like any other.

For Lise, this happened in the old building on the Türkenstrasse, in the cluttered laboratory, during the informal give-and-take of teachers and students. It helped that the subject was difficult, chosen only by a few. In Vienna, indeed worldwide, the number of physicists was small; nearly all were engaged in teaching and research, very few in business or industry. Physics was more a calling than a career.[47] Students who committed themselves to physics did so because they could not imagine a more fascinating way to spend their lives. By 1902, Lise Meitner knew she was one of them.

In her second university year, she began studying physics in earnest. Over the next six semesters, her *Meldungsbuch* lists analytical mechanics, electricity and magnetism, elasticity and hydrodynamics, acoustics, optics, thermodynamics, and kinetic theory of gases as well as mathematical physics each semester and a course in philosophy of science. A fairly typical curriculum, it was highly unusual in one respect: all of it was taught by just one person, the theoretical physicist Ludwig Boltzmann.

Fifty years later Lise Meitner would still remember Boltzmann's lectures as "the most beautiful and stimulating that I have ever heard. . . . He himself was so enthusiastic about everything he taught us that one left every lecture with the feeling that a completely new and wonderful world had been revealed."[48]

One can scarcely imagine a better teacher for the atomic world that lay ahead. In 1902, Boltzmann was fifty-eight years old, the famed theoretical physicist who had extended kinetic theory and established statistical mechanics, the leading "atomist" who tied the second law of thermodynamics to atomic theory by showing that the inherent irreversibility of natural processes arises from the statistical behavior of atoms in the aggregate. The notion of unseeable atoms with indeterminate behavior was more than some scientists could swallow. For years Boltzmann was forced to defend his work against the fairly widespread philosophy of scientific positivism that denied the value of scientific theory and the reality of anything that could not be directly observed.

A big man, heavy, very nearsighted, with curly brown hair and a full reddish beard that framed his broad face,[49] Boltzmann aroused admiration and affection in his students. He began his mechanics course in 1902 by offering his students "everything I have: myself, my entire way of thinking and feeling," and asking the same of them: "strict attention, iron discipline, tireless strength of mind. But forgive me if I [first] ask you for that which means most to me: for your trust, your affection, your love—in a word, for the most you have the power to give, yourself."[50]

Like many of the others, Lise was swept away. He was immensely engaging, she remembered, this famous professor whose lectures were models of clarity, this warmhearted *Hofrat* (Excellency) who would shrug at his title and laugh, "Ach, how dumb of me!" at his blackboard errors.[51]

> Boltzmann had no inhibitions whatsoever about showing his enthusiasm when he spoke, and this naturally carried his listeners along. He was fond of introducing remarks of an entirely personal character into his lectures. I particularly remember how, in describing the kinetic theory of gases, he told us how much difficulty and opposition he had encountered because he had been convinced of the real existence of atoms and how he had been attacked from the philosophical side without always understanding what the philosophers held against him. . . . I wonder what he would say about our huge machines and teamwork [today], when I remember how bitterly he complained . . . about the great extension of the subject matter of physics and

the resulting overspecialization. He stated categorically that [Hermann] Helmholtz was the last physicist who had been able to have an overall view of the whole subject.[52]

His relationship to students was very personal. . . . He not only saw to their knowledge of physics, but tried to understand their character. Formalities meant nothing to him, and he had no reservations about expressing his feelings. The few students who took part in the advanced seminar were invited to his house from time to time. There he would play for us—he was a very good pianist—and tell us all sorts of personal experiences.[53]

Boltzmann accepted women students as a matter of course. In 1872, long before women were admitted to Austrian universities, he met Henriette von Aigentler, an aspiring teacher of mathematics and physics in Graz. From their four-year correspondence we know of her desire to attend the university ("out of eagerness to learn and to qualify for teaching"), how she was refused permission to unofficially audit lectures (an administrator declared himself "delighted" to keep women out, since "the character of the university would be lost and the institution endangered" by their presence), that Boltzmann advised her to appeal (she did, successfully), and that when he proposed marriage, finally, he began, "It seems to me that a constant love cannot endure if the wife has no understanding, no enthusiasm for the endeavors of the husband, but is merely his housekeeper rather than the companion in his struggles."[54]

Lise may have heard some of this; she came to know his wife and daughters and considered their family life harmonious.[55] In any case, her university years were free of the obstacles she had encountered earlier and the difficulties that lay ahead. With his intellect and spirit, Boltzmann created a community to which she fully belonged. "He was in a way a 'pure soul,' full of goodness of heart, idealism, and reverence for the wonder of the natural order of things."[56]

All who were close to Boltzmann were also aware of his bouts of severe depression and his suicide attempts.[57] His students blamed it on the bitter controversy over whether atoms existed, in which Boltzmann gained many adherents among younger scientists but never the satisfaction of convincing his opponents. It was not that simple. Boltzmann himself jestingly attributed his rapid changes in temperament to the fact that he was born during the night between Shrove Tuesday and Ash Wednesday: he was, almost certainly, manic-depressive.[58] But he was also very sensitive. As

Meitner reflected, "[He] may have been wounded by many things a more robust person would have hardly noticed. . . . I believe he was such a powerful teacher just because of his uncommon humanity."[59]

Boltzmann's academic career was a series of wanderings. Born in Vienna in 1844, he graduated from the Akademisches Gymnasium, where Lise Meitner and also his daughter, Henriette, later took their Matura. At the University of Vienna, he was a student of Josef Loschmidt (1821–1895), who made reliable early estimates of molecular size and the number of molecules per mole,[60] and he was assistant to Josef Stefan (1835–1893), who devised an empirical formula for black body radiation that Boltzmann subsequently gave a theoretical basis.[61] Between 1869 and 1890, Boltzmann held appointments in Graz, then Vienna, then Graz again, a period during which he contributed to all branches of theoretical physics: electromagnetic theory, kinetic theory, the Maxwell-Boltzmann distribution, statistical mechanics. He went to Munich in 1890, returned to Vienna as Stefan's successor in 1894, left for Leipzig in 1900, and came back again in 1902. The university had kept his position open in the expectation that he would return.[62]

Boltzmann was torn between his attachment to Austria, especially Vienna, and the attractions of German universities. Meitner recalled that Boltzmann would tell how in Munich there was "wonderful equipment, but far fewer good ideas" than in Vienna and then hastily add, "One must not let the Austrian [education] ministry know that good work can sometimes be done with inferior equipment."[63] Of all universities, he most preferred Berlin, for its intense scientific atmosphere and the presence of Hermann Helmholtz, whom Boltzmann regarded as the greatest physicist of the nineteenth century. Yet in 1888 he refused the offer of a chair in Berlin, in part out of concern for his health,[64] in part, it was said, because he disliked the formality of the Prussian capital.[65] Later he would tell his students how much he regretted that decision. The position he refused went to a younger theoretical physicist, Max Planck.

A year after Boltzmann came to Vienna in 1894, he was joined on the faculty by one of his principal scientific adversaries, the formidable Ernst Mach. The leading proponent of the philosophy of scientific positivism, Mach argued that science can do no more than conduct positive—that is, direct—observations: while scientific theory may be of use for organizing such data, it must not create "pictures," as he called it, of underlying reality. Mach's impetus was antimetaphysical, a reaction against nine-teenth-century attempts to reduce all of physics to mechanical principles; he opposed the kinetic theory of gases, based as it was on molecular motion, and dismissed the existence of atoms in broad Viennese, "'Ave y'seen one of 'em?"[66] In the 1890s, according to Boltzmann, the attitude toward the gas theory was "malevolent,"[67] complete with angry debates at meetings, struggles for the allegiance of young scientists, fights over appointments to faculties and journals.[68] In central Europe especially, Mach attracted a sizable following, including "energeticists" led by the physical chemist Wilhelm Ostwald, for whom energy was the primary reality and the second law of thermodynamics superfluous. For his part, Boltzmann attacked positivism as a modern version of an "old aberration," going back to the philosophy of George Berkeley. In 1905 he visited a university in Cali-fornia whose campus he described as "the loveliest place one can imagine," except for its "philosophical aura":[69] "The name Berkeley is that of a highly reputed English [sic] philosopher who is famous for the greatest foolishness ever hatched by the human brain, philosophical idealism, which denies the existence of the material world."[70]

In 1898 Mach suffered a stroke, and he retired from teaching in 1901. When Boltzmann returned to Vienna in 1902, he claimed the philosophy of science course that had been Mach's for many years. Boltzmann's inaugural philosophy lecture in 1903 was thronged by the press, students, including Lise Meitner, and six hundred "sensation-seekers." With his predecessor in mind, Boltzmann confessed to his "dislike, even hate of philosophy," comparing it to "a hallowed virgin . . . [that] will remain eternally barren"[71] as long as it denies the existence of physical reality. Thus the hostilities between the atomic theorists and the followers of Mach went on.

But the dispute over the reality of atoms was nearing an end. The discovery of radioactivity in 1896 and the electron in 1897 transformed atoms from disputed specks of mass to complex structures that were divisible, measurable, packed with amazing amounts of internal energy, and composed of fundamental particles of electric charge. "No physicist today believes atoms are indivisible,"[72] Boltzmann told an audience at the World's Fair in St. Louis in 1904. That was probably true for those who believed in atoms, but not all physicists did, yet. The final blow came after Albert Einstein in 1905 and Jean Perrin in 1908 made detailed studies of

Brownian motion, the random movement of particles suspended in a liquid, visible under the microscope. By relating the movement of the suspended particles to the number and energy of the molecules in the liquid that were hitting them from one side or another, Einstein and Perrin obtained a value for the number of molecules in a mole that was consistent with other, unrelated experiments. The direct relationship between the heat energy of atoms and the mechanical energy of visible Brownian particles gave complete credence to Boltzmann's interpretation of thermodynamic laws. And somehow it made atoms nearly visible and very real. Ostwald was convinced and in 1908 conceded; it is uncertain if Mach ever did before he died in 1916.[73]

The controversy made plain to students that scientific endeavor is not coldly objective but relies on human judgment. From Boltzmann, Lise Meitner understood physics to be a passionate commitment of intellect, strength, and integrity. Many years later her nephew Otto Robert Frisch wrote, "Boltzmann gave her the vision of physics as a battle for ultimate truth, a vision she never lost."[74]

Meitner's goal in physics would be theoretical understanding; her means, nearly always, would be experiment. In the summer of 1905, her coursework completed, she began her doctoral research. In Austrian and German universities the thesis research for a doctorate generally took no more than a few months to complete. She chose an experimental project, under Franz Exner and his assistant, Hans Benndorf, undoubtedly because she wanted the laboratory experience but also, perhaps, because Boltzmann was lecturing in California that summer and had been quite ill before he left.[75]

In her research, Lise determined that Maxwell's formula for the conduction of electricity in an inhomogeneous solid also applies to the conduction of heat. Her inhomogeneous solid, an emulsion—a finely divided mixture—of mercury droplets embedded in fat, was layered between two horizontal copper plates, on which was laid a third copper plate that was insulated from the bottom two. The temperature of the bottom plate was kept constant by a stream of running water; when the top plate was heated by steam, three strategically placed thermometers in the emulsion measured the temperature gradient as the heat flowed through. Exner was pleased, commending her for a "not entirely easy" investigation that was brought to completion "not without experimental skill." Her dissertation,

"Test of a Formula of Maxwell's," was published under the title "Conduction of Heat in Inhomogeneous Solids" in the proceedings of the Vienna Physics Institute.[76] Experiment close to theory, it typified her later approach to physics. Personally, however, Exner's influence seems to have been slight: in her later memoirs she mentions him only in passing, never with the affectionate term *Doktorvater* that German-speaking students often use. Most probably Boltzmann's personality eclipsed all others. Both men took part in her *Rigorosen*, the oral examinations that she took in December 1905 and passed summa cum laude. She was awarded her doctoral degree on 1 February 1906.[77]

It was the middle of the academic year and Lise found herself somewhat at loose ends. At the time Paul Ehrenfest, a theoretical physicist her own age who had taken his doctorate under Boltzmann a few years before, was in Vienna. When he heard that Lise had taken detailed notes of all Boltzmann's lectures, he suggested they study his ideas as well as the work of others in analytical dynamics.[78] Ehrenfest had a gift for explaining theoretical physics; he called Lise's attention to Lord Rayleigh's scientific papers, in particular an article on optics that described an experiment the British physicist could not explain. Meitner not only explained it but also predicted some consequences, proved them experimentally, and described them in her report, "Some Conclusions Derived from the Fresnel Reflection Formula."[79] More than her thesis project, this investigation convinced her that she was capable of independent scientific work.[80]

While engaged in the optics study, she also decided to learn something of the experimental procedures used in the new field of radioactivity. She had taken an advanced seminar on the subject from Egon von Schweidler the year before; now she became acquainted with Stefan Meyer, an assistant in Boltzmann's institute who was only six years her senior and already quite well known in the field. At Meyer's suggestion, Meitner measured the absorption of alpha and beta radiation in foils of various metals. By June she completed the study,[81] having been introduced to several radioactive substances, the literature of radioactivity, and a new instrument, the leaf electroscope.

It was the summer of 1906, a time to assess her future. For the young woman who had just become Dr. Lise Meitner, the future was not much clearer than it had been at age fourteen. As the second woman to earn a doctorate in physics from the university, she knew of no prospects for a woman in physics;[82] it seemed entirely possible she might never work as a scientist. In Austria there had as yet been no female *Assistent*, the first position on the academic ladder; there were no women's colleges like those in America with positions for a few women scientists,[83] no great likelihood of a job in industry. Of course, Lise had heard of Marie Curie, who had won the 1903 Nobel Prize in physics with her husband, Pierre, and Henri Becquerel; if she also knew of Curie's enormous professional difficulties in Paris and how little the prize had alleviated them, she would not have been encouraged. At one point she wrote to Marie Curie about working in her laboratory, but there was no position available.[84] There seemed no choice but to follow her father's advice once again and obtain the credentials necessary to earn a living. She signed up for practice teaching at a girls' school.[85] Teaching did not appeal to her. Great, perhaps insurmountable, obstacles stood in the way for a woman in science. There was no path to follow.

Late that summer, on 5 September 1906, the physics community was shocked by the news that Ludwig Boltzmann had taken his life. In a tribute to his friend and scientific opponent, Wilhelm Ostwald described Boltzmann as a victim of the immense sacrifices of health and strength demanded of those who struggle for scientific truth.[86] Lise Meitner, more realistically, ascribed his suicide to "mental instability"; she never understood it.[87] But it seems likely that Boltzmann's death strengthened her determination to remain in physics, so that the spark he had kindled in her would remain alive.

In the fall of 1906, therefore, she continued working with Stefan Meyer, who temporarily took over Boltzmann's institute. By day she taught school; in the evening she returned to Türkenstrasse 3. During that year she became familiar with radioactivity research, although she had no particular intention of specializing in it.[88]

One of the earliest workers in the field, Meyer had been investigating the magnetic properties of various elements when polonium, radium, and then actinium were discovered in the Curies' laboratory in 1899.[89] With von Schweidler, he investigated the radiation emitted by the new elements; using a magnetic field to deflect the radiation, they discovered that beta radiation consists of particles with negative charge,[90] a discovery made at nearly the same time by Henri Becquerel in France and Friedrich Giesel in Germany. A year later Becquerel measured their charge-to-mass ratio and confirmed that beta rays—more correctly, beta particles—are identical to electrons. In a similar experiment, Meyer and von Schweidler studied the alpha radiation from polonium but were unable to observe a deflection; a few years later Ernest Rutherford would use stronger magnets to determine that alpha particles are positively charged.

Meyer remained in radioactivity. He recognized that the puzzling "induced activity" that early workers had found throughout their laboratories was due to radium, thorium, or actinium emitting a radioactive gas—Rutherford called it "emanation"—that diffused into the air and then decayed to a solid that coated objects all over the laboratory. The solid, also radioactive, was thereafter termed the "active deposit."[91] (For the radioactive decay series, see Appendix fig. 1.) In those early years of radioactivity research, Meyer also studied the physical effects of radiation, such as color changes in minerals.

In 1900, four years after the discovery of radioactivity, the number of recognized radioactive species stood at five: the elements uranium, thorium, polonium, radium, and actinium. When Lise Meitner began research in 1906, the number was over twenty and rising—most confusing, since at first every new radioactive substance was thought to be a new element (the existence of isotopes was not fully appreciated until about 1913, when it became clear that it was possible for different radioactive species to be chemically identical). In addition, the relationship between radioactive substances was not understood, and little was known about their radiation. And yet, chaotic as it was, the field was inviting, for a newcomer needed little more than a radioactive source and a measuring device to quickly discover a new activity or learn something new about its radiation.

In Vienna the preferred instrument for radiation measurements was the leaf electroscope. A simple device, it consisted of a very thin gold or aluminum leaf fastened to a metal rod whose end protruded from an enclosed container into which it was sealed. When the rod was given an electric charge, the gold leaf was repelled away from the rod; when a radioactive substance was brought near, its radiation would ionize the surrounding air, the charge on the metal rod and the gold leaf would leak away, and the leaf would descend to its resting position, the rate of descent being a measure of the strength of the radiation. Franz Exner had improved the electroscope for highly sensitive measurements of atmospheric radiation; Julius Elster and Hans Geitel had added a mirror, a scale, and a magnifying glass. The electroscope was a sensitive instrument, readily modified for alpha, beta, or gamma radiations; its disadvantages included the mind-numbing tedium of the measurements and the delicacy of the metal leaf. Karl Przibam recalled "how irritating it was when [the electroscope] was not carefully charged and the little gold leaf hit the top and got stuck, or even tore off!"[92]

In Meitner's first study, done in the spring of 1906, she measured the absorption in several metals of alpha and beta radiation emitted by the active deposits of thorium and actinium; she found, as others had, that alpha radiation exhibits a well-defined range in each metal, whereas the more penetrating beta radiation tapers off gradually. Although she pointed to the importance of absorption studies for understanding the nature of the radiation, she presented her measurements without speculating on their meaning.

Now, in late 1906, Meitner turned to the question of whether alpha particles are merely absorbed as they pass through matter, or whether they are also scattered to some extent. A number of scientists, including Marie Curie and Rutherford had found evidence of scattering, but W. H. Bragg disputed these findings. The question was of considerable interest for learning something of the nature of alpha particles and also the matter through which they passed; Rutherford had already noted that scattering "brings out clearly the fact that the atoms of matter must be the seat of very intense electrical forces."[93] Meitner devised an arrangement by which a beam of alpha particles was collimated—made parallel—by passing through a bundle of tiny metal tubes, then allowed to penetrate a metal foil, and then collimated again by another bundle of tubes some distance away.[94] The attenuation of the beam as it passed through the first and second collimators varied with the distance between them in a manner that could be explained only if the alpha particles were indeed scattered somewhat by the metal foil; Meitner found that the scattering increased with the atomic mass of the metal atoms. In a few years, alpha scattering would lead Rutherford to the nuclear atom; Meitner's was an early example of such experiments, cleverly designed and carefully executed.[95] She submitted her report to the *Physikalische Zeitschrift* on 29 June 1907.

Another year was over, and again Lise Meitner faced a decision. In Vienna her future appeared to hold nothing but teaching.[96] Behind her lay the three investigations she had independently completed. As she later wrote, "This gave me the courage to ask my parents to allow me to go to Berlin for a few terms."[97] Her parents agreed. Meitner regarded it as pure self-indulgence on her part and generosity on theirs. At the age of twenty-eight, she still depended on them for an allowance.

CHAPTER TWO

Beginnings in Berlin

We were young, contented, and carefree, perhaps politically too carefree.

Lise Meitner arrived in Berlin in September 1907. She expected to study there a few semesters; she would stay for more than thirty years.

She chose Berlin because it was a magnet for the German-speaking world, because Boltzmann had spoken of it with regret, and above all because she knew the name of Max Planck and had seen him when he was invited to Vienna as a possible successor to Boltzmann. She had not heard of his quantum theory, although it had been published in 1900, and she knew almost nothing about Berlin, not even that women were still excluded from Prussian universities.[1]

The university in Berlin, young by European standards, was founded under Friedrich Wilhelm III in 1809–1810. Berlin was large then but still amorphous, set on a sandy plain at the confluence of two minor rivers, the Spree and the Havel. With Prussia's growing dominance, Berlin developed into Germany's political, social, and intellectual center. Its focus was the tree-lined Unter den Linden, the landmark Brandenburger Tor, and an immense cluster of public buildings: the huge Reichstag, the Staatsoper, the Charité hospitals and clinics, museums of art, antiquity, and anthropology, churches, palaces, libraries, theaters. Lest Prussia's traditional source of strength be forgotten, the Brandenburg Gate was topped by the goddess Victory who gazed calmly over the cafés on Unter den Linden; to her left just beyond the Reichstag stood the Siegesäule, a victory column encrusted with ceremoniously gilded cannon barrels captured from Denmark, Austria, and, most satisfying, France. Taste was not Berlin's strong point, neither in women's fashions nor cuisine nor monumental architecture; it lacked the patina of Vienna, the grace of Munich, the solidity of Hamburg. But the city had wit, non-European energy—"Chicago on the Spree"—and a modern edginess that left room for nearly every form of human activity.[2] And Berlin's beauty could be surprisingly gentle, with warm summers and generous open space, waterways lacing the central city, and to the west a chain of lakes framed by thick pine woods as the Havel flowed south toward Potsdam.

By the turn of the century the Friedrich-Wilhelm-Universität ranked as one of the finest in Europe. Its main building opened onto Unter den Linden across from the Opera, with other buildings and institutes on the narrow streets between the hospital complex, the museums, and the Reichstag. The university's placement as one among many cultural emblems may well have emphasized its inaccessibility to women, who were expected to fill their designated roles of mother, *Hausfrau*, and guardian of virtue with less education than their counterparts elsewhere in Europe. At no time in the nineteenth century could German women attend universities as anything but unmatriculated auditors; their secondary education was correspondingly irregular. The state of Baden was the first to open its universities to women, in 1900; others slowly followed. Prussia, which did not admit women until the summer of 1908, was by no means the last.[3]

When Lise walked through the university gates, therefore, she entered a domain so decidedly male that she felt not just a stranger but an oddity. She reacted with a reserve so extreme that she herself described it as "bordering on fear of people."[4] One of the first people to speak to her was Gerta von Ubisch, a biology student attending classes as an auditor. Gerta, who was a native of Berlin, and her parents invited Lise for weekends in the country. "No one who knew you when you were so shy at first doubted that you would do something great," she told Lise forty years later. "My small contribution during those first months was to help you overcome your shyness with respect to your colleagues, which you would have done eventually on your own."[5]

Lise had to ask Max Planck for permission to attend his lectures.

> He received me very kindly and soon afterwards invited me to his home. The first time I visited him there he said to me, "But you are a Doctor already! What more do you want?" When I replied that I would like to gain some real understanding of physics, he just said a few friendly words and did not pursue the matter any further. Naturally, I concluded that he could have no

very high opinion of women students, and possibly that was true enough at the time.[6]

She was right. Planck's opinion was in fact on record, published in 1897 by a Berlin journalist who surveyed some one hundred professors, teachers, and writers for their views on higher education for women.[7] The responses varied widely. Impassioned supporters tended to be brief: "Keeping women from the universities is an injustice that has gone on far too long," "It is quite impossible for me to understand how a modern human being deserving of the name can deny women's right and ability for academic study," "What gives us men the right to always determine what women shall do?" Those strongly opposed argued at length: women would be too weak for surgery, frightful (schrecklich) in the ministry, untalented in history, a threat to the social and intellectual character of the university. Mathematicians unconditionally favored the admission of women: Felix Klein reported that his six current women students (fully matriculated in Göttingen under a trial program open only to foreign women)[8] were as capable as the men, while the Kiel mathematician G. Weyer listed twenty-one women mathematicians and astronomers from Hypatia to Maria Mitchell, including a biography of Sofia Kovalevskaia. Several respondents deplored the German feminine ideal of woman's place in nursery and kitchen, and its reverse, the caricature of the sexless learned woman (die gelehrte Frau); a few pointed out that since men did not object to women working in factories, their opposition to women in the professions was surely due to fear of competition. Hugo Münsterberg, a Freiburg philosophy professor who spent several years at Harvard, described the vitality of Boston society with its educated, intellectual, and interesting women; he noted, however, that their American college diplomas were by no means equivalent to a German university degree.[9]

Max Planck's response was well toward the conservative end of the spectrum.

> If a woman possesses a special gift for the tasks of theoretical physics and also the drive to develop her talent, which does not happen often, but does happen on occasion, then I consider it unjust, from a personal as well as an objective point of view, to categorically deny her the means to study as a matter of principle; if it is at all compatible with academic order I shall readily admit her, on a trial basis and always revocably, to my lectures and my practical courses. . . .

On the other hand, I must hold fast to the idea that such a case must always be considered an exception, and in particular that it would be a great mistake to establish special institutions to induce women into academic study, at least not into pure scientific research. Amazons are abnormal, even in intellectual fields. In certain practical situations, for example, women's health care, conditions might be different, but in general it can not be emphasized strongly enough that Nature itself has designated for woman her vocation as mother and housewife, and that under no circumstances can natural laws be ignored without grave damage, which in this case would appear especially in the next generation.[10]

Planck wrote this in 1897 when he was thirty-nine years old, already a professor and well-known scientist, husband and father of four. Unlike many of his contemporaries, he accepted the societal status quo without question and elevated his own family structure to natural law. Nevertheless Planck did not completely depersonify women: he was willing to admit exceptions. When Lise Meitner appeared in his office ten years later, he apparently recognized her as one of them. Perhaps he saw in her painful shyness the determination it took just to be there; he must have sensed how out of place she felt and invited her to his home.

Even with my first visit I was very impressed by the refined modesty of the house and the entire family. In Planck's lectures, however, I fought a certain feeling of disappointment at first. . . . Boltzmann had been full of enthusiasm . . . and he did not refrain from expressing this enthusiasm in a very personal way. . . . With this background, Planck's lectures, with their extraordinary clarity, seemed at first somewhat impersonal, almost dry. But I very quickly came to understand how little my first impression had to do with Planck's true personality.[11]

Lise soon realized that Planck's lectures would not occupy all her time, and she looked about for a place to do some experimental work. When she approached Professor Heinrich Rubens, head of the experimental physics institute, he offered her a place in his own laboratory.

Now it was quite clear to me then, as a beginner, how important it would be for me to be able to ask about anything I did not understand, and it was no less clear to me that I should not have the courage to ask Professor Rubens. While I was still considering how to answer without giving offense, Rubens added that Dr. Otto Hahn had indicated that he would be interested in collaborating with me, and Hahn himself came in a few minutes later. Hahn was of the same age as myself and very informal in manner, and I had the feeling that I would have no hesitation in asking him all I needed to know.

Moreover, he had a very good reputation in radioactivity, so I was convinced he could teach me a great deal.[12]

More than fifty years later, Otto Hahn would cite 28 September 1907 as the day he first met Lise Meitner.[13] Although the date, like other things Otto would later remember, is probably not quite correct, their meeting was clearly an important event for him, as it was for Lise. They would work under the same roof for the next thirty-one years, together at first and then independently, the closest of colleagues, the best of friends. Later, after Lise left Berlin, their collegiality would not survive the differences that grew between them. But their friendship proved irreversible, a constant in their lives.

Hahn was born in Frankfurt on 8 March 1879, four months after Lise, the youngest son of a well-to-do tradesman, a good but not especially diligent student who took his degree in chemistry from the University of Marburg. To learn some English and improve his prospects for a position in chemical industry, he went to London in 1904; Sir William Ramsay, the chemist famous for discovering argon and other noble gases, introduced him to the field of radioactivity. Hahn's research went so well—he promptly discovered a new radioactive substance, radiothorium (^{228}Th)— that he decided on an academic career as a radiochemist. He spent the following year in Montreal with Ernest Rutherford, discovered two more activities, thorium C (later revised to ThC′ [^{212}Po]) and radioactinium (^{227}Th), and in 1906 returned to Germany for a position as *Assistent* (assistant) in the institute of Emil Fischer, the great organic chemist in Berlin. At the time, very few physicists and even fewer chemists worked in radioactivity, none in Berlin. In Fischer's institute Hahn was surrounded by chemists who had no idea what he was up to; Fischer himself, who liked to say that no instrument was more sensitive than the human nose, found it hard to believe that an electroscope could detect far smaller quantities of radioactive material.[14] Hahn soon found another activity, which he named mesothorium, a mixture of MsTh1 (^{228}Ra) and MsTh2 (^{228}Ac), and in the spring of 1907 was ready for his *Habilitation*. On learning of Hahn's promotion, a department head sniffed, "It's incredible what gets to be a *Privatdozent* these days!"[15]

In the German academic hierarchy, the first rung on the academic ladder was that of Assistent, a research position that carried a small salary and no teaching duties. After a few years, the Assistent would prepare his

Habilitationsschrift, a written and verbal presentation of his independent work, receive the *venia legendi* (right to teach) at the university level, and be appointed Privatdozent (instructor), an unsalaried position that provided academic rank, research funding, and teaching assignments in return for fees paid by students. If things went well he could expect a *Ruf* (literally, "call") to an appointment as *ausserordentlicher Professor* (extraordinary professor) and if he was exceptional and a position was available, *ordentlicher Professor* (ordinary professor, or simply professor), a position most academics never reached.

The titles do not convey the same relative importance as comparable American titles: in Germany, the position of Professor was far more powerful than those below it because of the practice of having only one professor for a given subject area. Most universities, for example, had only two professors in physics, one theoretical and the other experimental (in Berlin, these were Planck and Rubens). Many physicists might work under them in various fields, but the number of professors remained constant unless major efforts were undertaken to convince the Ministry of Education that a new professorship was needed. Professors often headed their own institutes, were paid far more, and were the only faculty to have a say in hiring, budget, and policy decisions. Ordinary and extraordinary professors were government officials with civil service rank; very distinguished professors might be given the title *Geheimrat* (privy councilor; in Austria, *Hofrat*) or even ennobled with a "von" before their family name. It has been argued that with their power, privilege, and government rank, German academics of the period were a functional ruling caste—comparable to the military in Prussia—that substituted for a more traditional aristocracy.[16]

With few kindred spirits in the chemistry institute, Hahn regularly attended Professor Rubens's Wednesday physics colloquium, and it was there, most probably, that he was first introduced to Lise. They were delighted with each other from the start. Otto liked women in general; he had worked with at least one other woman physicist, Harriet Brooks, in Montreal, and he was genuinely glad to find another person—especially a physicist—with experience in radioactivity.[17] And Lise sensed that Otto's good-natured informality would make it easier for her to overcome her shyness. Each saw in the other something they lacked. Lise knew the physics and mathematics Otto had never studied, and she had the middle-class intellectual upbringing he had always admired from afar; Otto was charming and sociable to the tips of his Wilhelminian mustache, whose upturned ends signaled that a person "was somebody," or at least wanted to be. In each other's company both could escape some of the formalities of Berlin. For his outspokenness Hahn had already been described as "one of those Anglicized Berliners" (not intended as a compliment), and Meitner had been appalled at being asked to sign a paper saying she would behave "*standesgemäss*"—"according to her rank and station"—when renting a room.[18] In many ways Vienna and Frankfurt were closer to each other than either was to Berlin. Lise and Otto made plans to begin work together at once.

There was one problem. The Chemistry Institute was completely off-limits to women: Emil Fischer was afraid they would set fire to their hair, having once had a Russian student with an "exotic" hairstyle.[19] (He must have believed his beard to be flame resistant.) As a compromise, Lise was allowed to work in a basement room, formerly a carpenter's shop, which Otto had set up for measuring radiation; she was not to set foot in any other part of the institute, not even the laboratory upstairs where Otto did his chemical experiments. Fischer relented only because the wood shop had a separate outside entrance; to use a toilet Lise walked to a restaurant down the street. A year later, when women were legally admitted to Prussian universities, Fischer welcomed them, lifted his restrictions on Meitner, and installed a ladies' room.[20] Many of the chemistry assistants resented the change, and Lise remained essentially "nonexistent" in Fischer's institute.[21] Sometimes, when an assistant encountered Meitner and Hahn together, he would make a point of saying only, "Good day, Herr Hahn!"[22]

The physicists were much friendlier. At Rubens's physics colloquium Lise joined a group of young people that included James Franck, Gustav Hertz, Max Laue, Otto von Baeyer, Robert Pohl, Peter Pringsheim, and Erich Regener, and, later, many others who would be her lifelong friends. As she wrote, "Not only were they brilliant scientists, they were also exceptionally nice people to know. Each was ready to help the other, each welcomed the other's success."[23]

For their first investigation, Hahn and Meitner decided to survey all the beta-emitting radioactive sources at their disposal. Such a study was necessary, they believed, because earlier results, obtained by different scientists under varying experimental conditions, were proving difficult to interpret. Although no one doubted that beta particles were high-energy electrons, almost everything else about them was quite unclear, including their energy of emission and the manner in which they were absorbed in various materials.

On the whole alpha radiation presented a somewhat simpler picture. The nature of alpha particles was not known with certainty—Rutherford had not yet proven that they were helium atoms that had lost two electrons—but their energy and absorption characteristics were understood quite well. In 1904 Bragg had shown that each pure alpha source emits alpha particles with uniform energy and a sharply defined range (penetrating distance) that is sufficiently characteristic of the source to serve as a means of identification.[24] In Montreal Otto Hahn had used just this characteristic to find a new alpha source: the active deposit of thorium, already known to contain the alpha emitter ThB (later designated ThC [^{212}Bi]), emitted alpha particles with two distinct ranges, indicating the presence of a second activity he called ThC (later ThC′ [^{212}Po]).[25] This method was particularly useful for finding very short-lived substances. Hahn's ThC (ThC′), for example, was later found to decay with a half-life of 3×10^{-7} seconds—so fast that not enough could accumulate for detection by chemical means.

As a working hypothesis, Meitner and Hahn assumed that beta particles, like alpha particles, were emitted with uniform energy; they knew, however, that beta particles were far more penetrating and were absorbed only gradually as they traveled through matter. By 1907 there was a general consensus that beta particles of uniform energy would be exponentially absorbed.[26] If true, then any deviation from exponential absorption would indicate the presence of more than one beta source.

Hahn and Meitner measured the absorption characteristics of a number of pure beta sources and mixtures—mesothorium 1 and 2, the thorium and radium active deposits, uranium X (^{234}Th), radiolead (^{210}Pb), and radium E (^{210}Bi)—making chemical separations where possible, controlling physical parameters such as the thickness and area of the radioactive source, secondary absorption effects, and interference from alpha and gamma radiation. Their electroscope, a well-made brass design with aluminum leaves, was clamped in place with screw adjustments for varying the distance between source and instrument; the absorbing material consisted of thin aluminum foils—as many as fifty—layered over the source and held down with a metal ring. With one exception, each pure substance exhibited exponential absorption; mixtures did not. The lone exception was mesothorium 2 (^{228}Ac), whose nonexponential absorption Meitner and Hahn attributed to the presence of a yet-undiscovered substance; attempts to separate it chemically were unsuccessful. In April 1908 they submitted their results to the *Physikalische Zeitschrift*, a twelve-page article, Hahn noted, in a journal with unusually large page size. "Evidently we were very diligent in those days."[27]

Hahn and Meitner next turned to the active deposit of actinium, known to contain one beta emitter, actinium A, and its daughter, actinium B, thought to emit both alpha and beta particles. The beta radiation deviated considerably from exponential absorption,[28] so that Meitner and Hahn suspected the presence of another beta emitter; to find and characterize it, they chemically separated actinium B, measured alpha and beta decay simultaneously, and established that actinium B is an alpha emitter only, decaying to a new beta emitter they called actinium C. (AcA was later revised to AcB [^{211}Pb], AcB to AcC [^{211}Bi], and AcC to AcC″ [^{207}Tl]).[29]

The actinium work was completed in August 1908, just before the institute closed for the four-week summer vacation. In ten months Meitner and Hahn had accomplished far more together than either could have done alone, in part by sharing the tedious physical measurements, more fundamentally because radioactivity was by nature interdisciplinary, requiring the chemical separations that were Hahn's forte as well as Meitner's physical measurements and the mathematical and graphic skills she brought to the analysis of their data. In a field characterized by a profusion of strange new species and unexplained effects, their collaboration also benefited from their differences in scientific temperament: Hahn's patience and thoroughness inspired confidence that no detail was overlooked, whereas Meitner looked for bold generalizations that were essential for finding a way through the chaos. This was true in their first year's work together, and even though their assumption of uniform beta energies and exponential absorption would soon be proven incorrect, it provided a strategy for systematically exploring a large number of radioactive substances and even discovering a new one. When Lise went home to Vienna in August, she asked her parents to extend her allowance for another year.

In Vienna, Lise also attended to another matter. On 29 September 1908, she formally withdrew from the Jewish community in which her name had been registered at birth, and was baptized at the Evangelical (Protestant) Congregation. There is no record of why Lise took that step just then; she may have been prompted by the baptism of her sisters Carola (Lola) and Gisela, both as Catholics, earlier that year. Although she also never explicitly stated why she chose to be Protestant, she maintained a genuine interest in the ethical teachings of the religion all her life.[30] It may well be that her year in Berlin had something to do with it, especially her admiration for Planck, whose character and behavior exemplified the German Protestant ideal of "excellent, reliable, incorruptible, idealistic and generous men, devoted to the service of Church and State."[31] One can almost certainly rule out opportunism as a motive for Lise's conversion: professionally she was so thoroughly excluded for reasons of gender that religion played no essential role. And in Berlin, as in Vienna, everyone knew who was "really" Christian anyway.

That fall Hahn and Meitner turned a small discrepancy into an important new method for isolating individual radioactive species with exceptional purity. A number of scientists, including Stefan Meyer, had found minute amounts of what appeared to be actinium X (^{223}Ra) in the active deposit of actinium. This was puzzling, since actinium X decays to actinium emanation, a gas that escapes and then decays to form the active deposit some distance from its parent; no one could imagine how actinium X, a solid incapable of evaporation, could be transported to the active deposit:

$$\text{AcX } [^{223}\text{Ra}] \rightarrow \text{AcEm } [^{219}\text{Rn}] \rightarrow \text{Ac active deposit}$$

Toward the end of 1908, Hahn discovered that the effect did not occur unless radioactinium (^{227}Th), the parent of actinium X, was also present; this enabled him to explain the puzzle. When a radioactinium atom (^{227}Th) expels an alpha particle, he realized, it does so with such force that its daughter atom, actinium X (^{223}Ra), recoils, sometimes with enough energy to free itself from the solid and travel to a nearby surface some distance away:

$$^{227}\text{Th} \rightarrow {}^{223}\text{Ra} + \alpha$$

The phenomenon, no different in principle from the recoil of a rifle firing a bullet, has come to be known as radioactive recoil.

Hahn hastened to prepare his results for publication; preoccupied with actinium, he did not speculate on other systems. When Meitner read his manuscript, however, she immediately proposed that the recoil he had observed from fairly thick layers of actinium should occur far more readily from the extremely thin layers formed by active deposits.[32] Together they tested the active deposit of thorium and at once discovered a new radio-active substance, the beta emitter thorium D (now ThC″ [^{208}Tl]), which they could show was ejected from the active deposit by the alpha recoil of its parent, ThB (now ThC [^{212}Bi]):

$$^{212}\text{Bi} \rightarrow {}^{208}\text{Tl} + \alpha$$

The ^{208}Tl they obtained was exceptionally pure, so that they could determine its half-life of three minutes with unusually high precision for such a short period. In the radium active deposit they searched for evidence of beta recoil—expected to be much weaker—and indeed found minuscule amounts of radium C (^{214}Bi) that could only have come from the beta recoil of radium B (^{214}Pb). Finally they turned to the actinium active deposit, and from the alpha recoil of actinium B (now AcC [^{211}Bi]) they collected a pure sample and verified the half-life of actinium C (now AcC″ [^{207}Tl]), the species they had discovered a few months before.

The recoil method was so clean that their experiments were completed in a few days; in their report they emphasized "the great advantage of this physical separation method over chemical separations, not only with re-spect to purity but also the quantity of the preparations obtained."[33] Recoil continued to be a powerful method for separating and purifying radioactive substances; more than twenty years later Leo Szilard and T. A. Chalmers would use the recoil from hard gamma emission to separate radioisotopes produced by neutron irradiation.

In December 1908 Ernest Rutherford was awarded the Nobel Prize in chemistry; on their return from Stockholm to Manchester (where he had moved from Montreal in 1907), he and his wife visited Berlin for the first time. When he was introduced to Lise, he exclaimed in astonishment, "Oh, I thought you were a man!" (even though her first name was on every publication).[34] The visit was festive, Rutherford much amused about his sudden "transformation," as he called it, from physicist to chemist. He twitted Otto Hahn and later another "brother chemist," Bertram Bolt-wood of Yale, "I was very startled at my transformation at first but afterwards saw that it was quite in accord with the disintegration theory."[35] While Rutherford and Hahn had long talks, Lise accompanied Mrs. Rutherford, who spoke no German, on her Christmas shopping trips.[36] Lise could not have been pleased to be out shopping while the men talked shop.

Hahn made arrangements for a grand reception for his former professor, including visits to the physics and chemistry institutes and dinners and seminars in his honor.[37] At a meeting of the Deutsche Physikalische Gesellschaft, Hahn reported on the recoil of actinium X. Rutherford was most interested and, to Hahn's discomfort, told him that the effect had been described some years before in his laboratory in Montreal, when his student Harriet Brooks had observed what appeared to be "evaporation" of radium B from the radium active deposit. Noting that the effect occurred only when the active deposit was fresh and radium A (^{218}Po) was present, Rutherford attributed it to the recoil of radium B (^{214}Pb) from the alpha decay of radium A; he had, moreover, discussed recoil in several articles and in his 1904 textbook on radioactivity and its 1905 revision, which had been translated into German.[38]

Hahn countered, rather stiffly, that he had proved it was not evaporation but recoil by showing that radioactinium was required, and in any case he had done so with actinium, not radium. For all his charm, Hahn could be prickly about priorities, an expression, perhaps, of the insecurity felt by a chemist in a field dominated by physicists. In his memoirs fifty years later he was still defensive, insisting that Rutherford's "unexplained residual activities cannot be explained by radioactive recoil. The experimental proof of its existence was first furnished in the wood shop in Berlin."[39] Hahn's proof may have been more convincing, but one wonders why it was so difficult for him to acknowledge that his mind had been prepared for recoil by the concept Rutherford had explicitly and repeatedly proposed.

In the end there were no hard feelings.[40] Rutherford commiserated with Hahn over his lack of academic advancement and returned to Manchester enthusiastic about his reception in Berlin. Of Hahn's colleague he wrote, "Lise Mitner [*sic*] is a young lady but not beautiful so I judge Hahn will not fall a victim to the radioactive charms of the lady."[41]

In fact, it was true that Lise and Otto were colleagues and friends, nothing more. They took care to avoid the slightest appearance of impropriety, a necessity for a young woman and man who spent each day alone together in a single room. As Hahn described it,

> There was no question of any closer relationship between us outside of the laboratory. Lise Meitner had had a strict ladylike upbringing and was very

reserved, even shy. . . . [F]or many years I never had a meal with Lise Meitner except on official occasions. Nor did we ever go for a walk together [a "walk together" being one of the very few socially acceptable ways for an unmarried couple to spend some time alone]. Apart from the physics colloquia that we attended, we met only in the carpenter's shop. There we generally worked until nearly eight in the evening, so that one or the other of us would have to go out to buy salami or cheese before the shops shut at that hour. We never ate our cold supper together there. Lise Meitner went home alone, and so did I. And yet we were really very close friends.[42]

Warmhearted by nature, with a capacity for making and keeping good friends, Lise may at some time have wanted a closer relationship with Otto, or perhaps with one or another of the many young men she met. But there is no record of it, not even a hint, nor is there any indication that she regretted not having children of her own. She took a lively interest in the children and later the grandchildren of her relatives and friends and was always especially close to her nephew Otto Robert Frisch, "a darling little fellow[,] . . . exceptionally observant,"[43] who grew up to be a nuclear physicist, devoted to his Tante Lise. Many years later one of James Franck's daughters, Dagmar von Hippel, asked Lise why she never married, since she was "so beautiful" and there were so many young men around. "But Daggie, dear, I just never had time for it!" Lise exclaimed.[44] If she ever had the desire, she must have buried it early on.

Lise and Otto enjoyed each other's company, buoyed by the immediate success of their collaboration. With his relaxed Frankfurt accent and easy good humor, Otto helped Lise overcome her reserve. And her connections with the physicists meant that he was included in a congenial group that looked after each other and became good friends.

Radioactivity and atomic physics were then developing incredibly quickly; nearly every month brought a wonderful surprising new result from one of the laboratories working in these fields. When our work was going well we sang duets, mostly Brahms *Lieder*, which I could only hum, while Hahn had a very good singing voice. . . . If he was in an especially good mood he would whistle large sections of the Beethoven violin concerto, sometimes purposely changing the rhythm of the last movement just so he could laugh at my protests. . . . Both personally and scientifically we had a very good relationship with the young colleagues in the nearby physics institute. They often came to visit us, and sometimes they would climb in through the window of the wood shop instead of taking the usual way. In short, we were young, contented and carefree, perhaps politically too carefree.[45]

Disinterest in, even aversion to politics was traditional among German academics; they disdained its subjectivity and lack of consensus and certainly recognized that their own careers were best enhanced by joining the establishment, not changing it.[46] Although Lise grew up with the democratic ideals of Liberal Vienna, she took little interest in German issues, not even the struggle for equity in women's education and suffrage that was important at the time. Being a foreigner may have been part of it. And the physics was all-absorbing.

Each semester Lise extended her stay in Berlin. Her parents continued her small allowance. To supplement it, she occasionally translated scientific articles from English into German[47] and wrote quite regularly for the popular scientific periodical *Naturwissenschaftliche Rundschau* under the name "L. Meitner." Other such assignments were hard to come by. The editor of the *Brockhaus* encyclopedia, impressed by the *Rundschau* articles, decided to ask the "Herr Doktor" for an article on radioactivity for the encyclopedia and wrote to the *Rundschau* editor for "his" address. When the *Brockhaus* editor learned that L. Meitner was a "Fräulein Doktor," he replied with some heat that he "would not think of printing an article written by a woman!"[48]

Lise lived frugally, renting single rooms from a succession of landladies, never with private bath, occasionally with a piano she could use, or a telephone. She carefully listed her clothing—"7 blouses, 20 pair stockings, 4 underskirts . . ."—and accounted for every penny she spent. By eating very little she had enough for cigarettes, a daily newspaper, and concerts,[49] where she could be found high up in the cheapest seats—the section students called "Olympus"—often following the music with a full score.[50]

Not long after coming to Berlin, Lise met another young person on the city train, where their paths regularly joined on their way to the university. This was Elisabeth Schiemann, three years younger than Lise, who was studying botany in the Agricultural Institute (Landwirtschaftliche Hochschule). They quickly became close friends. Lise introduced Elisabeth to her circle of physicists, and Elisabeth's family, including her sister Gertrud, embraced Lise, who "came and went in [her] parents' house like a sister and was not permitted to miss a single family celebration."[51] On Sundays they often took excursions into the countryside around Berlin, so that Lise, who knew and loved the Austrian mountains, learned to appreciate the less spectacular beauty of the Mark Brandenburg. Their most ambitious *Wan-*

derung took place in the summer of 1913, when they made their way on foot and by train from Munich to Vienna, hiking with rucksacks through mountain meadows from hut to hut, climbing the Zugspitze, highest in the Bavarian Alps, passing through Salzburg, and arriving at last in Vienna for the annual meeting of the Society of German Scientists and Physicians (Gesellschaft Deutscher Naturforscher und Ärzte), which inaugurated the university's fine new Physics Institute on Boltzmanngasse.[52] It was on this trip, after years of friendship, that they first addressed each other with the familiar "Du," an occasion for celebration, toasts, and annual remembrances.[53] It would be another ten years before Lise and Otto Hahn would do the same.

With James Franck, such barriers did not exist. From nearly the first day they met in 1907, Lise said, "we both knew that we spoke the same language." After Franck married Ingrid Josephson in 1911, he would bring Lise home with him after the Wednesday physics colloquia; Ingrid, an accomplished pianist, would play the Brahms *Lieder* Lise loved, or Franck would accompany her on the violin. Friends for over fifty years, they never once disappointed each other.[54] "I've fallen in love with you," Franck teased when they were both in their eighties. "*Spät!* (Late!)" Lise laughed.[55]

Of all the people in Lise's circle in Berlin, the one she admired with near-reverence was Max Planck, the second great theoretical physicist whose extraordinary human qualities she came to know. Where Lise had been swept along by Boltzmann's exuberance, she loved and trusted Planck for his depth of character.

> He had an unusually pure disposition and inner rectitude, which corresponded to his outer simplicity and lack of pretension. . . . Again and again I saw with admiration that he never did or avoided doing something that might have been useful or damaging to himself. When he perceived something to be right he carried it out, without regard for his own person.[56]

His "inner rectitude" was tempered by kindness. What Planck once said of his friend, the great violinist Josef Joachim, Meitner applied to him: " '[He] was such a wonderful person, that when he entered a room, the air in the room got better.' " She added, "The younger generation of Berlin physicists . . . felt this very strongly."[57]

Planck believed physics to be inseparable from ethical values, because nothing less than complete honesty suffices to understand external reality;

Lise attributed his tolerance for ideas quite foreign to him, his sympathy for others, and his concern for justice to just this respect for reality. Despite a twenty-year difference in age and an immeasurable difference in status, they became "true friends," a steady influence in each other's lives for forty years.[58]

Planck lived in Grunewald, an attractive new suburb at the edge of the pine forest west of Berlin, in a villa, a big house with a large garden. Lise described the times she and her colleagues spent there:

> Planck loved happy, unaffected company, and his home was a focus for such social gatherings. The more advanced students and physics assistants were regularly invited to [his home on] Wangenheimstrasse. If the invitations fell during the summer semester, we played tag in the garden, in which Planck participated with almost childish ambition and great agility. It was almost impossible *not* to be caught by him. And it was obvious how satisfied he was when he had caught someone.[59]

As a young man, Planck had contemplated a career in music but recognized that some of his compositions sounded too much like music already known.[60] He was a pianist of great technical ability; away from the university, his life centered on music. Family and friends would gather for regular evenings of chamber music, Planck accompanying Josef Joachim or playing in a trio with Einstein. Or he might direct a choral work of Haydn or Brahms, with Otto Hahn a frequent soloist and a chorus of friends and neighbors that included Elisabeth and Gertrud Schiemann. Lise did not perform, but she was always in the audience.[61] She became good friends with Planck's twin daughters, Emma and Grete, who inherited his gift for music; she fit into the Planck family as one of the younger generation. There the relationship stood for several years.

Lise continued to visit her family several times a year, always for Christmas and usually for the Easter and summer holidays. There in Vienna, in her parents' home, a sense of belonging came over her that she felt nowhere else. "Surely wherever else I may be I shall always have to overcome a certain feeling of being a stranger," she reflected. "Nevertheless I know very well that leaving home was in certain respects a salvation for me—if I had stayed here I would, at least inwardly, have been destroyed."[62]

Lise never lost her nostalgia for Vienna, but the things that mattered most to her were in Berlin. Successful in her collaboration with Hahn, secure in her circle of good friends, her reticence gradually receded. With great trepidation, but with Otto's encouragement and prodding, she began giving talks, first to the German Physical Society and then, even more daunting, at the conference of German Scientists and Physicians that met in Salzburg in September 1909.[63] Later she would recall the conference primarily for her first meeting with Albert Einstein, to whom she was introduced by Anton Lampa, her former laboratory instructor from Vienna.[64] Einstein was thirty, a few months younger than she, and already a phenomenon. In his Salzburg lecture he made the revolutionary argument for a theory of light that would treat light as both particle and wave. Like most in the audience, including Planck, Lise was not ready for it; instead she was transfixed by Einstein's brief overview of the special theory of relativity and his derivation of the equivalence of mass and energy. Until then she had not heard of it (although Einstein had published it in 1905), and it was "so overwhelmingly new and surprising" that she remembered that moment all her life.[65]

Four years later Planck succeeded in bringing Einstein to Berlin, where he and his violin were soon a fixture for the evenings of chamber music in Planck's home. In this most revolutionary period in physics, there seemed no better place to be a physicist. Later Lise would describe the physics she experienced and the people who were part of it as a "magic musical accompaniment" to her life.[66]

But apart from the day-to-day absorption with work and friends, Lise's life was undefined. It was not so much that her living conditions were poor, or that she was still dependent on her parents; Otto too, in lieu of academic advancement, was still on an allowance (albeit considerably more generous than hers).[67] Far more troubling was her lack of position, the fact that she fit in nowhere, a permanent double exception: a woman who was a scientist, a scientist who was female. Physics was a personal need, a private indulgence; her pleasure in it was tinged with guilt. Shortly after the death of her father in 1910, she wrote to a friend,

> Sometimes I lack courage, and then my life, with its great insecurity, the constantly repeated worries, the feeling of being an exception, the absolute aloneness, seems almost unbearable to me. And what distresses me most is the frightful egotism of my current way of life. Everything I do benefits only me, my ambition and my pleasure in scientific work. It seems I have chosen a path which flies in the face of my most deeply held principle, that everyone should be there for others. By that I don't mean one must sacrifice oneself

for no reason, but that somehow our lives should be connected with others, should be necessary for others. I, however, am free as a bird, because I am of use to no one. Perhaps that is the worst loneliness of all.[68]

Although Lise added that she was usually far too busy for such negative thoughts, it is clear that physics alone was not enough: she needed a measure of acceptance, some sense of future. She did not dwell on it and immersed herself in work.

In her first years of collaboration with Otto Hahn, they published often—three major articles in 1908, six more in 1909—studied every known beta source, identified several new activities, and used radioactive recoil as a powerful separation technique for discovering more. They devoted much of 1909 to radium and its active deposit. Using the beta recoil of radium B (^{214}Pb), they collected a sample of radium C (^{214}Bi) pure enough to detect a slight variation in its rate of decay; this and the fact that its beta radiation was not exponentially absorbed led them to propose that radium C was followed by another beta emitter (radium C″, ^{210}Tl) with a half-life of 1 to 2.5 minutes. They also confirmed the presence of a penetrating alpha radiation, which they attributed to a very short-lived new activity (radium C′, ^{214}Po), also following radium C.[69] This completed their study of all three active deposits; in each, the nonexponential absorption of beta radiation had led them to new activities. Confident in their experimental technique, they were also quite convinced that their favorite hypotheses were true. "Pure substances emit uniform β-radiation, which is exponentially absorbed," they would write, often more than once, in every article. "Complex β-radiations correspond to a mixture of substances."

The assumption was simple, useful, and nicely resonant with chemical behavior: just as a pure chemical compound displays invariant chemical and physical properties, it seemed natural to suppose that a single radioactive source would emit radiation with uniform energy. Now Hahn and Meitner pressed the analogy one step further. Others had left open the possibility that a given source might emit both alpha and beta radiation, but in their work with the active deposits, they had not found this to be so.[70] "Each individual substance," they asserted confidently, "emits only one sort of homogeneous radiation, either alpha or beta radiation."[71]

They first applied their new premise to radium itself. In the course of their work with radium C, Meitner and Hahn had prepared an unusually pure sample of the element and found that radium (^{226}Ra), an alpha emitter, also emits a very weak beta radiation. They concluded that radium must contain another substance, which they denoted radium X. Although they cautioned that there had never been any indication that radium was complex, they were quite sure they were right.[72]

For months they tried to separate their "radium X" from radium but were unable to do so, either by chemical separations or by recoil. "Nevertheless we are convinced of the existence of radium X, but we cannot at the moment see any way to separate it," they reported wearily.[73] A year earlier they had found a similar weak beta radiation associated with radioactinium, another alpha emitter.[74] Soon they discovered weak beta radiation associated with still another alpha emitter, thorium X.[75] In each case they proposed a new substance they were unable to find. Although they drew some comfort from the fact that the effect appeared in all three radioactive series, it seemed that their means of detecting weak beta radiation had outstripped their ability to determine the source.

But this was not their only challenge. In mid-1909, William Wilson reported from Rutherford's laboratory in Manchester that the beta radiation from several pure beta emitters appeared inhomogeneous; when he created a homogeneous beam by deflection in a magnetic field, he found the beam was linearly, not exponentially, absorbed.[76] In response, Hahn and Meitner pointed out ambiguities in Wilson's experiment and suggested that it actually supported their own hypothesis.[77] Wilson did not agree.[78]

To resolve the conflict, Meitner and Hahn designed their own experiment with magnetically deflected beams of beta particles. When a moving charged particle traverses a magnetic field, its path is bent from a straight line into a curve; lower-energy, slower-moving particles are deflected more than faster-moving, higher-energy particles. The magnetic field, in effect, separates moving charged particles, in this case, electrons, by energy; beta particles of the same energy will be deflected equally and be found at the same position. To measure the position of the deflected beta particles, Wilson used an electroscope; Hahn and Meitner decided that a photographic plate would more clearly distinguish between the continuous spectrum of inhomogeneous beta radiation and the discontinuous spectrum they expected for radiation of uniform energy.

> As there were no magnets in the Chemistry Institute, we carried out these experiments with Otto von Baeyer in the Physics Institute. Hahn and I attempted to precipitate in as radioactively pure a condition as possible the

substances whose beta radiation we wished to investigate in the thinnest possible layers on very short lengths of very thin wire. The precipitation did not always work. We simply had to try, and, if our efforts were successful, we raced out of the Chemistry Institute as if shot from a gun, up the road to the Physics Institute a kilometer away, to examine the specimens in von Baeyer's very simple beta spectrometer.[79]

For very short half-lives, they arranged for a car to rush them from the Chemistry Institute on Hessische Strasse to the Physics Institute on the Reichstagsufer.[80]

For their first experiment they selected the active deposit of thorium. It had the advantage that it contained only two beta emitters, thorium A and thorium D (now ThB [^{212}Pb] and ThC″ [^{208}Tl]), whose energy they expected—if their absorption studies were at all meaningful—to be quite different. They immediately obtained the first magnetic line spectrum of beta radiation, which showed, just as expected, two strong lines.[81] A few weak lines appeared as well, which they attributed to secondary effects. The spectrum of pure radium E showed the expected single line, while mesothorium 2, whose beta radiation was not absorbed exponentially, had several lines in its spectrum. As an indication of the sensitivity of the photographic method, they easily recognized a strongly deflected line from thorium X, whose beta radiation was so easily absorbed that it had been detected electroscopically only with great difficulty,[82] and a similar line for "radiationless" radium D whose radiation had never been detected electroscopically at all.[83] The spectra were all discontinuous.

To improve the resolution of the line spectra, the three scientists used stronger magnetic fields and changed the position of the photographic plate. Suddenly Hahn and Meitner's hypothesis was in danger. The new photographs made it obvious that the high energy lines were not at all well defined but quite diffuse; the extra lines in the spectrum of the thorium active deposit were stronger and more difficult to explain.[84] The line spectrum of radium active deposit showed nine lines, of which five originated from a single source, radium B. Radium itself, an alpha emitter, had a beta spectrum of two lines.[85] Always they found more than one line from a given source; Hahn and Meitner were forced to concede that their hypothesis of homogeneous beta radiation was definitely wrong. Although they still thought that the exponential absorption they had observed for so many pure substances could be of use as an analytical method, they admitted that it "could not be a criterion for the homogeneity of the radiation, as Hahn and Meitner, in contrast to other scientists"—here they referred to Wilson—"have assumed."[86]

Together with von Baeyer, Meitner and Hahn continued to determine the line spectra of a number of other substances,[87] but in the absence of a unifying hypothesis their work consisted primarily of collecting data, much as scientists had done with optical spectra for many years. In fact, beta decay would prove far more complex, difficult, and, in the end, revealing than anyone first imagined: it would yield profound insights into the atomic nucleus. In this Lise Meitner would play a significant part, but it would take some fifteen years from the first photographic record of magnetically deflected beta particles. In 1911, the year Meitner and Hahn abandoned their simple hypotheses, no other ordering principle was evident: not from radioactivity, burdened with unexplained data; not from chemistry, its periodic table threatened by the profusion of radioactive species; not from physics itself, struggling to integrate the new discoveries.

Hahn and Meitner's adherence to simplicity had in fact held them back. Others had recognized the phenomenon of branching, in which one type of atom decays sometimes by alpha emission and sometimes by beta emission to yield two different daughters, a violation of any notion of atomic consistency. Meitner worked to catch up, with Hahn on beta radiation in the thorium active deposit,[88] with James Franck on gas-phase radioactive ions,[89] by herself on electrochemical classification schemes[90] and the difficult question of branching in the thorium decay sequence.[91] Her report on branching was submitted to *Physikalische Zeitschrift* on 17 June 1912, the last publication to emerge from the wood shop of the Chemistry Institute. After nearly five years, she and Otto Hahn were preparing to move on.

That summer the Kaiser-Wilhelm-Institut für Chemie was under construction on the flat farmland southwest of central Berlin. More than a new institute, it represented a new avenue for German science: academic research, privately supported, protected but not directly regulated by the state. The need arose from the rapid expansion of the sciences and the concern, expressed by scientists for some time, over the inability of universities to quickly incorporate important new fields of research. Radioactivity was one such field: in 1906 Otto Hahn just barely found a place in Emil Fischer's institute and was never subsequently offered a university position.[92] Many young German scientists were seeking better opportunities abroad; others were going into industry.

Beginning in 1905 a committee of prominent university and industrial chemists proposed the establishment of a Chemische Reichsanstalt, an independent research institute, analogous to the existing Physikalisch-Technische-Reichsanstalt, to be jointly endowed and administered by government and industry. Despite the amicable relationship between the two sectors, the negotiations for such a partnership proved so difficult that nothing came of it until 1911, when the state stepped in as financial guarantor through the Kaiser-Wilhelm-Gesellschaft zur Förderung der Wissenschaften (KWG; Kaiser Wilhelm Society for the Advancement of the Sciences). The Kaiser himself assumed the title of "Protector," persuaded by his close adviser on education, the theologian Adolf von Harnack, who argued that "military might and science are the two strong pillars of the greatness of Germany." Plans were made for the Kaiser-Wilhelm-Institut (KWI) für Chemie and a second institute, the KWI für Physikalische Chemie und Electrochemie, to open in 1912; others were to follow. The Kaiser donated crown lands for the institute sites and the suburb of Berlin-Dahlem that would develop around them.[93]

The KWI for Physical Chemistry and Electrochemistry was entirely underwritten by Leopold Koppel, an extraordinarily wealthy financier and chairman of the board of the Auergesellschaft, a Berlin firm with a strong interest in physical chemistry research. Koppel's designation for institute director was the physical chemist Fritz Haber, who had discovered the process for synthesizing ammonia from nitrogen and hydrogen and, with Carl Bosch, had made it industrially feasible. With its close ties to a single firm and Haber's practical expertise, the KWI for Physical Chemistry started out with a pronounced orientation toward industrial research.[94]

The KWI for Chemistry took a more academic turn, financed by a consortium of chemical industries but administered by its own parent organization, the Verein Chemische Reichsanstalt, chaired by Emil Fischer, whose leadership in shaping the chemistry institutes and the Kaiser Wilhelm Society made him virtually the "president of German science." The institute's first director was the distinguished physical-inorganic chemist Ernst Beckmann; the associate director was Richard Willstätter, an organic chemist famous for his studies of chlorophyll and other plant pigments.[95] By that time Fischer was convinced of the importance of radioactivity, and apparently at his request Otto Hahn was offered a junior position as scientific associate (*wissenschaftliches Mitglied*) in the new institute, with responsibility for a modest radioactivity section, the first such laboratory in Germany.[96] With it came the title of Professor and the very decent annual salary of 5,000 marks.[97] Lise Meitner was welcome too—as an unpaid "guest."

She had worked with Hahn for nearly five years, published more than twenty articles, and established her reputation as a scientist comparable to him and as a physicist, independent of him. Still she had no position, no income, and, it now appeared, no prospects. And no other place to go.

Years later, when Lise was in her seventies and many awards came her way, she always insisted that young people needed them more.

> I know very well that especially in one's developing years, one urgently needs the encouragement of external recognition, to know that one's chosen path is not a wrong one. I myself greatly needed such encouragement, and to this day I am filled with gratitude that I received it from all of you [in Berlin] in many different ways.[98]

Lise finally saw the first glimmer of external recognition just as she and Hahn were moving into the new institute in Dahlem. Late in 1912, Max Planck appointed her his Assistent—the first woman Assistent in Prussia, the first rung on the academic ladder, her first paid position. She graded his students' papers.[99]

Why Planck appointed her just then is not known. Most likely the move to the new KWI for Chemistry emphasized the inequity of Meitner's status; perhaps, with the death of her father in 1910, Planck was concerned that financial difficulties would force her to return to Vienna. In any event, Meitner always thought of her assistantship with Planck as a decisive turning point. Fischer, who had followed Lise's work with kindly interest, took notice: within a year she too would be a "scientific associate" in the institute, her position the same as Hahn's, the radioactivity section theirs.[100]

It made all the difference. "I love physics with all my heart," she wrote to Elisabeth Schiemann. "I can hardly imagine it not being part of my life. It is a kind of personal love, as one has for a person to whom one is grateful for many things. And I, who tends to suffer from a guilty conscience, am a physicist without the slightest guilty conscience."[101] She knew her chosen path was the right one after all.

CHAPTER THREE

The First World War

*At night I feel slightly homesick for physics, but during the day
I think only of the patients.*

The Kaiser Wilhelm Institutes were inaugurated on 23 October 1912, a
cold, wet day of considerable ceremony. The dignitaries came to the
Chemistry Institute in special trains and carriages; His Excellency Emil
Fischer spoke, the architect was praised, Kaiser Wilhelm II urged the
assembled scientists to devise a fire-damp detector (which the two institute
directors, Ernst Beckmann and Fritz Haber, soon did). During a tour of
the building, the Kaiser glanced at Richard Willstätter's chlorophyll crystals through a microscope and viewed Hahn and Meitner's mesothorium
glowing in the darkroom. At the Physical Chemistry Institute there was
more, and finally all were invited to the palace in Potsdam for a forty-
five-minute royal tea, the men in frock coats as instructed, the wives in
high-necked dresses and hats, also as instructed. In his memoirs Otto Hahn
always savored that day. Lise Meitner never mentioned it.[1]

Those who first planned Dahlem and its institutes envisioned a leafy
intellectual retreat, a German Oxford. In 1912 it was all very new: two
handsome institutes side by side on open fields, trees no bigger than sticks,
freshly paved streets numbered but still unnamed. To the south stretched
a broad plain dotted with picturesque windmills; to the west, the
Grunewald forest. The Berlin subway did not yet extend to Dahlem—one
took the Potsdam train and got off at Lichterfelde-West—and there was
not a restaurant or shop in sight. The few professors whose villas were
finished shared their gardens with rabbits and partridges.[2]

The streets soon acquired names: Van't Hoff, Hittorf, Boltzmann,
Landolt, Faraday Way, which Willstätter selected as his home address; the

Chemistry Institute faced Thielallee. For his studies of plant pigments, Willstätter planted fields of tulips, chrysanthemums, and dahlias near the two institutes, a gorgeous sight each summer until the war. He was awarded the 1915 Nobel Prize in chemistry for his chlorophyll work—a welcome first for the Kaiser Wilhelm Institutes—and soon after accepted a call to Munich; the inorganic chemist Alfred Stock took his place.[3]

Otto Hahn's new position and salary provided the security he needed to consider marriage. A year earlier he had met Edith Junghans, a pretty art student from a good family; shortly before the institute opened in 1912, he showed her around, and afterward, on their walk together in the Grunewald, they became engaged. They were married the following March.[4] Lise and her brother Fritz sent a telegram from Vienna:

> Outside the ramparts of science
> In life some new things are found
> We wish long life and high activity
> To you both in your new compound.[5]

The year 1913 was good to Lise Meitner as well. Late that year she was made a *Mitglied* (associate) of the institute, the same position as Hahn's, and although her salary was still considerably less, the radioactivity section was now officially theirs: the Laboratorium Hahn-Meitner.[6] How this came about is not entirely clear, but it is evident that Emil Fischer took a special interest in the young woman he had once barred from all but the basement of his chemistry institute. This called for celebration, and Lise had a big one, a splendid dinner party with friends in the elegant Hotel Adlon on Unter den Linden.[7]

For the first time, finances were not quite so tight. The mesothorium Hahn discovered in 1906 emitted gamma radiation sufficiently strong to be useful as a radium substitute ("German radium") in medical therapy. In Fischer's institute Lise had helped Otto purify several hundred milligrams of mesothorium, which Knöfler, a Berlin firm specializing in thorium products, eventually developed into a profitable process. In 1913, Otto suddenly received a few thousand marks, then 66,000 marks in 1914 (6,600 of which he gave to Lise), and again 40,000 in 1915, more money than either ever had before. Hahn called it a "blessing" (*Segen*), and like many other good things, it ended during the war.[8]

Professionally, Meitner's and Hahn's greatest good fortune was the cleanliness and space of their fine new laboratories. Their section consisted of four rooms in the north wing of the institute's ground floor; Willstätter and a dozen co-workers occupied the entire first floor, Beckmann and his co-workers the second. The library, colloquium rooms, darkroom, and basement cold rooms were shared. To prevent radioactive contamination, Hahn arranged to store and handle their strongest preparations in the old wood shop in Fischer's institute; years later a small "radium house" was built for the purpose on the grounds of the KWI for Chemistry.[9]

In the wood shop the background radiation from radioactive gases, spills, and dust had made it impossible to investigate weak activities. Meitner and Hahn devised rigorous measures to keep their new quarters free of such contamination. Chemical experiments and physical measurements were confined to separate rooms; people handling radioactive substances were required to follow uniform procedures, shield everything they touched—a roll of toilet paper hung next to the telephones and every door handle—and sit only on specially marked yellow chairs. No one ever shook hands in greeting, a suspension of the usual rules of German etiquette. Later, when they each had their own section and their research occupied most of the building, Hahn and Meitner systematically trained and tested every student, assistant, and visiting scientist who came to work with them. Throughout the institute Meitner was known for being especially strict and for insisting that strong activities be kept out of her physics section on the ground floor. Her precautions were effective: twenty-five years later her section could still be used for studying very weak activities.[10]

In their spacious new laboratories Lise and Otto embarked on an intriguing project, made difficult by its complexity and the absence of strong radioactivity: the search for the immediate precursor—the "mother substance"—of the element actinium. Extremely scarce, actinium was by far the least well understood of the radioactive elements: its atomic weight was unknown, its chemistry uncertain, its half-life, estimated at twenty-five years, still in dispute.[11] Actinium, a beta emitter, was known to head a decay series distinct from the uranium-radium decay series, but it appeared that actinium was itself descended from uranium: it was always—and only— found in uranium-bearing minerals, always in amounts proportional to the uranium present. With its relatively short half-life, actinium evidently required constant replenishment to prevent its complete disappearance, yet the link between uranium and actinium was a mystery: pure uranium had never been found to generate actinium, even though actinium was not especially difficult to detect. It was assumed that uranium and actinium were separated by a long-lived intermediate.[12] The problem, as Meitner and Hahn saw it, was "to find that substance which . . . forms the starting point for the actinium series, and to determine whether and through which intermediates actinium is derived."[13]

The search was made possible by fundamental new principles that had begun to clarify the entire field of radioactivity. For years people had struggled with a growing burden of seemingly disparate data: they had studied the three decay series and identified some thirty radioactive species, observed occasional patterns in genetic sequences and a few regularities in alpha and beta decay, and learned the chemistry of some species well enough to place them in the periodic table. There was always more that was uncertain than sure: the decay series were incomplete, the chemistry often questionable; some species, including uranium, decayed so slowly they were falsely considered "radiationless," others so fast it was difficult to purify them and impossible to learn their chemistry; there was no explanation for the growing list of radioactive species that seemed chemically identical—radiothorium with thorium, radium B and actinium B with lead, among others; no one knew how many elements to expect between lead and uranium, or what their chemistry was likely to be, or even whether the periodic system could be trusted at its upper reaches—the rare earth elements, ominously enough, did not fit into the periodic system and were grouped separately.

Gradually, however, certain patterns in alpha and beta decay became clear enough to be stated explicitly: alpha decay produces a daughter whose position in the periodic table is two places below the parent; beta decay produces a daughter one place higher. Known as the group displacement laws, the two generalizations were stated independently and nearly simultaneously early in 1913 by Kasimir Fajans, a Polish radiochemist working in Karlsruhe, and by Frederick Soddy, in Glasgow.

The new rules implicitly correlated charge with elemental position: emission of an alpha particle, of mass 4 and charge +2, produced a species two places down; emission of a beta particle (electron), with very small mass and charge -1, produced a species one place up. Mass, which until then was considered the determinant of an element's identity, seemed to play no role at all. This was most clearly seen in the common sequence of an alpha decay followed by two beta decays, which produced a species at the original position, chemically identical but lighter by 4 mass units. Soddy's term *isotope* ("same place") was universally adopted to describe species that appeared at the same position in the periodic table but were radioactively different; it was explicitly stated that they were chemically the same but differed in mass. The group displacement laws and the concept of isotopy had incorporated the profusion of radioactive species into the periodic system, tied radioactivity to chemistry, and substituted charge for mass as the distinguishing characteristic of a chemical element. It was part of a revolutionary change in the underlying understanding of atomic behavior.

By 1913, four radioactive elements had been placed in the periodic table, based on their chemical similarities with lighter elements. Radium, divalent and similar to barium, was placed in Group II; thorium, tetravalent, in Group IV; uranium, hexavalent, in Group VI. The placement of actinium was in doubt for some time, but in 1913 Soddy, Fajans, and Meitner and Hahn independently sifted the evidence and concluded that actinium most likely belonged in Group III.[14] The sequence, apparently homologous to the third row transition elements, had a gap in the Group V position between thorium and uranium; the missing element, presumably pentavalent, was expected to resemble tantalum in chemical behavior. (See Appendix fig. 2.)

According to the group displacement laws, only two choices were possible for an actinium precursor: a beta-emitting radium isotope or an alpha-emitting isotope of the unknown pentavalent element. Since actinium had never been found in the presence of radium, Soddy concluded in 1913 that the mother substance of actinium was a long-lived alpha-emitting isotope of the missing Group V element, which he designated "eka-tantalum."[15]

Earlier that year Fajans and Oswald Göhring discovered UX_2, daughter of the beta-emitting thorium isotope UX_1 (^{234}Th). Applying the group displacement laws that Fajans had just articulated a few months before, they assumed that UX_2 would be a member of Group V and chemically bolstered that assumption by precipitating it with tantalic acid to partially separate it from its thorium parent. They named their new element "brevium" for its very short half-life of about one minute but realized that brevium, a Group V beta emitter, decayed to an isotope of uranium (^{234}U): it could not be the mother substance of actinium. They searched for a long-lived alpha emitter with similar chemical properties but uncovered nothing. Meanwhile Meitner and Hahn modified Fajans and Göhring's procedures somewhat to obtain a better separation of UX_2.[16]

Within a few months the once-empty Group V position had acquired two members—the UX_2 quite certain, the actinium precursor a still hypothetical but potentially more interesting isotope whose discovery, it was hoped, would resolve the ancestry of actinium. By then there was new evidence for its existence. The uranium decay product UY had been discovered in Rutherford's laboratory by G. N. Antonoff in 1911; late in 1913 Antonoff, Soddy, and, most thoroughly, Meitner and Hahn verified that UY was a thorium isotope (now known to be ^{231}Th) and a beta emitter.[17] Its daughter was necessarily a member of Group V, but neither Soddy nor Göhring could find a trace of it.[18] This was just what one might expect of the actinium mother substance, whose slow rate of decay and correspondingly low energy alpha particles would be impossible to detect in the presence of stronger activities.

Encouraged, Meitner and Hahn set out to find the actinium precursor. It was essential, first of all, to separate sufficient quantities of it from other activities in order to monitor its weak but steady alpha radiation; to prove that it was the mother substance, it would also be necessary to demonstrate that it generated actinium. Low levels of beta radiation were difficult to detect, however, so Hahn and Meitner decided to monitor not actinium itself but the alpha radiation from its decay products, which included actinium emanation (^{219}Rn), AcA (^{215}Po), and AcC (^{211}Bi). Owing to actinium's half-life of some twenty-five years, this would be a long-term project.

To locate and separate the mother substance, Meitner and Hahn explored two possible starting materials: uranium salts that had been extracted from uranium ore years before and the untreated uranium ore pitchblende. From an "old" uranium nitrate salt, Hahn and Meitner separated the tantalum group using the same procedure as for UX_2, mounted several samples under fixed electroscopes, and prepared to monitor them for several years. They knew that precipitation of the tantulum group also coprecipitated small amounts of polonium, thorium, and various other uranium decay products whose activities would initially obscure the weak radiation from the actinium mother substance, but they expected these short-lived activities to fade away while the actinium, steadily supplied by its mother substance, would gradually increase.

For pitchblende, which contained much more mother substance but also more of every other uranium decay product, Hahn and Meitner needed a cleaner separation method. Early in 1914 they found that treating pitchblende with nitric acid left an insoluble residue that was largely silicon dioxide but also contained most of the tantalumlike substances and relatively little extraneous activity. Knowing that Soddy and Fajans were also looking for the actinium precursor, Meitner and Hahn kept this promising new method to themselves.

Meanwhile they engaged in a number of other studies. With Martin Rothenbach, his first assistant, Hahn estimated the half-life of the very weak beta decay associated with rubidium (^{87}Rb), a value later used for estimating the geological age of rubidium-strontium minerals.[19] Meitner explored the question of the final product of the uranium decay series—whether it ended with stable lead or decayed further to bismuth and thallium.[20]

Along with isotopy, the most exciting developments at the time were Rutherford's nuclear atom, the atomic theory of Niels Bohr, and the x-ray studies of H. G. J. Moseley; while theoretical attention focused on extranuclear electrons in atomic orbitals, there were indications that nuclei might be the source of radioactivity: alpha particles, and also the electrons emitted in beta decay.[21] Meitner began studying the relationship between gamma radiation and radioactive decay, taking a particular interest in Rutherford's suggestion that the discrete lines in magnetic beta spectra might be tied to gamma emission and the finding by James Chadwick, a student of Rutherford's, of a continuous beta spectrum when gamma emission was absent.[22] The problem, of fundamental importance and great difficulty, would be pursued by Meitner and by members of Rutherford's group for many years.

In the spring of 1914, Lise Meitner received an attractive offer from Prague: a position in the lower academic ranks, with the prospect of advancement. Lise may not have wanted to leave Berlin, but Max Planck made certain that Emil Fischer understood she was seriously considering the offer. It took several months, but the call had its effect: by summer Fischer doubled her salary, to 3,000 marks, and praised her work. She accepted Fischer's beneficence, thanking him for the increased pay and for extending it in such "honorable form."[23]

Meitner's letter to Fischer was written on 2 August 1914 and mailed from Vienna, where she had rushed to see her brothers off to war. Already much of Europe was involved. Having smoldered for generations, the conflict had been ignited, finally, in Sarajevo that June, by a group of Bosnian Serbs who assassinated Archduke Franz Ferdinand, nephew and heir to the aging Kaiser Franz Josef. Eager to crush the south Slavs once and for all, Austria-Hungary refused conciliation, declared war on Serbia on 28 July, and bombarded Belgrade the next day. Within a week the machinery of mutual alliances was locked into war: Russia, champion of the Slavs, mobilized along her Austrian frontier, giving Germany its long-desired rationale for declaring Russia the aggressor and invading France and Belgium, which required England, treaty-bound to defend Belgian neutrality, to enter the war. For the moment Turkey hesitated, and Italy, nominally an Austrian ally, declared itself neutral. Otherwise Germany and Austria had enemies on every side.[24]

The mood in Austria was jubilant. Early reports gave the impression of decisive victories. "Doing scientific busywork seems senseless to me now," Lise wrote to Elisabeth Schiemann. "My mother's apartment is directly across from the train station and each day I see the men going off to war with unbelievable enthusiasm. Those who remain behind outdo themselves in demonstrating love for those who are leaving, and the train station presents a festive, joyous sight all day long."[25] Lise followed the German success in Belgium and France with "honest admiration";[26] a few weeks later, as Austria suffered heavy casualties in Galicia (Austrian Poland), she took comfort in the thought that "the fate of the individual recedes behind the greater cause."[27]

Germans, too, exulted in a new sense of national purpose. Otto Hahn, James Franck, Gustav Hertz, Hans Geiger, all in the reserve, were called up and left at once, sure they would be home victorious by Christmas. "The die was cast, and hardly anyone had any doubt of our winning this just war," Hahn recalled. Although he witnessed the killing of civilians and the burning of medieval Louvain during the invasion of Belgium, the camaraderie of army life appealed to him, and being a soldier seemed quite pleasant at first, "rather like going for a stroll in an occupied country."[28]

It may be difficult for us, at the end of this savage century, to comprehend such eager acceptance of war, particularly among educated people whose outlook—at least professionally—was quite international. We must remember that for most Europeans in 1914 war was an abstraction, the most recent having been the Franco-Prussian War of 1870–1871, a conflict neither long nor bloody enough to discourage Germany from dreaming of further expansion or to keep France, humiliated by a huge indemnity payment and the loss of Alsace-Lorraine, from desiring a return engagement. In place of war, there had been forty years of nonpeace: economic belligerence, incessant trouble in the Balkans, rival alliances, an escalating arms race. For all its strength, Germany tended to feel underappreciated and insecure. It derided England as a nation of shopkeepers, France as decadent, Russia as a Slavic menace; believing they alone were infused with the depth of spirit and intellect they called *Kultur*, Germans could scarcely imagine why the French would refer to them as barbarians or the English as Huns (a designation, incidentally, that Kaiser Wilhelm II first regarded as positive in connotation).[29] Apart from a socialist minority, few in any nation were free of narrow nationalism; the press was hardly objective, universities even less so: academics took pride in avoiding politics, an activity they disdained in favor of an almost reflexive identification with the state. As it grew dominant in the heart of Europe, Germany felt encircled and confined, a condition seemingly unamenable to compromise, its only ally the disheveled Austro-Hungarian Empire, a cauldron of endless dissension. In the end, war seemed the only way to set things right. When it came, finally, it was greeted almost with relief.

In the first surge of national unity, professional and class distinctions seemed to fall away. Everyone wanted to do their share, and those not eligible for military service applied their science to the war effort. Max von Laue, rejected for medical reasons, spent the war investigating the physics of electronic tubes.[30] In Dahlem and elsewhere, scientists searched for substitutes for strategic raw materials—most urgently, nitrates from Chile—that had been cut off by trade embargoes and the British blockade. Emil Fischer and Fritz Haber took the lead in persuading industry to greatly increase the production of synthetic nitrates for munitions and agriculture; substitutes were also sought for many other materials, including copper, sulfates, rubber, fiber, and food. There had been no planning, in either the economy or the military, for a war lasting more than six months.

Fritz Haber was forty-six years old, above the age for active military duty; because of his Jewish background, he had been ineligible for commission as a reserve officer. As a physical chemist, he was already famous (together with Carl Bosch) for developing the Haber-Bosch process for the industrial synthesis of ammonia, a crucial starting material for the synthetic nitrates the military required for an extended war: without the Haber process, the war would have been over in a year.[31] For Haber, that was not enough: he was driven to prove his absolute devotion to Germany. In the first months of the war, he was asked by the military to find a motor fuel antifreeze to replace toluene, which was essential for TNT and already in short supply; he quickly determined that xylene and other petroleum derivatives would do. The project was minor. By December 1914, Haber had committed himself and his Dahlem institute to a larger and more dangerous aspect of the war effort: tactical military research, including alternative explosives, nonlethal chemical irritants, and poison gases.[32]

While radioactivity research was not a candidate for war service, Lise Meitner herself definitely was. In Vienna in August 1914, she immediately inquired of the Red Cross about the need for nurses, but the agency was disorganized and the nursing courses were all filled. When she realized there was nothing for her in Vienna, she contacted Planck about the Berlin university's war auxiliary and asked Ernst Beckmann whether the KWI for Chemistry might be converted into a military hospital. With all answers negative, she returned to Berlin in September 1914 and signed up for x-ray technician training and an anatomy course in the hospital in Lichter-felde.[33] Restless and distracted, she nevertheless maintained the labora-tory, completed a beta spectrum begun with Hahn and Otto von Baeyer before the war,[34] and finished her own study of the end product of the uranium decay series.[35] As a continuation of their most important long-term project, the search for the actinium precursor, Lise regularly mon-itored their samples and kept an eye out for Kasimir Fajans, who, she was sure, was doing the same.

For the duration, England and Germany each sequestered the other's nationals. A German glassblower who had worked in Manchester for years was interned; a move was made in the Royal Society to remove the physicist Sir Arthur Schuster from his position as secretary for no other reason than his German birth.[36] In Germany some five thousand British civilians were incarcerated for the duration of the war in Ruhleben, a deserted racetrack on the western outskirts of Berlin. Among the prisoners were James Chadwick, a student of Rutherford's who had come to Berlin to work with Hans Geiger, and Charles D. Ellis, a young engineering cadet who was trapped in Germany while on vacation in 1914. In Ruhleben the men suffered from cold, extreme congestion—six men to a two-horse stall—and bad food;[37] at the intervention of Meitner, Planck, Rubens, and others, Chadwick was allowed occasional visits and even some scientific equipment, but he remained a prisoner until the end of the war. Ellis's association with Chadwick in Ruhleben converted him to a career in physics; on his return to England he studied in Cambridge and later joined Rutherford's group. During the 1920s Ellis and Meitner maintained an active collegial relationship as they worked on similar problems of beta spectra.[38]

At the outset, everyone was sure the war would be brief and victory glorious. The English physicist H. G. J. Moseley was in Australia when the war began; anxious not to miss the great adventure, he raced home, insisted on being taken into the Royal Engineers, and then could hardly wait for combat.[39] No nation thought it necessary to preserve its scientific talent. Fritz Hasenöhrl, successor to Boltzmann and Austria's leading theoretical physicist, was inducted immediately. Both were killed in battle in 1915: Moseley in the Dardanelles expedition, Hasenöhrl at the Italian front. Theoretical physics in Austria did not recover for a generation.[40]

In a spirit of Austrian solidarity, Stefan Meyer and Lise Meitner maintained a steady correspondence. "It was very good of you," he wrote in September 1914, "to defend us against the [German] accusation that Austria has not yet defeated the Russians."[41] He was appalled at the internment of Chadwick; Robert Lawson, an English physicist in Meyer's Radium Institute, was not allowed to visit cafés or stay out late but was otherwise undisturbed.[42] Most news concerned colleagues in the military: Erwin Schrödinger at the Italian front, Arthur Boltzmann a balloonist in Serbia, Hasenöhrl a mechanic in a motor unit in Cracow, others in hospitals as medical and x-ray technicians. The fighting in Poland was especially fierce. "Poor Dr. Michl (alpha tracks on photographic plates) fell in Galicia, Kofler was killed there three months ago. Also O. Scheuer, who last worked with M. Curie."[43]

As the war continued, the mood in Germany turned somber. The world community strongly protested Germany's violation of Belgian neutrality, its atrocities against civilians and destruction of cultural treasures. Germans were pained and surprised: to them the war was purely defensive. When English newspapers suggested that a distinction be made between German culture and Prussian militarism, intellectuals issued the "Appeal to the Cultured Peoples of the World" (*An die Kulturwelt! Ein Aufruf*), proclaiming their solidarity with the German army, justifying the invasion of Belgium as a defense of German culture, and conjuring the "shameful spectacle . . . of Russian hordes . . . unleashed against the white race." The appeal was signed by ninety-three prominent artists and scientists, including Max Planck, Emil Fischer, Fritz Haber, Philipp Lenard, Walther Nernst, Wilhelm Ostwald, Wilhelm Röntgen, and Richard Willstätter. (A counterappeal for European unity and peace was signed by only four, one of whom was Albert Einstein.) Planck and Willstätter later regretted their early enthusiasm. "The outbreak of war overtook us like a natural disaster," Willstätter reflected in his memoirs. "The professors were convinced that Germany bore no responsibility for the war and that war had taken it by surprise." Willstätter concluded that educated Germans, himself included, had failed abysmally in their responsibilities as citizens. Planck, who welcomed the first flush of patriotism and saw two sons off to war, later did what he could to repair relations with the international scientific community.[44]

In December Otto Hahn was in heavy fighting in Belgium. Lise wrote constantly, trying for a cheerful tone, with scientific details, news of colleagues, reports on the well-being of her relatives in the military. When she learned that his assistant, Martin Rothenbach, had been killed in France, she could hardly tell him. "I knew it 10 days ago," she finally wrote in December 1914, "but could not get myself to write to you because I knew from the way it affected me, that it would also depress you greatly. But I think you should write to his parents."[45]

In January 1915 Hahn was recruited for a new assignment, a special unit for chemical warfare that Fritz Haber was organizing. To all objections, Haber countered that poison gas had already been used by the French; indeed, the French and British were experimenting with tear gas for driving soldiers from their trenches. Almost no one refused Haber: his close friend Richard Willstätter devised a gas mask without which a gas offensive would have been impossible; Hahn, Franck, Hertz, and other scientists joined Pioneer Regiment No. 36. After brief training in meteorology and the use of poison gases and protective devices, the "gas pioneers" began field experiments with chlorine in Belgium. But the first casualties of Haber's military research occurred in his own institute in December 1914, when a test mixture of explosive materials produced a huge detonation that killed the chemist Otto Sackur and caused another chemist, Gerhard Just, to lose his right hand.[46] Meitner described the funeral to Hahn: Haber and Willstätter wept uncontrollably. "One can only hope these dreadful sacrifices have not been in vain. If only things would move forward more quickly!"[47]

The chemical warfare project was secret, but Lise soon had a "fairly good idea" of what it was. "[I] can well understand your misgivings," she reassured Otto, "yet you are certainly justified in being an 'opportunist.' First, you were not asked [but ordered] and second, if you do not do it, someone else will. Above all, any means which might help shorten this horrible war are justified."[48] In April 1915 the Pioneer Regiment was transferred to Galicia, where a potent mixture of chlorine and phosgene was directed against Russian soldiers, with great suffering and loss of life. Although Hahn was shocked by the agony of the dying Russians, he stayed with the Pioneer Regiment for the remainder of the war, experimenting with poison gases and the techniques for delivering them. In his memoirs years later Hahn explained, "As a result of continuous work with these highly toxic gases, our minds were so numbed that we no longer had any scruples about the whole thing."[49]

The explanation was facile: Hahn was not an introspective man. As a participant, he was horrified and then suppressed his horror, perhaps a necessity at the time. But later, when the immediate numbing might have worn off, he spoke and wrote about his military experiences with considerable pride, as old soldiers do, never reflecting in depth on the use of poison gas, or the ethics of applying his science to war. During the Third Reich and World War II, the scale of horror would be greater, but Hahn and other Germans would find, even in the absence of toxic gases, that their minds were similarly numbed.[50]

Through the spring of 1915 Lise continued to work, managing the laboratory and purchasing their first electromagnet for beta spectra.[51] In February, two Turkish princes toured the institute. "Their royal highnesses assured me it was the first time they had the honor! (such flatterers) to meet a woman doctor. Naturally I am swollen with pride, which, thin as I am, I can certainly afford." From Brussels, Otto sent her a snapshot of himself and a collar of Belgian lace: "I really don't know which gave me more pleasure, perhaps your picture a little more."[52] In her letters Lise was by turn sister, friend, colleague, telling him of the whereabouts of their friends, scolding him when he did not write, planning for his thirty-sixth birthday to be celebrated at the front,[53] and always informing him, in detail, of the progress she was making with their research. And because German soldiers tended to disparage their Austrian counterparts, Lise made sure Otto understood that all the male members of her family— brothers and brothers-in-law—were fighting too: in Cracow, in the Polish offensive, in the heavy fighting at the San River in Galicia.[54]

By late spring, it was evident that the war would not soon be over. Fighting on the western front reached a stalemate, and Germany joined Austria in a major offensive against Russia in Galicia, from which Austria had retreated in the first weeks of the war. In May the Russians were forced back from the San; by the end of June the German-Austrian forces re- captured Lemberg (Lwow, Poland; now L'vov, Ukraine), capital of Gali- cia. The war widened. Turkey joined the Central Powers in 1914, and Italy finally declared war on Austria, its old enemy, in 1915 in the hope of redeeming Trieste, the Dalmatian coast, and the Südtirol (Trentino). Casualties mounted without end. That summer Meitner learned that Marie Curie had initiated an ambulance service with radiological equip- ment and was working in field hospitals with her eighteen-year-old daugh- ter, Irène, behind the lines in France.[55] In July 1915, Lise left for Vienna to volunteer as an x-ray nurse-technician with the Austrian army.

Within days Lise was accepted, trained, vaccinated, and assigned to a military hospital in Lemberg, not far from the Russian front. On 4 August she left Vienna with a unit of 220 men, 50 nurses, 10 physicians, and the complete facilities for a hospital, on an "endlessly long" train that slowly made its way to Budapest, skirted the Carpathians in northeastern Hun- gary, then wound north through the mountains to Poland. She wrote to Elisabeth,

> We were underway for 60 hours, but it didn't seem that long. Naturally it was a very old third-class car with narrow wooden benches, but the doctors and nurses each had their own bench so one could stretch out at night, and during the day there was so much that was new and interesting to see that time went quickly. In Miskolcz [Miskolc], a little Hungarian town, . . . a German doctor told us that Warsaw had been taken. . . . [Y]ou can imagine how happy we were. Near Mezo-Loborcz [now Michalovce, Slovakia], as you may remember, there was very bitter fighting. . . . There are houses that were shot to pieces, the front wall alone remains, sometimes even this is blown away, and the roof lies on the ground. . . . The surroundings . . . are just beautiful, on one side the plains and to the north the Carpathian

mountains with beautiful forests. From there until Lemberg we went through an uninterrupted war zone, always the same picture. All the railroad bridges were destroyed by mines, and the train crossings were temporary wooden bridges from which one could see the old bridge on the rocks below.

The suffering of the wounded came as a shock.

We are converting the local Technical Institute into a barracks hospital. Until now there was a field hospital here, with about 6,000 to 7,000 wounded who had to be transferred elsewhere as quickly as possible. Now as a barracks hospital at least some of the wounded can stay longer to make a recovery. . . . Oh, Elisabeth, what I have already seen—I never expected it to be as awful as it actually is. These poor people, who at best will be cripples, have the most horrible pains. One can hear their screams and groans as well as see their horrible wounds. Today we had a Czech who was severely wounded in his arms and legs who moaned in pain while tears ran down his face. . . . Since we are only about 40 km from the front we get only the most severely wounded here. I tell myself this for consolation. But one has one's own thoughts about war when one sees all this.[56]

Until the x-ray facilities were ready, Lise assisted in several operations each morning, cleaned operating tables and instruments, and bandaged the wounded.

27 August. Today we amputated the foot of a very young Hungarian, and it upset me that I could say nothing to him. . . . A young Polish soldier said quietly, "I know I will die," and he did.

24–26 September. It is already terribly cold. I work only with the very badly wounded. . . . It is impossible how much they suffer. We have so many wounded Hungarians, and far too few Hungarian nurses. Those with back wounds are often so badly wounded that nothing can be done for them, and they die very slowly. The Catholic priest here gives his time to Catholic, Protestant, and Jew alike. There are many good people here.

18 October. I have done over 200 x-rays. . . . The surgeon told me that x-rays have saved the life of at least one of the wounded, by identifying a hernia of the bladder and not a stomach wound as the doctors originally thought. . . . This is a small happiness among much that is very hard. I have seen young soldiers 18 or 19 years old who are operated on four or five times and die anyway.[57]

Writing to Otto, Lise omitted the medical details.

10 September 1915. Dear Herr Hahn! . . . We have many Russians among the wounded, and also prisoners who work for us. They are mostly good-natured, patient men, very caring to the sick ones. . . . It is difficult to understand them, of course. A few words that I picked up as a child, a few

Polish or Russian words that I learned here, is my whole vocabulary. Hungarian is even more difficult.

14 October. At night when I lie in bed and cannot sleep right away I feel slightly homesick for physics, but during the day I think only of the patients. . . . The gratitude that they show always makes me somewhat ashamed. My younger brother [Walter] is now a cadet at the Isonzo front [Italy].

28 November. The bitterness against the Italians is very great—the soldiers regard every single Italian as a personal enemy.

9 January 1916. I do sometimes have homesickness for physics, above all feel I hardly know what physics is anymore. . . . My health is always very good. I don't think I weigh as much as 50 kg [110 lb.], but after all I didn't come here to get fattened up. . . . When there are no x-rays needed, I help in the operating room—I have even done anesthesia, which I dislike—and in addition I am the hospital mechanic. I fix broken electrical cords and apparatus, make T-tubes, catheters, and so on.[58]

In Lemberg, Lise received a marriage proposal—perhaps not the first, or the last, but apparently the only one of which we have a record. Somehow she met a professor from Greece, who then wrote to her from Piraeus: "I would like to have the honor of marrying you. I admire you and the other Germans and your wonderful country. I hope you take my offer of marriage seriously. Also I would like your photograph. Please answer me. P.S. Greece is now all for the Germans."[59] One suspects Lise did not answer. But she did keep the note, flowery writing, purple ink, and all.

By early 1916 the eastern front reached a stalemate and there was very little to do.[60] Lise requested a transfer to the south, the scene of heavy fighting near Trient (Trento) in the alpine Südtirol, a region Austria would eventually lose to Italy. The transfer took months; she visited Berlin and worked halfheartedly in Vienna for a while in Stefan Meyer's Radium Institute.[61] Assigned to Trient in June, Lise was again mostly idle and asked to be sent "any place there is work."[62] In July she was in Lublin at the height of a renewed Russian offensive in Poland, but the doctors were themselves too exhausted or ill to treat the many wounded and Lise could do little on her own. "I feel superfluous," she wrote Elisabeth in August. "Without me things would go just as well. If this is true, then my duty is to go back to the Kaiser Wilhelm Institute. I say my duty because if I had followed my wishes, I would have gone back long ago."[63] In September: "My last card! I am coming back to Berlin. . . . I must work so I feel relieved already."[64]

Lise returned to Dahlem in early October 1916.[65] She found the KWI for Chemistry almost entirely converted to war research. After Willstätter left for Munich in March 1916, his section on the institute's first floor had been requisitioned by the aerial photography command (Luftbildkommando) and others from Haber's "military institute." In a transparent attempt to do the same for Meitner and Hahn's section, the Haber group offered Meitner a position with them, "with the special inducement," she noted sardonically, "that I would be paid. They evidently thought that I live here—shall we say—on a private income." She turned the Haber offer down, concerned about the fate of the radioactivity section she shared with Hahn.

> 25 October 1916. . . . I hope that the Lion [Haber] will not get his claws into our modest section, especially since our private laboratory with its physical apparatus would hardly be useful for the chemical studies in question. If I am not prevented from doing so, I will try to work. I would very much like to repeat Chadwick's work, namely, to count the spectrum of beta radiation in a magnetic field. . . . What do you say about [the German] victory in Dobrudscha [Dobruja, Romania]? One must be glad about it, if one still has the capacity to be glad.[66]

Like most Germans, Lise was exhausted by the war but by no means without hope for eventual victory.

Lise hastened to measure the preparations she had set aside before she left, in particular, the samples that had been monitored since 1913 for evidence of actinium precursor.

> 16 November 1916. Dear Herr Hahn! . . . The Haber people treat us of course like captured territory; they do not take what they *need* but what they *like*. . . . Who will guarantee that they won't come over here, and then everything will really be lost. I will do everything to prevent it; we have measurements here that have been in progress for so long . . . but they have the arrogance of victory. . . . It really is very thoughtful of you to be concerned about what I eat; you imagine [the shortage of food] to be worse than it actually is. . . . Last night I was at Plancks'. They played two marvelous trios, Schubert and Beethoven. Einstein played violin and occasionally made amazingly naive and really quite peculiar comments on political and military prospects. That there exists an educated person in these times who does not so much as pick up a newspaper, is really a curiosity.[67]

Einstein considered his views on German militarism and his hopes for the defeat of Germany to be so far removed from those of his friends that he did not discuss them, except to ask questions "in the Socratic manner, to challenge their complacency." He was quite aware that "people don't like that very much."[68]

In January 1917, Otto Hahn was in Berlin for several weeks. With Lise's permanent return, they were eager to resume their search for the actinium precursor. The tantalum group they had extracted from the old uranium nitrate salt in 1913 still gave not the slightest indication of actinium, its emanation or decay products. They abandoned it and turned to the silica residue they had first separated from pitchblende early in 1914. At the time they had determined that the residue included essentially all the tantalumlike substances in pitchblende and with them, presumably, the hoped-for "eka-tantalum," mother substance of actinium. Under their electroscopes three years later, the residue was giving off their first glimmer of success: a tiny but unmistakable amount of actinium emanation.

It was hardly the ideal time for a major project. In this, Germany and Austria's first "turnip winter," food was scarce, fuel almost nonexistent, equipment expensive and hard to get, pitchblende unavailable. Perhaps the greatest difficulty was that students, assistants, and technicians had long since disappeared into the military. Lise would have to work alone.

And as much as Lise favored the war effort, she and Hahn were eager to protect their work and their section from Haber's incursions. Support came, once again, from Emil Fischer: where Haber was glad to see the wall between academia and the military come down, Fischer was trying to shore it up. In January 1917, Lise Meitner was appointed head of her own section for physics, in effect dividing the Laboratorium Hahn-Meitner into the Laboratorium Hahn and the Laboratorium Meitner, with a pay increase of 1,000 to 4,000 marks (essentially equivalent to Hahn's 5,000 marks, which included a marriage supplement).[69] Perhaps more than any other appointment, Meitner took this as an expression of trust. Although the hiring and purchasing necessary to build her section would be deferred until after the war, she had been granted the administrative authority to direct her own work and the power to protect her section.

In February, Otto returned to his unit, impatient for news.

22 February 1917. Dear Herr Hahn! The pitchblende experiment is of course important and interesting but I cannot do it right now—don't be angry, please. . . . I have ordered the [platinum] vessels for our actinium experiment . . . and will begin as soon as they arrive. . . . Yesterday I gave

a colloquium. I thought of you and spoke loudly and looked at the people and not the blackboard, although under the circumstances the blackboard seemed far more appealing than some of the people. . . . Be well and *please don't be angry* about the pitchblende delay. Believe me, it's not due to lack of will, but really only lack of time. I can't very well do as much work alone as the three of us used to do together. Today I bought 3 meters of rubber tubing for 22M!! I got a real shock when I saw the bill.[70]

The equipment was ready, finally, in early March. The initial steps, several arduous treatments with boiling concentrated acids, were designed to separate the silica (SiO_2) from the pitchblende and the tantalumlike substances from the SiO_2; because the quantity of "eka-tantalum" was so minuscule, a tantalum compound was added as a carrier. Lise began by pulverizing 21 grams of pitchblende—nearly all they had—and boiling it in concentrated nitric acid, decanting the solution and boiling the undissolved remainder several times until an insoluble SiO_2 residue was obtained, which was filtered, washed several times, and dried. The residue weighed 2 grams, of which 1.5 grams was set aside as a control. The remaining 0.5 gram dissolved almost entirely after repeated treatment with hydrofluoric acid (HF) to which a few milligrams of potassium tantalum fluoride had been added; this reaction required platinum vessels, those Lise had ordered in February, since glass and most other metals dissolve in hydrofluoric acid. The HF solution was filtered, boiled down somewhat, and then evaporated to dryness in boiling concentrated sulfuric acid—of all the difficult steps involving hot concentrated acids, this was the most unpleasant—which left a solid material that dissolved almost entirely when repeatedly boiled in concentrated nitric acid. A tiny bit of solid did not dissolve: this contained all the tantalum and with it, presumably, the actinium precursor.

Meitner monitored the preparation for alpha radiation for several weeks. "According to our assumptions," Meitner and Hahn wrote later, "since the mother substance emits α-rays (presumably of shorter range than its decay products), our preparations ought to show a strong α-activity, which in view of the [25-year] half-life of actinium ought to increase only very slowly, despite its five α-emitting decay products. However, if one filters out the slow α-radiation of the mother substance of actinium, one should observe an activity which rises from almost zero at a much faster rate. (One cannot think of measuring the β-activity of such a weak preparation.)"[71]

Exactly as expected, the new preparation displayed a constant short-range α-radiation and a faster α-radiation whose intensity doubled in several weeks. The untreated 1.5 grams of SiO_2 showed the same.

The results could not have been better, but still they did not prove that the source of the alpha activity was in fact the actinium precursor and its decay products. Because Meitner and Hahn's preparations were too weak to separate the actinium decay products and measure them directly, they chose an indirect proof: eliminating from consideration any other activities with similar short-range alpha radiation and chemical behavior.

This drew on Hahn and Meitner's entire experience in radioactivity. In April, when Hahn was again in Berlin for a short period, he and Meitner narrowed the possibilities to two: ionium (^{230}Th), a long-lived alpha emitter, and radium D (^{210}Pb), which is a beta emitter but decays to ^{210}Bi and then ^{210}Po, an alpha emitter of fairly long half-life. To determine whether ionium or RaD were indeed present, Meitner and Hahn added to the silica residue measured quantities of their respective beta-emitting isotopes UX (^{234}Th) and ThB (^{212}Pb) and processed the silica as before. The complexity of the indicator experiment may be gathered from Meitner's letters to Hahn.

7 May 1917. Dear Herr Hahn! . . . Preparation #9 seems very promising. . . . [I]n any case the strong alpha activity cannot arise solely from ionium. . . . In my opinion, ruling out RaD is not as certain, since at first #9 had some ThB; the corresponding polonium [^{210}Po, derived from ^{210}Pb] must also be there. . . . I am also disturbed by #5; its α-activity has been constant since April 20; its beta activity, however, has continuously increased. According to your experiments this cannot be due to UX from uranium, so maybe it is due to the accumulation of RaE [^{210}Bi].[72]

15 May 1917. As it's hardly possible for you to retain all the preparations and their numbers in your head, I will very briefly remind you of the separations: no. 8 is what did not dissolve in HF [presumably containing no precursor]. No. 9 is the [one which did dissolve in HF and contains] Ta after H_2SO_4 evaporation [but which did not dissolve] in hot HNO_3 [this was expected to contain precursor]. No. 10 is the NH_3 precipitate from the nitric acid solution [the portion that did dissolve in HNO_3—no precursor expected]. No. 8 originally contained most of the ThB and UX [indicating that most RaD and ionium separated from the precursor]. . . . No. 9 also must have some ThB and a little UX. From the 5th to the 14th [of May] its beta activity declined exactly as UX [indicating no other beta impurities], its alpha activity shows much less decrease. There is, therefore, a constant alpha

activity, which in any case cannot be due only to ionium, as one can see from the relative amount of ionium and uranium X in no. 8 and no. 10. . . . In addition there is a not very penetrating beta radiation, possibly RaD or RaE. . . . No. 10 had (contrary to what you expected) only very little ThB and almost only UX, the ratio of these, as seen by decay measurements, is the same in no. 9, only no. 10 has much less constant alpha [less mother substance, as expected]. . . . No. 9 and no. 10 must have the same ratio of ThB to UX, but since no. 10 is much more strongly active than no. 9, it must have much more RaD than no. 9, if RaD is there at all. . . . No. 5 is the part we worked up (dissolved in HF, evaporated with H_2SO_4) [with no added indicators] from the 2 g of SiO_2 out of 21 g pitchblende. It shows, in addition to a quite constant alpha activity, a markedly increasing beta activity [possibly from actinium itself]. . . . You must only give me a little time—I mean you must not get impatient. The measurements all take a long time—I must also standardize [the electroscopes] exactly, very necessary with the weather we are now having.[73]

The intensity of the work consumed Lise; she hardly mentioned the war, and she forgot to ask about Otto's health or send her usual greetings to James Franck. The one intrusion came in her letter of 15 May, after four pages of science. Grete Planck died that day, after giving birth to a little girl. "It is terribly hard for me to bear," Lise wrote and did not mention it again. Perhaps she felt no right to express her own grief, not to Hahn after three years of war, not to the Plancks who had now lost two children, their oldest son, Karl, having been killed in France the year before. There was nothing one could say. Lise's letter ended, "You need hardly worry about others' eka-tantalum experiments. In Vienna they're not being done, I know that for certain, and I heard . . . that Fajans is doing war service."[74]

19 June 1917. I have good reports about our work. Preparation no. 9 seems really to be something, the alpha activity is apparently already constant. . . . I think we can count on having the substance in hand.[75]

For verification, it would be necessary to measure actinium decay products directly; for this, more starting material was needed. In Vienna at Easter, Lise had asked Stefan Meyer for some pitchblende, without telling him much about their work, and she had tried to buy some from the Joachimsthal mines. But, she told Otto pointedly, as a result of an embargo initiated by Germany, none could be shipped until after the war.[76]

So Meitner turned to Friedrich Giesel, the elderly industrial chemist who had discovered actinium in 1902 (independently of André Debierne but two years later) and was associated with the radium-producing firm of Buchler & Co. in Braunschweig. With considerable difficulty Lise obtained 100 grams of "double residue" (*Rückrückstände*)—pitchblende from which both uranium and radium had been removed—and with 43 grams, twice their previous trial, began the separations as before. But Giesel's pitchblende was different somehow, and the earlier method did not work. By the end of July, Lise was exasperated.

> 27 July 1917. Dear Herr Hahn! With all due respect for your reasons for not writing, don't you think that others besides yourself are also pressed for time? You are thrifty with everything, even with friendly words. Unfortunately, as you can see, I am not at the exalted place where I do not write to you. The extenuating circumstances are that I am not just writing to be friendly but also have some news to report. And so as not to offend the practical God of time I shall get to it right away. . . . With somewhat unjustified optimism I attempted to separate 43 g of Giesel's preparation A. It appeared to go quite well, but the final result was not very satisfactory.

After a detailed description of her attempts, Lise was interrupted.

> 6 August. I got this far when your wife visited me and took me swimming. . . . When I came home that evening I found, in addition to both your letters (the beginning of my letter is still justified since you only wrote for radioactive reasons!), a telegram from my brother telling me he would arrive the next morning. I spent 3½ days with him and did nothing in the institute. . . . Now that short pleasure is over. . . . Be well, and write, at least about radioactivity. I remember a time very long ago when you would once in a while send a line even without radioactivity.[77]

> 24 August 1917. Concerning our small personal misunderstanding, I am unconvinced by your objections. I am still of the opinion that a friendly line from time to time is not a great sacrifice for friendship—but I do not want to lead you to such old-fashioned excesses if you yourself do not feel a need for them. Anyway I am not angry now, nor was I before. Concerning the *abracadabra* [actinium precursor], the only news I can tell you is that I worked up 10 g of Giesel's preparation B. . . . No. 6 has increased significantly, no. 9 in beta electroscope is still decreasing slightly. . . . I hope you will not be angry with me if I do not try any further separations. First, I really believe it is of no use, as long as the preparations are decreasing and, second, to tell the truth, I would like to finally have a vacation. With all the things left to measure, I will not finish until the 31st, and then I would like to go to Vienna. Before I go I will set up some kind of system for collecting active deposits. As long as the strong preparations are still decreasing, it is difficult to take convincing measurements. . . . In the last issue of *Electrochem[ische]*

Z[eitschrift] Fajans . . . defends his brain-child "brevium" and defends himself
against the possibility that Soddy's "eka-tantalum" might have priority over
his "brevium." (He says that) Soddy proposed that eka-tantalum is a long-
lived mother substance of actinium that has *never been found* and for that
reason has never been named. So you see, Fajans does not have eka-tantalum!
. . . Now be well. Your wife told me you have obtained a bottle of schnaps
for my "spiritual" needs. Many thanks! It will come in handy as my last bottle
disappeared almost entirely into my brother's travel flask.[78]

When she returned to Berlin, Meitner monitored the old and new
preparations. "They are increasing [in activity] completely on schedule. I
have measured the curve [for actinium beta activity]; the increase has been
completely linear for six months now, which is very comforting."[79] For
Lise, all doubts were gone; every experiment afforded unequivocal evidence of actinium, its emanation, its active deposit, and its precursor. What
remained was to determine the half-life of the mother substance, by
quantitatively measuring the range of its alpha particles and the rate with
which its decay products formed.

For this a much stronger preparation was required. In November Lise
traveled to Braunschweig to ask Professor Giesel to perform the extraction
in quantity in Buchler's industrial laboratory.

16 November 1917. Tuesday I went to Braunschweig. The connections are
very poor and I spent two hours in Magdeburg . . . but I had a good lunch
which really counts for a lot these days. Giesel was very reserved for the first
15 minutes, but then thawed visibly and at the end was quite friendly. I told
him very frankly about our experiments and asked him to keep it confi-
dential. According to Giesel, a few hundred pounds of residues A and B are
available which contain a high percentage of pitchblende. But the factory
stopped radium production during the war so that those are the only residues
available. In addition, they had to give up their platinum vessels, and of
course one cannot boil sulfuric acid in lead vessels. My discussion with Giesel
was not easy because he is a somewhat absent-minded older gentleman who
finds it hard to follow someone else's thoughts and in addition seems to be
somewhat impractical, but we agreed on the following. Giesel will pretreat
1 to 2 kg of the [pitchblende] residue with HF and H_2SO_4. . . . I think the
quantity then will be small enough for me to do the evaporation of H_2SO_4
in our large platinum vessel. . . . I wrote the procedure down for Giesel, as
he is terribly forgetful. For example, he asked me three times why we did
not do the extraction with sulfuric acid. Each time I asked him if concentrated sulfuric acid can be filtered, and when he said no, I told him that
radium would precipitate in dilute sulfuric acid. He forgot this several times.

He certainly has gotten old, completely white and a little unsteady, but still he is quite a character. The death of his only son apparently dealt him a terrible blow. Personally he was very kind to me, insisted that I come to his house one evening where he and his intelligent and charming wife outdid themselves on my behalf. . . . The success in Italy is very good, and the events in Russia [revolution], however they may turn out, are in any case better for us than for the Entente.[80]

Victory for Germany and Austria still seemed possible.

As promised, Giesel partially processed 1 kilogram of pitchblende; in December, the third and most quantitative phase of the investigation began. Hahn, again in Berlin on leave, helped Meitner with the difficult evaporation of concentrated sulfuric acid, which they repeated several times until they obtained a fairly pure product that was intensely radioactive. After setting up samples for monitoring, Lise went to Vienna for Christmas and Otto returned to his unit a few days later.

> 17 January 1918. Take a deep breath before you begin reading, it will be a very long letter. . . . I received your card of 29 Dec. the day I left Vienna. Once I got here I wanted to finish some of the measurements so that I could tell what you want most to hear from me. And I shall tell you a variety of delightful things.
>
> Well, the strong Giesel preparations are quite all right . . . and all 3 have now increased strongly [in alpha activity]. . . . Also nos. 6, 9, 21, and 22 [from the first and second trials] have increased nicely. I have also determined the range. . . . The measurements at the end of the range, which are so important, take hours, for 10 days I left the lab only at 8:30 or 9 at night, but at least it was worth the effort. . . . The relatively large value of 3.32 cm (Ra has 3.13 cm) should not worry you. For the actinium series the relationship[81] $\log \lambda = A + B \log R$ yields a straight line whose slope differs from that of the uranium-radium series. . . . You can only compare within the same series. In the actinium series, though, there is the difficulty that either AcX or radioactinium don't correspond, so one gets two different slopes depending upon which is irregular. That also gives two different values for the half-life. If one takes the value of AcX as determining, the half-life of our eka-tantalum would be 1,200 years, the other one would give 150,000 years.[82] . . . In addition I have done the emanation measurements. The old preparation 25 [second trial] is more than double as strong, the new Giesel preparation [from 1 kg pitchblende] thirty times! stronger. The increase is so strong one notices it from one day to the next. That made me especially happy, I think you will also be glad. . . . Finally, I also collected actinium deposit from preparation 22 [second trial], and 10 days later repeated the experiment. Of

course the activity is very weak, but it can be measured with complete certainty and can be verified by its rate of decay. . . . In any case, we can now think of publishing very soon.[83]

This elegant piece of work proved the existence of a radioactive species that could not be observed directly but only by its radiations and those of its daughter and its decay products. The hunt for the elusive mother substance had taken five years. In 1913 Hahn and Meitner first looked for the mother substance in an undissolved silica residue from uranium ore; applying the concepts of isotopes and the group displacement laws, then very new, they used an isotope of the mother substance, UX_2, as an indicator of the correctness of their chemistry and the possible hiding place of the mother substance. Their suspicions were confirmed fours year later, in the winter of 1917, when the silica gave off a tiny bit of actinium emanation that had not been there before. In the spring of 1917 they tested the silica, accounted for the radiation from extraneous activities, and verified that some of that radiation was indeed coming from something new. Then in their larger trials, in the summer and fall of 1917, they concentrated the mother substance in earnest, let the extraneous activities fade away, and soon observed an increasing amount of actinium, which they detected by its emanation and the resulting decay products in the active deposit. Finally they characterized the mother substance, measuring the range of its alpha particles and estimating its half-life. This beautiful discovery called on every bit of expertise and intuition that Meitner and Hahn had acquired over the years.

On 16 March 1918, Lise Meitner and Otto Hahn submitted their paper, "The Mother Substance of Actinium, a New Radioactive Element of Long Half-life," to *Physikalische Zeitschrift*. With satisfaction they reported, "The supposition that pitchblende was the suitable starting material was indeed justified. We have succeeded in discovering a new radioactive element, and demonstrating that it is the mother substance of actinium. We propose, therefore, the name protactinium."[84]

Although Meitner had done nearly all the work, Hahn was the senior author on the protactinium paper. In matters of priority, Meitner was more than scrupulous with Hahn: she was intensely loyal, more sensitive to their collaboration than to their separate accomplishments, and aware that they shared a laboratory with equipment and materials they had acquired together. Thus, in January 1918, when she completed another study that she had done entirely alone over a period of many years, she still asked Hahn to "frankly" tell her if he wanted to co-author the paper. (He declined.)[85] Perhaps with protactinium she thought she owed him senior status, that to do otherwise might have been unfair to a soldier serving his country. Twenty years later, when the situation was reversed, when she was away from the laboratory and he was there for the day-to-day experiments, she might have expected a similar loyalty from him. It would not happen that way.

Once the protactinium article was submitted, what remained were the pleasant duties associated with a new element. Having told Stefan Meyer almost nothing all along, Lise hastened to give him the experimental details[86] and ask his advice in choosing a symbol. "You pose a terribly difficult question," he responded cheerfully. "I would prefer the names Lisonium, Lisottonium, etc., and I therefore propose the symbol Lo, but unfortunately these are unsuitable if one wants general acceptance. . . . In Pn a completely unimportant letter is brought to the fore, which leaves me most sympathetic to Pa: after all palladium for some reason is Pd. . . . Although I still prefer Lisotto, it is much more significant to have discovered . . . Pa or Pn than to come up with the most beautiful name."[87] For Meyer, at least, Lise came first. Lisotto & Co. decided on Pa.

There were also some delicate discussions with Kasimir Fajans. As the first to identify an isotope of "eka-tantalum," he and Oswald Göhring had the right to name the element, and they had chosen "brevium" for their short-lived UX_2. But common usage required that a radioactive element be represented by its longest-lived (that is, its most abundant) isotope, and "brevium" would hardly do for the mother substance of actinium. Fajans was known for his superior manner and aggressive treatment of his colleagues. "He feels as if he's the pope of radiochemistry,"[88] Stefan Meyer once commented, and he had taken to calling Fajans "Kasimir the Greatest."[89] But Fajans agreed to the name protactinium.

Fajans had not pursued "eka-tantalum," but Frederick Soddy had. In June 1918, Soddy and John Cranston reported that a high-temperature treatment of pitchblende yielded a sublimate that gradually generated increasing quantities of actinium. Their results appeared at about the same time as Hahn's and Meitner's; neither knew of the others' until after the war. Meitner and Hahn, however, had collected more actinium precursor and characterized it more completely, so the priority was theirs.[90]

While the discovery of protactinium identified the immediate parent of actinium, the more distant ancestry of actinium, in particular its relationship to uranium, was only slightly less murky than before. Although the sequence UY → Pa → Ac seemed correct, and although it appeared that UY was directly descended from uranium, it was still unclear just how this could be so, since the two known uranium isotopes, UI (now known to be ^{238}U) and UII (^{234}U), were both members of the uranium-radium series. If they also initiated the actinium series, this would involve branching of a completely novel sort: in every other known case, the branch diverged for only one generation and then immediately healed.[91] To resolve the question it would be necessary to determine the atomic weight of at least one member of the actinium series, the most promising candidate being the longest-lived, most abundant member—protactinium. This would occupy Hahn, and to a much lesser extent Meitner, for many years to come.

But in the summer of 1918, it was not possible to plan far ahead. For four years Germans had believed the rosy picture of the war; suddenly there was talk of collapse, exhaustion. Still the generals fought on, deceiving themselves and their country, wasting lives. No political faction would accept the burden of surrender and defeat. Civilians, desperately hungry, struggled to keep themselves and their families alive.

That June Otto suffered an attack of "weakness," probably mild phosgene poisoning, and was sent to a military hospital to recuperate. Some fresh food was still available for convalescing soldiers; when Edith came to visit, Otto sent Lise six eggs, "of which one is better than the other, or I should say was better, as all but two have become past tense. The last two I shall bestow upon the Francks."[92] Lise never complained of lack of food; only her gratitude reveals her hunger. Like everyone else, she carried on as best she could, completing a long-term study of the half-lives of several thorium products[93] and lecturing on the protactinium discovery.

> Did I write to you that I recently gave a colloquium on our work, and that Planck, Einstein, and Rubens told me afterwards how good it was? From which you can see that I gave quite a decent lecture, even though I was, stupidly enough, again very self-conscious. . . . I was glad you weren't there, you would surely have scolded me. This way, I quickly got over my shyness with a friendly jest from Planck and a very comforting psychological observation from Einstein. . . . I am optimistic enough to expect peace by autumn so I hope that we can again work together next winter.[94]

In Vienna there was chaos. "Concerning our living conditions, it is better to say nothing," Meyer wrote. "One can hardly tell which is at fault, lack of organization or actual shortage of food. As organizers, we are swarming with Reichsdeutschen [Germans, as distinct from German-Austrians]; the way they act—as if they cannot be wrong—and the way they pass judgment on others has made them beloved in all the world."[95]

The fighting continued through the summer and into the fall of 1918. On 20 October, Austria asked for an armistice; when it was signed on 3 November, revolution broke out throughout Germany. Sailors mutinied, troops left the front, radical socialists known as Spartacists—the only socialists to oppose the war from the start—organized workers' and soldiers' councils modeled on Russian soviets. On 9 November, the Kaiser fled; to forestall Spartacist leaders from proclaiming a soviet republic, moderate Social Democrats in the government joined with centrists and conservatives, declared a German republic, signed an armistice on 11 November, and ended the war.

In the cities and provinces of Germany rival factions fought, people believed nothing and everything, rumors flew. Stefan Meyer wrote, "[I am] almost more concerned now about Berlin than about Vienna, and with you the reverse seems to be true. . . . We are afflicted with influenza as much as you are in Berlin, and we don't know which is worse, influenza or the new majority of people who wish to strip us of everything German." He and his wife left their newborn son with a farm family for the winter; suicide and nervous breakdowns were commonplace. "One must now go about the world with blinders on, and without looking left or right try to get on and to work, as long as one's nerves hold out."[96]

For most of the war, Austria-Hungary had been a military satellite of Germany; at the end, as the empire fell into its national parts, many German-Austrians looked to a union with Germany as a fulfillment of their own national aspirations. As much as Lise Meitner loved Germany and favored such an *Anschluss*,[97] she disliked the trappings of German nationalism and was unsympathetic to her friends' attachment to the old order. In the first month of the revolution she tried to make Elisabeth understand that much of the change was for the better.

First, all rumors of houses being searched are completely untrue. . . . There may have been a few robbers who availed themselves of food from well-stocked houses. They were subsequently stopped from further "requisitions." So you really have no need to be concerned about your parents. Now, when I try to discuss some of the questions in your letter, please remember that I have always had very strong democratic inclinations. . . . I can certainly understand that you personally are very upset by the disappearance of the German nobility and especially the Hohenzollerns. But Elisabeth, the enormous mistakes they made in internal and international politics both before and during the war indicate very plainly that the system was a bad one; it is impossible for a single individual to control the fate of millions of people without answering to anyone but himself. . . . The last weeks of the Kaiser's regime once again showed this very clearly. He issued proclamations granting the German people broad democratic rights as if it were his own wish. But he would not negotiate with [Philipp] Scheidemann [one of two Social Democrats in the Kaiser's cabinet], whom he himself appointed. . . . Instead he manipulated the socialists so that they had to share the odium of the unfavorable cease-fire conditions. . . . You must not be unjust, Elisabeth, and say that the socialists sold our honor for bread, and that people concerned about honor would have secured better cease-fire conditions. . . . Bread means honor, life, and the existence of Germany right now. Because if we have starvation, the Bolshevists will rise up, which will not only result in civil war but the Entente will surely occupy the country, under the pretext of maintaining order. That, we hope, will be spared us. . . . The [Social Democrats] are not engaged in one-sided class politics, but want real democracy. . . . The main concern now is that dangerous fools like [Spartacist leader Karl] Liebknecht, Rosa Luxembourg, and the like do not acquire followers for their crazy ideas, and that on the other side some of the [right-wing] generals do not succeed with their plans.[98]

Meitner's politics were solidly Social Democratic; she did not share Elisabeth's upper-class distaste for democracy, or the general dismay of established academics over the disappearance of the monarchy and imperial Germany. In the laboratory she and Hahn had constant political discussions, in which "we almost always have differing opinions."[99] Perhaps because she felt politically isolated, she developed a greater appreciation for Einstein. In the fall of 1918 they briefly worked together, planning an experiment to verify an aspect of his light quantum hypothesis; unlike many other physicists at the time, Meitner believed in the hypothesis. Einstein's calculations were not quite right and Meitner never did the experiment, but she enjoyed the scientific and private contact with him: "I know of few people who have as strong and pure an individuality as Einstein."[100] For his part, Einstein began referring to Meitner as "our

Marie Curie"—great praise, although by then she could certainly have been compared not only to other women but to any physicist in her field.[101]

In Berlin the fighting continued. Socialist students occupied the university and seized the rector; Einstein, one of the few academics students trusted, was asked to negotiate.[102] For a time moderate socialists in the university quarter fought Spartacists just across the Spree from the Physics Institute; constant gunfire forced evacuation of the rooms facing the river. That winter Karl Liebknecht and Rosa Luxembourg were arrested and then murdered, the revolution was defeated, and a new German republic was created. The constituent assembly did not convene in Berlin, which was too dangerous, but in Weimar, birthplace and home of Goethe. Germany had been ruined by imperialism and militarism, and the choice of Weimar seemed an expression of hope.

There was no peace, but those who returned from the war were at least alive. Many physicists were demobilized in Berlin; away from the scientific literature for years, they requested a colloquium. Heinrich Rubens, unnerved by the fighting near his institute, asked Max Born, one of the young theoretical physicists, to make the arrangements. And so the Wednesday colloquium resumed, the men bundled in their old army coats, intent on physics, forgetting the cold for a while and the gunfire outside.[103]

Otto Hahn returned to Dahlem, and James Franck took a position as a department head in Fritz Haber's institute. Lise Meitner also headed her own section: with her appointment in January 1917 came the responsibility for establishing an independent physics section in the KWI for Chemistry. Scientifically, the independence could not have come at a better time. For a physicist, radioactivity had reached a certain completeness, far less interesting in itself than for what it might reveal of its nuclear origins. Radioactivity was evolving into nuclear physics, a term and a field that did not yet exist. Almost nothing was known of the atomic nucleus; Lise was ready for it. Her most important years of research lay just ahead.

CHAPTER FOUR

Professor in the Kaiser-Wilhelm-Institut

We spent several hours firing questions at Bohr, who was always full of generous good humor, and at lunch Haber tried to explain the meaning of the word "Bonze" (bigwig).

Germany in 1919 was exceptionally bleak: defeated, divided, poor. By summer people knew they faced a fourth winter of hunger and cold. In Austria the situation was equally grim. "We are already freezing and going hungry," Stefan Meyer wrote to Lise Meitner in October. "Milk and meat are unknown words, coal is nonexistent, wood insufficient and expensive."[1] He and his wife left their children in the mountain village where in better times they had spent their summers. In the countryside some food was available, but Vienna was a city of starvation and death.

For Germans there was the added fear that the harsh terms of the Versailles treaty would make recovery impossible. Even Einstein, who had always blamed Germany for the war and welcomed its defeat, was moved. "Those countries whose victory I had considered during the war by far the lesser evil I now consider only slightly less of an evil." Having hoped for a peace "that did not conceal a future war," he regarded the Allies' failure to assist the German people and their new democracy as heartless and shortsighted.[2]

The war brought disaster, but peace was a catastrophe and the government was blamed for it. Weimar satisfied almost no one, neither the Right nor the Left, certainly not industry or the military or cultural elite.[3] In the universities many of the senior professors were demoralized and bitter; Max Planck was one of the few to devote himself to rebuilding German scientific institutions. On several occasions military factions seized parts of Berlin, and trade unions retaliated with general strikes; to staff essential utilities, Berlin scientists organized an emergency technical service (*technische Nothilfe*), which assigned Otto Hahn, James Franck, and others to gas and electricity plants as all-night stokers.[4] For Lise Meitner, the disturbances were "quite depressing, not so much because . . . we were without gas, light and for a time even without water . . . but because the general situation makes me fear once again for the health of the Reich."[5] The situation included a vitriolic anti-Semitic campaign against Einstein. Planck, anxious not to lose Germany's chief scientific emblem, was instrumental in keeping his friend from leaving permanently for a post abroad.[6]

In the summer of 1919, friends invited Lise to Sweden for a short respite. There she enjoyed the serenity of a country that had not seen war for a century; her letters to Otto reveal an intense—and for Lise, quite uncharacteristic—interest in food. "In Göteborg Frau Franck's sister [Ingrid Franck was Swedish] came for me, and gave me a wonderful breakfast (beefsteak with eggs, sausage, milk, butter, etc.)." On the Swedish coast she was the guest of Emma Jacobsson, an Austrian physicist Lise had met before the war, and her husband, Malte, a Swedish engineer. "I am eating an unbelievable amount, drink at least a liter of milk a day, am eating eggs, butter, bacon, puddings, in short everything good, and am very spoiled by my friends. I shall return to Berlin fat and brown."[7] Lise probably never weighed more than 50 kilograms (110 lbs.), but that summer she did gain a new title, that of Professor in the institute. She had little taste for titles, "but," she said, "what gave me pleasure was the real pleasure my friends took in it." On her return to Berlin she found flowers in her apartment, more flowers in the laboratory.[8]

Whether she appreciated it or not, it was a significant event. In all likelihood, she was the first woman in Germany, certainly in Prussia, with the title Professor. And although her status in the institute did not immediately change, the title, especially in Prussia, was helpful in attracting students and assistants, obtaining grants, and working with officials. And it must have provided yet another defense against the feelings of insecurity that occasionally still overcame her.

The Kaiser-Wilhelm-Gesellschaft may have functioned well enough to dispense Professor titles, but working conditions in the institute were poor for quite some time. For more than a year after the protactinium discovery, Meitner and Hahn did little but tie up loose ends, publishing an improved value for the half-life of actinium,[9] speculating on the branch point of the actinium series,[10] describing protactinium's chemical properties in some detail.[11] In every publication Hahn was first author, to which Meitner apparently did not object, even though she had done much of the work.

In 1919 the Verein Deutscher Chemiker (Association of German Chemists) awarded its Emil Fischer Medal to Hahn, citing essentially everything he had done in radiochemistry, including radioactive recoil, magnetic beta spectra, and the discovery of protactinium;[12] the Verein also voted to present Meitner with a *copy* of Hahn's medal, an odd attempt to acknowledge her contribution without really recognizing it. The fact that Meitner did not share in Hahn's award seems to have disturbed him, but not her. "I find it *very gratifying* and very justified that you receive it," she assured him. "To me it seems less justified that I should also come away with something. Prof. [Theodor] Diehl wrote to me, namely, that the Verein at the same time resolved to recognize my collaboration [*Mitwirkung*] and present me with a copy of the medal; he invited me to the festivities on September 7. I don't have much desire to make the trip [she was on vacation] but would like your opinion. Of course *you* must go."[13]

Why would Meitner find it "very justified" that Hahn alone have the award, and "less justified" that she be recognized at all? Was this a matter of turf, a physicist reluctant to join chemists celebrating the accomplishments of chemistry, in which context her role would, almost by definition, be regarded as subsidiary?[14] Very possibly this was an important factor, for unlike Rutherford, who in 1908 was much amused by his Nobel Prize in chemistry, Meitner still needed to demonstrate her independence as a physicist. What is surprising, however, is how far she distanced herself from protactinium, almost as if she had willed herself to believe that chemistry was the key to the discovery and her own contributions hardly mattered.

At the heart of Lise's ambivalence was her relationship with Otto Hahn. Scientifically, their partnership had been most productive, but she knew that others would never see a woman working with a man as anything but his subordinate—even though it should have been obvious that for a substantial part of their collaboration, with the beta spectra, for example, she was the leader and Hahn the one at the margin, a chemist in the domain of physics. The difficulty was that personally she was indeed dependent on him: grateful, utterly loyal, more in need of his friendship than he of hers. The asymmetry was evident in their careful formality—still "*Sie*" after all those years—and in Meitner's deference, especially visible during the war when she ran the laboratory, did the work, and put Hahn's name first on every publication. To free herself, at least in science, it appears that she put protactinium, prize of their work together, out of mind. That this required some artful rethinking on her part may be seen in the sparse and, for Meitner, unusually faulty memoir she wrote in 1964: "I worked from mid-1915 to autumn 1917 [*sic:* 1916] as a radiologist in Austrian hospitals at the front. However, Hahn . . . often came to Dahlem, while I was able to get leave of absence from my voluntary position frequently enough for us to be able, even before the end of the war, to point conclusively to protactinium."[15] She was remembering Hahn's role better than her own, dissolving her year of intense and often lonely effort into a part-time venture by scientific comrades-in-arms.[16]

What Meitner did remember accurately and always cited with pride was her assignment, in 1917, to establish a physics department within the institute. She took it as a sign of recognition, trust, and professional coming-of-age: it was, above all, her own. It took a year or two to get ready, but by 1920 she and Hahn were well along their separate scientific paths. Her work with him had been prelude, radioactivity no longer an end in itself but preparation for the work ahead.

Physicists knew that these were the most exciting years any scientist could wish for. Only a generation or so before, bright students, including Max Planck at age sixteen, were advised to seek intellectual adventure somewhere else: physics was nearly complete, it seemed, its treatment of energy, motion, radiation, electromagnetism so brilliantly perfect that nothing worthwhile seemed left to discover.[17] And if little was understood of the nature of matter—well, that sort of thing was best left to chemists. Then came a rash of discoveries and everything changed: x-rays were unexplained; atomic spectra were a mystery; radioactivity was strange; electrons were everywhere, in atoms and electricity and beta decay and spectra. And despite overwhelming evidence for radiation as a continuous wave, new results implied just the opposite. In 1900, Planck found it necessary to call upon *quanta*, discontinuous energy corpuscles, to interpret blackbody radiation; in 1905, Einstein characterized radiation itself as a discontinuous medium composed of energy quanta, which were essential for explaining the photoelectric effect and other phenomena. Also that year Einstein transformed space and time with relativity and deduced the equivalence of matter and energy. There were more questions about matter and energy than ever before, but physics was making progress.

Radioactivity supplied some of the ammunition, the energetic alpha particles Rutherford called his "pets." Based on alpha-scattering experiments of his assistant, Hans Geiger, and his student, Ernest Marsden, Rutherford in 1911 proposed a planetary atomic arrangement: electrons orbiting a tiny, massive, positively charged nucleus. The nuclear atom had a grave defect, however: it did not account for the stability of atoms. According to everything that was known about moving charges, electrons should have been radiating energy and spiraling into the nucleus.

The revolution came in 1913. Niels Bohr took hydrogen, boldly declared its energy states to be nonradiating, quantized them, and calculated a set of discrete electron orbits in perfect agreement with the hydrogen spectrum. It was unexpected and *right;* it made quantum theory the basis for atomic physics, and the nuclear atom the model for everything radioactivity, spectroscopy, and chemistry had to offer. The nuclear atom embodied the primacy of charge over mass, something already indicated by radioactivity's group displacement laws and isotopes. Now it was also evident that alpha particles, being positive, must originate in the nucleus; that an element is characterized by its nuclear charge, which fixes the number and energy levels of its orbital electrons; that isotopes of an element have the same nuclear charge but different mass; that chemical reactions involve orbital electrons, not the nucleus. The site of beta decay was briefly debated. At first Rutherford and Fajans thought beta electrons were orbital, but after Bohr listed isotopic beta emitters that are radioactively different and Soddy pointed to the elemental change associated with beta decay, the conclusion was inescapable: beta decay is a nuclear process.

In 1914, James Franck and Gustav Hertz experimentally showed that when energetic electrons collide with mercury atoms in the gas phase, energy is transferred from an electron to a mercury atom only in distinct quanta. This indicated that quantization applies to all atomic processes, not just the absorption and emission of radiation, and to all elements, even though Bohr's theory was exact only for one-electron atoms such as hydrogen.[18]

The experimental field that advanced most symbiotically with the new atomic theory was spectroscopy. From the start the discrete lines in atomic spectra were understood as electron transitions between quantized energy levels: optical spectra as transitions of outer electrons, x-ray spectra as those closest in. Compared to optical spectra, x-ray spectra tended to be simpler and, because inner electrons are so close to the nucleus, more characteristic of the element used as the x-ray source. In 1914, when Moseley measured the characteristic x-radiation emitted by a set of consecutive elements, he found that the frequencies of the x-ray lines were a function of atomic number (the ordinal position of an element in the periodic system). He concluded that the physical basis for atomic number must be nuclear charge. Moseley's work illuminated the entire periodic system, affirmed the Rutherford-Bohr atomic model, and replaced atomic weight with nuclear charge as the mark of chemical identity. In turn, the quantized atom made possible a detailed interpretation of x-ray spectra, which became a powerful tool for studying inner electrons, for detecting new elements, and for chemical analysis. Although optical spectra were far more difficult to unravel, they would eventually reveal properties of all orbital electrons and spur the advance of atomic theory.[19]

Meitner, meanwhile, was engaged with spectra of a different sort, the magnetic beta spectra that she, Hahn, and von Baeyer had collected over a period of years. Meitner came to beta spectra with an interest in beta radiation and persisted because so little about the spectra—and the decaying nuclei that produced them—was well understood. During the war she made little progress, but after it was over she was head of her own section and eager to strike out on her own. Nuclei were as uncharted as atoms before Rutherford, their structure, energy, and even composition unknown: beta spectra offered a window into the nucleus as promising as optical and x-ray spectra were for the atom as a whole. In fact, beta spectra would prove more demanding and in many respects more significant than anyone expected; with them Meitner would make her mark as one of the leading physicists of her day.

The first beta spectrum was taken in 1909 by William Wilson to resolve a dispute with Meitner and Hahn over the homogeneity of beta radiation.[20] Meitner expected the spectrum of each beta source to be a single line, the image of electrons expelled with uniform energy in beta decay; when instead every spectrum displayed several lines, she had to concede that beta radiation is not homogeneous. The advent of the nuclear atom and the recognition that beta decay is a nuclear process permitted a revival of her former view: just one line would be formed by the primary decay electrons, and all other lines would be secondary, formed by electrons expelled from their orbits outside the nucleus.

In 1914, however, James Chadwick found what appeared to be a continuous beta spectrum for radium B + C (a mixture of ^{214}Pb and its daughter, ^{214}Bi), leading Rutherford to suggest that all beta lines were secondary and the primary spectrum continuous. This Meitner could not accept.[21] Convinced that nuclei, like atoms, must be quantized, she did not believe that identical nuclei would emit primary electrons of variable energy. She was still guided by the analogy between alpha and beta decay that she had always favored; considering the uniform energy always found for alpha particles, the presence of discrete beta lines, and the triumph of early quantum theory, her view was not unreasonable. She thought she could prove it by analyzing beta spectra for their primary and secondary components.

It was not immediately obvious how to do this, so Meitner first looked for beta spectra with no primary line at all. Before the war she and Hahn had found that the alpha emitters radium (^{226}Ra), radioactinium (^{227}Th), and radiothorium (^{228}Th) exhibited beta spectra with well-defined lines.[22] Before concluding that these spectra were entirely secondary, Meitner wanted to be sure these three alpha emitters were not primary beta emitters also.[23] Such dual alpha and beta decay, known as branching, had been observed for several species; if present, the respective beta decay products would have been ^{226}Ac, ^{227}Pa, and ^{228}Pa.

Hahn and Meitner searched for these products and found nothing.[24] The beta spectra in question were, therefore, entirely secondary. The result, Meitner believed, applied to primary beta emitters as well, for it "proved without doubt that . . . there exist β-particles of defined energy which do not originate in the nucleus; they are thus of secondary origin; i.e., they must come from the electron rings. From this one would expect that part of the beta spectrum of a typical beta emitter . . . is likewise of secondary origin."[25]

What drove these secondary electrons out of their electron orbits? Collisions with alpha particles would not do it: from the position of the lines, Meitner noted, it was evident that the energy of the secondary electrons was far too great to result from collisions between alpha particles and orbital electrons.[26] Gamma radiation was the only other possibility: it was of nuclear origin, very high in energy, and nearly always accompanied alpha and beta decay. "All three substances have a relatively easy-to-detect γ-radiation," she noted, "and this makes it seem likely that linked to the alpha radiation in some way is a γ-radiation which is emitted from the nucleus and gives rise to the observed [secondary] β-radiation."[27]

A connection between gamma radiation and secondary beta spectra had been recognized by Rutherford in 1914, after he and his co-workers investigated radium B (^{214}Pb), a lead isotope that emits beta and gamma radiation. When they enclosed RaB in lead foil, they obtained a beta spectrum that necessarily came from electrons in the foil, since RaB's own beta particles could not penetrate the lead enclosure. Surprisingly, the lead foil spectrum and the natural beta spectrum of RaB appeared identical. Rutherford concluded that both spectra must be secondary, both incited by the gamma radiation of RaB.[28]

In another experiment, Rutherford and E. N. daC. Andrade found that RaB emitted soft gamma radiation seemingly identical to the characteristic L x-rays of lead (x-rays emitted when an electron falls from a higher level to a vacancy in the L shell). The existence of an orbital vacancy in the L shell was another indication that orbital electrons were expelled from their orbits during radioactive decay.[29] In 1917, Rutherford suggested that nuclear gamma radiation might expel secondary beta particles, whose energy might be used to determine the wavelengths of the gamma rays by means of the quantum relation. At the time, however, the calculation could not be done, because the energy required to remove the orbital electrons was not well known, nor was Rutherford confident that the quantum relation was valid at such high energies.[30]

In the spring of 1921, Lise Meitner spent several weeks in Sweden as a visiting professor at the University of Lund. The invitation came from Manne Siegbahn, a Swedish x-ray spectroscopist, who asked her to give colloquia on radioactivity and to instruct assistants and technicians in the techniques of beta and gamma spectra. Essentially no radioactivity research was done in Sweden, and Meitner was soon teaching a full course in radioactivity, including theory and laboratory preparations, to physicists and chemists. By then she had conquered most of her inhibitions about lecturing, but she was still somewhat taken aback at the end when, as she later wrote, "the students gave me a lovely bouquet of flowers and a nice thank you-farewell speech which, according to Swedish custom, ended with three hurrahs. I, poor thing, was totally unprepared for this and had to thank them for it."[31]

Meitner had also come to Lund to learn something about x-ray spectroscopy. The field had matured rapidly since its beginnings around 1910, when C. G. Barkla first detected the characteristic x-rays emitted by different elements. X-ray diffraction, discovered in 1912 by Max Laue (Max *von* Laue after his father acquired a hereditary title in 1913) and developed by W. H. Bragg and W. L. Bragg, made it possible to determine x-ray wavelengths with precision. By measuring the wavelengths of the x-ray spectra of a series of elements, Moseley identified atomic number with nuclear charge; others began unraveling the characteristic x-ray spectra of the elements and mapping the energies of their inner electron shells. Siegbahn, who would receive the 1924 Nobel Prize in physics, was known for the precision of his x-ray measurements and for extending these to a large number of elements. By 1921 the innermost K shell energies and also the next higher L, M, and N shell energy values for the heaviest metals were quite precisely known.[32] These would be essential for Meitner's analyses of beta spectra.

In Siegbahn's laboratory Meitner met Dirk Coster, a young Dutch spectroscopist, and his wife, Miep, a student working toward her doctorate in Indonesian languages and culture. They became immediate friends, Lise enjoying Dirk's enthusiasm for science and the couple's warmth and concern for others. Politically involved and committed to social equality, the Costers were quite apart from the academic norm; Lise's evenings with them were filled with "exciting talk about every possible subject." By day Miep and the baby sunned themselves on the steps outside the institute, while Coster demonstrated for Meitner the instruments and techniques of x-ray spectroscopy.[33]

Returning to Berlin in May, Meitner prepared for a fresh look at beta spectra. She had learned enough about spectroscopy to understand the limits of its precision: she would use spectroscopic data and her own magnetic beta spectra to determine the energy of the gamma radiation, from which she was prepared to determine which beta electrons were secondary and which were primary.

From each beta source, she expected a single primary line; as there was no way to directly recognize which one that was, it would be necessary to first identify the secondary lines. These, she assumed, were produced by an internal photoelectric process: a gamma quantum leaving the nucleus would be absorbed by an orbital electron, which would use some of the gamma energy as ionization energy to free itself from the atom and convert the remainder into kinetic energy. The deeper the electron's shell, the more tightly it was bound to the nucleus, and the less kinetic energy it would have at the end, but no matter from which shell an electron came, its ionization energy and kinetic energy, added together, would equal the energy of the gamma quantum.

Meitner's experimental apparatus was similar to the one used in her earlier studies with Hahn and von Baeyer. A beta source was deposited on a thin wire, and its electrons were passed through a slit, deflected into a circular path by a magnetic field, and deposited onto a photographic plate. From the position of the lines and the known strength of the magnetic field, the kinetic energy of the electrons was readily calculated.

For her first study, Meitner selected the lead isotope thorium B (^{212}Pb) for its strong gamma radiation and relatively simple spectrum of one weak and two strong lines. To determine which lines were secondary, she encased ThB in a thin lead tube that absorbed ThB's beta particles but not its gamma radiation. In the resultant secondary spectrum, formed by orbital electrons from the lead case, the weak line was absent, but the two strong lines had the same position and relative intensity as in the natural spectrum, an observation quite similar to Rutherford's several years before.

The difference in energy between the two lines corresponded to the difference in ionization energies (known from x-ray data) of the K and L_1 shells of lead, indicating that the slower electrons came from the innermost K shell and the faster ones from the L_1 shell. The energy of the gamma quantum was obtained by adding the kinetic energy of each electron to the ionization energy of the shell from which it came: the two values agreed within 1 percent. As a check, Meitner enclosed ThB in platinum, again obtained a two-line secondary spectrum, then used platinum ionization energies to calculate a gamma energy in good agreement with the previous values.[34] The beta spectra were, therefore, gamma spectra also: from them one determined which lines were secondary, from which shells the electrons came, and the energy of the gamma photon. The gamma wavelength followed at once from the quantum relation, $E = hc/\lambda$, where E = energy and λ = wavelength. Since h (Planck's constant) and c (the speed of light) are known constants, the energy can be calculated from measured wavelength, and vice versa.

One thing remained—to identify the primary line. To assign it merely by elimination seemed unsatisfactory: Meitner reached for an explanation to tie everything together. She postulated that the energy of some, possibly most, decay electrons was converted into gamma radiation before they left the nucleus; unconverted primary electrons would form a line in the beta spectrum corresponding to the gamma energy. (This mechanism for the genesis of gamma radiation was too farfetched to be true, and Meitner eventually abandoned it.) In the ThB spectrum, Meitner did find a weak line at precisely the gamma energy. To verify it, she took the spectrum again, using an experimental arrangement that gave greater resolution, developed before the war by the Polish physicist Jean Danysz. The three previous lines appeared, along with a very weak fourth line that corresponded perfectly to a secondary electron from the L_2 energy level. It all fit; Meitner considered her hypothesis valid.[35]

Implicit was the sequence primary (decay) electron first, then gamma emission, then secondary electrons. From this Meitner raised a question others had not fully considered: do the secondary electrons come from the parent atom or the daughter? To Meitner it seemed evident that once a decay electron left the nucleus, the parent no longer existed and all subsequent events took place in the daughter atom. To calculate gamma energies, therefore, it would be correct to use ionization energies not of the beta source (in this case, lead) but its decay product (bismuth). In practice, however, Meitner used the values for lead. The x-ray data for its ionization energies were better known, and within the precision of the experiment the difference was negligible.[36]

Meitner then reviewed the beta spectrum of radium D (^{210}Pb) taken by Danysz in 1914, a few months before he was killed in the war.[37] Calculating the gamma energy as before, she found a beta line at just that energy and attributed it to the primary electron. As a check she reexamined the spectra for radium (^{226}Ra) and radiothorium (^{228}Th) and determined their gamma energies: as expected for alpha sources, all lines were secondary. Meitner concluded that "typical beta emitters release from their nuclei a beta particle of a single energy. These beta particles are partly converted to gamma radiation of the same energy, which ejects secondary beta particles from the electron orbits, at speeds determined by the speed of the primary beta particle and the ionization energy of the orbital electrons. The greatest speed corresponds to the primary beta radiation and its measurement gives at once the wavelength of the gamma radiation."[38]

A few months after Meitner began her ThB study, Charles D. Ellis published the beta spectrum of radium B (^{214}Pb), calculating gamma energies in the same way.[39] This was the same Ellis who had been interned along with James Chadwick in Ruhleben during the war; he had given up a career as an artillery officer in favor of physics, and now he and Chadwick were working with Rutherford in Cambridge. Ellis was at first less interested in beta spectra themselves than in using them to determine gamma wavelengths, which were in general too short to be measured by crystal diffraction.[40] The spectrum of RaB was quite complex, in part because it was always contaminated with its decay product RaC (^{214}Bi), also a beta source; to be certain of his measurements, Ellis used four different metals as shields. In accordance with Rutherford's suggestion some years before, Ellis regarded the line spectrum as entirely secondary; to account for twelve lines, he calculated six gamma energies. In her own analysis of Ellis's data, Meitner assigned two lines to primary beta particles and others to RaC, requiring only two gamma energies to account for the rest of the RaB spectrum.[41]

Soon afterward Ellis noted an arithmetic relationship among the frequencies of the RaB gamma rays, an indication that gamma rays are emitted in transitions between quantized nuclear energy levels. Like Meitner, Ellis presumed a link between gamma and primary beta decay, but his sequence was the reverse. Based on Rutherford and Andrade's 1914 finding of soft RaB gamma rays apparently identical to the L x-rays of lead,[42] Ellis presumed that gamma emission must precede beta decay: "It would appear that the γ-ray is emitted, travels out to the L ring, from where it may eject an electron, this electron goes clear of the atom and another electron falls into a vacant place in the L ring, and the nucleus has still not disintegrated." Gamma emission not only preceded beta decay, Ellis believed, but triggered it: an electron in an upper (nuclear) energy level "passes to a [lower] energy level with the emission of a γ-ray. Finally, [the electron] will arrive in one of the states which is connected with instability, and the nucleus disintegrates with the emission of an electron."[43]

Ellis thought primary electrons were not of uniform energy but instead formed a continuous spectrum, as Chadwick had found for RaB + C in

1914. Working in Hans Geiger's laboratory in the Physikalisch-Technische Reichsanstalt in Berlin, Chadwick counted electrons using an ionization chamber at a fixed radius while varying the magnetic field. A relatively small number of electrons were at the lines; the great majority—presumably the primary disintegration electrons—formed a continuous spectrum so diffuse it barely darkened a photographic plate.[44] From the beginning the continuous spectrum presented a dilemma: how could identical nuclei emit electrons of continuously variable energy and still produce identical daughter nuclei? The notion was at odds with a quantum view of the atom and defied the first law of thermodynamics.

Ellis had postulated nuclear energy levels to explain additive gamma energies but then blurred the idea to accommodate the continuous spectrum: "One would not anticipate that all the radium B nuclei were absolutely identical."[45] He struggled to find an explanation that did not violate the law of conservation of energy.

> The electron arrives in a stationary state in which it is not permanently stable and it flies out from the nucleus. The kinetic energy of the electron must be considered to depend on other factors besides those of the stationary state, and the variable kinetic energy is possibly connected with the two facts that the nuclear field must vary considerably in distances comparable with the diameter of the electron and that the electron cannot be considered as rigid under these conditions.[46]

Ellis criticized Meitner's data and repeated her ThB spectrum with somewhat different results.[47] Strongly opposing Meitner's "simple analogy between α- and β-decay,"[48] he raised the "most serious objection" that her theory "gives no possibility of explaining the general [continuous] spectrum of β-rays, in fact appears to deny its existence. This general spectrum certainly does exist, and no theory which disregards it can be correct."[49]

In response, Meitner repeated several of her own and Ellis's measurements with improved apparatus. Some spectra contained so many closely spaced lines that a slight variation in experimental conditions or ionization energies led to a different assignment. In each case Meitner defended her interpretation, ascribing discrepancies to Ellis's use of outdated ionization energy values and spectral data taken in Rutherford's group before the war.[50]

She reserved her sharpest criticism, however, for his hypothesis that the emission of nuclear gamma radiation would precede the primary decay electron.

> I must confess that these assumptions appear very hard to reconcile with the customary view of radiation processes, because in general it is found that by emitting radiation the electron falls to a more stable energy state. [Mr. Ellis] presumes that just the opposite occurs—the electron by emitting gamma radiation ends up in a more unstable state. In addition, the claim that the electron may leave the nucleus with any allowed speed, and not with a single speed defined by the energy state of the nucleus, is hardly an explanation for the existence of a continuous spectrum, it is only a description of it.[51]

Meitner was simply not impressed by the continuous spectrum, despite Chadwick's reputation and Rutherford's endorsement. It had been done just once, by one person, for only one species; she may have thought it anomalous. While Chadwick thought the photographic image falsely exaggerated discrete lines,[52] Meitner considered Chadwick's experimental arrangement too imprecise to resolve closely spaced lines, noting that individual beta lines are always somewhat inhomogeneous due to electron collisions. "In any case," she ended, "one cannot conclude from Chadwick's experiments that a primary continuous β-spectrum exists."[53] Her attack on the continuous beta spectrum—part criticism, part denial—was communicated before publication to Chadwick, who voiced his objections to Hahn in a roundabout way when Meitner was in Vienna for Easter. Meitner was irritated: Chadwick ought to have contacted her directly; after all these years he apparently still regarded Hahn as her superior. And she thought it was "really quite unscientific not to believe someone's work before reading it *thoroughly*. He surely wouldn't have done that for a work of [Hans] Geiger's, for instance; what is being expressed here is a lack of respect for me as a woman, and that upsets me a little." But still she wanted to know, "Did he say anything about Ellis's new results?"[54]

The controversy between Meitner and Ellis was now quite sharply drawn.[55] In person they were friendly and collegial—they had known each other since Ellis's internment in Berlin—and they corresponded amiably and often. But in print they were unsparingly contentious. Certain that his data were superior, Ellis relied heavily on experiments done by others in Rutherford's group, even when they were difficult to interpret; his aversion to theorizing was typical of Rutherford and his co-workers. Meitner also reflected her immediate scientific milieu. Although she considered herself an experimentalist first and foremost, she was guided by theory and not at all reluctant to criticize or ignore experimental data she thought made no sense.

For their next investigations, Meitner and Ellis each chose experiments that most strongly supported their own hypotheses. In the summer of 1922, Ellis and Chadwick repeated Chadwick's 1914 determination of the RaB + C spectrum, using an electroscope to measure ionization. As before, the spectrum was predominantly continuous, its magnitude "roughly what would be expected if each disintegrating atom contributed one electron to the continuous spectrum. It appears to us that the simplest way of viewing these facts is to suppose that the continuous spectrum is formed by the actual disintegrating electrons."[56]

Meitner countered Ellis's contention that gamma radiation triggers beta decay by citing two beta emitters, RaE (^{210}Bi) and ThC (^{212}Bi), for which no gamma radiation was evident. Another example was UX_1 (^{234}Th), a thorium isotope with well-defined beta lines but very soft gamma radiation. Meitner determined that its secondary electrons originated in the L, M, and N shells and its gamma energy corresponded to the characteristic K_α x-rays of thorium; she was sure the radiation was not of nuclear origin.[57]

To be certain no higher-energy gamma radiation had gone undetected, Meitner decided to measure the UX_1 gamma energy by absorption, independent of the beta spectrum. With Hahn's help, she extracted UX_1 from a large quantity of uranium nitrate. "Since UX has a half-life of 24 days we really worked intensely," she wrote to her brother after it was over, "and of course, processing 40 kg in the laboratory to get an invisible trace of UX is not exactly an easy job."[58] The absorption curves proved that the only gamma radiation present was of low energy, of the magnitude of characteristic x-rays.[59] The absence of nuclear gamma radiation in a beta emitter, Meitner noted with satisfaction, "cannot be reconciled with the views of C. D. Ellis" since his mechanism for beta decay required gamma radiation as a trigger.[60]

According to Meitner, the primary process was simply the emission of a decay electron from the nucleus. In UX_1, she believed, there was no nuclear gamma radiation at all. Instead, the decay electron directly ejected a K shell electron, an L electron dropped into the vacancy, and the resultant K_α radiation was mostly reabsorbed to eject L, M, or N electrons from their orbits, all in the same atom. The possibility of multiple transitions without the emission of radiation had been discussed theoretically;[61] Meitner was the first to observe and describe such radiationless transitions. Two years later, Pierre Auger detected the short heavy tracks of the ejected secondary electrons in a cloud chamber, and the effect was named for him. It has been suggested that the "Auger effect" might well have been the "Meitner effect" or at least the "Meitner-Auger effect" had she described it with greater fanfare, but in 1923 it was only part of a thirteen-page article whose main thrust was the beta spectrum of UX_1 and the mechanism of its decay.[62]

If Meitner thought she had proved that UX_1 did not emit nuclear gamma radiation, C. D. Ellis did not. "It is perfectly straightforward," he wrote, "to see in this a γ-ray emitted from the nucleus, like all other cases, and it is a mere coincidence that its energy happens to be near that of the K_α radiations. . . . As far as we can see, uranium X_1 appears to be a perfectly normal body, emitting a soft γ-ray from the nucleus like radium D."[63]

While Ellis denied that beta decay might occur without nuclear gamma radiation, Meitner denied the validity of the continuous beta spectrum. By 1923, however, she realized she would have to explain—or at least address—the issue. She could not deny that some inhomogeneity was associated with nearly every beta line, and she found it difficult to explain the occasional lines that were so diffuse (*verwaschenen*) that they were more correctly termed bands. In the UX_1 spectrum Meitner had assigned the primary beta particles to such a diffuse band,[64] and Ellis quickly pointed out that "the diffuse band, as Meitner states, probably consists of the disintegration electrons, and we now see that it constitutes a very good example of the varying velocity which these electrons always appear to possess."[65] Meitner ascribed inhomogeneity to extranuclear events such as collisions between orbital and primary electrons.[66]

In 1923, Meitner enlisted the just-discovered Compton effect to support her view. Adopting a completely quantum approach, A. H. Compton of the University of Chicago treated the observed scattering of x-rays (and gamma rays) in matter as a two-particle collision between x-ray quanta and electrons: when a quantum collided with an electron, the energy could be lost by the quantum and gained by the electron in amounts that varied continuously with the angle of recoil. Compton's close agreement between experiment and theory provided the convincing proof for Einstein's light quantum hypothesis that Einstein and others had long sought.[67] Meitner was a believer in the light quantum hypothesis—she had briefly worked with Einstein on an experiment to verify it in 1918[68]—and she quickly applied Compton scattering to her analysis of beta spectra. The Compton effect, Meitner thought, might explain the continuous beta spectrum. Although the Compton effect was first described for free electrons, Meitner applied it to "almost free electrons," the outer, least strongly bound, orbital electrons: on being hit by a quantum of gamma radiation, orbital electrons would recoil at various angles and with varying energies, forming the nearly continuous background of the beta spectrum. "The shorter the wavelength of the γ-ray, the more it is true that even inner electrons of the atom seem loosely bound. . . . In the region of fast beta radiation in general, therefore, one would expect only very weak lines against a continuous background. This is exactly what is observed."[69] In her next paper Meitner discussed the magnitude of the Compton effect and again concluded that it "makes itself evident, as I have shown before, in the continuous β-spectrum of radioactive substances."[70]

Having once again explained the continuous spectrum to her satisfaction, Meitner returned to the question that interested her most: the beta-then-gamma sequence of decay. If secondary electrons originated in the daughter atom, as she believed, then gamma energy calculations would agree better if ionization energies of the daughter, rather than the parent, were used. Meitner turned to an alpha source first because parent and daughter differed by two units of charge, making it easier to distinguish the possibilities.

In the spring of 1924, she retook the beta spectrum of radium (^{226}Ra),[71] technically the most difficult of all beta spectra. Again Otto Hahn assisted; great care was required to prevent radon (^{222}Rn) from fogging the photographic plates and contaminating the apparatus and the entire laboratory.

The spectrum showed three lines, with energy differences consistent with origin in the K, L, and M shells. The gamma energies were calculated as before, but the radium spectrum was quite blurred, and since the ionization energies of radium and its daughter, radon, had never been measured but had to be estimated, the experimental error was such that neither possibility could be excluded.

Nevertheless, the experiment was not a complete failure. By determining the gamma energy of an alpha emitter more precisely than had been done before, Meitner showed that it was monochromatic and of sufficiently high energy to prove its nuclear origin, "which again supports the position I have emphasized, in opposition to C. D. Ellis, of the analogy between α- and β-decay."[72]

In a deeper sense, her analogy extended atomic quantization into the nucleus. She was the first to clearly state that nuclear gamma emission was entirely analogous to optical and x-ray spectra.

> When an α- or β-particle flies out of the nucleus, a reorganization of the remaining nuclear particles must set in. . . . Quantum changes in nuclear configuration may occur which give rise to monochromatic γ-rays. . . . The appearance of γ-rays is, so to speak, a measure of the magnitude of the disturbance in nuclear configuration, which has been incurred by the ejection of α- or β-particles. . . . The greater the disturbance of the nucleus, the more numerous the possible energy transitions, i.e., the gamma spectrum shows more lines, and in general extends to shorter wavelengths.[73]

However reasonable Meitner's analogies, actual proof of the decay sequence still eluded her. In 1924, Ellis and H. W. B. Skinner measured the spectrum of RaB (^{214}Pb) again and found better agreement using ionization energies of the daughter atom (bismuth) rather than the parent (lead).[74] Ellis struggled to fit the new data to his old interpretation, suggesting that somehow the slowest radiation was "held until after the disintegration," perhaps because the "electronic levels had the power of holding absorbed radiation for a finite time before emitting the electron."[75] Meitner found this "very hard to imagine. One can avoid this difficulty by recognizing that all γ-radiation follows beta decay, and therefore all secondary electrons come from the daughter atom."[76]

Turning to another alpha emitter, she measured the beta spectrum of radioactinium (^{227}Th), whose decay products actinium X (^{223}Ra) and actinium active deposit were always present. The spectra were extremely complex. To be certain all lines were visible, it was necessary to use strong preparations, while taking care to prevent actinium emanation (^{219}Rn) from fogging the plates. With Hahn's assistance she recorded nearly one hundred spectra under varying conditions; the radioactinium spectrum consisted of 49 lines, actinium X had 21.[77]

It took Meitner a year to analyze the spectra, but when she was done, her calculations of gamma energies showed significantly better agreement when ionization energies of the decay product rather than the parent were used. She broadcast her success in the title of her article: "The γ-Radiation of the Actinium Series, and the Proof that γ-Rays Are Emitted Only after Radioactive Decay."[78]

Meitner submitted the article on 20 October 1925 and sent a copy to Ellis. He had just obtained similar results of his own. He wrote,

> I am so sorry to have been so long in answering your letter, but every day I hoped to have some new reprints to send you. I particularly wanted to send them because I know you would be pleased with the results. We have been at considerable trouble to settle the question of whether the gamma-ray sometimes preceeds [*sic*] the disintegration as I deduced from Rutherford's and Andrade's measurements, or whether the simpler standpoint advanced by you that the disintegration always happens first was correct. We have had three different ways of testing this and they all show that the gamma rays come out afterwards, so you were right![79]

In a new and exceptionally precise measurement of several RaB lines, Ellis and his student, W. A. Wooster, had confirmed that Meitner's sequence was true for beta decay as well.[80]

> I cannot help feeling a little annoyed that we were led astray by that old experiment, and I don't know yet what the error was. I have seen Andrade's old plates and measurements and everything looks all right but yet there is little doubt that everything was really all wrong. I am very pleased that this removes one "Streitfrage" [controversial question] between us, but the other one over the nature of the continuous spectrum still remains. . . . The continuous spectrum does present very difficult problems, because I cannot see how secondary effects like Compton scattering, excitation of characteristic radiation, or emission of general radiation can produce a large enough effect. . . . The position might be summed up as follows, we both agree that once the disintegration electrons are outside the parent atom they are already inhomogeneous in velocity. We both agree that a quantized nucleus ought to give disintegration electrons of a definite speed, but whereas you think various subsidiary effects are sufficiently large to produce the observed inhomogeneity, I think they are much too small. That seems to be the only point of difference.[81]

For Ellis, the continuous beta spectrum was a *Streitfrage*, but to Meitner it did not seem very controversial, or even very important. In her radio-actinium paper she dismissed it in a single paragraph: "All photographs . . . show a background. As radioactinium is an α-emitter, this definitely cannot be due to continuously distributed primary beta particles . . . but could be Compton scattering of the gamma radiation by orbital electrons, as I have stated in an earlier work."[82] Even before completing that study, she began new experiments designed to probe the nucleus from a quite different perspective.

With her work on beta and gamma spectra Meitner joined the first rank of experimental physicists. For physics, the period was so rich in discovery that it was, even at the time, referred to as a "golden age." Berlin was one of the great centers for atomic physics, Meitner a leader in the newer and smaller field of nuclear physics. It was her work, Otto Hahn would later note, more than his own, that contributed to the growing international reputation of their institute.[83]

Hahn stayed with the refinement of radiochemical techniques throughout the 1920s, a time when the field was widely thought to have outlived its usefulness for fundamental insight into atomic or nuclear behavior.[84] Although Meitner's and Hahn's direct collaboration ended soon after the protactinium discovery, they remained close colleagues and good friends. The bonds became even stronger when Edith and Otto's only child, Johann Otto, was born in 1922. "I am already incredibly eager to see your son," Lise wrote Otto from Vienna. "I am really very happy that the little boy has arrived so nicely, and has not made it difficult for his mother. . . . I send her a kiss, and very hearty greetings to you."[85] Here—after almost fifteen years—Lise was using the familiar *Du*. Now little Hanno, as he was called, was her godson and Otto was her *Fachbruder*, her colleague-brother.[86]

If the Hahns had become almost family for Meitner, the physics community was her home. Her capacity for friendship was very strong. She liked people of all ages and sorts: employees, students, colleagues, their wives, children, and eventually grandchildren; she enjoyed the company of families that were successful and inclusive, such as the Francks and Gustav and Ellen Hertz, but she had no trouble adjusting to the much less traditional marital arrangements of Annemarie and Erwin Schrödinger.[87] She often formed individual friendships with the wives of her physicist friends: with Hedi Born, a talented poet, with Annemarie Schrödinger and Margrethe Bohr. When she was with Max von Laue she would talk about Goethe, ancient Greece, or his beloved car; with Max and Magda von Laue, she would go to the movies and occasionally on weekend trips. With Arnold Berliner, the highly cultured editor of *Naturwissenschaften*, she would discuss music and modern physics; with Planck, everything. Among her close women friends, besides Elisabeth Schiemann, was Eva von Bahr-Bergius, a Swedish physicist who had spent some years in Berlin before the war working under Heinrich Rubens and who would later be Meitner's closest confidante in Sweden. In her travels and at conferences, Meitner made new friends and kept them, grateful to physics for the people who were part of it. She was far from the shy, almost fearful young woman who first arrived in Berlin; although she never forgot her difficult start and occasionally relived her old insecurity, she had left it mostly behind her. The mature Lise Meitner was a confident and increasingly assertive person who was exactly where and what she wanted to be: a physicist, among friends, in Berlin.

One of Berlin's attractions was the famous Wednesday colloquium, held in the old university physics building by the Spree. For many years it was organized by Laue, the front row traditionally taken by the great professors—Einstein, Planck, Laue, Nernst, Haber—and the rows behind filled with others not quite so distinguished—Meitner, Baeyer, Franck, Geiger, Hertz—and then students, physicists from industry, and visitors from Germany and abroad.[88]

Niels Bohr first came to a Wednesday colloquium in April 1920. It was a grand event; Bohr had never been to Berlin. He stayed with Planck, whom he had not met before, and he met Einstein in person for the first time. Afterward Einstein expressed his joy in Bohr's "mere presence," Bohr his gratitude for his talks with Einstein, the "greatest experience" he ever had. George de Hevesy reported that he had "never experienced an ovation similar to that given to Bohr in Berlin. Young and old celebrated him with complete conviction and enthusiasm."[89]

At the colloquium Bohr lectured on atomic spectra and the correspondence principle. Some of the younger physicists, including Lise Meitner, left his talk feeling they had understood very little and also somewhat annoyed that the professors had dominated the discussion.[90]

> In this half-depressed and half-playful spirit, we decided to invite Bohr to spend a day in Dahlem, but not to include in the party any physicists who were already professors. That meant that I had to go to Planck and explain to him that we wanted to invite Bohr, who [was staying] with Planck, but not Planck himself. In the same way [James] Franck had to go to Professor Haber—because, after all, if we were going to have Bohr in Dahlem for the whole day we wanted to give him something to eat—and ask Haber for the

use of his clubhouse for our discussion "without bigwigs" (*bonzenfrei*), again stressing that we did not want to invite Haber himself, as he was already a professor. Haber was not the least put out. Instead he invited us all to his villa—this, you must remember, was the very difficult period after Germany had lost the war and to get something to eat was rather difficult in Dahlem. Haber only asked our permission to ask Einstein to lunch as well. We spent several hours firing questions at Bohr, who was always full of generous good humor, and at lunch Haber tried to explain the meaning of the word "*Bonze*" (bigwig).[91]

A year later Meitner was in Copenhagen for the first time, to lecture in Bohr's Institute for Theoretical Physics on her studies of beta-gamma spectra. Despite its name, Bohr's institute always had an experimental section, and, especially at that time, the unity of theoretical and experimental work was emphasized. It was one of the many ways in which Bohr and Meitner thought alike; on this visit they became close friends. Lise was full of admiration for Bohr and charmed by Margrethe; she could speak to them "about everything under the sun, whether grave or gay." And, she told Margrethe, it was enormously reassuring to know that Bohr valued her work, for it helped her overcome the insecurity that still sometimes afflicted her. Their friendship would be intense and permanent, Lise never forgetting the "magic" of their first meeting, "a magic which was only enhanced" on her many subsequent visits to Bohr's institute and home.[92]

In Germany, meanwhile, the political situation was poor, and the economy disastrous. Following the war the value of the mark declined steadily; in 1922 it suddenly plunged by a factor of fifty. "In the last few months prices have risen unbelievably fast," Lise wrote to her mother in October. "The winter will be very hard. . . . We are already freezing in the institute and at home."[93] She was tired of her student room,[94] but an apartment and some furniture were luxuries she could not afford.

No aspect of professional or private life was exempt from the inflation. Scientific meetings were often poorly attended because participants could not afford to travel to another city; as the inflation worsened, some lacked even the carfare to cross Berlin.[95] The finances of the Kaiser-Wilhelm-Gesellschaft crumbled, institute budgets became meaningless, supplies and equipment were borrowed or bartered.[96] "Things seldom go as one would like," Lise wrote to her brother Walter in March 1923. "Since Christmas my paper on the beta spectrum [of UX$_1$] has been finished, but I am waiting to do a gamma study [of UX] with Hahn, because the two papers would go well together."[97] The problem was obtaining enough uranium for the strong UX preparations needed to measure gamma absorption; for months Lise and Otto went around to factories asking for donations. In May 1923, "we finally got 40 kg of uranium nitrate. Hahn and I have been working, even on Sundays, to extract the UX."[98] The UX_1 papers were finally published that summer, and Meitner attempted no further experiments that year. Instead, she reviewed theoretical studies of the interaction between gamma radiation and electrons and devoted two articles to the importance of the Compton effect for inhomogeneities in beta spectra.[99]

In 1923 the mark plummeted by a factor of a thousand million. By summer salaries were paid every week, later almost daily. Even so, Lise's salary was astronomical. In August her paycheck "again came to thousands of millions, which I again—anxiously—used up. I don't think we need to worry about the Communists, but altogether things look very bad. It would be nice if we could have a little money left over, instead of nervously calculating every little purchase. One cigarette costs 50,000 marks."[100]

A month later prices were ten times higher.

> There really is enough to eat, although the prices are completely crazy; a kilogram of margarine costs 30–40 million, one egg costs 1½–2 million, etc. Personally I feel fine. . . . I myself don't mind if there is less to eat. It's much worse for ill people and children—no milk, hardly any butter. In the laboratories the workshops are open only from 9–11 and 5–7 because it's so cold.[101]

That was in September: as always Lise was more distressed by cold than lack of food. Every two weeks her mother sent a small package—coffee, butter, sometimes a nutcake. "You shouldn't send so much. I still have coffee, which I drink on Sundays. . . . Sometimes I wish I had my own apartment but of course at the moment that is out of the question. When one is over forty, a little student room is not quite the right thing." Nevertheless, she assured her mother, "I am hardly suffering."[102]

It required effort just to cope with the volume of paper currency. Planck and other professors could be seen leaving the university cashier with their salaries stuffed into rucksacks and suitcases.[103] In the institute Meitner and Hahn "devoted most of [their] time to acquiring and paying out salaries which constantly increased in amount. Then all the employees would run out to spend it as quickly as possible, as the buying power of the money could decline markedly within a few hours."[104] In November, the worst month of all, food riots broke out, the government in Berlin was threatened by Communists, and National Socialists in Munich attempted a putsch against the Bavarian government. Their leader was imprisoned just long enough to write *Mein Kampf* and emerge a hero to his followers.[105]

That autumn Dirk Coster arranged Meitner's first visit to the Netherlands, a lecture tour to Delft, Eindhoven, Haarlem, Utrecht, and Amsterdam. It provided an honorarium in stable currency and escape from the chaos in Germany. The Costers took her everywhere, introducing her to the tranquil landscape and superb museums, Dirk always the "affectionate 'Impresario.' . . . [A] film star could not have been more pampered."[106] Everywhere Dutch physicists were discussing how best to counter the French, who still insisted that German scientists be excluded from international meetings.[107] In Delft, Adriaan Fokker, a nuclear physicist whose work was close to Meitner's, handed her some money for her students and assistants, a reminder that "some scientific solidarity still exists in the world."[108] That Christmas, after Meitner was back in Berlin, Coster sent packages of food in her name to many of her friends in Germany, a kindness she never forgot.[109]

Late in 1923, austerity measures were introduced and the mark was stabilized. In 1924, the situation had improved enough for Meitner and Hahn to concentrate on the technically difficult investigation of radium; by June 1924, the beta spectrum and Meitner's analyses were complete. Soon afterward she began her study of radioactinium, from which she finally proved the sequence of radioactive decay in 1925.

Hedwig Meitner died in December 1924, fourteen years, almost to the day, after her husband, Philipp.[110] After her mother's death, Lise's ties to Vienna were not quite as strong; although her sisters and brother Walter were still all there (her brother Fritz had died in a mountaineering accident some years before), she no longer went "home" to Vienna for every vacation. In 1927, her favorite nephew, Otto Robert Frisch, fresh from his physics doctorate in Vienna, came to work at the Physikalisch-Technische Reichsanstalt in Berlin. He was a good young (23 years old) experimentalist with a talent for instrumentation and had managed to get the position without a reference from his Tante Lise, who had scrupulously refused to provide one. Otto Robert took a flat near his aunt's apartment, and although he was a good pianist and she was not, they sometimes played duets together, mostly the slow movements, translating *Allegro ma non tanto* as "Fast, but not auntie." Until then, most of Otto Robert's music education had come from his mother, Auguste (Gusti), a concert pianist and teacher; now Lise introduced him to symphony concerts under Wilhelm Furtwängler and Bruno Walter, to chamber music, and to more of the varied and splendid musical offerings of Berlin.[111]

Nuclear physics was still beginning, but atomic physics was moving remarkably fast and would, before long, be important for the nucleus as well. Just as Bohr's theory of the hydrogen atom began with a fusion of quantum theory and spectroscopy, atomic physics progressed with an ongoing interplay between experiment and theory.

Soon after Bohr's theory of the hydrogen atom, Arnold Sommerfeld generalized the Bohr atom from circular to elliptical electron orbits. This quantized three atomic properties: the principal energy level, the orbital shape, and the spatial orientation of the orbit in an external magnetic field. The Bohr-Sommerfeld theory accounted for the fine structure of the hydrogen spectrum, and space quantization was experimentally confirmed by Otto Stern and Walther Gerlach in 1921.

The spectra of higher elements were extremely difficult to interpret, but in time some yielded, with help from theory and sometimes chemistry, guided by Bohr's correspondence principle, which permitted a judicious application of classical physics to quantum processes. In 1920, Bohr set out to determine the electron configuration of every element in the periodic table, an enormous undertaking that drew on every possible source: optical and x-ray spectra, atomic theory, chemical properties, the arrangement of the periodic system itself. Within two years Bohr found the regularities in electron configuration that underly the periodic system; at once some puzzles were solved. One concerned the rare earth elements, for which there had been no place in the periodic system; Bohr showed that they introduce a new subshell, the $4f$. This clarified the position of the elements immediately following the rare earths, permitting him to predict that element 72, which had not yet been found, would be located directly below zirconium, with virtually identical chemistry. In Copenhagen in 1922, Dirk Coster and George de Hevesy investigated zirconium ore using x-ray spectroscopy and almost immediately found the spectral lines of the new element; they telegraphed the news to Bohr in Stockholm just in time for him to include it in his Nobel lecture, which made his lecture, with the announcement of this new element, among the most dramatic ever. The discovery was a stunning confirmation of Bohr's theory of the periodic system, acceptable to both chemists and physicists. Coster and Hevesy named the element *hafnium* in honor of Copenhagen (Latin: *Hafnia*) and Bohr's institute.[112]

With a clearer understanding of atomic structure, some spectral effects became obvious which existing atomic theory could not explain. In 1925, Wolfgang Pauli, a young Austrian theoretical physicist, proposed that every electron must possess not three but four quantum numbers, with the added restriction, now known as the Pauli exclusion principle, that no two electrons in an atom may share the same four quantum numbers. With this, Pauli accounted in a most satisfactory way for the maximum number of electrons each energy level was known to contain. The fourth quantum number lacked physical meaning, however, until two young Dutch physicists (so much good physics was done by young men in their early twenties that this period of physics has been dubbed *Knabenphysik*—boys' physics), Samuel Goudsmit and George Uhlenbeck, suggested in 1925 that electrons rotate about their own axis and thus possess magnetism, or spin, whose orientation (up or down) is two-valued. The model of a spinning electron was resisted at first as being relativistically impossible and too mechanical for an effect Pauli had called "classically nondescribable," but these objections were soon overcome by experimental evidence for intrinsic electron magnetism.

Together the new developments dramatically illustrated the inadequacy of existing atomic theory. Immensely successful up to a point, it seemed exhausted and incapable of further extension. The problem seemed to lie with the inherent contradiction of Bohr's theory: the orbiting electron was treated classically except for the arbitrary imposition of a foreign quantum condition. Without a physical basis, quantization appeared no more than a formalism introduced for the sole purpose of achieving agreement with observation.

Theorists looked for a more natural theory, one based on physically observable properties of matter and energy. Wave-particle dualism— unamenable to classical interpretation—became the basis for a new quantum theory. Einstein had proposed his light quantum hypothesis early on, but most physicists found it counterintuitive to think of light as both continuous wave and discrete particle, although they recognized that the quantum relationship $E = hc/\lambda$ already embodied the contradiction by defining the energy of a quantum in terms of the length of its wave. In 1925, the French physicist Louis de Broglie concluded that waves and particles are aspects of the same reality; using the special theory of relativity—which introduces its own dualism, the equivalence of matter and energy—de Broglie extended wave-particle dualism to matter, proposing that waves must be associated with material particles.

The experimental proof came very soon, in 1927, when Clinton J. Davisson and Lester H. Germer at Bell Laboratories in New York and George P. Thomson in Aberdeen demonstrated that electrons are diffracted as they pass through a crystal, a phenomenon very similar to x-ray diffraction that physicists had traditionally interpreted as interference of waves. By then, Erwin Schrödinger had incorporated the matter-wave concept into a new theory. Representing orbital electrons as standing waves, Schrödinger in 1926 devised a wave equation for the hydrogen atom whose solutions yielded the hydrogen spectrum; quantum numbers came out automatically as an entirely natural consequence of the wave properties of the electron.

A very different approach came out of Göttingen the year before. Max Born, Werner Heisenberg, and Pascual Jordan had formulated another theory based on physically observable quantities only, which emphasized the position and momentum of a particle, avoiding orbits and other unobservable representations. However, its matrix-based mathematics proved daunting, even to physicists, while the form of the Schrödinger wave equation was familiar to anyone who had studied wave motion. Meanwhile, Paul A. M. Dirac, a Cambridge mathematical physicist, reformulated matrix mechanics in a more general way consistent with classical mechanics for macroscopic systems. In 1926, a number of physicists, including Born, Schrödinger, and Carl Eckart at the California Institute of Technology, demonstrated that matrix mechanics and wave mechanics are mathematically equivalent and therefore can be regarded as complementary aspects of wave-particle duality. From the new quantum mechanics, Heisenberg derived his famous uncertainty principle, the impossibility of knowing both the position and momentum of a particle-wave with perfect precision. The limitations inherent in the uncertainty principle raised profound questions about the effect of the observer on the subject being observed. It appeared that the experimental conditions determined which of the dual aspects of energy or matter would appear:

wavelike, as in electron diffraction or interference effects of light, or particulate, as in electron beams or the Compton effect. The pace of discovery was dizzying. The physicist Victor Weisskopf would reflect that there was hardly a time in the history of science "in which so much has been clarified by so few in so short a period."[113]

The new quantum mechanics was incomplete, however. Neither matrix nor wave mechanics yielded more than three quantum numbers for the electron; the fourth quantum number, required for spin, was missing from all treatments and had to be separately introduced. In 1928, however, Dirac incorporated relativity into wave mechanics; because relativity requires equivalence of the four coordinates of space and time, his theory yielded four quantum numbers, with spin the natural consequence of a relativistic electron. Dirac's theory also yielded the correct numerical values for the physical properties associated with electron spin, evidence that a true quantum theory had been achieved.[114]

There was good reason to expect quantum theory to apply to nuclei as well: monoenergetic alpha emission and monochromatic nuclear gamma radiation were obviously quantum effects. And since electrons outside the nucleus were quantized, there was every reason to expect the same for the electrons inside.[115]

But the continuous beta spectrum raised doubts about nuclear quantization—and even about the conservation of energy in nuclear processes. Lise Meitner, firm in her quantum outlook, finally conceded that the continuous spectrum was real, but she refused to believe the inhomogeneity could be anything but secondary. Nevertheless, the energy of primary electrons had never been unambiguously determined. In an attempt to do this, in 1926 Meitner used a Wilson cloud chamber to measure the range of beta particles from radium D (^{210}Pb), a difficult undertaking because electrons are readily scattered. Of the thousands of electron tracks in four hundred photographs, Meitner found only eighty-three that were straight enough to measure. These fell into two distinct sets of energy corresponding to the two secondary lines in the beta spectrum of RaD. A monoenergetic group of primary electrons was not in evidence.[116]

While Ellis admitted that he had no explanation for the continuous beta spectrum, he staunchly believed that primary electrons were inhomogeneous at the instant of decay. Late in 1925, he cited new evidence to support his premise. Radium E (^{210}Bi), which emitted essentially no gamma radiation and thus had no secondary beta lines, was found to have a continuous beta spectrum. Without gamma radiation, the continuous spectrum could not be due to Compton scattering. If primary RaE electrons become inhomogeneous by directly ejecting secondary electrons from their orbits, one would expect to find several electrons for each primary decay. Ellis could rule this out also: in the Cavendish laboratory, K. G. Emeleus measured the total number of electrons from RaE and found just over one electron per decay.[117]

A student of Meitner's, Nikolaus Riehl, repeated Emeleus's experiment with about the same results but, like Meitner, did not regard it as proof of primary inhomogeneity. Nevertheless, Meitner had become quite uncertain of her former view.[118] Under her direction two of her assistants, K. Donat and K. Philipp, tested recoil nuclei from the beta decay of thorium B (^{212}Pb) for evidence of the variable energy distribution one would expect if primary electrons were emitted inhomogeneously. The results were too inexact for any conclusion to be drawn.[119]

Ellis, meanwhile, had decided to directly measure the heat produced by the decay of RaE. He devised a calorimeter with walls thick enough to absorb all beta particles: since RaE emitted no gamma radiation and no secondary electrons, the heat measured would be the energy of the primary electrons only, whether or not that energy was divided into smaller components by secondary processes. If all RaE electrons were emitted with identical energy but lost some energy to secondary processes such as collisions, then the energy measured per electron would correspond to the upper limit of the continuous spectrum, about 1,000 kiloelectron volts (keV). If, however, the energy of decay varied continuously, as Ellis contended, the energy per electron would be closer to the average energy of the continuous spectrum, about 390 keV. It took over two years for Ellis and Wooster to complete their measurements. In 1927, they concluded that the primary electrons were continuous. They had measured an average energy per electron of 344 ± 40 keV, very close to the mean energy of the continuous beta spectrum.[120]

The result had serious consequences for nuclear quantization and implied that energy was conserved only statistically and not exactly in nuclear processes. Meitner, indeed all physicists close to theory, were deeply disturbed.[121] The problem was so important that Meitner thought it would not be superfluous to repeat the experiment. Before doing so she congratulated Ellis on his "beautiful results," thanked him for sending his "beautiful paper," and made arrangements for her first visit to Cambridge, where she would give the opening lecture to a congress on beta and gamma spectra.

> I received an invitation from Mr. Chadwick as well, and I am writing to him today that I would very much like to come to Cambridge. [Meitner's disagreement with Chadwick was no longer in evidence; they maintained a very cordial relationship for many years.] A few months ago I began preparations for a calorimetry experiment with an assistant of Prof. Nernst, but so far the apparatus is not finished. The question is of fundamental importance, however, because the idea that the atomic nucleus might have energy states that are only statistically defined would be something that is new in principle.[122]

Meitner improved Ellis and Wooster's procedure in some respects: where they had determined the number of decay electrons of RaE indirectly by measuring the change in the rate of heat evolved, Meitner measured the number of decay electrons independently of the calorimetry experiment by using a standardized preparation of RaE. With Wilhelm Orthmann, Nernst's assistant, she also built an improved calorimeter. Their value for the average energy per beta particle was 337 ± 20 keV.[123]

"We have verified your results completely," she wrote to Ellis in July 1929. "It seems to me now that there can be absolutely no doubt that you were completely correct in assuming that beta radiations are primarily inhomogeneous. But I do not understand this result at all."[124]

The Meitner-Ellis controversy was over, but the data were irreconcilable with existing theory. Meitner was particularly disturbed because recent work by George Gamow and by R. W. Gurney and E. U. Condon had provided a theoretical foundation for the monoenergetic emission of alpha particles from the nucleus. To Meitner it seemed natural that the energy of nuclear electrons should be similarly monoenergetic. In classical theory an alpha particle could never escape the nucleus because its energy is always substantially less than the energy binding the nucleus together, but a wave mechanical alpha particle has a finite escape probability that increases with energy. This "tunnel effect" provided a theoretical basis for the empirical Geiger-Nuttall rule, the inverse relationship between the energy of an alpha particle and the half-life of its source. In a more general sense, Gamow's theory of alpha decay proved that quantum mechanics could be extended to nuclear processes.[125]

But beta decay remained a mystery. At issue was the validity of the principle of conservation of energy. "Every theory that is consistent with the principle of conservation of energy leads to the supposition that the continuous beta spectrum must be accompanied by a continuous γ-spectrum so that the energy emitted by each individual nucleus is, in total, the same."[126] Although gamma radiation had never been observed for RaE, Ellis had not explicitly shown that it was absent. Meitner searched again. It was not there. Resigned, she ended her calorimetry paper, "It is quite certain that radium E does not possess any continuous γ-radiation whose energy might compensate for the inhomogeneity of the primary beta radiation."[127] To Ellis she wrote, "We searched carefully for the existence of continuous gamma radiation; the gamma that is there is much too weak. And yet, if one does not wish to ignore the first law of thermodynamics, there is no theory that would not require a continuous gamma spectrum to compensate for the continuous beta spectrum. Also, according to quantum mechanics, such a gamma radiation should be there, but it is not there, neither for RaE, nor for ThC. I am anxiously awaiting the solution to this enigma."[128]

The continuous beta spectrum was but one of a constellation of problems associated with electrons in the nucleus[129]—all most disturbing, since nuclear electrons were required, it seemed, for beta decay and also for nuclear charge. A nucleus of mass number A and atomic number Z required A protons (for mass) and A-Z electrons (to adjust the nuclear charge down to Z).

In addition to the beta spectrum, it appeared that something was wrong with electron spin within the nucleus. Experimentally it was known—primarily from atomic spectra—that electrons and protons each have a spin of ½. A nucleus with an odd number of constituent particles was therefore expected to have half-integer spin; an even number of constituent particles would have integer spin. But for those nuclei whose spins were known—again from spectra—the predicted spin was correct when A-Z was even (that is, an even number of nuclear electrons) but failed for ^{6}Li and ^{14}N, whose A-Z was odd. With respect to spin, the nucleus behaved as if nuclear electrons were not there.

Related discrepancies surfaced for the symmetry of the wave function and the corresponding statistics of nuclei. Particles with half-integer spin obey the Pauli exclusion principle, are governed by Fermi-Dirac statistics, and have antisymmetrical wave functions, while particles with integer spin (alpha particles, for example) are not subject to the exclusion principle, are governed by Bose-Einstein statistics, and have symmetrical wave functions. By 1930, enough experimental data had accumulated to show that not only spin but also symmetry and statistics were incorrectly predicted for nuclei with odd A-Z.

Thus decay, mass, and charge considerations required both protons and electrons in the nucleus, but spin, symmetry, and statistics behaved as if electrons were entirely absent. Bohr spoke of the "remarkable passivity" of electrons in the nucleus; there were also theoretical reasons for concern. According to the uncertainty principle, a particle as light as an electron cannot be confined to a region as small as a nucleus; expressed another way, the de Broglie wavelength of the electron is much larger than the radius of a nucleus.[130]

Perhaps most disturbing, however, was the continuous primary beta spectrum, a threat to the universal validity of conservation of energy. Bohr believed a "new physics" might be required for nuclear electrons and suggested that energy might be conserved only statistically in the nucleus, much as the second law of thermodynamics is valid only statistically for the behavior of atoms.

Wolfgang Pauli strongly disagreed. In December 1930, he sent an open letter to physicists attending a meeting in Tübingen, proposing a "desperate remedy"—a new particle he called the neutron (later called the neutrino). The letter, in mock formal style, was delivered by a friend and addressed to Lise Meitner and Hans Geiger.

> Dear Radioactive Ladies and Gentlemen! As the bearer of these lines, for whom I ask your gracious attention, will explain to you in more detail, I have, faced with the "false" statistics of the N-14 and Li-6 nuclei as well as the continuous β-spectrum, stumbled upon a desperate remedy. Namely, the possibility that in the nucleus there could exist electrically neutral particles, which I will call neutrons, which have a spin of one-half and obey the exclusion principle and in addition also differ from light quanta in that they do not travel with the speed of light. The mass of the neutron must be of the same order of magnitude as the mass of the electron and in any case not larger than 0.01 proton mass. The continuous beta spectrum would then be understandable assuming that in β-decay a neutron is emitted along with the electron in such a way that the sum of the energies of neutron and electron is constant. . . .

At the moment I don't trust myself enough to publish anything about this idea and turn confidently to you, dear radioactives, with the question of how one might experimentally prove such a neutron, if its penetrating ability is similar or about 10 times that of γ-radiation. I admit that my remedy may at first seem only slightly probable, because if neutrons do exist they should certainly have been observed long ago! But, nothing ventured, nothing gained, and the gravity of the situation with the continuous beta spectrum is illustrated by a statement of my respected predecessor in this office, Herr [Peter] Debye, who told me recently in Brussels: "Oh, it is best not to think about it at all, like the new taxes!" Thus one should seriously discuss every means of salvation. Therefore, dear radioactives, test and decide! Unfortunately I cannot appear in Tübingen in person, since I am indispensable here due to a ball which will take place the night of December 6 to 7 in Zurich. With many greetings to you all, your most humble and obedient servant, W. Pauli.[131]

Although the neutrino was particularly important for explaining the continuous beta spectrum, Pauli was also thinking of the spin and statistics of stable nuclei and was, therefore, proposing a new nuclear inhabitant.

Beginning with Hans Geiger and Lise Meitner, the neutrino was widely and favorably discussed,[132] but Pauli did not publish his hypothesis for some time. His neutrino did not resolve other problems of nuclear electrons, including the questions raised by the uncertainty principle. During this period, George Gamow wrote a nuclear physics textbook in which he stamped every reference to nuclear electrons in his manuscript with a skull and crossbones (they appeared as tildes [˜] in the printed book).[133] A decade of work with beta spectra had illuminated the problems, but the solutions were still completely hidden from view.

CHAPTER FIVE

Experimental Nuclear Physics

We were struck by the surprisingly large number of almost straight line electron tracks emanating from the source.

In the 1920s it was possible, indeed essential, for a scientist to have a detailed overview of the field of nuclear physics. In this Lise Meitner was typical, closely following the experiments and theoretical work of others, writing review articles, speculating on the significance of new developments. Even at the height of her controversy with C. D. Ellis, she undertook a variety of other experiments, kept an eye out for new instruments, and altogether maintained the versatility needed in a field that tended to sprout surprises in unexpected places. In addition to beta and gamma spectra, she investigated long-range alpha particles, the absorption and energy of beta particles, nuclear scattering interactions, the scattering and absorption of high-energy gamma radiation, and artificial nuclear reactions.[1]

Already a titular professor in the KWI for Chemistry, Meitner began scaling the conventional academic ladder in 1922. Elsewhere in Germany women were granted the right to Habilitation in 1918, in Prussia in 1920; Meitner was *habilitiert* and received her venia legendi in 1922, which qualified her as the first woman physics Privatdozent (in her case, *Privatdozentin*) in Prussia, the second in all Germany.[2] In support of her candidacy, Max von Laue wrote, "As Fräulein Meitner is one of the world's best-known scientists in radioactivity, her Habilitation is completely consistent with the interests of the faculty. I propose that she be granted permission to present a trial lecture and colloquium. . . . My only reason for not proposing that these be waived (on grounds of special merit) is to give her the opportunity to demonstrate her extremely thorough understanding of other fields of physics."[3]

With more than forty publications to her name, Meitner was not required to submit a Habilitation thesis; her inaugural university lecture in 1922 was titled "Die Bedeutung der Radioaktivität für kosmische Prozesse" (The Significance of Radioactivity for Cosmic Processes).[4] Such lectures were semipublic and often appeared in print. An academic press, evidently a bit rattled, requested permission from the new Privatdozentin to publish her lecture on the "significance of radioactivity for cosmetic (*kosmetische*) processes," a quote she relished for many years.[5]

The venia legendi officially qualified Meitner for university teaching. From 1923 until 1933, when her teaching privileges were rescinded, she taught a colloquium or tutorial at the university in Berlin nearly every semester and supervised the doctoral research of university students in her own section in the KWI for Chemistry.[6] In 1926, soon after her proof of the sequence of radioactive decay, the university promoted her to the adjunct position of *nichtbeamteter ausserordentlicher Professor*, making her the first woman university physics professor in Germany.[7] One recognition followed another: the Prussian Academy of Sciences' Silver Leibniz Medal in 1924, the Vienna Academy of Sciences' Ignaz Lieben Prize in 1925, and the American Ellen Richards Prize, which Meitner shared with the French scientist Ramart Lucas in 1928.[8] Each year she became more prominent, and her physics section larger: a permanent Assistent, a sizable group of doctoral students, visiting scientists from Germany and abroad.

Throughout the 1920s the nucleus was still very much a newcomer to the atomic scene. The enormous development of theory and experiment that revealed so much of the atom had done little for the nucleus except create a list of discrepancies. Nuclear electrons in particular wreaked havoc with theories that worked perfectly outside the nucleus. Far from being a discouraging influence, however, the severity of the theoretical problems made this a splendid time for experiment: it seemed that something new just had to come along that would cut through the difficulties at a single stroke.[9] There was no way to tell whether this would result from the steady application of work in progress, or as a bolt from the blue. As it happened, the two crucial discoveries—the neutron and the positron—would come from each of these directions, but not until 1932.

Until then, radioactivity remained the prime source of nuclear data. Its larger picture was contradictory, alpha decay and gamma emission being consistent with nuclear quantization and conservation of energy while the continuous beta spectrum implied the opposite. But individual radioactive species, their transformations and patterns of instability, provided insight into nuclear behavior that was unavailable from any other source. Besides, the radiations themselves were important: alpha particles were used as energetic projectiles in scattering experiments and artificial nuclear reactions, beta particles for studying the interactions of electrons with matter, gamma rays for the high-energy radiation only radioactive disintegrations provided. So important was radioactivity to nuclear research that Rutherford, Chadwick, and Ellis used the title *Radiations from Radioactive Substances* for a treatise covering nearly all of nuclear physics in 1930.[10] Reflecting the absence of a comprehensive theoretical foundation, the authors were authoritative in their presentation of experiments and data but curiously hesitant in discussing the implications, giving the impression, on the whole, that much was known and little understood.

Scattering experiments had a long and distinguished history. In 1906, Rutherford observed that alpha particles deviate slightly from their original trajectory in air—Meitner studied the effect in 1907—and related it to encounters between the fast-moving alpha particles and the molecules in their path. Rutherford's co-workers Geiger and Marsden found similar small-angle scattering for alpha particles passing through thin metal foils; their observation of a few alpha particles hitting the foil and bouncing off backward was at odds with all existing ideas of atomic structure and led Rutherford to his theory of the nuclear atom in 1911.[11] Scattering continued to be a useful probe for nonradioactive nuclei and for observing single nuclear events. In 1920, James Chadwick used alpha particle scattering to determine the absolute charge of platinum, silver, and copper nuclei, the first direct verification of Moseley's conclusion that atomic number is identical to nuclear charge.[12]

In similar experiments with light nuclei as targets, Rutherford observed the first artificial nuclear reaction in 1919. When he enclosed an alpha source in a container filled with nitrogen gas, he found that high-speed hydrogen nuclei were produced and concluded that the nitrogen nucleus "disintegrated under the intense forces developed in a close collision with a swift alpha particle, and that the hydrogen atom which is liberated formed a constituent part of the nitrogen nucleus."[13] This confirmed what had long been assumed: that the hydrogen nucleus is a fundamental particle common to all nuclei. Following Prout's century-old hypothesis that all atoms are built up from hydrogen atoms or "protyles," Rutherford named the particle "proton" in 1920. At that point, the question of nuclear composition seemed to be settled: a nucleus of mass A and charge Z required A protons for mass and A-Z electrons to bring the nuclear charge down to Z.

To account for the stability of multiple positive charges confined to the exceedingly small dimensions of the nucleus, Rutherford suggested in 1920 that the nucleus consists of a core of highly stable alpha particles surrounded by the remaining protons and electrons; his proposal that protons and electrons might pair up into tightly combined "neutral doublets" launched the search for a "neutron." In 1921, Lise Meitner looked for clues to nuclear behavior and structure in patterns of radioactive instability and modes of decay. Noting that alpha decay is often followed by two beta decays, she proposed that an alpha particle neutralized by two electrons might exist as another relatively stable nuclear subunit.[14] Rutherford incorporated Meitner's neutral alpha particle and his own "neutral doublet" into models of nuclear structure that he developed and refined throughout the 1920s.[15]

There was hope that artificial nuclear reactions would be the starting point for a nuclear chemistry of sorts, as valuable for clues to nuclear structure as chemical reactions had been for atomic structure. Over the next several years a number of artificial reactions were induced, primarily by Rutherford and Chadwick in Cambridge, in elements ranging from boron to potassium; in every case alpha particles were the projectiles, and protons were the only product detected. At first Rutherford referred to the reactions as disintegrations, believing that the energetic alpha particle simply knocked protons out of the larger nucleus, but in 1924, P. M. S. Blackett photographed an alpha-nitrogen collision in a Wilson cloud chamber and saw no trace of the alpha particle after collision—only a proton and a heavy nucleus. The reaction was, therefore, not disintegration so much as synthesis: nitrogen captured the alpha particle and expelled a proton, which meant that the resultant heavy nucleus had to be oxygen:

$$^{14}N + {}^{4}He \rightarrow {}^{17}O + {}^{1}H$$

Alpha particle capture was intriguing but difficult to understand. The energy of the alpha particles was known—and in theory it was too small to overcome the coulomb repulsion of the nucleus. It was necessary to assume that a new, hitherto unknown strong attractive force predominated at very close range. Moreover, the incident alpha particles were monoenergetic, but the ejected protons varied widely in energy; while it was possible that compensating gamma radiation was simultaneously emitted or that the new heavier nuclei varied substantially in mass, this was difficult to ascertain experimentally. In all, few conclusions regarding nuclear structure could be made.[16]

In any case, it was possible to induce artificial nuclear reactions in only a few light elements: nuclei beyond potassium always repelled and scattered alpha particles before capture could occur. Attention again turned to scattering experiments. Meitner followed these closely, summarizing the results in a long review article.[17] Close encounters between alpha particles and lighter nuclei revealed marked departures from coulombic forces, an indication that nuclear structure was complex. From such noncoulombic scattering, Rutherford and Chadwick made estimates of nuclear size, shape, and charge distribution and speculated that electron-proton pairs, Rutherford's so-called neutral doublets, might form the outer shell of the nucleus. In Stefan Meyer's institute in Vienna, the Swedish physicist Hans Pettersson suggested that noncoulombic scattering might be explained by induced polarization of the nuclei; others thought magnetic effects might play a role. On the whole, scattering experiments were difficult to perform and more difficult to interpret—even more so because there was no guarantee that classical electrodynamics was valid for nuclear events.[18]

For her own scattering experiments, Meitner built a Wilson cloud chamber, the first in Berlin.[19] In the cloud chamber, the tracks of alpha, beta, and other charged particles could be clearly seen and photographed, with the great advantage that individual collisions and other rare events could be directly observed. With her student Kurt Freitag, Meitner studied the vast majority of alpha particles that never actually collide with another nucleus but instead brush past the outer electrons of the atoms in their path, lose energy, and come to a stop at a distance, or range, characteristic of their initial energy. From the large number of individual tracks made by alpha particles from ThC (^{212}Bi) and ThC' (^{212}Po), they determined the ranges with great precision. Meitner was primarily interested in testing a theoretical formula of Bohr's that related statistical variations in range to the mechanism by which alpha particles lost energy to the atoms they encountered: if Bohr's formula were correct, something of the electron arrangement of the atoms or possibly the structure of the alpha particle itself might be inferred. Meitner and Freitag measured ranges in several different gases, but, as often happened with scattering experiments, agreement with theory was neither good nor bad, so that one could hardly decide which was more flawed, the formula or the experiment.[20]

For Meitner, the greatest benefit may have been the experience she gained with the Wilson cloud chamber, the single most useful instrument for observing individual nuclear events. Invented by C. T. R. Wilson near the turn of the century for studying the physics of clouds and mist, its operation was in principle quite simple: a sealed chamber filled with moist air is quickly expanded, producing a super-saturated water vapor that condenses along the trajectory of a charged particle into a visible trail of tiny foglike droplets. Meitner's cloud chamber, which she built in 1924, was about 21 centimeters in diameter, sufficiently large to display the entire range of the alpha particles in question, and flat enough to photograph the tracks without optical distortion; she used a stereoscopic camera triggered by the mechanism that expanded the chamber. Over the next several years, Meitner and her co-workers systematically studied the effects of gases other than air in the expansion chamber, as well as condensates other than water vapor.[21]

In her work with Freitag on ThC + C′, Meitner's most interesting discovery was that among the nearly one million alpha tracks in some three thousand cloud chamber photographs, there were a few hundred tracks with two distinct ranges that were each considerably longer than the normal alpha ranges from ThC and ThC′. The most reasonable interpretation was that alpha emission was not absolutely monoenergetic, a fact that had been suspected for some time. Rutherford had observed some long-range alpha particles several years before,[22] and as early as 1922 Meitner had observed small variations in the range of radium alpha particles and suggested that the gamma radiation associated with radium alpha decay might account for the difference.[23] In a similar vein, George Gamow proposed in 1930 that slight variations in the measured ranges of alpha particles from ThC required the daughter nucleus, ThC″ (^{208}Tl), to emit gamma radiation to equalize the energy differences.

To test Gamow's proposal, it was necessary to correctly assign the various gamma frequencies to their nucleus of origin, an experiment requiring considerable ingenuity since ThC (^{212}Bi) branched, emitting alpha particles to form ThC″ (^{208}Tl, a beta emitter) as well as beta particles to form ThC′ (^{212}Po, an alpha emitter), the end product in both cases being ThD (^{208}Pb, stable lead). Meitner, together with her assistant Kurt Philipp, focused on the ThC → ThC″ → ThD branch. From the alpha recoil of ThC, they isolated pure ThC″ and obtained its secondary beta spectrum and gamma frequencies for the ThC″ → ThD step. Comparing these to the gamma spectrum of a ThC + C″ mixture, they could assign some gamma frequencies to the alpha decay of ThC → ThC″ alone: these corresponded to Gamow's prediction.[24] This meant that the emission of energy in alpha decay was analogous to similar processes for atomic spectra: one ThC nucleus might lose energy by alpha emission alone, another by alpha and gamma emission combined, but no matter the mechanism, it appeared that energy was always strictly conserved for each individual nucleus undergoing decay. At a time when the continuous beta spectrum threatened the sanctity of energy conservation in nuclear processes, these findings upheld it.

During the 1920s, physicists tried to glean what they could from radioactive decay and the abundant collection of natural nuclear reactions. In 1926, Lise Meitner again noted empirical correlations between the relative stabilities of radioactive species and their modes of decay. From these she predicted that the atomic mass of protactinium, which was still unknown, would be less than 234, which meant that the origin of the actinium series, also still unknown, could not be uranium-238. Meitner favored the idea that uranium-234 headed both the uranium-radium and the actinium series but noted that others had suggested a still-undiscovered third isotope of uranium for the precursor of the actinium series.[25] In 1929, a mass spectrographic analysis of the lead isotopes in a uranium mineral revealed that in addition to ^{206}Pb, the known end product of the uranium-radium series, there was a trace of ^{207}Pb. Immediately this was hailed as a possible end product of the actinium series, whose first member would then be ^{235}U. The existence of ^{235}U remained uncertain for several years, however, in part because the isotopic data were imprecise and also because the possibility could not be ruled out that some other process produced ^{207}Pb in uranium minerals. Finally in 1934, protactinium was purified and its mass determined to be 231, which established the mass of every member of the actinium series and proved the existence of ^{235}U.[26]

For Meitner, alpha decay was always simpler and clearer than beta decay. In 1928, when she had nearly accepted the reality of the continuous primary beta spectrum but was deeply disturbed by its implications, she investigated a number of alpha sources in the hope of correlating their gamma emission and nuclear properties. Her study of the beta and gamma spectra of protactinium (^{231}Pa) did little more than demonstrate that its gamma radiation was, as expected, monochromatic,[27] but the corresponding beta-gamma spectra of radiothorium (^{228}Th), another alpha emitter, were more interesting.[28] Because its secondary electrons originated in the L, M, and N shells and its gamma radiation corresponded to a K_α transition (the transition of an orbital electron from the L down to the K shell), Meitner was quite sure that nuclear gamma radiation was absent. The usual mechanism for the expulsion of secondary electrons is internal conversion, in which the nucleus expels a gamma quantum (photon) that is then absorbed by an orbital electron; this was viewed as *radiative* coupling between nucleus and electrons, with gamma radiation serving as an intermediary. In the absence of nuclear gamma emission, however, it appeared that the nucleus was losing energy *directly* to an orbital electron. If true, this was a very intriguing radiationless transition between nucleus and electron, demonstrating that orbital electrons are capable of penetrating the nucleus.

Meitner's interest in radiationless transitions went back to 1923, when she first recognized and described the process among the orbital electrons of UX$_1$.[29] Already in 1922 the Viennese theoretical physicist Adolf Smekal had proposed that quantum effects of nucleus and orbital electrons be treated as inseparable; he objected to the entire notion of internal conversion as inherently unobservable. Perhaps because Meitner found his positivistic argument distasteful, she did not refer to Smekal in the discussion of her radiothorium findings in 1928 and only briefly alluded to a "very close" (*sehr enge*) nucleus-orbit coupling as a possible explanation.[30] In the next few years, several groups, including C. D. Ellis and his co-workers in Cambridge, studied radiationless transitions in a number of beta sources; by 1930, they were convinced that such radiationless transitions between nuclei and orbital electrons were possible. Meanwhile the idea that subatomic particles might penetrate the potential barrier of the nucleus found a different expression in Gamow's 1928 theory of alpha decay; attempts to formulate a related theory of beta decay were unsuccessful.[31] Meitner continued to study radiationless transitions, rather ingeniously adapting Millikan's oil drop method to determine the probability of multiple ionizations caused by alpha particle collisions with light atoms.[32]

Among the most valuable new instruments of the postwar period was F. W. Aston's mass spectrograph, capable of measuring atomic masses to a precision of one part in 1,000. One of Aston's first measurements showed that neon is a mixture of atoms of mass 20.00 and mass 22.00, proving that isotopic mixtures are not an exclusive feature of radioactive nuclei but are also present in stable elements of low atomic mass. In itself the existence of stable isotopes was of great interest, but the most important finding was that the mass of each individual isotope is in every case very close to a whole number: chlorine, for example, was clearly shown to be a mixture of isotopes, 76 percent of mass 35.0 and 24 percent of mass 37.0, its atomic weight of 35.45 being the isotopic average. The finding of near-integral nuclear masses removed the last obstacle for accepting the proton of unit mass as a fundamental particle present in all nuclei.[33]

Mass spectrographic data also revealed marked differences in the abundance and masses of stable isotopes. In the earth's crust, elements of even atomic number had many more isotopes and were far more abundant— presumably an indication of stability—than elements of odd atomic number. In 1926, Meitner noted that all known isobars—atoms with the same mass number but different atomic number—were of even atomic number: isobaric pairs, for example ^{40}Ar and ^{40}Ca, skipped a possible middle isobar, in this case ^{40}K, which was either missing entirely or present in amounts too small to be detected. This reminded Meitner of several radioactive sequences where two successive beta decays produced three isobars, of which the middle isobar was usually the most unstable. She concluded that the missing ^{40}K might also be unstable, possibly the source of the very weak beta radiation associated with potassium; this was later found to be true.[34] The fact that even atomic number favors stability, she added, may indicate that protons in the nucleus tend to form even-numbered complexes.[35] Such empirical correlations of stability with atomic number and mass number anticipated the "magic numbers" that eventually formed the basis for the shell theory of nuclear structure.[36]

Meitner was also interested in using isotopic distributions for estimating geological and cosmological ages. The distribution of lead isotopes in ordinary lead was known to be quite different from that in uranium or thorium minerals, from which it was assumed that the lead isotopes in radioactive minerals were the stable end products of radioactive decay. From the relative amounts of lead and the known rates of uranium or thorium decay, the age of the minerals could be deduced; in general these were found to be several billion (10^9) years old, providing a minimum age for the hardening of the earth's crust. In 1933, Lise Meitner and Otto Hahn discussed the possibility that ordinary lead was also derived from radioactive decay but had separated from its radioactive antecedents before the crust congealed. In particular, they attributed the sizable fraction of ^{207}Pb in ordinary lead to ^{235}U; from the proportions of the lead isotopes and the decay rates of their uranium and thorium antecedents, Meitner and Hahn arrived at an age of some 10^{10} years for the formation of the earth.[37]

From this they placed the minimum age of the sun at 10^{11} years, which in turn raised new questions. As Meitner noted, no known process on earth, including radioactive decay, could continuously emit energy for longer than some 20 million years. "A more abundant energy source must exist in the interior of stars, and we have such a source in the conversion of mass into radiation."[38]

For a century scientists had supposed that elements are built up from atoms of hydrogen; since the early days of spectroscopy it was known that young stars consist primarily of hydrogen and helium and that the spectral lines of heavier elements are more intense in stars that are older. Not until the 1920s, however, did physicists understand that the process of forming heavier elements must also be the source of stellar energy. Once again, mass spectrographic data played an essential role: even the earliest measurements indicated that atomic masses are in every case somewhat less than the total mass of the individual protons and electrons of which they are composed.[39] The missing mass, or "mass defect," was recognized as the basis for both stellar energy and nuclear stability: at the instant a nucleus forms, its excess mass is converted into energy and radiated away; the new nucleus is stable with respect to decomposition, being lower in mass and energy than the particles from which it is formed.

When Aston's mass spectrograph was improved to a precision of one part in 10,000 in 1927, some mass defect trends became obvious. The proportion of mass lost in forming nuclei—the so-called packing fraction—varies slightly but systematically through the periodic table, rising gradually to a distinct maximum for atomic masses near 60, then declining for larger atoms. This meant that energy could be released in two very different sorts of nuclear reactions: the fusion of light to heavier nuclei below mass 60 and the disintegration of very heavy nuclei to lighter ones above mass 60. As Meitner pointed out in a 1931 review, examples of both processes had been observed in the laboratory, in artificial nuclear transformations, on the one hand, and radioactive decay, on the other. Neither she nor others foresaw that a third process, the splitting of a heavy nucleus, was also possible, although Meitner would, before long, be instrumental in its discovery.[40]

From the beginnings of radioactivity, scientists were tantalized by the prospect of vast new stores of energy. Rutherford had often pointed to the enormous energy latent in the atom, estimating the energy released by the decay of radium to be many millions of times greater than its weight in coal; artificial nuclear transmutations, despite their limited applicability, also released far more energy than comparable chemical reactions. Yet a simple calculation showed that the energy emitted in the fusion of hydrogen into heavier nuclei was greater still, leading Aston to imagine that some day "the human race will have at its command powers beyond the dreams of science fiction."[41]

Until then, energy of that order of magnitude had been detected only in the penetrating rays known as cosmic radiation. In 1929, Meitner reviewed the field, in particular the origins of cosmic radiation that had intrigued physicists for many years. More penetrating than radiation from any terrestrial source, cosmic rays seemed to come from all directions of outer space and were believed to consist of electrons and ultra high energy gamma radiation. (Not until the 1930s was it discovered that cosmic "radiation" is not electromagnetic but consists of charged particles.) Walther Nernst attributed cosmic radiation to the presence throughout the universe of transuranic elements—elements beyond uranium—that disintegrated with far more energy than ordinary radioactive decay. But the American physicist Robert Millikan, who coined the term "cosmic rays," thought these were the energies released when hydrogen fuses to form helium and other nuclei; with a knack for the memorable phrase, he described cosmic rays as "birth cries of the elements" and "music of the spheres."[42] To estimate their energy, Millikan and his co-workers conducted absorption measurements and found three absorption coefficients, indicating three distinct wavelengths, which he believed corresponded to the formation of helium, oxygen, and silicon. Meitner was very interested, particularly in the idea that "cosmic radiation actually possesses defined wavelengths, that is, it consists of two or more monochromatic rays, because this indicates that the origin of this radiation is somehow linked to fundamental atomic processes, in fact to as many different atomic processes as there are groups of monochromatic radiation."[43] Although Millikan's hypothesis was weakened by the improbability of hydrogen nuclei finding each other in the vastness of interstellar space, his absorption measurements and wavelengths were at least open to experimental test. Meitner had available a sensitive new instrument and a strong new theoretical approach for doing just that.

Wavelengths of very high energy radiation such as gamma or cosmic radiation were too short to be measured directly. Instead, their absorption was measured and a formula applied which permitted the wavelength to be calculated as a function of absorption coefficient. Such formulas were developed with a specific absorption mechanism in view; for very high energy gamma radiation the only known absorption mechanism was Compton scattering. Two formulas, one derived by A. H. Compton and another by P. A. M. Dirac, had been in use for some time. In 1928, Oskar Klein and Yoshio Nishina published a new formula, based on the relativistic electron theory introduced that year by Dirac, which gave quite different results for radiation of very high energy. "Which formula corresponds best to reality," Meitner decided, "and therefore which wavelengths can be ascribed to cosmic radiation, can only be determined with more experimental data on the scattering of γ-radiation of very short wavelength."[44]

In 1929, Meitner began a series of experiments designed to test the Klein-Nishina formula. As a practical matter, it was important to know whether Klein-Nishina was more reliable than the older formulas for the very short wavelengths associated with high-energy light quanta (photons); more fundamentally, a test of Klein-Nishina was also an implicit test of Dirac's new theory.

When first introduced, relativistic electron theory was a great success: it accounted for the electron's intrinsic spin and magnetic moment in an entirely natural way without the need to introduce these properties separately as had been the case for earlier, nonrelativistic theories. On closer examination, however, Dirac's theory revealed a number of peculiarities. Most striking was the possibility that electrons could exist in a state of negative kinetic energy, a bizarre concept that implied the existence of negative mass. Late in 1928 Oskar Klein pointed out that an electron encountering a very strong potential gradient could lose so much energy that it could, according to Dirac's theory, pass through the potential barrier and emerge with negative kinetic energy.[45] As potential gradients of this magnitude were considered necessary to contain electrons within the nucleus, the so-called Klein paradox was taken as yet another argument against intranuclear electrons. In any event, electrons with negative energy had never been observed, even in regions with extreme potential gradients; the strangeness of negative energy together with the lack of experimental evidence for it raised serious doubts about Dirac's theory.

Dirac boldly claimed that negative energy is unobservable because the "normal" world is in fact immersed in an infinite sea of electrons of negative energy: because nearly all negative energy states are already occupied, ordinary electrons are barred, by the Pauli exclusion principle, from making the transition from positive to negative energy. While this argument resolved the Klein paradox by declaring it impossible, it presented new problems. Nothing prevented a negative-energy electron from acquiring enough energy to bring it into the realm of positive energy, where it would appear as an ordinary electron. This, however, would leave a vacancy in the sea of negative energy, which, according to Dirac's equations, would behave like a positively charged electron with positive energy. But positive electrons had never been observed. Dirac thought the holes might be protons, but the difference in mass created great difficulties. To Bohr, the infinite electron sea was a "fatal" objection; Pauli thought Dirac's theory "hopeless in the face of its consequences."[46]

To relate Dirac's theory to a physical process more amenable to experimental test, Klein and Nishina applied it to Compton scattering in 1928. Compton had treated the scattering of gamma radiation in matter as a collision between photon and electron; with this mechanical model he found that the wavelength of scattered radiation increases with angle of scattering; that is, the more a photon is deflected from its original path, the more energy it loses. Because the increase in wavelength is independent of the wavelength of the incident radiation, the Compton effect becomes significant only at the very short wavelengths of hard x-rays and gamma radiation, where scattering produces a relatively marked change in wavelength. Compton's formula for the intensity of scattering as a function of wavelength agreed with experiment fairly well, at least to a first approximation. For radiation of fairly long wavelength the Klein-Nishina formula agreed with the Compton formula, but for shorter wavelengths it differed substantially.[47]

The scattering process could be studied in several ways. One could measure the intensity of the scattered radiation or the recoil electrons as a function of scattering angle; this had been attempted without clear results. Meitner decided to determine the scattering coefficient, a measure of the absorption of the incident gamma beam as it passes through matter.

The investigation suited her: for its theoretical significance, for the convergence of cosmology with nuclear physics, and because it called for just the sort of experiment she was prepared to do. Familiar with absorption measurements from her earliest studies of beta and gamma radiation, she had access to gamma sources, the techniques for handling them, and a new and highly sensitive measuring instrument that she was eager to use.

> To obtain exact measurements of scattering coefficients one must use monochromatic radiation in a parallel, narrowly collimated beam, so that one can be certain that one's measurements are not affected by stray radiation. The narrow collimation that is necessary reduces the intensity of the already weak preparations by a factor of about $3 \cdot 10^4$ to 1. One must therefore use a very sensitive measuring instrument, and the Geiger-Müller counter is especially well suited for this.[48]

In fact, the new counter that Hans Geiger and Wilhelm Müller had just developed was so sensitive that when Meitner asked a student to try it, he complained that stray radiation from an adjoining room was affecting his measurements.[49]

At the time there were only two known mechanisms, each quite distinct, for the interaction of gamma radiation with matter. In a photoelectric process a gamma photon loses its energy to an electron and the photon disappears entirely. In Compton scattering, photons collide with electrons and are scattered in all directions. In both processes, however, the primary gamma beam appears to be absorbed; that is, the intensity of the incident radiation decreases as it travels through the material.

The photoelectric effect occurs most readily when the energy of the gamma photon is only a little greater than the ionization energy of the electron it encounters, while Compton scattering dominates for photon energies much higher than ionization energies. Meitner's experiments were designed for high-energy gamma radiation, where the observed absorption of the primary beam could be attributed entirely to scattering. By measuring the intensity of the gamma radiation as a function of the thickness of scattering material, she would obtain a scattering coefficient she could compare with the value predicted by the Klein-Nishina formula.

For the first set of measurements, Meitner, together with her student H. H. Hupfeld, used gamma radiation from ThC″ (^{208}Tl) and measured its scattering coefficient in thirteen different elements ranging from carbon to lead. ThC″ was chosen for its monochromatic gamma ray of very short wavelength, 4.7 X.U. (1 X.U. = 10^{-11} cm). The apparatus was quite simple: a large block of thick iron plates, a gamma source, and a Geiger-Müller counter. In the center of the iron block was a slender shaft into which the gamma source was inserted; the shaft collimated the radiation into a parallel beam that passed through varying thicknesses of scattering elements at the shaft opening and was registered by the counter several meters away. The apparatus was designed to record only the primary beam, excluding stray effects from scattering or reflection.

For the light elements Meitner and Hupfeld found Klein-Nishina in excellent agreement with experiment, much better than the older formula,[50] but for heavier elements they immediately found discrepancies. According to Klein-Nishina, the scattering coefficient per electron would depend only on the gamma wavelength, not on the scattering material. Meitner and Hupfeld's scattering coefficients agreed with Klein-Nishina for elements up to magnesium, rose distinctly in aluminum, and kept increasing with atomic number. "Compared to the Compton formula, the scattering coefficient agrees best with Klein and Nishina, but there are definite deviations that increase with atomic weight. . . . One must consider the possibility that a hitherto unknown effect is present."[51]

Meitner and Hupfeld measured the scattering coefficients again, using lower-energy 6.7 X.U. gamma radiation from RaC (^{214}Bi). If the excess absorption were due to an unexpected photoelectric effect, the deviation would be larger for radiation of lower energy. Exactly the opposite was true, and Meitner ruled out photoelectric absorption as a cause.[52]

Under other experimental conditions, however, the Klein-Nishina formula was fully verified. New measurements of the angular distribution of recoil electrons were just what the formula predicted, and Meitner's own experiments showed good agreement for scattering in the lighter elements. This meant that the formula was valid to the extent that it described scattering but that some other process became important with increasing nuclear charge and higher gamma energy. In an effort to characterize this unknown effect, Meitner set out to find the missing radiation.

She thought the nucleus itself might be scattering radiation.

> Since there can be no doubt about the correctness of the Klein-Nishina formula, it seems likely that the observed deviation is due to a nuclear scattering process. If one takes into account the fact that the wavelengths are long compared to nuclear dimensions, one can take this as an analog to Rayleigh scattering of light, and expect it to show spherical symmetry, with no change in wavelength.[53]

In a series of experiments, Meitner and Hupfeld did find some radiation scattered through an angle of 90° with no change in wavelength, in contrast to Compton scattering, which increases the wavelength.

In England, L. Gray and G. Tarrant were finding something quite different. Using gamma radiation that was not monochromatic, they too measured scattering in excess of the Klein-Nishina formula. Against a background of scattered radiation they always identified the same two wavelengths, no matter what radiation or scattering material they used. From this they proposed that incident gamma radiation was absorbed by a nucleus, then preferentially reradiated in two characteristic wavelengths by alpha particles within the nucleus, a process they called nuclear fluorescence.

Meitner was unable to confirm their observations. Varying the type of radiation and the scattering material, she always found some radiation scattered with wavelength unchanged but not enough to account for the entire deviation from the Klein-Nishina formula.[54] The search for the missing radiation spanned a period of almost three years, from 1930 to 1933.[55]

Meitner had come to this investigation with an interest in cosmic radiation and the desire to test Dirac's relativistic electron theory. Ironically, the deviation from Klein-Nishina was a manifestation of the theory's most incomprehensible consequence, and the explanation for it would begin with seemingly unrelated studies of cosmic radiation. Only later would it be clear that the Klein-Nishina experiments provided just those conditions—hard gamma radiation in the vicinity of heavy nuclei—that favor the creation of positive electrons, those "holes" in the infinite sea of negative energy that were incompletely predicted by Dirac. Unfortunately for Meitner and others working on Klein-Nishina, their experiments created positrons but excluded the means for detecting them. The positron was first recognized, directly and without benefit of theory, in a cloud chamber in 1932.[56]

For physics, 1932 was a "miracle year," the likes of which had not been seen since Einstein's revolutionary theoretical work in 1905. First came the neutron, discovered by James Chadwick in February. This particle was long awaited, especially in Cambridge, from Rutherford's first mention of a "neutral doublet" in his Bakerian lecture of 1920; in his studies of scattering and artificial nuclear reactions, Chadwick had the neutron in mind and on at least one earlier occasion specifically tested for its presence.[57] Although there was no trace of it then, Rutherford steadfastly incorporated the neutral doublet and other neutral subunits such as Meitner's neutral alpha particle into his hypothetical models of nuclear structure, primarily to explain the emission and scattering of alpha particles.[58] Even after Gamow's theory of alpha decay made these nuclear models obsolete, the idea of neutral subunits survived. In their 1930 treatise, Rutherford, Chadwick, and Ellis weighed the possibility of neutral doublets: "It may be that electrons in a nucleus are always bound to positively charged particles and thus have no independent existence within the nucleus." But it was only speculation, an attempt to evade the difficulties associated with free electrons in the nucleus. After ten years they lacked "convincing evidence on any of these points, for we have little if any definite information on the internal structure of nuclei to guide us."[59]

The information provided by artificial nuclear reactions had been meager. In all, only thirteen artificial reactions had been induced, in elements from boron to potassium. In every case only a single proton was emitted; there was no evidence of other products. Elements beyond potassium did not react, presumably because of coulomb repulsion between alpha particle and nucleus. Unaccountably, several light elements also failed to react, including lithium, beryllium, carbon, and oxygen.

Then in 1930, in Heidelberg, Walther Bothe and Herbert Becker discovered that when beryllium was bombarded with polonium alpha particles, radiation was emitted, more penetrating than any ever seen from radioactivity or artificial transmutations. Under similar conditions lithium and boron did the same, leading Bothe and Becker to conclude that these elements did undergo reaction after all but that the only product was gamma radiation of unusually high energy. In January 1932, Irène Curie and her husband, Frédéric Joliot, reported that this penetrating radiation expelled protons from hydrogen-containing substances such as paraffin, cellophane, and water; they attributed the effect to Compton collisions between the ultra high-energy gamma radiation and protons in the hydrogenous materials. Chadwick immediately realized this could not be true: the frequency of the scattered radiation did not obey the Klein-Nishina formula, and it was inconceivable that a nuclear reaction would generate gamma radiation energetic enough to set in motion a particle as massive as a proton. With the neutron in mind, Chadwick charted a series of experiments designed to prove its existence, exposing not just hydrogen but several other elements to the penetrating radiation and measuring the energy of the recoil atoms. Three weeks later Chadwick reported that the penetrating "radiation" was not radiation at all but a particle as heavy as a proton whose penetrating power is enormous because it has no charge.[60]

Neutrons were produced by the collision of beryllium and alpha particles (helium nuclei) according to the reaction:

$$^{9}\text{Be} + {}^{4}\text{He} \rightarrow {}^{12}\text{C} + {}^{1}\text{n}$$

The fact that boron and lithium underwent similar reactions with alpha particles indicated that the neutron is common to all elements.

Physicists rushed to confirm Chadwick's results. Meitner learned of the neutron just as she was repeating the Curie-Joliot experiments in a cloud chamber. Together with her assistant Kurt Philipp, she proceeded to estimate the relative penetrating ability of the lithium, beryllium, and boron neutrons by measuring the range of protons after collisions with neutrons; based on cloud chamber photographs, they calculated a "surprisingly large" cross section, or probability, for effective collisions between neutrons and protons and suggested that it might be based on mutual attraction between the two. Later they established that neutron-proton collision cross sections increase for slower-moving neutrons, one of the first indications that slower neutrons are more readily captured.[61] When Chadwick visited Berlin in June 1932, he and Meitner had a great deal to discuss. As he left she brought him a loaf of his favorite German bread. "Thank you so much for your kind gift of pumpernickel," Chadwick wrote when he was back in England. "It was so very nice of you. I like pumpernickel so much and we can't get it here."[62]

Meitner was particularly intrigued by a report that neutrons may initiate nuclear reactions. In the cloud chamber Norman Feather had observed neutrons reacting with nitrogen, the products being boron and an alpha particle:

$$^{14}N + {}^1n \rightarrow {}^{11}B + {}^4He$$

Meitner and Philipp confirmed this, and observed a similar reaction with oxygen:

$$^{16}O + {}^1n \rightarrow {}^{13}C + {}^4He$$

For both reactions stereoscopic cloud chamber photographs showed that the tracks of reactants and products were coplanar, indicating that momentum and energy were conserved. From the geometry of the tracks, the kinetic energy of the neutrons, and the approximate masses of the reactants and products, Meitner and Philipp estimated the energy emitted in the course of reaction. Franco Rasetti, a member of Enrico Fermi's group visiting Meitner's laboratory from Rome, established that the alpha particle-beryllium reaction that produces neutrons also emits gamma radiation, an important consideration in analyzing the mass-energy relationships of reactants and products.[63]

The work centered on the essential question of the neutron's identity: is it an elementary particle in its own right, or a closely combined proton-electron complex? The question was crucial to understanding the nucleus, for an elementary neutron would resolve the great problems associated with electrons in the nucleus while a complex neutron might not.

Chadwick's initial concept of the neutron was no different from Rutherford's: "We may suppose it to consist of a proton and an electron in close combination, the 'neutron' discussed by Rutherford in his Bakerian lecture of 1920." He thought the proton-electron complex might be arranged as a dipole, or with the proton embedded in the electron (the popular description was "dumbbell" or "onion"), and suggested detailed study of neutron collisions for information about the neutron's structure and field.[64] A more decisive approach to the identity question was to determine the mass of the neutron: if it was less than 1.0078 atomic mass units (the mass of a proton plus an electron), the neutron might well be complex, stabilized by its "mass defect"; if greater, the neutron had to be elementary. Chadwick's first measurements gave a mass of about 1.0067 atomic mass units (amu), but the margin of error was considerable. In Berkeley, E. O. Lawrence, M. Stanley Livingston, and G. N. Lewis used their 27-inch cyclotron to bombard various targets with deuterons ("heavy" hydrogen nuclei consisting of a proton and neutron) and came up with an almost shocking low value of 1.0006 amu, while in Paris Curie and Joliot used a different set of reactions and assumptions to obtain the exceedingly high mass of 1.012 amu. The high value was so large that the neutron would have disintegrated spontaneously into a proton and electron, whereas the Berkeley low value would have required the proton to collapse into a neutron and a positron—neither very promising for nuclear stability.[65]

The matter was the subject of intense discussion at the seventh Solvay conference in Brussels, 22–29 October 1933, whose topic overall was "Structure and Properties of the Atomic Nucleus." Lise Meitner attended, as did the protagonists in the neutron-mass debate, all of whom remained convinced of their conclusions.

After the conference, Meitner, together with Kurt Philipp, analyzed a number of nuclear reactions in the Wilson cloud chamber in an attempt to determine upper and lower limits for neutron mass. For the process

$$^9\text{Be} + {}^4\text{He} \rightarrow {}^{12}\text{C} + {}^1\text{n} + \gamma$$

the kinetic energy of ^{4}He, the polonium alpha particle, was known; the masses of the Be, He, and C nuclei were known from mass spectroscopic studies; and the neutron kinetic energy was determined from cloud chamber photographs of protons set in motion by neutron-hydrogen collisions. The gamma energy and neutron mass, however, were both unknown. Meitner and Philipp's determination of 1.0053 amu was calculated without considering gamma energy and thus represented a maximum value for the neutron mass.

Similarly, for the process

$$^{14}\text{N} + {}^1\text{n} \rightarrow {}^{11}\text{B} + {}^4\text{He} + \gamma$$

Meitner and Philipp obtained a minimum value for neutron mass of 1.0056 amu.[66]

These results, like Chadwick's, were consistent with a complex neutron, but they were imprecise, due primarily to uncertainties in the atomic masses. Nevertheless, Meitner and Philipp concluded that Lawrence's much lower value "does not seem possible," and within a few months the Berkeley group discovered that their value was an artifact of contaminated apparatus and retracted it.[67] But the question of elementary versus complex neutron was, for the moment, still unresolved.

Theoretical physicists were making their way through the new data without a firm sense of direction. An elementary neutron eliminated the problems of intranuclear electrons but created some theoretical problems of its own; there were new values for the magnetic properties of the proton to explain and unresolved questions of the origin of beta decay. With no compelling basis at first for favoring one type of neutron over the other, theorists drifted toward a hybrid approach. By late 1934, however, Chadwick had arrived at a neutron mass of 1.008; in 1935, experiments using different reactions and improved atomic masses, as determined by mass spectroscopy, confirmed the result. The slight mass difference between neutron and proton—less than the mass of an electron—indicated that *both* were elementary particles and neither was unstable with respect to the other: theory came to regard neutron and proton as alternate states of the same elementary particle.[68]

In 1934, Fermi proposed a theory of beta decay in which the neutron was elementary and the electron no longer inhabited the nucleus. Instead, the neutron was transformed at the instant of beta decay into a proton, an electron, and a neutrino, which explained the continuous beta spectrum without forfeiting conservation of energy and accounted for nuclear spin and statistics. So successful was this theory that the "neutrino," Fermi's diminutive for the "neutron" Pauli first suggested in 1930, was accepted as real, although it would remain undetected by experiment for another twenty-two years.[69]

In 1932, Meitner and Philipp were studying beryllium-polonium neutrons in the cloud chamber when they observed still another effect.

> We were struck by the surprisingly large number of almost straight line [i.e., very high energy] electron tracks emanating from the source (Be and Po in a 2 mm thick brass chamber). When the Be was removed from the chamber the number of electron tracks was much smaller and only a small fraction of them were straight.[70]

Like others studying neutrons, Meitner considered these energetic electrons to be of secondary origin, ejected by the high-energy gamma radiation that accompanied the neutrons produced by the beryllium-alpha reaction.[71] In fact, this was the second time positrons were making an appearance in her experiments, this time in a much more obvious way.

In contrast to the neutron, which was elusive but quickly detected because Chadwick had been looking for it, the highly visible positron was ignored—in fact, explained away—for quite some time because it was almost completely unexpected. For years scientists had observed thin, beady electron tracks in cloud chamber photographs of cosmic radiation and ascribed them to ordinary electrons traversing the chamber in an upward direction. This was peculiar, given the origin of cosmic rays, and some attributed the tracks to protons traveling downward.

A strong research program in cosmic radiation had been developed by Robert Millikan at the California Institute of Technology in Pasadena. In August 1932, Carl D. Anderson, one of Millikan's former students, photographed a thin track crossing a lead plate inserted into a cloud chamber. The particle's downward direction of travel could be seen from the increased curvature of the track after crossing the plate; its deflection in a strong magnetic field proved that it was positively charged. But a particle capable of penetrating lead that was also appreciably deflected in a magnetic field had to be much lighter than a proton. Anderson was certain he had found a positive electron, never before identified in ordinary matter on earth, and by the end of 1932 confirmed his discovery with many more photographs. In Cambridge, P. M. S. Blackett and G. Occhialini set up a triple-coincidence Geiger counter arrangement that enabled them to photograph a shower of cosmic ray electrons and positrons.[72]

Meitner remembered the "surprisingly large number of straight line electron tracks" emanating from the beryllium-polonium source in the cloud chamber. In March 1933, she and Philipp reported,

> We have photographed these tracks in a magnetic field. . . . Most striking was the frequent appearance of electron tracks whose curvature was the reverse of that expected for negatively charged electrons. As we are considering only rays which [one can see] come *directly* from the source, there can be no doubt about the original direction of travel of the rays. . . . On 14 photographic plates we measured 9 negative and 8 positive curves. Of

these, 4 of each (altogether 8) had [very high] energies between 3.4–4.4 MeV. With Po alone as a control no positive curves were observed.[73]

In Paris, Curie and Joliot had observed electron tracks with "reversed curves" but attributed them to electrons returning to the source; Blackett and Occhialini suggested that these were positive electrons.

In our experimental arrangement, as we have already emphasized, there is no doubt that these positively curved tracks come from the source. . . . It seems obvious that these particles are in essence identical to the "positive electrons" observed in cosmic radiation by Anderson and by Blackett and Occhialini.[74]

With this, Meitner and Philipp were the first to identify positrons from a noncosmic source and to show, moreover, that positrons appear together with negative electrons in pairs.[75]

Now the connection to Dirac's theory was obvious. According to the theory, an energy of 1.02 MeV (1 MeV = 10^6 electron volts), equivalent to the mass of two electrons, was necessary to raise a negative-energy electron out of the sea of negative energy, that is, to create a positive-energy electron and a hole, an electron-positron pair. Meitner and Philipp had already seen that the relatively modest 0.9 MeV gamma radiation from Po alone was insufficient, but they had not yet determined whether positrons were produced by the neutrons from the Be-Po reaction, or by an interaction between its very high-energy gamma radiation and the surrounding brass enclosure. When they enclosed a sample of ThB + C + C″ in a heavy lead cylinder that permitted only the 2.6 MeV gamma radiation of ThC″ to pass through, they observed nine fast positive electrons in 160 cloud chamber photographs; the energy of the positrons, as determined by their magnetic deflection, was in agreement with Dirac's predicted value. "These positive electrons clearly owe their origin to the interaction of the energy-rich γ-radiation [of ThC″] with atomic nuclei of lead. In our study of the release of positive electrons [along] with the induction of neutrons and γ-rays in beryllium, these electrons originated from the brass enclosure, that is, from copper or zinc nuclei [in the brass]." But Meitner and Philipp were not the first with these results. In Paris just a few weeks before, Curie and Joliot found that positrons are produced by the intense gamma radiation of the Be-Po source, not the neutrons. The formation of electron-positron pairs associated with ThC″ gamma radiation had been reported a few weeks earlier by Carl Anderson as well.[76]

The deviation from the Klein-Nishina formula that Meitner and her students had pursued for so long was finally explained: the "missing" gamma radiation had been transformed into positron-electron pairs. The process, later known as "materialization," required gamma radiation in close proximity to a nucleus and was enhanced, as Meitner and her students had found, with increasing gamma energies and greater nuclear charge. Her last paper on the Klein-Nishina deviations was in press when this discovery was made. In a note added in proof, Meitner's assistant Max Delbrück attributed the deviations from the Klein-Nishina formula to "the positive electrons that are formed in various elements of middle to high atomic number by the gamma radiation from ThC″ and even more readily by the harder gamma radiation from [the reaction of α particles with] Be."[77]

Pair formation was an astonishing confirmation of the most controversial aspect of relativistic electron theory. At the time of his discovery, Carl Anderson had not known that Dirac's theory implied the existence of the positron. This is hardly surprising, since Dirac himself first thought the positive particles would be protons and thus did not anticipate the positron as clearly as he foresaw the process by which it is created. As Dirac predicted, an electron in the sea of negative energy may absorb a high-energy gamma photon to become an ordinary electron with positive energy, leaving a hole in the sea of negative energy. The electron and its hole are matter and antimatter: the electron-positron pair. In many ways the positron resembles the ordinary negative electron, but when they meet, both disappear: the electron has fallen into the hole. This is annihilation, the reverse of pair formation, in which the mass of the positron and electron is entirely converted into energy. The two recurring gamma frequencies detected by Gray and Tarrant may well have been annihilation energies: two photons of 0.51 MeV or one of 1.02 MeV are the energy equivalent of the mass of two electrons.

In the years 1932 and 1933, Meitner investigated beta-gamma spectra, alpha particle fine structure, cosmology, gamma scattering and the Klein-Nishina formula, neutrons, and positrons. Her work spanned almost all of experimental nuclear physics; she had the equipment, resources, and coworkers to quickly step in as each amazing new finding was announced.

Years later, Otto Frisch noted that some of her publications were brief, an indication that Meitner and her collaborators were trying to keep up with the headlong pace of discovery.[78] In 1933, it was also a reflection of the political situation in Germany, which forced Lise Meitner to think about many things that had nothing to do with physics.

CHAPTER SIX

Under the Third Reich

Nichtarisch.

Lise Meitner began 1933, as always, with a new diary and some notes for the coming weeks. She would buy *Max und Moritz*, the tale of two rascals, for ten-year-old Hanno Hahn; in early February she would chair a meeting at which Werner Heisenberg and Max Born were scheduled to speak; Otto Hahn was leaving Berlin for America at the end of February. In the entry for New Year's Day she wrote, "Plus ça change, plus c'est même chose"— The more things change, the more they stay the same.[1] We cannot know what Lise was thinking on that day. The saying was as untrue for physics after the "miracle year" of 1932 as it was for the events that were about to overtake Germany in 1933.

Of course, everyone knew that the political situation was dreadful. It had been for years: depression and deep unemployment, constant elections, endless power struggles between the Right and Left, rising anti-Semitism, the Nazis ever stronger in the Reichstag and more violent on the streets. But what could one do? Life went on, and despite the very difficult economic times, so did physics.[2]

Lise's circle of physicists had expanded most pleasantly when a fellow Viennese, Erwin Schrödinger, came to Berlin in 1927. Then forty years old, Schrödinger succeeded Max Planck in the chair of theoretical physics; Planck had retired from the university and assumed the presidency of the Kaiser-Wilhelm-Gesellschaft. Erwin and Annemarie descended on Berlin with a breezy Viennese sociability and many happy parties that did not, unfortunately, reflect the state of their marriage. At the end of January 1933, as the couple planned their annual February Wiener Würstel-Abend (a Viennese sausage feast and costume party), they did not foresee the changes that were almost upon them. They invited their friends, in verse, "from near and far," to their home on Cuno Street, "the Hotel $\Psi\Psi^*$." ($\Psi\Psi^*$, psi-psi-star, indicates electron density according to the Schrödinger wave equation.) The invitation announced that the sausages would be white that year instead of brown, but the Schrödingers were not considered to be political and no one took it as commentary.[3]

Nevertheless, politics had people worried, enough so that on a Monday morning at the end of January, Annemarie Schrödinger sought out some company and came over to Lise's apartment to listen to the radio. At noon on that fateful day, 30 January 1933, the two women heard Adolf Hitler sworn in as chancellor of the German Reich.[4] It was caricature, as ugly and surreal as a cartoon by George Grosz: a chancellor who made no secret of his contempt for democracy, whose private militia, the SA (Sturmabteilung: storm troops), was far larger than the German army, a man whose followers spilled blood on the streets while his Nationalsozialistische Deutsche Arbeiterpartei (NSDAP), the largest single party in the Reichstag, reduced parliamentary deliberations to the level of a sustained brawl. The NSDAP had never achieved a parliamentary majority, however. Hitler was made Reichskanzler because the nationalist Right thought they could rein him in, use him to form a ruling coalition, and consolidate their own power.

True to his word, Hitler lost no time in destroying constitutional government and all those who stood in his way. The Reichstag was dissolved, new elections were scheduled for March 5, and tens of thousands of SA brownshirts were turned loose, in Berlin and throughout Germany, in a bloody campaign to suppress the opposition. A week before the elections, the Reichstag building went up in flames. The Communists were blamed and thousands of them were jailed, a state of emergency was decreed, and the civil rights guarantees of the Weimar constitution were suspended. The elections took place amid press censorship, arrests, beatings, and murders, with local police joining in or looking away. Even so, the NSDAP failed to secure an outright majority.

"The political situation is rather strange," Lise wrote to Otto Hahn a few days after the election, "but I very much hope it will take a calmer, more sensible turn." Hahn was scheduled to spend the entire spring semester as a guest professor at Cornell University, and Meitner was taking his place as temporary director of the institute. She wrote on letterhead stationery and chose her words carefully, expecting Otto to read between the lines. "Today the accounting office ordered us to estimate the cost of our national flag, because it is to be replaced with a black-white-red one which the KWG will pay for."[5]

The black-gold-red flag of Weimar was gone. Brownshirts roamed the streets, physically removed uncooperative officials from government posts, hauled Jewish lawyers and judges out of courtrooms, and jailed political opponents. In Dachau, near Munich, and in Oranienburg, outside Berlin, the first concentration camps were made ready. "Everything and everyone is influenced by the political upheavals," Lise wrote two weeks later. She gave no details. It took ten days for a letter to reach America; surely Otto would read the newspapers. And it was more prudent to write only about the institute. "Already last week we were notified by the KWG that along with the black-white-red flag, we must also display the swastika. . . . It must have been very difficult for Haber to raise the swastika."[6] Presumably it was not entirely easy for Meitner, either.

The ceremonial opening of the newly constituted Reichstag was held on 21 March in Prussia's great shrine, the old Garrison Church in Potsdam, where Bismarck convened the first Reichstag and Frederick the Great was buried. As Lise described it to Otto, "Mrs. Schiemann and Edith were here today to hear the radio broadcast of the Potsdam ceremony. It was harmonious and dignified throughout. [President] Hindenburg said a few short sentences and then yielded to Hitler, who spoke in a very moderate, tactful, and conciliatory way. Hopefully it will continue this way."[7] Like many others, Lise was encouraged by the change in tone. In the evocative setting, Hitler had deferred to the old president, humbly alluding to the union of Germany's "old greatness and new strength." All of it, including the radio broadcast, was staged by Joseph Goebbels. As Hitler spoke, thousands of SA and SS massed outside the church, and the concentration camps filled with political prisoners.[8]

Two days later the Reichstag put an end to parliamentary democracy in Germany. With all the Communist deputies and many of the Social Democrats arrested or in hiding, the remaining deputies granted Hitler the unlimited power to rule by decree. Thus Hitler succeeded, behind a facade of legality, in seizing absolute power. In the Reichstag the NSDAP deputies stretched their arms in the Nazi salute and sang their party anthem, the Horst Wessellied: "SA marching . . . Jew blood in the streets . . ."[9] The Third Reich was under way.

Unofficially it had begun weeks before, but now the National Socialists were free of political and legal constraints. One of their first acts was to call for a nationwide boycott of Jewish businesses and offices on Saturday, 1 April. Many citizens wanted nothing to do with the boycott, and a few openly opposed it—Edith Hahn, for example, made a point of visiting her Jewish dentist that day—but it was obvious that neither the police nor the courts would protect Jews or their sympathizers.

But isolating and terrorizing individual Jews was not Hitler's first priority. His immediate aim was to purge them from public life: from government, from the medical and legal professions, from education and the arts. One of the first confrontations involved Albert Einstein, the world's most famous scientist and Germany's most prominent Jew. The incident revealed the breadth of pro-Nazi sentiment and the absence of effective resistance among Germany's most distinguished scientists.

Einstein was in California when Hitler came to power; from the newspapers he learned that he and other intellectuals had been singled out for attack. Having been the target of virulent anti-Semitism throughout the 1920s, Einstein did not underestimate the danger. On 10 March, just before sailing for Europe, he publicly announced that he would not return to Germany where "civil liberty, tolerance, and equality of all citizens before the law" no longer existed; en route he issued a statement objecting to the "transfer of police powers to a raw and rabid mob of the Nazi militia."[10] From Belgium on 28 March he resigned his position with the Prussian Academy of Sciences.

The Prussian minister of education and culture, Bernhard Rust, had hoped to highlight the day of the *Judenboykott* by expelling Einstein from the academy; Einstein's preemptive resignation infuriated him. A secretary of the academy denounced Einstein for "atrocity-mongering" abroad and declared that the academy had "no reason to regret Einstein's resignation." When Einstein objected to this "deliberate distortion," he was accused of failing to defend Germany against the "flood of lies which has been unleashed against us." Einstein responded that "such testimony . . . would have contributed, if only indirectly, to moral corruption and the destruction of all existing cultural values. Your letter only shows me how right I was to resign my position with the Academy."[11]

Max von Laue was appalled at the academy's position but could not muster a show of support for Einstein. Even Max Planck, who had brought Einstein to Berlin and urged him to stay on several previous occasions, agreed that Einstein's "political" statements made it impossible for him to remain in the academy.[12] Einstein vowed never to set foot in Germany again. "The conduct of German intellectuals—as a group—was no better than the rabble."[13]

On 1 April, the day of the Judenboykott and in the midst of the furor over Einstein, Lise spent the evening with Max and Magda von Laue. The next day she guardedly wrote to Otto, "[Laue] is of the opinion that under the circumstances we shall not be able to do without you for very long. The university vacation has been officially extended to 1 May, obviously to provide time to arrive at a position on various questions."[14]

What was coming was the "Law for the Restoration of the Professional Civil Service" (Gesetz zur Wiederherstellung des Berufsbeamtentums), enacted on 7 April. "Non-Aryans" and political undesirables were to be purged from all government agencies, including the universities, a "non-Aryan" being defined as a person with at least one Jewish grandparent. There were some exemptions: Jews who had been appointed to their positions before World War I, or who had fought at the front, or whose father or son had been killed in the war.

As an adjunct university professor, Lise Meitner was required to fill out a questionnaire giving dates of employment, type of war service, and—ultimately the only question of importance—the *Rassezugehörigkeit der 4 Großeltern*, the "racial membership" of her four grandparents.[15] Meitner had never concealed her Jewish descent. Still, to write the word *nichtarisch* for the first time, black on white, for her grandparents and thus herself was clearly a step into the unknown.

It was unclear, however, how the law would affect Meitner's position in the institute, since the KWI for Chemistry, unlike the universities, was funded and administered by an industry-government partnership that had never been under direct government control. Then, too, she had been employed as an assistant to Planck before the war, she had served in the military, and, being Austrian, she was not sure she was subject to a German civil service law. Again and again Meitner discussed her situation with Max von Laue. Would things get worse? Or would this madness be brief, as many predicted, and decency assert itself? If dismissed, where would she go? If permitted to stay, should she resign anyway, on principle? On 26 April, Carl Bosch, director of the I. G. Farbenindustrie, a major sponsor of the KWI for Chemistry, assured Meitner that "special regulations are being intensively sought for deserving men and women of science who are subject to the Civil Service Act. Therefore I entirely welcome the advice that Professor von Laue has given you, not to entertain thoughts of resigning at this time, but to continue your very meaningful work at the Kaiser Wilhelm Institute."[16]

But in the universities the new law created havoc. In 1933, there were some 600,000 Jews in Germany, about 1 percent of the population. But their representation in the academic community was much higher: about 20 percent in the sciences overall, over 25 percent in physics. In April and May 1933, Jews from Privatdozent to Professor were quickly and systematically dismissed. From their students and colleagues came not the slightest public protest—no expressions of outrage, no student or faculty strikes on their behalf, no demands for academic freedom or appeals to fairness and decency. On the contrary, the universities had been the breeding grounds for right-wing ideology for years.[17] As National Socialist student groups flexed their new political muscle, their teachers did the same, from young Privatdozents to prominent professors. Even the rector of the University of Berlin, one Ludwig Bieberbach, wore a brown shirt: a mathematician, he was one of the leading proponents of the Nazi *deutsche Mathematik* and eventually edited a journal specializing in "Aryan mathematics."[18]

Those who were exempt from dismissal due to war service faced difficult decisions. Should they stay and fight? Should they resign in protest? Or should they just get out? James Franck was a Nobel laureate, director of the Second Physics Institute in Göttingen, and, until Hitler came to power, the likely successor to Walther Nernst in the University of Berlin and eventually to Fritz Haber as director of the KWI for Physical Chemistry in Dahlem. Franck had served at the front—with distinction—but on 17 April 1933 he submitted his resignation:

> We Germans of Jewish descent are being treated as aliens and enemies of the Fatherland.... Whoever was in the war is supposed to receive permission to serve the state further. I refuse to make use of this privilege, even though I also understand the position of those today who consider it their duty to hold out at their posts.[19]

Franck's resignation was considered sensational and was widely published; Edith Hahn was almost envious of the Francks "for being Jews and thus having justice entirely on your side, while we bear the disgrace and inextinguishable, irreparable shame for all time."[20] Among Franck's Göttingen colleagues, forty-two instructors accused him of inflaming anti-German propaganda with his public resignation and condemned it as "equivalent to an act of sabotage."[21]

Max Born, head of the Institute for Theoretical Physics, left Göttingen quietly. His military service provided grounds for appeal, but he knew he did not have the stamina for a fight. The example of his friend Einstein was not encouraging.

> On 25 April 1933 the newspaper published a list of civil servants dismissed ... [including] my name and those of [mathematician Richard] Courant and other men of Jewish origin. Though we expected this, it hit us hard. All I had built up in Göttingen during twelve years' hard work, was shattered. . . . I went for a walk in the woods, in despair, brooding on how to save my family. When I came home nothing seemed to have changed. Hedi kept her head and showed no sign of desperation. Then visitors began to drop in to express their sympathy. . . . But there were others who got on our nerves. . . . The zoologist [Alfred] Kühn . . . tried to console me with the consideration that we were in crisis like war; just as one man is killed in battle, while the other survived, so it happened that I belonged to the casualties. . . . [I]t made me irrationally angry.[22]

Surveying the damage, Einstein wrote to Born at the end of May, "You know, I think, that I have never had a particularly favorable opinion of the Germans (morally and politically speaking). But I must confess that the degree of their brutality and cowardice came as something of a surprise to me."[23]

Some Jewish professors with exemptions tried to stay on but soon realized their mistake. Colleagues spread defamatory rumors; students jeered and disrupted their lectures. Students were in no mood for learning anyway. Campaigning against the "un-German spirit," the German Students Association wanted all books by Jewish authors marked "translated from the Hebrew" and organized public book burnings to take place throughout Germany on the night of 10 May. In Berlin the event was held in the large square next to the Opera, across Unter den Linden from the main entrance to the university. Some twenty thousand books went up in flames that night—Einstein, Freud, Kafka, Marx, Heine, and others—as thousands of students and professors carried the books in by the armful, hurled them into the fire, and stayed for the pleasure of watching them burn.[24] This was the image of the new Germany, portending the destruction that lay ahead.

But to the great majority of Germans, it did not matter that books were burned or newspapers were shut down or that Jews, Communists, and Socialists were thrown out of work or put in jail. Most Germans welcomed the new national spirit and the chance that their economic situation might improve—and even those who disliked the harshness of the new regime accepted its legality. Among Jews, few understood that the Nazis would use any means—not just "laws" but deception, threats, and violence—to remove them from German life. It would take time before it was clear that no Jew, no matter how far from Judaism or how devoted to Germany, was safe. For these reasons the fate of Fritz Haber came as a surprise and a revelation.

Haber, chemist, Nobel laureate, and director of the KWI for Physical Chemistry and Electrochemistry since its inception, was a passionate German patriot. The Haber process for the synthesis of ammonia had been essential to the wartime production of nitrates for munitions; he had organized and led the poison gas unit during the war and then had spent years unsuccessfully trying to extract gold from seawater in the hope of paying off Germany's crippling war reparations. A baptized Jew, he dissociated himself from Judaism—his children were baptized and he attended church occasionally—but he was surrounded by Jews nevertheless: his best friends, his two wives, and, inevitably, many talented associates who could not find work elsewhere because of the entrenched anti-Semitism in academia and the chemical industry.[25]

In mid-April 1933, Haber was informed by the Prussian Ministry of Education that the number of "non-Aryans" in his institute was intolerable. (Unlike other Kaiser Wilhelm Institutes, Haber's was under direct government control, so that he and his associates were civil servants.) Because he had served at the front and had been employed prior to 1914, Haber himself was not threatened with dismissal, but the ministry was demanding that he carry out a purge: his three section leaders, several assistants, technicians, and secretaries, including Irene Sackur, the daughter of Otto Sackur, the chemist who was killed in a test of military explosives in 1914. Desperate to protect his institute and personnel, Haber turned to his friends for advice, including Richard Willstätter, Max von Laue, and Lise Meitner. At one time Meitner had described Haber as wanting to be "both your best friend and God at the same time." Now he simply needed his friends. "I remember our talks with Haber," Laue wrote to Meitner years later, "when he sought advice and wrestled with the decision of how to deal with the different, conflicting demands from the Nazis concerning the reopening of his institute after the Easter vacation. The spiritual suffering that this great man endured is unforgettable."[26]

Haber finally realized he had no choice: he had been stripped of his authority within his institute and his standing outside. He submitted his resignation on 30 April: "My tradition requires of me that in my scientific position I select my collaborators on the basis of their professional qualities and their character, without questioning their racial condition."

For a week Haber heard nothing. Then Bernhard Rust, Prussian Minister of Education, mentioned in a public address that a well-known Jew, a chemist, insisted on choosing his collaborators on the basis of their qualifications. This was completely unacceptable, Rust said. "We must have a new Aryan generation in the universities or we will lose the future." Haber was out, brushed away like an insect. He was already in poor health; it almost broke his spirit. He felt useless, bitter, and alone; he thought he had lived too long. He hoped to turn his institute over to a worthy successor, but that too would be denied him.[27]

Max Planck had been vacationing in Italy since the beginning of April, but when Meitner wrote to him of Haber's dismissal he hurried back to Berlin. As president of the KWG he was prepared to intercede on Haber's behalf, but he had not paid detailed attention to the news from Germany, and the idea of challenging the government bewildered him. Years later Meitner recalled that when they were discussing Haber, Planck "once said, truly desperately, 'But what should I do? It is the law.' And when I said: But how can something so lawless be a law? he seemed visibly relieved."[28]

Planck met with Hitler on 16 May 1933. According to Planck's account, published fourteen years later, Hitler ranted against Jews in general, working himself into such a rage that Planck could do nothing but wait in silence until he could take his leave.[29] From contemporary reports, however, it appears that the meeting was not so discouraging. For much of the summer of 1933, Planck, Heisenberg, and Laue were fairly optimistic that the more prominent "non-Aryan" physicists might be allowed to stay—most of the older physicists had military exemptions—and they urged Max Born and others to delay taking positions elsewhere. They were willing to see some Jews go, but for the sake of German science, not all; it was, at the very least, a form of collaboration with the Nazi regime. As Heisenberg expressed it to Born, "Since . . . only the very least are affected by the [civil service] law—you and Franck certainly not, nor Courant—the political revolution [i.e., the Nazi takeover] could take place without any damage to Göttingen physics. . . . Certainly in the course of time the splendid things will separate from the hateful." Heisenberg's optimism, especially his suggestion that it was possible to sacrifice the "very least"— the young scientists—and still salvage physics and other "splendid" German things, must have seemed to Born as heartless as the battlefield rationale that so infuriated him the day he was dismissed. Neither Born, nor Franck, nor Courant returned to Göttingen; they would not, of course, have been able to stay. The mass dismissals of scientists—the young and promising along with the older and more prominent—damaged German science for years to come.[30]

Otto Robert Frisch had been in Hamburg since 1930, working in the institute of Otto Stern; in 1933 Stern and most of his collaborators, including Frisch, were dismissed. Stern eventually went to the Carnegie-Mellon Institute in Pittsburgh; Frisch, then twenty-nine years old, had the advantages of youth: optimism and a willingness to go almost anywhere. Although he had a Rockefeller Foundation fellowship for a year with Enrico Fermi in Rome, the grant was rescinded because he did not have a permanent position to return to (a Rockefeller statute that effectively excluded nearly all dismissed "non-Aryans"). But there were other opportunities; the British were helping displaced scholars through the Academic Assistance Council, of which Rutherford was president, and Bohr was doing all he could to find positions for refugees, especially for young scientists, in his own institute and elsewhere.[31] In late 1933, Frisch went to London for a year with Patrick Blackett at Birkbeck College and then to Copenhagen where he stayed until 1939.[32]

In the KWI for Chemistry, Lise Meitner's position seemed safe for the moment, but the old ties among the *Mitarbeiter* were subject to a new political alignment. Hahn's chief assistant, Otto Erbacher, an active party member, was the party steward (*Vertrauensmann*) for the institute; Kurt Philipp, Meitner's chief assistant, followed enthusiastically. One of her students, Gottfried von Droste, regularly wore his brown shirt after joining the SA at the urging of another student, Herbert Hupfeld, whose views were so extreme that he soon left the institute. Perhaps most threatening was Professor Kurt Hess, who headed an independent section for organic chemistry, the so-called guest section, on the third floor of the institute building. Known to all as a "fanatic" Nazi, Hess had long coveted Otto Hahn's position as director of the KWI for Chemistry, an ambition fueled by the events of 1933. Hess was also Meitner's neighbor: he lived in the apartment adjoining hers in the institute villa on Thielallee.[33]

It was difficult to concentrate that spring. For Lise, there were worried discussions with friends, sleepless nights, a feeling of helplessness and isolation; for the Mitarbeiter, new power, party meetings, jostling for status within the party hierarchy.[34] When student groups staged rallies outside the institute, the laboratories would empty. Once Philipp returned, excited and proud, "Now I can sing the *Horst Wessellied* as well as Professor Hess." Meitner was speechless for a moment. Philipp was no boy but a man nearly forty years old. "I'm not quite sure that's what you're being paid for," she finally said.[35]

Four thousand miles away, Otto Hahn was comfortably insulated from the events in Germany. Like many academics, he had always been apolitical; now he found it difficult to believe what he read in American and Canadian newspapers.[36] In an interview with the *Toronto Star Weekly* in early April, Hahn reflexively defended Germany and suggested that any persecution of Jews was incidental to the suppression of communism; referring to Hitler's ascetic personal habits, Hahn said he "lived almost like a saint." The interview was published under the headlines, "He Defends Hitler; Denies Man Who 'Lives Like a Saint' Is Guilty of the Atrocities Charged."[37] Hahn's instinctive nationalism was evidently more pronounced than his concern for civil rights. In this he was not alone: Planck and many other "good" Germans were willing to overlook the Nazis' methods at first in the hope that something "splendid" would come from the renewed sense of national unity.[38]

Hahn became alarmed, finally, when he learned from Edith and Lise that many of their friends and colleagues had been dismissed. On May 3, about the time of Haber's dismissal, Lise urged Otto to return quickly.

> I do think it would be better if you return here after your lectures are over, and don't go West as planned. It is hard for Edith, too, that you are not

here. . . . You know how depressed she can be at times, and I am not exactly suited just now to make her feel better. A rather large Nat[ional] Soc[ialist] cell has formed in the institute, it is all quite methodical. The universities are starting classes again, now that the prescribed leaves of absence [dismissals] have taken place. . . . I am quite tired and . . . somewhat useless right now.[39]

In her next letter Lise apologized for not writing more often—"for 25 years you have known of my strong inhibitions against speaking of emotional things"—and again asked Otto to give up his trip to California and return to Berlin. "That is the opinion of all our scientific acquaintances and colleagues and not just my personal wish. . . . Haber's resignation is of course perceived as very painful. Also not very pleasant is the zealous activity of all those who believe that an outspoken show of obsequiousness will result in some personal advantage." Nevertheless, Meitner assured him, her relationship with his Mitarbeiter and hers continued to be very good, "and that makes everything easier."[40]

Otto prepared to return to Berlin as soon as he finished his lectures at Cornell. "It will take some time before you really have the full picture," Lise warned. "With such great upheavals the life of an individual plays a subordinate role, but even here it is obvious who values people and who does not. And I was not surprised that my house-partner [Kurt Hess] is of the latter type. But one must admire Planck, Laue, [Wilhelm] Schlenk, and many others. . . . Franck was in Berlin once; it is really very hard for him not to be allowed to set foot in his [former] institute, and this, unfortunately, is true for many." She saw the actions against Jews as a symptom of a deeper malaise. "Anti-Semitism is only *one* problem; there are problems as serious or even more serious, and anyone who cares about Germany should be worried about what will become of it all."[41]

The Nazis were consolidating their grip on every aspect of society. In education they emphasized political indoctrination—"character development"—and expressed their contempt for scholarship or scientific research. In this context, intellectuals and scientists who opposed the regime were overcome by a feeling of powerlessness. Hahn encountered this soon after returning to Berlin, when he proposed that a large number of well-known "Aryan" professors might protest the treatment of their "non-Aryan" colleagues. "But Planck (and others) advised against it: 'If today you assemble 50 such people, then tomorrow 150 others will rise up who want the positions of the former, or who in some way wish to ingratiate themselves with the Minister.' I therefore did not attempt anything."[42] Because there was no group protest, individuals stood alone. Wilhelm Schlenk, a prominent chemist who was Emil Fischer's successor, spoke out against the Nazis and continued to defend Haber; he was forced to vacate his Berlin professorship.[43]

Planck, Hahn, and others in positions of influence believed there was little they could do except protect themselves and save what remained of the universities, learned societies, and institutes. But survival came at a price: immediate submission on some matters, protracted compromise on the rest. Under Planck's presidency, the Kaiser-Wilhelm-Gesellschaft displayed the swastika, used "Heil Hitler!" in its correspondence, and routinely praised the Third Reich in its annual reports; in exchange, the KWG retained a few "non-Aryan" scientists for a time and a tenuous measure of scientific independence.[44] There was no question, however, that its moral and professional authority was gone. When Haber left his institute in the summer of 1933, Hahn stepped in as interim director. In that position he followed the orders Haber had refused to carry out: he dismissed nearly all personnel and dismantled the institute, paving the way for the new Nazi-appointed director, Gerhart Jander, who arrived in the fall. Jander would have never even been considered for such a position by any pre-Nazi standards; his only research, ironically, had been with chemical weapons, and the Ministry of Defense hoped to convert Haber's former institute into a research facility for poison gases.[45] By the end of 1933, the KWI for Physical Chemistry was a full-fledged "model institute" staffed entirely by party members, a warning to potential dissidents that they were indeed utterly replaceable, just as Planck had suggested to Hahn. As to his own role, Hahn would later claim he had done his best with an "unpleasant and thankless task."[46] One could also say that he had done the Nazi regime's dirty work and lent them his good name in the process, receiving nothing for it but the vague hope that he and his institute would be left alone. It was a poor bargain, and now the Nazis knew his price.

Members of the academic elite had neither the ideological strength nor the political experience necessary to fight the Nazis. They were disarmed by the promise of a national revival: they had, after all, traditionally disdained the divisiveness of party politics for the greater good of serving the state. The idea of serving one's country by opposing its government was foreign to them. By the summer of 1933, active resistance—or even protest—seemed impractical. Leo Szilard, a young Hungarian physicist who worked briefly with Meitner, sensed that attitude even before Hitler came to power.

> Many . . . people took a very optimistic view of the situation. They all thought that civilized Germans would not stand for anything really rough happening. The reason that I took the opposite position was . . . [that] I noticed that the Germans always took a utilitarian point of view. They asked, "Well, suppose I would oppose this thinking, what good would I do? I wouldn't do very much good, I would just lose my influence. Then why should I oppose it?" You see, the moral point of view was completely absent, or very weak. . . . And on that basis I reached in 1931 the conclusion that Hitler would get into power, not because the forces of the Nazi revolution were so strong, but rather because I thought that there would be no resistance whatsoever.[47]

Some who opposed the regime considered emigration. It was an agonizing decision: should they abandon hope for change and leave their country to the Nazis? Or should they stay, lending their talents to the Third Reich, signaling to the world that fascism was tolerable? Werner Heisenberg "almost envied the friends whose basis for living in Germany was removed from them by force so that they knew they had to leave our land."[48] Heisenberg did not leave, despite attractive offers from abroad; his bonds to Germany and German culture were too strong. To justify his decision, he imagined that he had embarked on an "inner exile," preserving an island of German culture for the future, preserving the true, "hidden" Germany from the onslaught of nazism. It was a fantasy: he could not separate the two, either in his personal actions or in his scientific work.[49]

For Planck, Heisenberg, Laue, and other leaders of the scientific establishment, the most appalling aspect of the new regime was not the injustice or terror but the threat to German science and their own independence. They may have regarded the treatment of Jewish friends and colleagues as distasteful, even wrenching, but for them it was not intolerable; their focus was not on morality or individual rights but on their own profession. Thus in the summer of 1933, Planck and Heisenberg tried to keep prominent Jewish scientists in Germany but did not defend the rest; later they sought the best replacements they could find for Born and Franck and countless others without worrying about the morality of replacing people who had been unethically dismissed, or the fact that the continued success of German science might make the Nazis look respectable to the outside world. In fact, science in Germany declined during the Nazi period. It has been argued that German science was damaged less by the dismissal policies (although the loss was significant, especially in physics) than by the isolation of the remaining German scientists from the international scientific community. As a result of their moral weakness and narrow nationalism, Germans were cut off from the international contacts vital to the progress of science, not just for the duration of the Third Reich but for decades after, not only by displaced Jewish colleagues but by scientists everywhere.[50]

Schrödinger, who was not Jewish, reacted instinctively: he found the Nazis disgusting, and he left. His departure was a rebuke to the regime; Planck took it as a blow to German science, which it was, and worried about the political damage it might inflict on those left behind. He asked Schrödinger to ask for a leave of absence and to cite reasons of health. Schrödinger departed for Oxford. The fact that Schrödinger could do as he pleased does not, of course, mean that others who were less prominent could do the same. Schrödinger was at the height of his fame and mobility: in 1933 the British scrambled to find a place for him. He won the Nobel Prize that year, and when he left Oxford in 1936, he quickly found a position in Austria.[51]

In 1933 many of the elder statesmen of science, including Planck, believed that the Nazis would soon become more responsible. "Take a pleasant trip abroad and carry on some studies," Planck advised a worried colleague. "And when you return all the unpleasant features of our present government will have disappeared."[52] According to Max Born, Planck "trusted that violence and oppression would subside in time and everything return to normal. He did not see that an irreversible process was going on."[53]

Neither did Lise Meitner. Both Planck and Hahn urged her to stay, and that was what she wanted to hear. Emigration was hard: the world was gripped by depression and positions were scarce. Lise could not bring herself to leap into the unknown, to relive her early days in Berlin, to be a frightened outsider again, a stranger in a foreign land. She clung to her physics section: "I built it from its very first little stone; it was, so to speak, my life's work, and it seemed so terribly hard to separate myself from it."[54] In November 1933, Niels Bohr obtained a Rockefeller grant for her to work for a year in his institute; she went through the motions and then turned it down after Planck suggested that her absence might jeopardize her return to Berlin.[55] When she learned that Swarthmore College in Pennsylvania was seeking a physicist, she made it clear that she would not consider a position until she knew more about "laboratory space, how many assistants are available, whether it is certain that one can obtain co-workers in addition to assistants, how well one is assisted with mechanics and workshops and whether it is possible to work with relatively large quantities of radium or other radioactive substances"—conditions the undergraduate college could not begin to meet.[56]

Although Meitner demanded perfection as a prerequisite for emigration, she easily found reasons to stay in Berlin. The political situation was bad but would surely improve. Until then the institute was her haven, her position safe. It would be wrong to go abroad and take a job from someone who had been forced to leave.[57] And her relationship with her staff was something "quite exceptional" in the political climate of the day: "There was really a very strong feeling of solidarity between us, built on mutual trust, which made it possible for the work to continue quite undisturbed even after 1933, although the staff was not entirely united in its political views. They were, however, all united in the desire not to let our personal and professional solidarity be disrupted."[58]

If her Mitarbeiter were willing to work with a "non-Aryan," then she was willing to overlook their politics. Students and assistants might wear brown shirts and interrupt work to attend rallies, but it was perfectly understandable: young people found it necessary to cooperate a bit with the Nazis—"*ein bisschen mitmachen*," they called it—that was all.[59] Physics was the important thing, and there was no better place to do it than in her own physics section. And so she stayed, "only too willing to let myself be persuaded by Planck and Hahn."[60]

Lise Meitner would not leave until she lost everything and was driven out. Until then Germany gave her what she thought was essential: her work was untouched, her position was the same, most of her friends were still there. She had Planck, the father figure she admired without reservation: she overlooked his political stumbles, seeing only his inner integrity and uncompromised moral strength. She had her Fachbruder Otto, who was not only her friend and colleague but now, in his position of director of the KWI for Chemistry, her protection. She had Gustav and Ellen Hertz for company, the Laues for movies and car trips on weekends, and her friend Elisabeth; she had the rich musical life of Berlin, and her sisters and brother in Vienna were only a day's train ride away. In short, she had a life she loved and work that was hers, and if things were not quite as good as before, she chose not to dwell on it. She was not prepared to give up what she had for the unknown. She had seen firsthand the enormous displacement of her exiled friends and the tragedy of Haber.

But Meitner's professional activities outside the institute were over. During the summer of 1933, Meitner's name appeared on a list of faculty to be dismissed from the University of Berlin; although exemptions were provided for those who had been *planmässige* (regular) employees prior to the war, there was some question whether Meitner's assistantship with Planck was planmässig since she did not undergo Habilitation until 1922. In last-minute pleas, Planck stressed the importance of Meitner's work and Hahn argued that her assistantship would have been planmässig had women been permitted Habilitation before the war.[61] The Prussian ministry was not moved. On 6 September 1933, Meitner's venia legendi was rescinded: "*Sofort!* [At once!] Based upon the third paragraph of the Law for the Restoration of the Civil Service of 7 April 1933, I revoke herewith the right to teach at the University of Berlin." The reasons: "§3: 100% non-Aryan; military service as x-ray technician *not at the front;* assistant 1912, hab[ilitated] 1922, *unprotected;* vote: to be dismissed."[62]

The loss of her external professorship at the university had no direct effect on Meitner's work in the institute or her salary. But in the university she was a pariah, unable to teach or take on new doctoral students. And she never again attended Max von Laue's Wednesday physics collo-quium.[63] That was hard, but the colloquium was not the same either. Einstein, Haber, and Schrödinger were gone; more than one-third of the Berlin physicists, including many young people, had been dismissed or no longer attended, and far fewer visitors chose to come from abroad.[64]

Not just the universities but every public organization was required to adopt Nazi goals and methods, a process known as *Gleichschaltung*, or alignment. Scientific societies fell into line, barring Jews and surrendering organizational control to a Nazi leader responsible only to higher authorities, an arrangement known as the "Führer principle." Meitner was no longer permitted to give reports at scientific meetings; eventually she stopped attending entirely. By 1936, even her name was suppressed. Hahn noted that "one constantly hears essentially only my name for an investigation in which Lise Meitner has participated at least as much as I have."[65] Had Meitner not continued to publish, she would have entirely disappeared from view.

Although Meitner retreated into her institute, she was not entirely insulated from the hostility outside. In 1934, Hermann Fahlenbrach, a young *Dozent* whom a co-worker later described as a "troublemaker in the nastiest way," tried to bring charges against her at the instigation of his Dozent association leader. Thirteen years later, when his Nazi past was under investigation, Fahlenbrach begged Meitner to exonerate him, explaining that he had been just an immature twenty-five-year-old "Westphalian hothead" who resented working for a woman.[66] True or not, Meitner knew this was irrelevant, for it was race, not gender, that made her a target for any opportunistic troublemaker who came along.

At times racial attacks were disguised as scientific criticism. In 1935, when Meitner and her theoretical assistant Max Delbrück published a monograph on nuclear structure,[67] a reviewer for *Naturwissenschaften* saw his chance to inflict harm. "One cannot in good conscience recommend this book for further distribution," he warned, because it contained a "grave error which might cause great confusion in wider circles." The error? A photograph taken by Curie and Joliot was mistakenly attributed to Chadwick, Blackett, and Occhialini. Unwilling to publish the biased review, the editor of *Naturwissenschaften* sought advice from Arnold Sommerfeld, who agreed that the "completely unfounded invective against our dear Meitner should not be printed."[68]

Nazi activists in the scientific community clamored for conformity in all of science, including scientific publications. They were furious that *Naturwissenschaften* continued to accept articles from Jews—Meitner, for example, published almost nowhere else between 1933 and 1935—and in retaliation they boycotted the journal so severely that it faced extinction by the end of 1933. Its editor, Arnold Berliner, pleaded for articles from friends. "I am now in fact living with the journal from hand to mouth," he wrote to Sommerfeld in October 1933. And a month later, "*Naturwissenschaften* is doing very badly. . . . Every article I now receive is truly help in need. . . . It has been gently called to my attention [by the publisher] that I have too many 'non-Aryans' among the authors. That will gradually come to an end of its own accord."[69] It did, but that was not enough to save the journal, for Berliner was himself a Jew. Hugo Dingler, a philosopher of science, regarded it as "beneath my dignity for obvious reasons" to have his books reviewed in *Naturwissenschaften*.[70] Two longtime promoters of a racist "Aryan physics," Johannes Stark and Ernst Gehrcke, accused Berliner of conspiring with Planck to suppress free discussion in the sciences by "propagating Einstein's teachings for decades, to the exclusion of every other opinion."[71] The effect of such attacks was augmented by the silence of the great majority. *Naturwissenschaften* did poorly, and in August 1935, its publisher, Julius Springer, summarily dismissed Berliner. Berliner had founded *Naturwissenschaften* and been its editor for twenty-two years. "The form of my dismissal was truly an affront," he noted sadly, "not because it lacked fairness, but because of its clumsiness and cowardice."[72] On his card he wrote under the printed *Dr. Arnold Berliner:* "had to leave *Nw*, his lifelong work, on [13 August] because he had become unbearable for the publisher."[73]

For Nazi ideologues, the purge of individual Jewish scientists was not enough: they wanted to expunge Jewish influence from science itself. As a discipline, physics was particularly threatened, not just because a disproportionate number of talented physicists had been dismissed but also because modern theoretical physics was under attack as alien to the Aryan soul. A new gospel of German or Aryan physics declared the superiority of German spirit over decadent materialism, experiment over theory, truth over logic, intuition over incomprehensible mathematics. Underlying it was the belief that intellect was a function of race, products of the Nordic mind being inherently superior to anything Jewish. (Curiously, Aryan physics was also acclaimed for its modesty.) Begun in the early 1920s with the vicious campaign against Einstein and relativity, Aryan physics was fostered by two physicists with seemingly impeccable credentials, Philipp Lenard and Johannes Stark. Both were Nobel laureates, both difficult men, embittered by scientific and personal slights, resentful of social and scientific changes they could not follow. Extreme nationalists, they saw Germany's defeat in the war as betrayal, and they despised the Weimar democracy; scientifically, too, they were unable to grasp relativity and quantum theory.[74] The international acclaim for the pacifist Einstein and other theoreticians was more than they could bear. They distilled their nationalism, envy, and racism into "Aryan physics."[75]

As science, Aryan physics was so empty that few took it seriously at first. Politically, however, it posed an immediate threat to scientific institutions already weakened by the dismissal policy. Its advocates attacked theoreticians such as Werner Heisenberg for teaching relativity and quantum theory and denounced the Nobel awards for falling prey to a Jewish conspiracy.[76] It appeared that the vacuum created by the dismissals would soon be filled by mediocre people whose scientific talents were not nearly as strong as their party loyalty.

Under these circumstances, a number of scientists rallied to the defense of their profession. As a group, they had been largely silent when German democracy and political freedom were destroyed; silent, too, when racial persecution drove their colleagues into exile. They saw these as "political" issues outside their sphere of responsibility, but they were willing to take a stand when their own influence and the future of their scientific disciplines were at stake. Their principal spokesman was Planck, whom Meitner revered less for his political acumen than for his "unflinching moral probity and desire for justice."[77] Planck tirelessly appealed for unfettered scientific research, arguing that scientific quality was vital to Germany's strength and its image abroad. To a large extent, this argument succeeded. Despite its contempt for all things intellectual, the Third Reich supported scientific research; once the universities had been purged of Jews, the regime left most scientists, including known anti-Nazis, relatively free to carry on their work.

The Kaiser-Wilhelm-Gesellschaft prospered. It managed to avoid a thoroughgoing Gleichschaltung, retained its non-Nazi directors, and kept a number of Jewish professors for several years. Such independence, however, required a demonstration of political allegiance. The KWG and its member institutes adopted a policy of voluntary self-alignment, or *Selbstgleichschaltung*, that required virtually every move to be weighed on a highly sensitive political balance.[78]

In the KWI for Chemistry, Selbstgleichschaltung permitted Otto Hahn and Lise Meitner to remain in charge but gave their assistants, Otto Erbacher and Kurt Philipp, both party members, considerable influence in running the institute.[79] Selbstgleichschaltung meant that the chemist Fritz Strassmann, an outspoken anti-Nazi, could stay but that he could not be promoted or paid a decent salary. It meant that the institute closed its doors to any more Jews or anti-Nazis, at a time when Jews and dissidents could find few other places to work as scientists. And although Hahn and Meitner could exercise some choice in co-workers—Hahn accepted one doctoral student, W. Seelmann-Eggebert, because he was the only one who did not say "Heil Hitler!" during the interview—they eventually excluded anyone who was politically tainted in the Third Reich. Thus in 1935, Martin Nordmeyer, a young physicist who had been blacklisted for his anti-Nazi views, appealed to Planck, "Frau Prof. Meitner has informed me that you fear that my employment could be a burden to the institute.... The KWI is the only possible solution for me, just because it is not under university regulations. . . . It means the difference between working in scientific research or giving up physics." Nordmeyer was not taken on in the institute, even as an unpaid volunteer.[80]

In practice, then, Selbstgleichschaltung meant virtually complete political collaboration in exchange for noninterference with science as usual. Scientists who opposed the regime were convinced this was the best they could do: they were buying time, preserving German science by training the next generation. Determined to avoid confrontation, they left politics to the state and a show of moral outrage to others. Political defiance, they were sure, would only be crushed and subservience rewarded, as shown by the swift destruction of Haber's institute and the new, utterly mediocre Nazi model institute that took its place.

But by representing themselves as professionals working in the national interest, scientists became, to varying degrees, servants of the Fascist state. Within Germany, adherence to the principle of science-not-politics made it impossible for scientists to withhold their expertise from the Nazis, especially in time of war. To the outside world, their ability to conduct science as before could be taken as a sign that the Nazi regime was tolerable.

Scientists only gradually became aware of this dilemma. Meitner found that she was "constantly vacillating between the necessity of doing nothing, which went against one's conscience, and the responsibility for the institute and the Mitarbeiter." Even so, years went by before she fully understood that her decision to stay in Germany after 1933 had been "very wrong, not only from a practical point of view, but also morally. Unfortunately this did not become clear to me until after I had left Germany."[81]

For most scientists, however, the morality of working under the Third Reich was never an issue. On the contrary, they convinced themselves that they resisted the Nazis most effectively by keeping science free of moral or political questions. Heisenberg objected strongly when the Nazis interfered with theoretical physics, but after they left him alone, he was ready to work in any scientific capacity and eventually directed the German fission project. Planck detested the Nazis but acquiesced to compromises that sustained German science in general and the KWG in particular.[82] Hahn anxiously tried to compartmentalize his private and professional lives, fearing that his anti-Nazi views might jeopardize the KWI for Chemistry or his own position as director.

> One noticed that there was much that I disagreed with. I never participated in Mayday celebrations! The presence of . . . L. Meitner did not make the situation better. Thus at the yearly meetings of the KWG, I was always seated in a less prestigious place at the dinner table than was appropriate for my position and my age and my length of service with the KWG.[83]

Long after the war, Hahn would still recall these minor slights as "painful experiences,"[84] although his institute never actually suffered. Obsessed with his institute and frightened for the future, he would not complicate matters with moral reservations.

As the Nazis consolidated their hold, organized resistance was compromised or eliminated in every segment of society. Trade unions disappeared into the Nazi Labor Front; political parties other than the NSDAP were banned; education, the press, and the arts came entirely under Nazi control. In a religious version of Selbstgleichschaltung, the Vatican hastened to seal a Concordat with the Nazis, renouncing political involvement in return for religious protection. Protestant churches, to an overwhelming extent, gave the Third Reich their blessing. Some, like Max von Laue, expected the army to resist, but when Hitler assumed the dual role of president and supreme commander in the summer of 1934, he did so with the support of the military high command. From that time on, every organized state function was under Hitler's direct control. Rearmament and conscription followed in 1935, a violation of the Treaty of Versailles that the European powers chose to ignore. Emboldened, Hitler intensified his offensive against the Jews. The Nuremberg laws of 1935 completed their legal and social isolation by stripping Jews of their citizenship and placing them, without rights or protection, at the mercy of the Nazi state.

Among scientists the only instance of organized protest centered around Fritz Haber, who died in Switzerland of heart failure on 29 January 1934, a few months after leaving Germany. No official notice was taken of his death, although a few individuals spoke out. In an obituary for *Naturwissenschaften*, Laue compared Haber to Themistocles, the Greek military hero who was banished from Athens: "Themistocles did not go down in history as an exile . . . but as the victor at Salamis. Haber will go down in history as the genius who discovered the means of binding nitrogen and hydrogen."[85]

At Laue's urging, Planck arranged for the KWG to hold a memorial service on the first anniversary of Haber's death.[86] The Prussian minister of education, Bernhard Rust, was enraged by this "provocation of the National-Socialist State" and forbade anyone under his jurisdiction to attend; the Society of German Chemists did the same.[87] Planck found it painfully difficult to defy the government. Finally, the evening before, he told Lise, "I will do this service unless the police haul me not."[88]

The service was dignified and well attended. Max Planck and Otto Hahn were the only speakers. No university professors, not even Max von Laue,[89] dared attend; their wives came in their stead. Scientists from industry, military representatives, and diplomats filled the auditorium, and Lise Meitner, Fritz Strassmann, Elisabeth Schiemann, and Max Delbrück were there.[90]

Planck ended his speech with the words, "Haber was true to us, we shall be true to him."[91] According to Hahn, the Haber memorial showed that "during the early years of the Hitler regime some resistance—minor, to be sure—was still possible."[92] But the Haber memorial was the only concerted act of protest by the scientific community. As resistance, it was little more than a gesture. It did not call for action or inspire the resolve to carry on the fight. Rather, the experience seemed to exhaust the participants. Hahn feared he had "distinctly weakened" his institute.[93] Like most of his colleagues, he avoided further protest from then on.

There were some who had more courage. One was Fritz Strassmann, who risked his career and even his life rather than follow the National Socialist majority. Strassmann had come to the KWI for Chemistry in 1929 with a doctorate in analytical chemistry, expecting that a year with Otto Hahn would improve his chances for a position in industry. Finding the work and the people to his liking, Strassmann extended his stay even after his small stipend expired in 1932. By 1933, he was thirty pounds underweight, severely malnourished, and suffering from fainting spells, yet he turned down a lucrative industrial offer because the company required political training and prior membership in a Nazi association. Another of Hahn's assistants took the position instead, and Meitner prevailed on Hahn to pay Strassmann 50 marks a month—barely enough for food—out of the director's fund for special circumstances.[94]

By refusing to join National Socialist organizations, Strassmann was unemployable in academia or industry. He resigned from the Society of German Chemists in 1933, shortly after it became part of the Labor Front, a Nazi-controlled public corporation with compulsory membership for employers and employees. The society let him know he would be black-listed: "We shall undertake to place an appropriate notation in your file." Strassmann did not back down: "It would interest me very much to know if that sentence . . . is intended to influence me in a particular direction."[95] Strassmann had no professional future except in the KWI for Chemistry. In 1935, Hahn and Meitner found an assistantship for him, at half-pay.[96] He considered himself fortunate, for "despite my affinity for chemistry, I value my personal freedom so highly that to preserve it I would break stones for a living."[97] When he married in 1937, his wife, also a chemist, agreed. During the war Fritz and Maria Heckter Strassmann concealed a Jewish friend in their apartment for months, at enormous risk to themselves and their three-year-old son.[98] A photograph of Strassmann taken just after the war shows a man exhausted by years of deprivation and anxiety.

In 1934, Meitner and Hahn asked Strassmann to join them in their investigation of elements beyond uranium, a project that eventually culminated in the discovery of fission. Politically isolated in the institute, Meitner and Strassmann became close during their long hours in the laboratory. He looked to her as the intellectual leader of their team, while she in turn drew strength from the trust of her younger colleague.[99]

Outside the institute Meitner's closest confidante was Max von Laue. When the Nazis first came to power, Laue felt powerless to oppose them. He believed scientists should not actively engage in politics and, after fighting Einstein's expulsion from the Prussian academy, chided him for his public statements. "Political battles call for different methods and purposes from scientific research. Scholars are usually chewed up by them." In reply, Einstein cited scholars from Spinoza to Alexander von Humboldt who had taken political stands; Laue's response was so pedantic that Einstein decided not to answer.[100]

But during the summer of 1933, Laue changed. He fought bitterly to keep Johannes Stark out of the Prussian academy.[101] In September 1933,

Laue opened a physics conference by referring to Galileo, condemned by the Church for "a theory that aroused opposition just as the relativity theory has done in this century. . . . Power may obstruct knowledge for a time, but knowledge will prevail. . . . Even under oppression scholars can take heart with the victorious knowledge . . . 'and still it moves!'"[102] And again, in the spring of 1934, Laue praised Haber in an obituary for *Naturwissenschaften*. The Galileo speech and the obituary provoked reprimands from the Ministry of Culture. "Someone there must have felt a need to do something for my amusement," Laue commented.[103]

Max von Laue had achieved spiritual independence. This quality was rare. While others anxiously weighed every word and deed, Laue was free to follow his conscience. As Einstein wrote, "The spinal cord plays a far more important role than the brain itself. . . . We really should not be surprised that scientists (the vast majority of them) are no exception to this rule and *if* they are different it is not due to their reasoning powers but to their personal stature, as in the case of Laue. It was interesting to see the way in which he cut himself off, step by step, from the traditions of the herd, under the influence of a strong sense of justice."[104]

On 4 August 1934, the day Hitler seized control of the army, Laue believed Germany had received a "final, mortal dagger in its back."[105] He left the political arena and devoted himself to helping individuals. As Arnold Berliner became increasingly isolated, depressed, and ill, Laue was his most faithful friend. In 1936, together with Planck and Heisenberg, Laue tried to force a Nobel Prize for Lise Meitner and Otto Hahn, in the hope that the award would give them some political protection.[106] (After 1936, when the Nobel Peace Prize went to Carl von Ossietzky, an imprisoned pacifist, Germans were forbidden to accept Nobel awards.) Occasionally Laue was able to warn colleagues who were about to lose their positions; often he smoothed their way abroad with inquiries and testimonials. "Because mail was censored, these letters went by more secure routes over the border. One time I brought someone who was being followed over the Czech border in my car. . . . All this was done as secretly as possible."[107] Most of what he did will never be known. According to his friend James Franck, Laue "was not a daredevil, blinded against peril by vitality and good nerves; he was rather a sensitive and even a nervous man who never underestimated the risk he ran in opposing Nazidom. He was forced into this line of conduct because he could bear the danger thus incurred better than he could have borne passive acceptance of a government whose immorality and cruelty he despised."[108]

Laue was proud of his obituary for Haber but wary of purely symbolic protest. He did not attend the Haber memorial service: "I could not . . . go against Rust's prohibition. Shortly before, I went over to Harnack-House and established that my special friend Ludwig August Sommer was lying in wait, quite certainly for the purpose of denouncing me, if at all possible."[109] In 1937, Laue sent his son to the United States, so that he would not be forced to fight for Hitler. But Laue himself would not leave. He wanted something to survive of the Germany he valued, and he wanted, above all, to witness the destruction of the regime he hated.[110] During a visit to the United States in 1937, a nervous Laue told Einstein, "I hate them so much I must be close to them. I have to go back."[111]

During those years when Meitner was professionally isolated outside her institute, Laue came regularly to the KWI for Chemistry to talk; they came to rely on each other's integrity and judgment. After the end of World War II, Meitner was not surprised to find that Laue was among the very few who took responsibility for what he had done, and failed to do. He wrote to her in 1958,

> I am not of the opinion that we were all free of guilt and only "those on top" were responsible. We all knew that injustice was taking place, we did not want to see it. . . . Came the year 1933 I followed a banner that we ought to have torn down at once. I did not do it and therefore must also bear responsibility.

During those years Laue had looked to Meitner for courage and strength.

> At that time, in spite of everything, you tried to understand us, guided us with such tact that I must admire it more and more. . . . One day [in May 1933] the doorbell rang. It was James Franck. One could see what lay behind him. Even in the depths of our institute he had no peace, he feared his presence would jeopardize it. When he left—much too soon—you spoke with us very calmly about the man and the scientist Franck. Quite incidentally you mentioned his Iron Cross I [a World War I military honor]. Did you realize how deeply your words affected us? . . . [Y]our goodness, your consideration had their effect. The notion of humanity acquired substance. For this I am grateful to you. . . . I was saved from things for which I would never have been able to forgive myself.[112]

As long as Meitner remained in Berlin, she had no premonition of physical danger and until 1938 freely traveled abroad for interludes of normalcy. In October 1933, Meitner attended the Solvay conference devoted to the sensational findings of 1932–1933: the neutron, the positron, and the deuteron (the heavy isotope of hydrogen), which together reshaped nuclear physics. The meeting may have convinced Meitner not to leave her laboratory, where she was engaged with so many of the new discoveries. A year later she traveled to Leningrad for the centennial celebration of Mendeleev's birth, and she was often in Copenhagen for Niels Bohr's physics conferences. She visited her friends Dirk and Miep Coster in Holland and welcomed visitors from abroad. It may be that Meitner sensed how urgently she would need her friends later on.

Meanwhile, physics gave her strength: its objectivity and universal validity seemed to repudiate the turmoil outside the laboratory. These words, from a speech given by Planck in 1935, held special meaning for her: "The scientific incontrovertibility of physics leads directly to the ethical demand for veracity and honesty. And justice is inseparable from truth. . . . Just as the laws of nature work consistently and without exception, in great things as in small, so too people cannot live together without equal justice for all."[113]

CHAPTER SEVEN

Toward the Discovery of Nuclear Fission

I found these experiments so fascinating that as soon as they appeared I talked to Otto Hahn about resuming our direct collaboration.

The journey to the discovery of nuclear fission was a jumble of paths and blind alleys—"Wege und Irrwege," Lise Meitner would later say—that misled physicists and chemists, experimentalists and theoreticians for years; when fission was recognized, finally, it was a sensational surprise that would change the history of its time. Had fission been born into a world at peace, its energy might first have been used to provide light and heat for people's homes. Had fission been discovered in a world free of racial persecution, it might well have been the crowning achievement of Lise Meitner's career.

The journey began with the discovery of the neutron in 1932. When Meitner attended the seventh Solvay conference in Brussels in October 1933, the essential nature of the neutron was still unknown, although experiments were under way to determine its mass. Positrons, also new in 1932, were bizarre in some ways but, on the whole, better understood; the formation of positron-electron pairs was consistent with theory and demonstrated that new particles can be created by the direct conversion of energy into matter. Returning to Rome from the Solvay conference, Enrico Fermi made use of this idea, incorporating Wolfgang Pauli's proposed neutral particle, the *neutrino*, into a theory of beta decay that accounted most satisfactorily for the continuous beta spectrum. Nuclear physics was maturing rapidly, and almost anything seemed possible.[1]

In January 1934, yet another discovery astonished the physics community. In the process of bombarding light elements with alpha particles, Irène Curie and Frédéric Joliot found that new radioactive isotopes were produced—the first instance of artificial radioactivity. Aluminum, for example, reacted with alpha particles to produce radioactive phosphorus,

$$^4\text{He} + {}^{27}\text{Al} \rightarrow {}^{30}\text{P} + {}^1\text{n}$$

which decayed to stable silicon by emitting a positron,

$$^{30}_{15}\text{P} \rightarrow {}^{30}_{14}\text{Si} + {}^{0}_{+1}\text{e}$$

The half-life of ^{30}P was about three minutes.[2]

"The significance of these extraordinarily beautiful results is certainly very far-reaching," Meitner commented. At one stroke, artificial radioactivity enlarged the number of known nuclear reactions, extended radioactivity to the entire periodic table, and demonstrated that positrons can be emitted in the course of radioactive decay. Meitner quickly confirmed the Curie-Joliot findings by photographing phosphorus decay positrons in a cloud chamber some time after the alpha particle–aluminum reaction had been interrupted.[3] A few months later she showed that the energy spectrum of the phosphorus decay positrons was analogous in every respect to the beta spectra of negative electrons.[4]

For several years Fermi and his group had been preparing to reorient their research from atomic to nuclear physics; to learn the experimental techniques, they made extended visits to various laboratories in Europe and the United States. For radioactivity, Franco Rasetti went to Meitner's laboratory in late 1931 and again in 1932, just after the discovery of the neutron. On his return to Rome, Rasetti was the "moving spirit" in the preparatory work, building a number of Geiger-Müller counters and a large cloud chamber, based on Meitner's model in Berlin-Dahlem; from Meitner, he had also learned how to prepare neutron sources by evaporating polonium onto beryllium. With this, Fermi had the idea of using neutrons rather than alpha particles to induce artificial radioactivity.[5]

Fermi rightly supposed that neutrons would be more effective than alpha particles for initiating nuclear reactions: because they had no charge, neutrons were more likely to reach and penetrate a target nucleus. The difficulty, however, was that most neutron sources were very weak: thousands of alpha particles might bombard a light element (usually beryllium) and produce only a few neutrons. Here the Rome group was fortunate. They had access to a large quantity—over a gram—of radium, which produced copious amounts of radon, a gas readily separated from its radium parent. When mixed with powdered beryllium in a sealed capsule, the alpha-emitting radon was a much stronger neutron source than the polonium-beryllium mixtures then in use.[6]

Starting with hydrogen, Fermi and his co-workers began working their way through the periodic table, systematically irradiating one element after another with neutrons. Fluorine was the first to display artificial radio-activity; aluminum was next. On 25 March 1934, Fermi reported these results in a letter to the small Italian journal *La Ricerca Scientifica*, chosen for its rapid publication time.[7] A few weeks later Fermi listed twenty elements more, up to lanthanum.[8] To inform their colleagues quickly, the Rome scientists mailed preprints of the papers to forty of the most active nuclear physicists worldwide.[9]

Thus Lise Meitner was among the first to learn of the neutron irra-diation experiments in Rome. She was interested from the start, as we know from a note she wrote to Fermi on 16 May 1934 asking him to send "once more the reprints of both of your articles about neutrons. Unfortunately I misplaced the reprints you so kindly sent and cannot find them again."[10] That week Meitner reported to *Naturwissenschaften* that she had verified the artificial radioactivities Fermi had found from the neutron irradiation of aluminum (Al), silicon (Si), phosphorus (P), copper (Cu), and zinc (Zn) (the first three also in a Wilson chamber) and measured the half-lives of the Al, Si, and P activities.[11]

The Rome experiments attracted wide attention. Rutherford jovially congratulated Fermi on his "escape from the sphere of theoretical phys-ics."[12] In Niels Bohr's institute in Copenhagen, Otto Robert Frisch was one of the few who could read Italian: "When each new copy of the *Ricerca* arrived I found myself the centre of a crowd demanding instant translation of Fermi's latest discoveries. And what an exciting time it was!"[13]

On 10 May 1934, Fermi and his co-workers submitted their third report to *La Ricerca Scientifica*.[14] They had arrived at the last element, uranium. On neutron bombardment, uranium yielded several new beta activities whose chemistry appeared different from uranium (element 92) or any nearby element down to radon (86). Fermi cautiously suggested the "spon-taneous hypothesis that the active substance of U might have atomic number 93"—the first element beyond uranium. The idea of a new ele-ment—a synthetic new element—attracted wide attention. One Italian newspaper hailed it as proof that "in the Fascist atmosphere Italy has resumed her ancient role of teacher and vanguard in all fields"; another fantasized that Fermi had presented a small vial of element 93 to the queen of Italy.[15] Beyond the aberrations of the popular press, scientists were equally fascinated.

A few weeks later, Fermi noted that the observed neutron-induced reactions fell into three categories:

1. Emission of an alpha particle, an (n,α) reaction. For example,

$$^{27}\text{Al} + {}^{1}\text{n} \rightarrow {}^{24}\text{Na} + {}^{4}\text{He}$$

2. Emission of a proton, (n,p):

$$^{28}\text{Si} + {}^{1}\text{n} \rightarrow {}^{28}\text{Al} + {}^{1}\text{H}$$

3. Neutron capture with gamma emission, (n,γ):

$$^{127}\text{I} + {}^{1}\text{n} \rightarrow {}^{128}\text{I} + \gamma$$

In each case, the new artificially radioactive nucleus differed from the original nucleus by no more than two atomic numbers. Theoretically this made sense: a particle as modest as a neutron was expected to inflict only minor damage.

In addition, each new radioactive nucleus was, without exception, a beta emitter. For example,

$$^{128}_{53}\text{I} \rightarrow {}^{128}_{54}\text{Xe} + {}^{0}_{-1}\text{e}$$

Fermi also noted that the (n,α) and the (n,p) reactions were found only for the lighter elements, while heavier elements favored neutron capture (n,γ). Thus it was not only possible but plausible that uranium would capture a neutron, then decay with beta emission to element 93:

$$^{238}_{92}\text{U} + {}^{1}_{0}\text{n} \rightarrow {}^{239}_{92}\text{U} \rightarrow {}^{239}_{93}? + {}^{0}_{-1}\text{e}$$

But there was more: under neutron bombardment, uranium produced at least four new beta activities. This suggested to Fermi a sequence of beta decays, perhaps to element 94 or 95. He cautioned, however, that it was "premature to form any definite hypothesis on the chain of disintegrations involved."[16]

At this point Meitner spoke to Otto Hahn: "I found these experiments so fascinating, that as soon as they appeared in *Nuovo Cimento* and in *Nature* I talked to Otto Hahn about resuming our direct collaboration, after an interruption of several years, in order to resolve these problems."[17] Fermi's investigations, she remembered later, were of "consuming interest to me, and it was at the same time clear to me that one could not get ahead in this field with physics alone. The help of an outstanding chemist like Otto was needed to get results."[18]

But it took Lise several weeks to get Otto interested.[19] What sparked his interest, finally, was a publication by Aristide von Grosse, a former student who had since emigrated to the United States. For several years von Grosse and Hahn had been embroiled in a controversy over the discovery and properties of protactinium, an unpleasant dispute that had twice spilled onto the pages of *Naturwissenschaften*.[20] Now in the 1 August 1934 issue of *Physical Review*, von Grosse argued that the chemical behavior of Fermi's proposed element 93 was closer to that of protactinium.[21] At this point Hahn took notice.

> After the appearance of the articles by Fermi, etc., came an article by v.
> Grosse and Agruss, according to which it was not at all certain that the Fermi
> activities were element 93 or so, but that it was most likely element 91:
> ekatantalum [protactinium]. After these publications L. Meitner and I de-
> cided to repeat Fermi's experiments and test Grosse's assumptions.[22]

So Hahn remembered it in 1945 and in all his many reminiscences and autobiographies.[23] While it may be true that this reflected his own first engagement with the uranium experiments, it certainly was not Meitner's: the issue of *Physical Review* with von Grosse's note would have arrived in Germany only toward the end of August, weeks after Meitner first approached Hahn, months after she herself first became interested in Fermi's work. Indeed, in their first joint uranium publication, Hahn and Meitner clearly state, "In the course of several other experiments"—here they cite two earlier neutron studies by Meitner[24]—"we have now undertaken a thorough investigation of these uranium processes."[25]

Before Meitner and Hahn could begin their investigation, they attended the great celebration in Leningrad for the centenary of the birth of Dmitri Mendeleev. For her conference lecture, Meitner spoke on "The Atomic Nucleus and the Periodic System." Concerned as always with the interplay between theory and experiment, she surveyed the dramatic changes brought by the neutron, cited the importance of neutron mass for understanding nuclear stability, and discussed Heisenberg's new theory that neutrons and protons were bound by exchange forces that transformed one into the other. A key to the nucleus, the neutron was also the source of new reactions and new radioactive nuclei. It was entirely possible, Meitner concluded, that Fermi's recent neutron experiments had extended the periodic table with the creation of two new elements, 93 and 94.[26]

On her return to Berlin, Meitner prepared for the uranium experiments by investigating neutron sources. Leo Szilard and T. A. Chalmers had just reported that radium gamma radiation acting on beryllium generated neutrons that were captured by iodine, the same reaction Fermi had observed with neutrons from a radon (α) + beryllium source. When Meitner repeated the Szilard-Chalmers experiment, she found that the gamma + beryllium neutrons were captured by iodine, silver, and gold but did not react with the lighter elements sodium, aluminum, or silicon. Meitner surmised that the gamma + beryllium neutrons were substantially lower in energy than those generated by radon (α) + beryllium and were therefore "more suited for [capture processes], which are just those processes favored by Ag, I, and Au. In contrast, the processes involving Na, Al, and Si require higher neutron energies because they split off an alpha particle or a proton, which must acquire substantial energy in order to be able to leave the nucleus with finite probability."[27]

Meitner was thus one of the first to suggest that neutron energy affects the course of reaction, in particular that slower neutrons are more readily captured. She submitted her note to *Naturwissenschaften* in mid-October. A few days later, on 22 October, Fermi and his co-workers observed a dramatic increase in the neutron-induced activities of silver, copper, and iodine when paraffin, water, or some other hydrogen-containing material was placed between the neutron source and the target. They quickly established that such hydrogenous substances enhanced neutron-capture reactions only, "a possible explanation of these facts [being] that neutrons rapidly lose their energy by repeated collisions with hydrogen nuclei." They sent their results to *La Ricerca Scientifica* that night.[28]

Efficient as always, the Rome group mailed this newest preprint to their colleagues. Meitner responded immediately, writing to Fermi on 26 October, "Enclosed is a small notice currently in press in *Naturwissenschaften*, from which you can see that by quite different means I arrived at similar conclusions . . . to yours, as your latest, so very beautiful communication shows."[29]

By then Meitner and Hahn had begun their own experiments with the neutron irradiation of uranium. Of the four activities found by Fermi's group, with half-lives of 10 seconds, 40 seconds, 13 minutes, and 90 minutes, the two longer-lived substances were easiest to study. In December 1934, Meitner and Hahn reported a separation procedure based on the expected chemical similarities of transuranic elements to rhenium and platinum. Using a Rn(α) + Be source similar to Fermi's, they irradiated several grams of a uranium salt with neutrons, then dissolved it and added potassium perrhenate and platinum chloride to the solution. The addition of sodium hydroxide quantitatively precipitated elements 90, 91, and 92; elements beyond 92 were expected to remain in solution. Hahn and Meitner found the 13- and 90-minute activities in the solution and concluded that neither was an isotope of elements 90, 91, or 92. When the solution was acidified and saturated with hydrogen sulfide, first platinum sulfide and then rhenium sulfide precipitated, carrying the 13- and 90-minute activities with them. Therefore, it seemed likely that these were elements beyond uranium, with chemistry similar to the transition elements rhenium and platinum.

As a specific test for protactinium, Hahn and Meitner added the known protactinium isotope UZ (^{234}Pa) to the irradiated uranium solution. By various precipitations UZ was completely separated from the 13- and 90-minute activities. Hahn and Meitner noted with satisfaction, "It has thus been directly shown that these two substances are not isotopes of element 91, as A. v. Grosse has claimed." In fact, their chemical reactions ruled out all elements down to mercury (80), "which again makes it very probable that the 13- and 90-minute substances are elements beyond 92."[30]

Fermi, Meitner, and Hahn, indeed all who worked in the field, were guided from the outset by two prevailing principles. The first was a generalization of all previous experience in nuclear physics: in every known nuclear reaction, including neutron-induced reactions, the changes were always minor: the product nucleus never differed from the original by more than a few protons or neutrons. This observation corresponded to one of the few successful applications of quantum theory to the nucleus, George Gamow's theory of alpha decay, which showed that only small particles— alpha particles, protons, neutrons—have a chance of tunneling through the nuclear barrier and escaping the nucleus.[31] Thus, if the 13- and 90-minute activities were not elements just below uranium, it was consistent with both experiment and theory to suppose they would be elements just above. The thought that a uranium nucleus might shatter, split, or otherwise drastically change when hit by a neutron was never entertained by either experimentalists or theoreticians until fission was finally discovered four years later. With one exception: in September 1934, Ida Noddack, chemist and co-discoverer of rhenium, criticized Fermi's chemistry and his assumption of small nuclear changes. "It is conceivable," she wrote, "that when heavy nuclei are bombarded by neutrons, these nuclei break up into several *larger* fragments, which would of course be isotopes of known elements but not neighbors of the irradiated elements."[32] No one explored the idea, not even Noddack herself. In Rome her article was read but ignored, and in Berlin it was never seriously considered.[33]

The second guiding principle was that elements beyond uranium would be higher homologues of third row transition elements. (See Appendix fig. 2.) This, too, was based on induction: because the known elements Ac, Th, Pa, and U chemically resembled La, Hf, Ta, and W, it was assumed that the elements beyond U would resemble Re, Os, Ir, Pt, and so on. In some respects it is difficult to understand why this assumption went unchallenged for so long, since chemists did have a theory of the periodic table, first proposed by Bohr in 1921–1922, in which elements beyond uranium were expected to form a second rare-earth series and thus resemble the lanthanides—the elements following La—more than transition elements. But Bohr was unable to predict just where this second rare-earth series would begin (see Appendix fig. 3), and for the most part, Bohr's proposal was set aside, even by Bohr himself. An exception was Aristide von Grosse, who noted in 1935 that the transuranium elements "might belong to a second group of rare-earth elements in N. Bohr's sense." But the evidence was not there. When the first true transuranium elements were found, they did indeed resemble lanthanides and not transition elements, and in 1944 Glenn T. Seaborg, the discoverer of several transuranium elements, first proposed that the second rare-earth series begins with actinium[34]—entirely homologous to the lanthanides—and so it is found on the modern periodic table today. (See Appendix fig. 4.)

But for several years the two erroneous principles, the one from physics and the other from chemistry, gave credence to each other. Physics claimed that the neutron irradiation of uranium would produce only minor changes. Chemistry confirmed that the resulting products were "transuranes," chemically so much like third row transition elements that they were designated ekarhenium, ekaosmium, and so on, to the elements beyond.

Early in March 1935, Hahn and Meitner reported that the 13- and 90-minute activities were not isotopes of each other. Modifying their separation procedure somewhat for speed, they irradiated a uranium compound with neutrons, dissolved it in acidic hydrogen sulfide, then added rhenium and osmium compounds as carriers to selectively precipitate their sulfides. Uranium, its decay products, and all other species down to element 85 remained in solution, the 90-minute activity precipitated with osmium sulfide and the 13-minute activity with rhenium sulfide. With improved separation of the two species, it was evident that the 90-minute substance contained another, considerably longer-lived activity.[35]

The uranium investigation was growing more complex. Although it would always be driven by questions of nuclear physics, it relied, at least at first, on radiochemical data. As director of the KWI for Chemistry, Hahn could not devote his full attention to research, and in the spring of 1935, he and Meitner asked Fritz Strassmann, one of his assistants, to join them. Strassmann, a physical-inorganic chemist, was expert in chemical analysis; he had already helped Meitner and Hahn with their first transurane separations. Scientifically, the three were remarkably suited to one another and to the project; politically and personally, their affinity was also very strong. By 1935, Meitner had become quite isolated professionally, Hahn was known for his anti-Nazi views, and Strassmann had made himself unemployable in academia or industry by refusing to join any Nazi-affiliated professional organizations.[36] For Meitner and Strassmann particularly, it was far more congenial to work together than with others in the institute. In their first joint publication that August, they verified Hahn and Meitner's previous findings, confirming the 13-minute activity as ekarhenium and the 90-minute activity (revised to 100 minutes) as ekaosmium. For the new long-lived activity, also a transurane, they obtained a half-life of about 3.5 days.[37]

Probing a radioactive mixture for tiny amounts of unknown activities was no simple task. Because the neutron-induced activities were always much weaker than the natural radioactivity of uranium, the Berlin team always began by separating the transuranes from the irradiated uranium. Once they established that the transuranes precipitated quantitatively from a solution of uranium and its decay products, they ignored the filtrate and pursued only the precipitate, using ever more refined techniques to disentangle its multiple activities.[38] They felt justified in doing this, since they had established at the outset that precipitation separated the activities of interest—the presumed "transuranes"—from elements 90, 91, and 92. The approach was based on traditional analytical methods of chemical separation and precipitation. As it turned out, their failure to investigate the filtrate critically restricted the scope of their investigation.

From cloud chamber photographs, Meitner determined early on that the transuranes emitted only beta particles.[39] The beta decay was measured with Geiger-Müller counters, and a decay curve—logarithm of intensity versus time—was plotted. For a single radioactive substance, the decay curve is a straight line from whose slope its characteristic half-life is obtained; for a mixture, the shape of the decay curve may indicate the number of activities and their sequence. A short-lived parent decaying into a long-lived daughter produces a decay curve that drops rapidly, then levels off to the more gradual slope of the long-lived daughter; by subtracting the contribution from the longer-lived species, the shorter half-life can be determined. But if the parent is longer-lived than the daughter, the decay curve is a straight line determined by the parent and the short-lived daughter can only be detected by other, usually chemical, means. When still more activities are present, the decay curve becomes increasingly difficult to interpret. This was particularly true for the neutron irradiation products, where several simultaneous reactions each initiated a sequence of disintegrations. Chemical separations were attempted whenever possible, but these were always difficult owing to the minuscule quantities and uncertain chemistry of the transuranes.

To disentangle the "genetic" links—the parent-daughter relationships—among the various activities, it was essential to know whether the neutron irradiation of uranium initiated more than one primary reaction. Early in 1935 Fermi and his collaborators reported that neutrons slowed by paraffin or water increased the amounts of some activities but not others, an indication that at least two different primary reactions were taking place.[40] The Berlin group was unable to verify Fermi's observations and began their own studies of the effects of slow and fast neutrons and of variations in the length of neutron irradiation.

Compared to other groups working on the uranium products, the Berlin team had the advantage of laboratories that were uncontaminated by radioactivity. To confine the spread of radioactivity in case of accident, the various procedures were carried out in separate rooms, all in Meitner's section on the ground floor. Neutron irradiations were done in one room; chemical separations were performed in a laboratory several doors down, just across the hall from a third room where the activities were measured. Most of the physical apparatus was simple and self-made: neutron sources consisting of radium and beryllium sealed into small glass tubes, paraffin blocks in several sizes; Geiger-Müller tubes, batteries, counters, and amplifiers; and an assortment of lead blocks for shielding and lead vessels to enclose the active samples.[41]

In March of 1936, Hahn, Meitner, and Strassmann submitted their first comprehensive report to a chemical journal, reviewing the entire development of transurane research and describing in detail their own chemical separations and radioactivity data. They listed a total of ten radioactive species and their half-lives:

U: 24 minutes
ekaRe: 16 (formerly 13) minutes
ekaOs: 12 hours
ekaIr: 3 days
U: 10 seconds
ekaOs: 59 (formerly 90–100) minutes
Th: 4 minutes
U: 40 seconds
ekaRe: 2.2 minutes
Pa: very short

The species were identified with varying degrees of certainty. The only truly unambiguous assignment was the 24-minute U, which was isolated and chemically identified as uranium. The 10- and 40-second activities were too short-lived for chemical identification but were attributed to uranium because they were beta emitters that preceded the two activities with the expected chemical behavior of ekarhenium. Such reasoning was the basis for the ekaOs and ekaIr assignments as well. Thus, even when genetic sequences were difficult to evaluate and chemical data absent or shaky, the assignments were considered firm if the radioactivity and chemistry did not contradict each other and were consistent with the expected characteristics of transuranic elements. As a result, the chemistry report was in part based on circular reasoning.

> For the 2.2-minute and the 16-minute substances, there is no option but to assign them to ekarhenium; not only the genetic relationships [to the 10- and 40-second activities ascribed to U because they preceded the supposed ekaRe] but also the chemical characteristics leave no doubt that they are element 93.

And on the whole, it exuded a confidence unjustified by its individual data.

> For the representatives of the elements 94–96 we are certain from their general chemical properties to which groups they belong [even though] we cannot yet give satisfactory chemical separation methods for the individual species.[42]

Hahn was senior author for the chemistry report, published in *Chemische Berichte*. Appearing at the same time was the physical interpretation, published in *Naturwissenschaften*, for which Meitner was chiefly responsible (her name appeared first on the publication, and Strassmann's was absent).[43] It was up to Meitner to incorporate the various species and their genetic sequences into nuclear reactions that made sense.

The existence of three uranium isotopes required three different primary neutron-uranium reactions, each initiating a series of beta decays:

Neutron capture (n,γ):

$$^{238}U + {}^{1}n \rightarrow {}^{239}U \ (10 \ sec) \rightarrow {}^{239}ekaRe \ (2.2 \ min) \rightarrow$$
$$^{239}ekaOs \ (59 \ min) \rightarrow {}^{239}ekaIr \ (3 \ days)$$

An (n,2n) reaction:

$$^{238}U + {}^{1}n \rightarrow 2{}^{1}n + {}^{237}U \ (40 \ sec) \rightarrow {}^{237}ekaRe \ (16 \ min) \rightarrow$$
$$^{237}ekaOs \ (12 \ hr)$$

An (n,α) reaction:

$$^{238}U + {}^{1}n \rightarrow \alpha + {}^{235}Th \ (4 \ min) \rightarrow {}^{235}Pa \ (short) \rightarrow {}^{235}U \ (24 \ min)$$

At this point there was no direct physical evidence for any of these reactions. On the contrary, a search for the alpha particles emitted in the (n,α) reaction had been unsuccessful. From the physicist's perspective, the situation was quite unsatisfactory. Not only were (n,2n) processes unknown for other heavy nuclei but the (n,α) reaction was effective only with slow neutrons, something Meitner found "remarkable," since theory and all previous experience predicted that energetic (fast) neutrons would be needed to expel an alpha particle. Thus the chemistry and radioactivity data that inspired such confidence in the chemists led to a physical interpretation that was decidedly weak. If in 1934 Meitner thought she could not get anywhere without a chemist of Hahn's ability, by 1936 she must have reflected that the relentless accumulation of new species had cast chemistry in the role of sorcerer's apprentice.

There was no choice but to go on. While Hahn and Strassmann refined the chemistry and genetic relationships, Meitner devised experiments to determine the mechanism of the reaction processes. In May 1937, they again issued two parallel reports, one in *Zeitschrift für Physik*[44] with Meitner as principal author and the other in *Chemische Berichte* with Hahn as senior author.[45] Hahn presented what appeared to be iron-clad chemical evidence for the existence of the transuranes. In contrast, Meitner's work demonstrated that the uranium problem had become much larger, more difficult, and more significant than the discovery of new elements alone.

Their new results called for a substantial revision of the neutron-uranium reactions they had proposed the year before. Now the three processes were all neutron capture (n,γ) reactions:

1. U + n → U (10 sec) → ekaRe (2.2 min) → ekaOs (59 min) → ekaIr (66 hr) → ekaPt (2.5 hr) → ekaAu (?)

2. U + n → U (40 sec) → ekaRe (16 min) → ekaOs (5.7 hr) → ekaIr (?)

3. U + n → U (23 min) → ekaRe (?)

The 23-minute (formerly 24-minute) U of process 3 was unambiguously uranium. The extremely short 10- and 40-second activities could not be confirmed as uranium, but, as before, genetic relationships seemed to exclude any other possibility: both were beta emitters that apparently decayed to an ekarhenium of atomic number 93. In addition, a new 2.5-hour ekaPt was found, and the half-lives of the 12-hour ekaOs and the 3-day ekaIr were modified to 5.7 hours and 66 hours, respectively. With improved tests, the activities were consistent with the chemical and physical properties expected of the higher homologues of rhenium, osmium, iridium, and platinum.

The genetic relationships in processes 1 and 2 were established by several methods. In a few cases activities could be separated chemically so that the decay of the parent and growth of the daughter could be observed. More often the sequences were determined from decay curves taken with variations in neutron irradiation time from a few seconds to several hours. In every case genetic relationships were consistent with chemistry for the sequence U → ekaRe → ekaOs → ekaIr → ekaPt. For example, decay curves indicated that the 16-minute activity decayed to the 5.7-hour activity; chemically, the 16-minute activity had the properties expected for ekaRe and the 5.7-hour activity, for ekaOs. Thus independent data from radioactivity and chemistry were mutually supporting. Like interlocking pieces of a puzzle, they made a compelling case for the existence of transuranic elements. Moreover, the striking parallelism of the two decay sequences was reassuring, for the evidence presented itself not just once but twice.

From a chemical viewpoint, the transuranes were a spectacular feat: new elements were synthesized, periodic behavior was confirmed, the periodic table was launched into the unknown. As each new transurane fit the sequence, it gave credence to those before. Chemists were convinced. Summarizing the similarities and differences between the transuranes and their lower homologues, Hahn ended his report emphatically: "In general, the chemical behavior of the transuranes . . . is such that their position in the periodic system is no longer in doubt. *Above all, their chemical distinction from all previously known elements needs no further discussion*" (emphasis in original).[46]

As the physicist on the Berlin team, Meitner's task was to incorporate the data from chemistry, radioactivity, and her own physical experiments into a coherent explanation of the nuclear processes. Her report in *Zeitschrift für Physik* clearly shows how difficult that was. Uranium was different somehow from other heavy nuclei, reacting with neutrons in three distinct ways, giving three sets of products. Equally disturbing was the length of the decay sequence in processes 1 and 2 and the parallels between them.

To find a physical distinction between the three processes, Meitner studied the energy of the effective neutrons for each. Her radium (α) + beryllium source produced fast neutrons of energy 10^5 to 10^7 electron volts (eV), which could be slowed by enclosing the source in paraffin. When thermal neutrons (the slowest possible, of energy 0.025 eV) were desired, the neutron source and the uranium target were both enclosed in paraffin; for neutrons that were slow but not thermal, only the source was enclosed in paraffin. For fast neutrons, no paraffin was used.

We have done such experiments in two directions. First, we have investigated the energy of the effective neutrons for the three uranium transformation processes, and second, we have attempted to see if the *relative* intensity of processes 1 and 2 may be influenced by various filters.[47]

Meitner noted that in other elements the capture of slow neutrons was a resonance effect, occurring only for neutrons of characteristic energy. Cadmium, for example, captured thermal neutrons but was almost transparent to neutrons higher than 1 volt, silver had a known resonance absorption at 2.5 eV, gold at 4.5 eV, and so on. These and other elements were used as filters to control the energy of the neutrons used.

To determine the extent to which the reactions in processes 1 and 2 were the result of slow neutrons, irradiation was done inside a large paraffin cylinder with and without the cadmium filters that absorbed thermal neutrons. Meitner found that about 90 percent of the activation was due to thermal neutrons and 10 percent to fast neutrons and that both processes were induced in the same proportion. And no matter which neutrons were selected for irradiation—fast or thermal, filtered or not—the activities produced in processes 1 and 2 were always the same. The significance of this, Meitner concluded, was that the same reaction mechanism operated with both fast and slow neutrons, for both processes.

As it was inconceivable that slow neutrons would induce (n,2n) or (n,α) reactions, it appeared that the only possible mechanism for processes 1 and 2 was neutron capture (n,γ). In 1937, neither Meitner nor any other physicist imagined splitting or any other massive disintegration—the question was never discussed. Moreover, the existence of transuranes, so convincingly proved by the chemistry, not only supported neutron capture but required it.

Meitner considered the possibility that processes 1 and 2 might each originate with a different uranium isotope. Uranium was known to contain ^{238}U (99.3%), ^{235}U (0.7%), and a trace of ^{234}U. After measuring the absolute yield of the 16-minute ekaRe (process 2) with fast neutrons, Meitner calculated the cross section—an index of probability—for the capture of fast neutrons by uranium. The cross section was large, which excluded from consideration isotopes other than the most abundant, ^{238}U. Since the yield for process 1 was similar, its cross section and effective isotope had to be the same. "Since processes 1 and 2 are induced by fast neutrons, and the cross section for ^{238}U is 1.6×10^{-25} cm^2, one can easily see that the two isotopes 235 and 234 are out of the question as starting isotopes for these processes, as they would have cross sections for the capture of fast neutrons that would be impossibly large."[48] Fast neutrons therefore reacted with ^{238}U only—and since fast and slow neutrons induced exactly the same activities, Meitner concluded that slow neutrons *also* reacted only with ^{238}U. Thus processes 1 and 2 both began with ^{238}U and both proceeded with fast and slow neutrons. The mystery remained: there was no difference in the reaction mechanisms of processes 1 and 2.

Process 3 was clearly different. The 23-minute uranium was only weakly produced by thermal neutrons and not at all by fast neutrons, an indication that process 3 was primarily a resonance capture of neutrons with an energy of a few volts. To determine the energy of the resonance neutrons, Meitner measured their absorption in boron and compared it to the absorption of thermal neutrons in boron, obtaining a resonance energy of 25 ± 10 eV. Finally, she determined the effective isotope for process 3 by measuring the reaction absorption coefficient in uranium and again obtained a cross section compatible only with ^{238}U. Thus three processes—all very different—seemed to start with neutron capture by ^{238}U.

The result was incomprehensible. Meitner reached for a new explanation. For the parallel processes 1 and 2, induced by neutrons of the same energy, she proposed that

> two "isomeric" uranium nuclei with atomic weight 239 might be created, which despite their identical nuclear charge and mass might decay with different decay constants and give rise to two radioactive series. To the extent that the two series are parallel, they must exhibit two comparable series of "isomeric" nuclei. These "isomeric" nuclei must differ only in their level of excitation, and their beta decay must occur before gamma emission de-excites them from one energy level to the other, which means one state must be metastable.[49]

The notion of isomeric nuclei was not entirely new. In 1922, Otto Hahn had discovered uranium Z, identical in nuclear charge and mass to UX$_2$ (^{234}Pa) but with a longer half-life; more recently there had been experimental evidence for isomeric nuclear states in bromine. Hahn's discovery of isomerism may have predisposed Meitner to apply isomerism to the transuranes; for a theoretical interpretation, she worked closely with Carl Friedrich von Weizsäcker, a young physicist who had been her assistant in the summer of 1936 and had since moved over to the nearby KWI for

Physics in Dahlem. In Weizsäcker's theoretical model, two isomeric nuclei would have almost the same energy but different angular momentum and thus have different half-lives.[50]

But the existence of a third isomer, the ^{239}U of process 3, weakened the argument. "The supposition of three isomeric nuclei involves considerable difficulty for von Weizsäcker's suggested model," Meitner noted, and she suggested that nuclear isomerism might instead be analogous to molecular isomerism. But it was just speculation. She ended her report abruptly. "The processes must be neutron capture by uranium-238, which leads to three isomeric nuclei of uranium-239. This result is very difficult to reconcile with current concepts of the nucleus."[51]

It seemed the Berlin team had gone as far as it could go. The chemistry was consistent, the genetic sequences sensible, the physical measurements impeccable. The evidence for the 23-minute U was particularly strong. And because it captured neutrons like other heavy nuclei and underwent beta decay to produce the next higher element, it added to the credibility of the neutron capture and transuranes of processes 1 and 2. If the Berlin group had looked for the decay product of the 23-minute U, they might have found the true element 93 and learned from its chemistry that the 16- and 2.2-minute activities were not element 93 at all. But they did not do it. Their neutron sources were too weak, the uranium filtrate was difficult to probe, and they had transuranes enough, or so it seemed, in processes 1 and 2. No one suggested anything else. To the extent that the Berlin experiments were repeated by others, the results were the same. Neither Irène Curie in Paris nor Philip Abelson in Berkeley nor the other scientists in Vienna, Cambridge, Ann Arbor, and Zurich came to a different conclusion. By 1936, Fermi and his co-workers were so sure of the transuranes that they gave names to the first two: ausonium and hesperium.[52] Everything about the transuranes was most impressive. Only a physical foundation was lacking.

The difficulty lay not in the methods or the data but in the underlying assumptions that framed the investigation from the start—the small nuclear changes ordained by physics, the mistaken chemistry of the transuranes. The difficulty was compounded when these two false principles unserendipitously dovetailed, each lending the other an appearance of truth, each inhibiting a critical test of the other. In Berlin and elsewhere, false confidence restricted the experiments to the transurane precipitate and prevented a search for other activities. Later Meitner recalled,

> During irradiations with fast neutrons our precipitations were always done in such a way as to leave U, Pa, and Th in the filtrate, from which we drew a certain support for the transuranic nature of the precipitated elements. For this reason—and this was our mistake—we for a time never examined the filtrate after precipitation, not even in the trials with slowed neutrons.[53]

That was a remarkable oversight, because the Berlin group did find that the 23-minute U is formed only with slow neutrons and should have suspected that other new activities would be formed by slow neutrons as well. But they had little desire to search the filtrate—not when the transuranes were so accessible and far more interesting. There was something dazzling about these new elements, a mix of scientific adventure and general appeal. For Meitner, Hahn, and Strassmann, as politically vulnerable as any group still in Germany, the international attention that came with the transuranes was the best protection they could hope to secure for themselves and their institute.

But their failure to thoroughly search the filtrate also reflected the nature of their teamwork. For all their collective expertise, they maintained a pronounced disciplinary divide between chemistry and physics. Standing on the institute stairs, Meitner could be heard to say, "Hähnchen, Geh' nach oben, von Physik verstehst Du nichts!" (Hahn dear, go upstairs, you understand nothing of physics!),[54] and Hahn did not deny it. Nor did Strassmann. At one time in 1936, Strassmann thought he had evidence of barium among the uranium products; when she shrugged it off, he let it go.[55] As the team's leader and only physicist, Meitner was responsible for the project as a whole, more aware of its difficulties and more troubled by those aspects she did not fully understand. Yet she was completely dependent on Hahn and Strassmann in matters of chemistry. Although their assignment of the transuranes was the basis for her measurements and her physical interpretation, she did not—could not—question their analytic methods, and even when she had grave doubts about the meaning of the results, she was unable to convince them to search the filtrate for products of slow-neutron irradiation. She accepted their judgment that it was too difficult. In an interview years later, she remembered,

> Of course what we did was wrong, Hahn and I. I really think our misfortune was that we didn't search the filtrate. We couldn't search it because uranium was in it, you know, we couldn't see anything. Our neutron sources were too weak. . . . The chemists absolutely didn't want to, I begged them to while

I was there because I was so disturbed by it. Just because I understand too little chemistry, I naturally was always worried about what wasn't done.[56]

Both fields contributed errors, but chemistry provided no means for detecting them. Reports in *Chemische Berichte* seamlessly answered questions of chemical interest: Do transuranes exist? Yes. Are they higher homologues of rhenium and the platinum metals? Yes. Chemistry had no further questions. But physics required an understanding of the underlying nuclear reactions. Did the irradiation of uranium with neutrons give rise to transuranes? Apparently. Did the neutron-induced reactions make sense? No. It did not make sense that thermal neutrons *and* fast neutrons would initiate the same reactions in not just one but two different processes. And despite experimental precedent and theoretical justification for isomerism, *triple* isomerism was unexplained and *inherited* isomerism—for five generations—incomprehensible.

Apart from isomerism, Meitner was troubled by the length of the beta decay sequences in processes 1 and 2: she knew that the capture of just one neutron by ^{238}U ought not produce a nucleus so unstable that five or more successive beta emissions were required to relieve it. Later she recalled, "I was always unhappy about it because I couldn't understand, 'How can the atomic number keep rising with the same mass?' That's what I kept asking Weizsäcker: How is that possible? You see, I was never satisfied with our experiments before fission."[57]

Meitner's studies of the effects of neutron energies on reaction processes, described in her May 1937 article, provided no obvious solution. The Berlin group turned to the neutron irradiation of thorium, in the hope that insight might come from another direction, or possibly, as Strassmann wryly remarked, "to recover from the horror of the work with uranium."[58] Hahn and Meitner had worked with thorium before. In 1935, they found two distinct neutron-induced processes in thorium.[59] One, quite strongly enhanced by slow neutrons, yielded a 30-minute beta activity with the chemical properties of thorium, a classic example of neutron capture:

$$^{232}Th + {}^{1}n \rightarrow {}^{233}Th \ (30 \ min) \rightarrow {}^{233}Pa \ (?)$$

^{233}Th was the first example of a nucleus of mass $4n+1$, unknown among the natural radioactive species. Meitner and Hahn emphasized the novelty in their report, submitted in early May 1935. Later that month, Irène Curie, Hans von Halban, and Peter Preiswerk independently reported the same discovery, without citing Hahn and Meitner. Hahn, who counted on scientific recognition to bolster his shaky political position, was "a little angry" with Curie.[60] A polite but pronounced rivalry developed between the groups in Paris and Berlin.

With fast neutrons, Hahn and Meitner found 1- and 11-minute activities, both beta emitters. They proposed an (n,α) reaction, followed by two beta decays:

$$^{232}\text{Th} + {}^{1}\text{n} \rightarrow \alpha + {}^{229}\text{Ra} \ (1 \text{ min}) \rightarrow {}^{229}\text{Ac} \ (11 \text{ min}) \rightarrow {}^{229}\text{Th} \ (?)$$

The fast-neutron reaction was by no means certain: no (n,α) process had been found for other heavy nuclei, nor, for that matter, were alpha particles actually detected.

Later in 1935, Irène Curie and her co-workers identified the 1-minute substance as radium. They also found a second actinium isotope with a 3.5-hour half-life, and in 1936, Elizabeth Rona and Elisabeth Neuninger in Vienna reported a third actinium isotope with a 12-hour half-life. As with the uranium studies, it was assumed that irradiation with neutrons produced nuclei near thorium, and lower homologues of the suspected products were used as carriers for separation and precipitation. Thus an activity that precipitated with barium compounds was assumed to be radium, and one that followed a lanthanum carrier was attributed to actinium.

When the Berlin team returned to the thorium reactions in 1937, Fritz Strassmann was responsible for most of the chemistry.[61] Compared to uranium, work with thorium was inherently more difficult because its decay product radiothorium (^{228}Th) tended to mask the weaker neutron-induced activities. Here the Berlin team had the advantage of a thorium preparation that Hahn and Meitner had purified over the years by regularly extracting mesothorium (^{228}Ra), the precursor of radiothorium; unreplenished, the radiothorium initially present had almost disappeared.[62] Even so, the work was not simple. Unlike the transuranes, which could be quickly and cleanly separated from uranium and its decay products, the artificial thorium activities—presumably radium, actinium, and thorium—were isotopic with the natural products of thorium decay. Thus the thorium experiments were always more tedious and the results less certain.

In May 1938, Meitner, Strassmann, and Hahn reported that the reaction of fast neutrons with thorium was considerably more complex than they had previously thought. Altogether they found three radium and three actinium activities, all beta emitters.[63] From this they proposed three different (n,α) reactions, initiating three parallel decay series:

1. ^{1}n + ^{232}Th → α + ^{229}Ra (<1 min) → ^{229}Ac (18 min) → Th?

2. ^{1}n + ^{232}Th → α + ^{229}Ra (15 min) → ^{229}Ac (3.5 hr) → Th?

3. ^{1}n + ^{232}Th → α + ^{229}Ra (4 hr) → ^{229}Ac (20–30 hr) → Th?

Here, as in the uranium studies, radioactivity and chemistry offered apparently unassailable results that had never been seen for other nuclei. The three parallel decay series reinforced the uranium results to some extent but were no easier to explain. And an (n,α) process was unknown for any other heavy element.

To see if a fast-neutron (n,α) reaction with thorium was at all feasible, Meitner made a rough theoretical calculation and found that it was, but just barely. With 2.5 MeV kinetic energy for the incident neutron, 6 to 7 MeV for its binding energy, and 3 to 4 MeV for alpha emission, she estimated a total energy of 12 to 13 MeV, which, according to the Gamow formula, corresponded to alpha decay so fast that it could indeed take place before energy was lost in the form of gamma radiation. If the incident neutrons were any slower, however, alpha decay would take longer and gamma radiation would be emitted first. Just this was observed in the slow-neutron (n,γ) capture reaction that formed ^{233}Th.

Accounting for the three parallel (n,α) reactions was a greater problem.

> With fast neutrons . . . three isomeric radium nuclei are formed which are the starting point for three isomeric series. Since thorium possesses only *one* isotope, ^{232}Th, we are forced to conclude that three isomeric nuclei have been formed. The simplest assumption for the origin of these isomeric nuclei is that identical intermediate nuclei (^{232}Th + n) eject one of three types of alpha particles, with the result that three [radium] isomers are formed, of which two are metastable.[64]

But the high energy required for rapid alpha decay meant that the three types of alpha particles could not differ significantly in energy, which made it hard to imagine how three distinct isomeric series might result. Moreover, the alpha particles themselves were elusive. In Zurich, A. Braun, Peter Preiswerk, and Paul Scherrer reported some high-energy alpha particles, but they did not find three distinct groups of alpha particles.[65]

Triple isomerism was hard to justify and inherited triple isomerism—through several transformations—still a complete mystery. "The occurrence of isomeric nuclei means that a nucleus is not uniquely characterized by its mass and charge, but that another defining property is necessary. v. Weizsäcker has tentatively suggested mechanical angular momentum as such a property. . . . This hypothesis meets with considerable difficulty when, as has been found for thorium and uranium, isomerism persists over one or even several transformation processes. . . . The differences in momentum would then be so great as to become very improbable." In the end, it made no more sense for thorium than for uranium. Again Meitner concluded abruptly, "Perhaps one must search for a more general basis for the existence of isomeric nuclei,"[66] and referred to an article by Niels Bohr and Fritz Kalckar, published in 1937, on the collective behavior of tightly bound intranuclear particles, a liquidlike nuclear model described as the "liquid-drop." This was closer to the answer than she or anyone else imagined at the time.[67]

While the Berlin team was working on thorium, Irène Curie and her co-worker Pavel Savitch in Paris took a new approach to the uranium reactions. Instead of precipitating the transuranes with hydrogen sulfide, Curie and Savitch measured the activities without chemical separation, using an arrangement of paired counters and copper filters to compensate for the natural activity of uranium and its decay products. With this arrangement, they found the known 40-second, 2-minute, and 16-minute activities and in addition a new, surprisingly strong 3.5-hour activity. The activity was intense and, like the transuranes, was produced by fast neutrons and enhanced by slow neutrons; it had not been detected before because it remained in the filtrate after the transuranes were precipitated. The 3.5-hour activity could be separated from uranium; it was, therefore, neither a transurane nor uranium. In late 1937, Curie and Savitch postulated that it was an isotope of thorium.[68]

The Berlin group was startled to think they had missed such a strong activity. In their early experiments they had proposed that thorium was a product of the neutron irradiation of uranium but subsequently proved it was not. They tended to be skeptical of Curie's chemistry, thinking her method of examining the entire radioactive mixture without separation was liable to gross misinterpretation. In any event, Meitner was sure that slow neutrons could never react with uranium to give thorium: that would be a slow neutron (n,α) reaction, and there were strong theoretical arguments against it. Nevertheless, Meitner asked Strassmann to search the filtrate for thorium. She later wrote, "Unfortunately, we repeated the experiment of the French scientists only so far as to look for a thorium isotope in the filtrate; we established with certainty that it was not there."[69] Convinced that Curie and Savitch were wrong, they did not try to locate or identify the 3.5-hour activity, an omission Meitner would sorely regret.

The Paris and Berlin groups were competitors, viewing each other with considerable respect, occasional irritation, and some condescension. In Berlin, Hahn and Meitner were of the opinion that Curie's chemistry was not nearly as good as theirs. Meitner and Curie knew each other's work and had met at conferences, but there was no special rapport. They maintained a definite collegiality, however, and rather than contradicting Curie in print, Meitner decided to inform her privately of their failure to find thorium in the filtrate. On 20 January 1938, Meitner wrote to Curie, described their experiments, and suggested a retraction.[70] In their next publication Curie and Savitch acknowledged Meitner's letter and retracted the thorium isotope; they had meanwhile precipitated the 3.5-hour substance with lanthanum and concluded that it was actinium.[71] According to Strassmann, at this point Meitner "lost interest in the situation. She could rightfully point to the fact that it would be tremendously improbable for slow neutrons to knock out α-particles *and* protons from uranium . . . especially since alpha particles could never be observed."[72]

Curie and Savitch persisted. In May 1938, they reported that the 3.5-hour substance, after precipitation with a lanthanum salt as carrier, "was put through a fractional separation process which was used by Marie Curie for concentrating the activity of lanthanum mixed with actinium. . . . The 3.5-hour substance separated itself entirely from actinium," behaving more like lanthanum (element 57). Unwilling to suppose that it might actually be lanthanum, Curie and Savitch concluded that it "cannot be anything except a transuranic element, possessing very different [chemical] properties from those of other known transuranics, a hypothesis which raises great difficulties for its interpretation."[73] In Berlin they looked at it with a jaundiced eye and called the 3.5-hour substance "Curiosum."

By then, Meitner had more immediate worries. The political situation in Germany, grim since 1933, had taken a turn for the worse.

CHAPTER EIGHT

Escape

One dare not look back, one cannot look forward.

On 12 March 1938, German troops poured over the border into Austria. Not a shot was fired. Delirious crowds greeted Hitler at every village along his route; in Linz that evening a crowd of one hundred thousand cheered wildly as he addressed them as "German racial comrades." Spurred by the ecstatic welcome, Hitler proclaimed the total *Anschluss*—annexation—of his native Austria, reducing once-mighty Österreich to a province of greater Germany. On 15 March, hundreds of thousands of Viennese, the largest crowd in Austrian history, thronged the Heldenplatz to welcome their *Führer*. The police were already in brown shirts, and the churches, bells ringing, were draped with swastikas. The celebrations over, Viennese turned on Jews, hauling them out of their homes and beating and robbing them with a vicious enthusiasm that startled even German observers. Many Jews were killed; many committed suicide.[1] The unqualified success and ease of the Anschluss emboldened Nazis and their sympathizers throughout greater Germany and abroad.

Overnight Lise Meitner lost the thin protection of her Austrian citizenship. The most immediate threat came from Kurt Hess, the avid Nazi who headed the small "guest" institute on the top floor of the KWI for Chemistry. The day after the Anschluss he denounced her: "The Jewess endangers the institute."[2] Word reached Georg Graue, an assistant in the nearby KWI for Physical Chemistry and Electrochemistry. Haber's former institute was a Nazi "model institute" staffed entirely by party members, but Graue had been a student of Otto Hahn's and was still quite attached to his Doktorvater.[3] On 14 March, Graue reported Hess's de-

184

nunciation to Hahn, who discussed it with Meitner the same day.[4] According to Graue, her case had already come to the attention of Rudolf Mentzel, head of the Reich Research Council.[5]

The tension was too much for Hahn. "At that point I too lost my nerve somewhat and spoke to [Heinrich] Hörlein [treasurer of the Emil-Fischer-Gesellschaft, sponsor organization for the KWI for Chemistry] about Lise Meitner and the new foolish situation since the incorporation of Austria."[6] As yet the regime had done nothing, but Hahn was anxiously trying to head off trouble.

On 17 March, Hahn noted in his pocket calendar, he went by train to Elberfeld "to speak with Prof. Hörlein about Hess, about Lise." The next morning at 9:30 the "discussion about Hess and Lise" took place.[7] He gave lectures that day in Elberfeld and nearby Leverkusen; Lise meanwhile saw a lawyer, Dr. Erich Leist, to discuss her changed legal status. When Hahn returned to Berlin on the afternoon of 20 March and told her of his meeting with Hörlein, Lise was stunned. "Hörlein demanded that I leave," she recorded in her diary. "Hahn says I should not come to the institute anymore." The next morning she went to the institute anyway to begin writing up the results of their neutron irradiation experiments with thorium. But she was in shock, "very unhappy" about Hahn. "He has, in essence, thrown me out."[8]

The following day, 22 March, was Otto and Edith Hahn's twenty-fifth wedding anniversary. So as not to spoil the day for Edith, Lise went through the motions of congratulating the couple in their home that morning. Later that day Hörlein telephoned Hahn. He had changed his mind: "'We want to do the right thing.'" The anniversary celebration did not go well. "7 P.M. Hahn called, sent for me to tell me what Hörlein said. [We talked for] 10 minutes. Then with Edith, [Otto von] Baeyer and Hanno to the Germania-Palace, silver wedding anniversary. Later home by car. I very upset."[9] In her calendar she added, "Nothing but Vienna waltzes (runarounds): enough to drive one to despair."[10] For Otto, too, "the evening was quite depressing, because Edith and Hanno knew nothing of our conversation earlier. . . . Lise was very unhappy and angry with me that now I too had left her in the lurch."[11]

On 23 March, Lise asked her friend Ellen Hertz to come to the institute. "We talked about Hörlein. But I could not tell Ellen the truth about Otto."[12] The truth, too painful to articulate, was that he had been ready to sacrifice her to save his institute—and himself. Instinctively, he had distanced himself from her.

Lise turned to Paul Rosbaud, one of the few people she knew with reliable inside information. They had been friends for several years, beginning in the late 1920s when Paul, a fellow Austrian, was a physics student in Dahlem and Berlin. Witty and gregarious, Paul and his wife, Hilde, were part of a lively circle of scientists, theater people, musicians, and athletes; his brother Hans was conductor of the Frankfurt Radio Orchestra. In the early 1930s, Rosbaud took the position of scientific adviser to Springer Verlag, publishers of scientific reference works and periodicals; after Arnold Berliner was forced out in 1935, Rosbaud assumed much of the editorial responsibility for *Naturwissenschaften*. He traveled widely for his work and cultivated contacts among scientists in academia, industry, and the military; he had friends who hated the Nazis as much as he did, and he also had friends well placed within the regime, some with very low party numbers.[13] On the afternoon of 26 March, Lise talked to Rosbaud, and together they visited her lawyer. They decided she should ask Ernst Telschow, General Director of the Kaiser-Wilhelm-Gesellschaft, to clarify her status, including her pension rights should she be dismissed.[14]

Telschow was a party member who had replaced the non-Nazi Friedrich Glum as general director of the KWG in 1937. Hahn had welcomed the change. Glum, he thought, had tried to placate the regime by slighting the KWI for Chemistry in favor of the Nazi-dominated institutes, while Telschow, a chemist who had been one of Hahn's first doctoral students, could, despite his party affiliation, be relied on to support the institute. Telschow may have been quite loyal to his Doktorvater Hahn, but he was an enthusiastic and ambitious Nazi who diligently enforced the regime's statutes, with no sympathy for the few Jews who remained and (also in line with Nazi precepts) a distinct antipathy to professional women.[15] Meitner met with Telschow on 31 March. Hahn was present. "Telschow does not want me to leave," she noted. Carl Bosch, president of the KWG since Planck retired in 1937, had insisted that she stay.[16]

The assurances did little to calm Meitner's fears. Five years under a lawless regime had taught her that things could change from one day to the next, that she had no rights whatever, that the KWG, however well intentioned at some levels, was largely ineffectual. Control was in the hands of the government and its education ministries. With respect to Meitner, however, they had been silent.

It was impossible to assess how seriously her position was threatened. If she were dismissed, she had to leave at once. But where would she go? Where could she work? Emigration required preparation: contacts, letters, visas, setting her scientific and personal affairs in order, activities likely to jeopardize the position she still hoped to salvage. For weeks Lise was paralyzed with indecision.

Her friends abroad understood that she was in trouble. Two days after the Anschluss, Paul Scherrer wrote from Zurich: "We will almost certainly have a Congress here this summer. In case it doesn't take place, I would like to invite you to give a lecture in any case on April 23. We have a colloquium every Wednesday from 3–5 P.M."[17] The letter served several purposes. It provided Meitner with a professional (in this case, fabricated) reason for leaving Germany quickly and being admitted to Switzerland. The invitation was specific enough in the short term but extended into the summer. And it provided evidence that scientists outside Germany were concerned about her welfare.

A few weeks later a similar letter came from Niels Bohr.

> The local Physical Society and the Chemistry Association have delegated me to ask you if you would give their members the great pleasure and generous instruction by holding a seminar in the near future concerning your very fruitful investigations of the artificially induced new radioactive families of radioactive elements. As to the date, we can entirely accommodate ourselves to your convenience. It would suit us particularly well, however, if you were able to arrange to come here some time in the first half of the month of May. It goes without saying that the Physical Society and the Chemistry Association will cover all your travel expenses, and you would give my wife and me special pleasure if you would live with us during your stay in Copenhagen.[18]

Bohr had been helping refugee scientists since 1933. Like Scherrer's, his letter was expertly composed. Under the letterhead of his internationally known institute, Bohr mentioned twice that two Danish academic societies were interested in Meitner, took care not to specify a date but urged her to come soon, and anticipated possible currency restrictions by promising that all her expenses would be paid.

Early in the Hitler regime such evidence of foreign interest sometimes had an effect. But in the spring of 1938 one could no longer be sure. Over the years Hitler had successfully appeased foreign protests, and the Anschluss had provoked barely a murmur from his European neighbors. For Meitner, the "invitations" from Scherrer and Bohr were insurance, nothing more. She did not want to leave her institute or her circle of friends and colleagues; she clung to the hope that she could keep her position and stay in Berlin. Hörlein had said she should leave but then changed his mind; Bosch had urged her to stay. Never did she think of physical danger. Her greatest fear was that she would be forced from her institute, her only hope that she could somehow retain her status as an exception. But the news from Austria that April was very bad. In Vienna especially, where the proportion of Jews in academia was high, the civil service law was carried out with a vengeance. Stefan Meyer preempted his dismissal by resigning as director of the Radium Institute, but other "non-Aryans" were simply thrown out. Their former associates and colleagues eagerly vied for their positions.[19]

During Easter vacation, Max and Magda von Laue took Lise with them for a few days by car in the spring countryside. When they returned, Lise signed Laue's *Autogästebuch* (auto guest book) with some lines from Goethe: when troubled days come into our lives, God gives us sunshine, friendship, and beauty. She thanked the "dear Laue couple" for the beautiful Easter outing.[20] It would be their last trip together for a very long time.

Toward the end of April, Meitner learned that as a consequence of Hahn's talk with Hörlein, the Ministry of Education was considering her case. It was an ominous sign. On Friday, 22 April, she met Carl Bosch at the Hotel Adlon, where he stayed when he visited Berlin. She wrote in her diary, "He related what Hörlein told him about the conversation with Hahn, said he must find out [from the Ministry] what the truth is. Promises are of no use, they are not kept. Possibilities narrowing."[21]

From that moment Lise Meitner met friends and colleagues nearly every day, endlessly sifting the meager news available, weighing the alternatives. In particular, she sought out visitors from abroad as they relayed messages that could no longer be entrusted to the mail. Lest her behavior attract attention among the Mitarbeiter in the institute, she continued her normal schedule of work. In her diary she recorded,

April 23. Planck's 80th birthday. [Torkild] Bjerge [from Bohr's institute] came to the institute early, [Max von] Laue came for Hahn and me and took us to Planck's. Speeches, addresses from the Academy. Planck answered, wonderfully fresh. I gave him a photograph album. Then to the institute, had discussions with Hahn concerning Hörlein-Bosch. Noon, Bjerge and

[Arnold] Flammersfeld [one of Meitner's Mitarbeiter] with us for lunch. Bjerge very nice, open, told us that in Copenhagen there are too few people for the large apparatus. [Dutch physicist A. D.] Fokker came to the institute and then got his very nice wife, showed pictures of his children (a girl and a boy) from his first marriage. Evening with Hahns to the Planck celebration. Sat with Fokker at dinner.

April 24. Afternoon, [Annemarie and Erwin] Schrödinger [visiting from Graz] for tea.

April 25. Afternoon, tea with the Fokkers. Letter to [James] Franck [in Baltimore].

April 26. Afternoon, [Annemarie] Schrödinger. Evening, with Nelly Planck at Hertz's.

April 27. Tea at Plancks. First talked to Pl[anck] alone, he was very tired. Schrödinger visited Pl[anck] for 3 hours without saying anything about himself.

April 28. With Gertrud [Schiemann] to concert. Very beautiful Chopin.

April 29. Evening with Elisabeth [Schiemann].

April 30. Evening, [Hilde] Rosbaud with me. I wrote to Lola [Allers, Meitner's sister] and J. F. [Franck].

May 5. With Hahn early to Bosch. . . . Hahn asked B[osch] whether he had told Hörl[ein] that Hahn was very sorry about his talk [with Hörlein]. Bosch answered no, there was no point to it. Evening with the Laues.

May 7. Afternoon, tea with Planck. He wanted to know what Bosch said and [before leaving on a trip to London] took greetings for Lotte and Waltl [Lise's brother and sister-in-law who had emigrated to England].[22]

In the nearly two months since the Anschluss, Meitner slowly grasped the inevitable: she had to get out. She informed her sister Lola, who had left Vienna a few months before and was living in Washington, D.C.,[23] and she also wrote to James Franck, who was about to leave the Johns Hopkins University for the University of Chicago. As soon as he received Lise's letter, Franck submitted an affidavit, the first step for U.S. immigration. As required, he pledged to support her if necessary so that "she never could become a public charge in the United States," listed his annual salary at Johns Hopkins ($6,000) and his expected salary in Chicago ($10,000), and provided details of his life insurance policies and his own application for American citizenship.[24]

But Meitner never seriously considered going to the United States. America was too distant, too foreign. Except for her sister Lola, Franck, and a few other friends, all refugees themselves, her American connections were slight—and the country, recovering from the depression, had few positions for émigré scientists. Bohr's institute was more to her liking, and Bjerge had told her that more people were needed to operate the cyclotron. She knew Bohr's invitation was only temporary, but Copenhagen would be a haven, as it had been for many refugee scientists, until she could find a permanent position somewhere else. She loved Bohr, his family and his institute. And her favorite nephew, Otto Robert Frisch, was working there. On 9 May, she made her decision. As recorded in her diary, "O[tto] R[obert] called [from Copenhagen], whether I could come. Said yes."[25]

But at the Danish consulate the next day, she was refused a travel visa: with the Anschluss, she was told, Denmark no longer recognized her Austrian passport as valid. This was a severe and unexpected setback. At the Hahns' that evening Lise discussed the situation with Gustav and Ellen Hertz but went home "very dissatisfied" with their "superfluous talk, beating about the bush."[26]

The next morning, 11 May, Meitner "telephoned Bosch early, whether he could talk. Waited with Hahn in the [Hotel] Adlon for over an hour. Bosch very friendly, wanted to inquire with a certain Dr. Gatmien about passport regulations."[27]

Again Meitner turned to her friends and foreign colleagues.

May 12. Swedish mineralogist Prof. Quensel visited Hahn. I remembered that I met him in 1919 in Stockholm, he remembered too and asked Hahn about me right away. Knows Eva v. Bahr [Bergius] and Eva Ramstedt very well, took greetings from me also to Niels [Bohr]. In the evening Hahn left for Rome. I met Elisabeth and she spent the evening with me.

May 13. Dr. Gatmien telephoned from Vienna. After dinner Rosbaud came and then with [Vilhelm] Bjerknes from Oslo.

May 15. Dinner with Plancks. They told about their trip [to London], about Lotte and Waltl [Meitner] and Gisela [Gisela Lion-Meitner, Lise's sister, also in London].

May 16. Went to Dr. Gatmien. Was made to wait almost 1 hour. Then spoke briefly with him, quite friendly. He had promised to give me advice, which he did not do.[28]

The delay was not only maddening but frightening. The regime was in no hurry to issue Meitner a new passport, and perhaps had no intention of doing so. In Rome for a scientific meeting, Otto was quietly but urgently seeking a new position for her, but without valid travel papers she could not go anywhere. Very worried, she met again with Carl Bosch on 20 May. He decided to openly request permission for Meitner to leave Germany and wrote to the minister of the interior, Wilhelm Frick, that day.

> Honorable Herr Reichsminister! In my position as head of the Kaiser-Wilhelm-Gesellschaft, I have concerned myself with an assignment which, in our opinion, only you can decide. It concerns Frau Prof. Meitner, who works scientifically in the KWI for Chemistry. Frau Meitner is non-Aryan, but with the agreement of the Ministry of Culture has been permitted to work, as she possesses great scientific experience, especially in her capacity as a physicist. During continuous scientific work with Professor Hahn she has solved many large problems. Frau Meitner has Austrian citizenship. With the return of Austria she has become a [German] citizen, and it may be assumed that the question of her separation will sooner or later become acute. Because Frau Meitner is well known in international scientific circles, I regard it as desirable for the interests of the institute to make it possible to find a solution for this situation. Frau Meitner is prepared to leave at any time to assume a scientific position in another country. She has received proposals of this kind. It is only a question of obtaining for Frau Meitner, who has an Austrian passport, notice that she may return to Germany, otherwise travel abroad for purposes of employment is impossible, or that Frau Meitner be issued a German passport. I would, honorable Herr Reichsminister, be very grateful if, in the interests of the institute and its scientific concerns, you could put me in the position of settling this situation. Heil Hitler! C. Bosch.[29]

The cards were now on the table. Meitner could do little more than wait, not knowing how long it would take or whether the authorities would respond at all. The tension was very great, and it affected even those not directly involved. Edith Hahn had always been prone to nervousness and depression; in May, she became unusually disturbed.

> May 22. I was to go to Edith; she called, I shouldn't come, Hanno had written dumb letters, etc. Hahn returned from Rome in the evening.
>
> May 31. Main meeting of the KWG. Edith wanted to come to me in the evening, to tell me things about Hanno. I refused, as I had the feeling that talking would increase her agitation.
>
> June 1. I telephoned Edith in the afternoon, she sounded very agitated. . . . 9:30 at night, Dr. Wolff called, Edith had said very strange things on the telephone. I called Edith again, spoke to her 1 hour, tried to talk her out of some of her ideas, and told her to telephone me.
>
> June 2. Edith did not call, so I called her. She sounded distant, said she

wanted to sleep. To my frightened question "not now" she answered . . .
in the evening. Spent the whole evening with her. She lay on the sofa, said
strange things about Otto, how much they loved each other, but he is so
unpsychological. . . . Said two or three times, I don't want to go crazy. . . .
She also spoke about our relationship. I comforted her and we kissed when
I left. "For the first time in 20 years we could speak about everything!" she
said.

June 3. Called Edith, she seemed quieter. I proposed that we go with Frau
Rosbaud to the movies. . . . Edith laughed a lot, then apathetic again.

June 4. Morning: I called Edith, calm. Half an hour later Frau Rosbaud
bicycled over, said Edith kept repeating on the telephone, "Lise, Lise, the
great misfortune has happened." I took Hanno with me. Afternoon: Otto
came. After much telephoning, Prof. Zutt took Edith, completely bewildered, to his clinic.[30]

Edith Hahn had suffered a severe nervous breakdown and would remain
hospitalized for months.

Now Lise's own difficulties came to the fore. Weeks had gone by since
Carl Bosch had requested permission for her to leave Germany; with each
passing day, the outlook grew bleaker.

News spread to Lise's friends and colleagues outside. In May, Dirk
Coster wrote from the Netherlands, urging her to spend the summer with
him and his family in Groningen. A most unusual academic couple, Dirk
and Miep Coster lived by their socialist convictions, always ready to extend
a hand to others. Although they had four children and Dirk was often in
poor health, they had given help and a temporary home to a number of
German-Jewish refugees since 1933.[31] Knowing this, Lise wrote, "I would
accept your loving invitation without hesitation, but my situation at the
moment is such that I cannot travel at all, or at least am very uncertain it
would be possible."[32]

From Zurich, Paul Scherrer wrote again, this time more forcefully.

I return to my letter at the start of the semester in which I invited you to
lecture to our colloquium or seminar. We were all very disappointed that
you didn't come after you said you would. . . . Next Friday is our last seminar
about nuclear questions. . . . So gather yourself together and come this week,
by airplane it is only a short hop. You could give your lecture Wednesday
or Friday, 5–7 P.M.[33]

But even the most insistent "invitations" were of no use; Meitner could
not enter Switzerland or any other country without valid travel documents.

On 6 June, Niels and Margrethe Bohr passed through Berlin. Lise met them for lunch, then went with Bohr to see Peter Debye. "D[ebye] told Bohr there is time, there is not great hurry [for me to leave]."[34] Debye, a Dutch physicist who had spent most of his professional life in Germany, was director of the Kaiser Wilhelm Institute for Physics in Dahlem; the dedication ceremonies for the new institute had taken place just a week before. Perhaps the afterglow had left Debye with an artificially rosy view of the political situation.[35]

Bohr, however, was alarmed. On his return to Copenhagen he began seeking a position for Meitner in the Scandinavian countries and asked the Leiden theoretical physicist H. A. Kramers to notify Dutch physicists that she was in urgent need of help.[36] Kramers immediately contacted Coster in Groningen and Adriaan Fokker in Haarlem. "Lise Meitner will probably be thrown out of Berlin-Dahlem shortly," Coster wrote to Fokker on 11 June. "It would be splendid if she could work in Holland for a time," perhaps with a subsidy from the Lorentz Foundation.[37] Fokker had seen her in April in Berlin. "I told you already a month ago that Lise Meitner is very unstable in Dahlem," he reminded Coster. "At the time I doubted we could find the means to offer her a place here. I am glad that with the attention of Hans Kramers, we can now be more hopeful."[38] With that, the two physicists began an intense campaign to find support for their friend and bring her out of Germany.

Fokker and Coster both knew that university positions were virtually unavailable for foreigners. Laboratory space was not a problem, however, either in Groningen or Haarlem. "Perhaps we can tap colleagues for regular contributions," Coster suggested. "I am prepared to commit myself for 5 years to an amount between f.50 and f.100 per year. If Lise Meitner could work in Groningen, there would also be a grant of about f.500 per year out of Groningen [University] funds. She would have outstanding working conditions here, as we have a neutron generator of 400,000 volts, i.e., equivalent to more than 100 grams of Ra + Be. I would like to have her here, but would not make my personal commitment dependent upon that." It was essential, Coster added, that they move quickly. "I have given my word that if I should get the impression that there is nothing for L.M. in Holland I shall let Bohr know in a week so that he can seek help in Denmark or Sweden. But I would regret it very much if we couldn't get her to Holland."[39]

Coster hoped that large industry could afford to be more generous than the universities, which were chronically short of funds.[40] To A. Bouwers, head of the x-ray division of N. V. Philips in Eindhoven, he wrote, "I have heard that Lise Meitner will most probably be dismissed. As long as she remains in Germany she will receive her full pension but . . . if she leaves Germany she loses all her pension rights. Bohr is trying to find something for her in Denmark or Sweden, but is not at all sure he can. He would be very grateful if we could offer her a five-year stipend here in Holland . . . but it seems to me f.2000 a year is the absolute minimum she needs."[41] In a similar letter to Professor G. J. Sizoo in Amsterdam, who had worked in Meitner's laboratory for a time, Coster cautioned him "not to write to L.M. in Berlin since I understand that the privacy of mail is not respected in Germany."[42]

Fokker set a goal of f.20,000, enough to support Meitner for five years, and immediately began contacting colleagues for advice and donations. Some weeks earlier he had asked Johanna Westerdijk, director of the phytopathology laboratory in Baarn, to contact the International Federation of University Women on Meitner's behalf. The federation's London office maintained a small fund to help well-known refugee women, Westerdijk informed him at the time, but she had heard it was exhausted.[43] In June, Fokker wrote to Westerdijk again.

> It would be of great value to our country if we could offer her a place to work. Perhaps I was too skeptical before, thinking it was not possible. I would be happy to offer her work here . . . and Coster would be equally glad to have her in Groningen. But we need special sources for her salary, say three thousand for living expenses and a thousand for scientific materials. . . . Bohr thinks that if we can't get it here we should perhaps go to the Rockefeller Foundation.[44]

Westerdijk doubted that they would garner much institutional support: "From the budget committee of the International Federation of University Women I have learned that there are 30 applications . . . for support from Austria alone and that they have 100 pounds in their account. They will meet to make their decisions next week."[45] Fokker was optimistic, however. "I can't complain," he told Westerdijk on 17 June. Several colleagues had already responded with pledges of assistance and money.[46] "Of the necessary twenty thousand, I have already gotten three. I hope that when we arrive at 15 thousand we can ask the Rockefeller Foundation for the rest."[47] As for the Lorentz Fund, he informed Coster, "We should try first to get money from living people. . . . Lise Meitner's work is experimental, and the statutes of the Lorentz Fund explicitly state that it is to be used for theoretical research."[48]

Coster and Fokker soon realized that one week's time was not enough to tell Bohr what they could do.[49] The initial response had been prompt and quite encouraging, but the effort was daunting. Fokker began to wonder how serious Meitner's problems actually were, "whether it is certain that Lise Meitner will be dismissed, and whether she is living badly or in fear. I would gladly write to Otto Hahn about this," he told Coster, "but I don't know whether his letters are opened; if it became known that he is trying to find work for her abroad she might be dismissed at once. I think if the Nazis let her keep her position, then we should not try to get her here."[50] Fokker had heard that Otto Warburg, a Jew, had been allowed to work undisturbed in Germany; "to be sure our attempts to get L.M. are really needed," he wanted to know whether or not the Nazis would leave her be.[51]

Even before those words were written, Meitner had the definitive answer to that question. On 14 June, she learned of new restrictions on emigration from Germany. Hurriedly she noted in her diary, "Go . . . for information. Hear that technical and academic [people] will not be permitted to leave. The same from Laue, from the legal faculty." The next day, 15 June, she heard "the same from Bosch."[52]

The information had the ring of truth, but until the new restrictions were officially announced, Meitner could hope that she at least might still get out. Bosch's appeal on her behalf had been sent weeks before; there was still a chance she would be granted permission to travel. That slim expectation was laid to rest two days later, on 16 June, when Bosch received a response from the Ministry of Interior. Meitner met Bosch at the Hotel Adlon and transcribed the letter in shorthand onto hotel stationery.

> Per instructions of the Reichsminister Dr. Frick, I may most humbly tell you, in response to your letter of the twentieth of last month, that political considerations are in effect that prevent the issuance of a passport for Frau Prof. Meitner to travel abroad. It is considered undesirable that well-known Jews leave Germany to travel abroad where they appear to be representatives of German science, or with their names and their corresponding experience might even demonstrate their inner attitude against Germany. Surely the K.W.G. can find a way for Frau Prof. Meitner to remain in Germany even after she resigns, and if circumstances permit she can work privately in the interests of the K.W.G. This statement represents in particular the view of

the Reichsführer-SS and Chief of the German Police in the Reichsministry of the Interior.[53]

Plainly before her, distilled onto a single page, was everything Meitner had feared. Her "resignation" was a foregone conclusion. She was forbidden to leave. And she had lost anonymity: her case had come to the attention of the Reichsführer of the SS, Heinrich Himmler. Bosch prepared a direct appeal to Himmler, but Meitner knew she had to get out—fast. Otto Hahn asked Paul Rosbaud to try to get her a forged passport.[54] Meanwhile, Rosbaud telephoned Scherrer, who sent a telegram on 17 June: ARE YOU COMING FOR A "PHYSICS WEEK"? 29 JUNE TO 1 JULY.[55]

But with her invalid passport, she could never get into Switzerland. Holland and Sweden were considered more lenient.

At this point, Peter Debye became Lise Meitner's critical contact to friends outside. He was politically adept, non-German and head of a major institute, altogether less likely than others to arouse suspicion with his extensive international correspondence. Even so, when Debye wrote to Bohr on 16 June, he described Meitner's situation with the utmost caution, never once mentioning her name: "When we last spoke [6 June] . . . I assumed everything was quite all right, but in the meantime it has become clear to me that circumstances have substantially changed." Trusting Bohr to read between the lines, Debye went on, "I now believe it would be good if something could happen as soon as possible. Even a very modest offer would be considered and followed up if only it provides the possibility to work and to live. That is how the situation was represented to me, and it was emphasized that a poorer but earlier offer would be preferred over one that is better but later. I have taken the responsibility of writing all this myself, so that you can see that I too concur with the opinion of the concerned party." To be sure Bohr understood that Meitner's dismissal was no longer conjecture but about to be forced, Debye stressed that "even the most dispassionate observer of the situation would not come to a different conclusion."[56]

Bohr understood. On 21 June, he sent Fokker a copy of Debye's letter. Believing that "it even may be necessary for her safety to leave Germany at the earliest opportunity," Bohr asked for a reply by return post "exactly how [the] matter stands in Holland and what proposal you and Coster are able to make at the moment." To his regret, Bohr wrote, he could not offer

Meitner a position in Copenhagen, it being "quite impossible to obtain the necessary permission of the authorities on account of the great number of foreigners working already in this Institute," among them Meitner's nephew, Frisch. Nevertheless, Bohr believed, "a person with her unique qualifications should hardly [have] difficulties to find some . . . solution for the long run, if only she can get out of her for [the] present most precarious situation."[57] There was a chance that Manne Siegbahn could take Meitner into his new physics institute in Stockholm, but "the possibilities there are only small, and it will therefore be most desirable if you and Coster continue your endeavours in Holland without delay."[58]

Bohr's letter reached Coster and Fokker just as they had become quite discouraged. "At the moment I still have no results," Coster had written a few days before. There was nothing for Lise Meitner in Amsterdam, he expected little help from other physicists, he had heard nothing from industry. "I shall try to get something from the institutions in Groningen. . . . That would be limited to a few hundred gulden a year, an amount scarcely worth counting."[59] After ten days of constant effort, they had collected only f.4,000, far short of their goal of f.20,000.[60]

And Fokker had just learned from an Amsterdam colleague, Professor D. Cohen, that getting Meitner into the Netherlands would "not be entirely easy."[61] Like many countries, the Netherlands had erected major barriers to immigration.

> Only those are admitted for whom it can be shown that their presence delivers important benefits to the Netherlands. The way to go about it is to direct a request to the Minister of Justice for the admission of Mevr[ouw] Meitner. The letter will go to [the Ministry of] Education for advice, and you should at the same time inform Education, preferably in person, and convince them that her presence in the Netherlands is of great importance. I fear that even then there will be difficulty, in that the regulations in fact apply to Germany and not to Austria. But if you and your friends push hard enough it seems to me you can show that this is a very special case. . . . I hope from my heart that you succeed.[62]

Fokker asked W. J. de Haas of Leiden, a friend of Meitner's, to telephone the Ministry of Justice in the Hague for specific advice.[63] To help Fokker with the difficult and unfamiliar task of petitioning the government, Coster traveled to Haarlem on Friday, 24 June. As required, their request to the Minister of Justice took the form of a highly stylized resolution. They appealed to national honor: "Whereas the measures of the German government have already forced the expulsion of many scientific scholars and others of the first rank, who have found positions in France, England, Belgium, Denmark and America, and whereas the same misfortune now awaits Mevrouw Prof. Dr. Lise Meitner . . ." They appealed to the Dutch reputation for scientific excellence: "Whereas it would be of great value and also esteem to the development of physics in the Netherlands if a scientist of the quality of Mevrouw Meitner could work in this country . . ." They indicated that Lise Meitner would be welcome in Holland: "Whereas this proposal has the consent of the Royal Academy of Sciences in Amsterdam. . . . Whereas she is prepared to work in laboratories in Haarlem as well as Groningen . . ." Perhaps most important, they explained that Meitner's presence would not cause economic distress to a single Dutch citizen: "Whereas the money [for Meitner's support] has been donated thus far by private parties . . . she would not deprive any Dutch scientist of employment opportunities." After testifying to her mental and political stability—"Whereas Mevrouw Meitner has never been mentally ill, nor taken part in political actions or propaganda"—Coster and Fokker concluded, "These reasons commend themselves to your Excellency with the respectful request that Mevrouw Meitner be granted her request to be admitted to the Netherlands."[64]

On 28 June, the petition was sent to the Ministry of Justice. It was accompanied by a less formal but more detailed appeal to the Ministry of Education.[65] According to de Haas, Justice would view the matter most favorably if Education could make the case that the admission of Lise Meitner served Dutch interests; for this it would be best if she had a definite university position.[66] As foreigners were not permitted to work for pay, the unsalaried position of *privaat-docente* (lecturer) was the only option.[67] This required faculty consent, normally a lengthy process. It happened that a meeting of the faculty in Leiden was scheduled for 28 June; de Haas and a Leiden chemist, Dr. A. E. van Arkel, saw to it that the position was quickly approved.[68] After assuring Education that Groningen, too, would be honored to have Meitner, Coster formally requested that she be admitted into the Netherlands as a privaat-docente who would maintain "close contact with physics and chemistry students . . . in Groningen and Leiden . . . directly supervising their research and giving lectures in the universities."[69] Privately Coster warned the ministry that there was "periculum in mora"—danger in delay.[70]

There was still the problem of money. University faculty had little to spare. "I don't know how I can be of service to you," one physicist wrote from Amsterdam, "but I shall gladly be available."[71] Van Arkel expressed his willingness to help by finding students for Meitner in Leiden, but "financially I cannot do much."[72] Funds would have to come from wealthier individuals and from industry. Coster obtained the addresses of four Jewish industrialists and asked each for f.1,000. Gilles Holst, director of research at N. V. Philips in Eindhoven, pledged a personal contribution and thought the company might underwrite some of Meitner's expenses, especially if they could arrange for her to give lectures there each year. "I suggested f.500 per year, but he could not at the moment give me an answer."[73] Fokker appealed to A. F. Philips directly, referring to Meitner as "one of the pioneers in radioactivity" whose work in "our land, with our students" would be a "strong inspiration," reminding him that Meitner had in the past lectured all over Holland, "including your physics laboratory in Eindhoven."[74]

It was to little avail. For five years, Europe and America had absorbed refugee scientists from Nazi Germany, sometimes gladly, often with difficulty;[75] in 1938, the number of refugees rose sharply and resources were spread thin. "We understand the unhappy circumstances of so many unfortunate people," one industrialist wrote to Fokker, "but we have so many responsibilities to fulfill that it is no longer possible to take part in your fund."[76] Another regretted that "owing to the dire circumstances of relatives and friends in Germany and now also in Austria, the demands on me are such that I am no longer free to offer philanthropic help to people outside my own circle."[77] "Economically things are bad," one businessman responded, citing the "American debacle of 1929" and Roosevelt's "anti-capitalistic politics."[78] P. F. S. Otten, director of N. V. Philips, responded coolly: the company did support scientists but only indirectly through grants to universities. "Internal budgetary considerations as well as considerations of a political sort" made it impossible to help.[79] Fokker spoke to another Dutch industrialist, in charge of a large concern in Germany. "I could not persuade him. So many unbelievable things have happened [in Germany], he said. He asked me L.M.'s age, and said many people there were forced to retire who are much younger than she is. It all sounds very hard."[80]

Even colleagues who knew and liked Meitner found it hard to believe her situation was truly desperate. Sizoo thought he was acting in her best interests by warning Coster, "Since Lise Meitner will lose her pension rights upon leaving Germany, the responsibility one assumes by offering her a position here is very great."[81] Another friend of Meitner's, Professor H. R. Kruyt of Utrecht, declared himself "of course gladly willing" to use his influence with the government, but only "if the question really becomes acute"—and then cautioned Fokker and Coster "to think ahead carefully about what [you are] doing, especially since [the loss of pension rights] apparently cannot be reversed."[82] Clearly it was difficult for people to comprehend the injustice of life without work, to understand that Meitner could not remain in Germany a despised outcast, even with a pension from her forced "retirement." And almost no one, including Meitner herself, was able to truly "think ahead carefully" and foresee what lay ahead for the Jews of Europe.

After three weeks of constant effort, Coster and Fokker had collected only enough to support Meitner for one year; at times even they thought she might be better off staying in Berlin. They knew she was about to be dismissed and they knew of her passport and visa problems, but Debye's letter of 16 June had been so guarded that they were unaware that she had been expressly prohibited from leaving Germany and thus they did not realize how essential it was for her to get out quickly, before the prohibitions were strictly enforced. Communications were poor in both directions. "Out of fear of party spies" Fokker and Coster had never directly written to Meitner or Debye that they could support her for a year; they did inform Bohr but heard nothing in response.[83]

Toward the end of June, Coster thought it best to go to Berlin to see for himself how bad things were and if necessary to bring Meitner back with him.[84] Before he left, Fokker offered some ambivalent last-minute advice. "Don't panic! Don't let your presence in Berlin . . . make L.M. leave too hastily. Let her calmly conclude her business and pack her suitcase; remember she cannot travel alone as well as with you. Forgive me for saying this, but don't fall victim to the masculine protective instinct. . . . Think of Warburg, who says he is left undisturbed and makes no trouble. There is no axiom that says you must bring L.M. to me. Also you must

let her calmly make the decision *herself.*" Unaware of Meitner's increasingly dangerous situation, Fokker still thought that perhaps "L.M. can get
an ordinary passport, without swearing that she will never come back to
G[ermany]." His ambivalence reflected his fear: "It is always possible that
the two of us will *not* get the money!"[85]

On 27 June, Coster sent Debye a short, coded message: he was coming
to Berlin to look for an "assistant" (Meitner) to fill a one-year appointment. That day, Ebbe Rasmussen, one of Bohr's close associates in Copenhagen, came to Berlin with news for Meitner of another offer: a position
with Manne Siegbahn, whose new institute in Stockholm would soon be
finished, the first in Sweden devoted to nuclear research. Meitner met
Rasmussen in Peter Debye's home; Max von Laue was also present. Together they weighed the alternatives between Holland and Sweden.[86]

Meitner's ties to the Netherlands were much closer, but she was by no
means a stranger to Sweden, having visited several times, and she had a
number of good friends and colleagues there; although she did not know
Siegbahn well, they had been acquainted for some twenty years. (It was at
Siegbahn's institute at the University of Lund that she first met Dirk
Coster in 1921.) Moreover, Siegbahn's new institute was part of the Royal
Swedish Academy of Sciences, which, it appeared, would make it easier for
her to be admitted to Sweden and obtain permission to work.[87] For
Meitner, however, the deciding factor was that experimental nuclear phys-
ics was undeveloped in Sweden: she believed she could be of use. She
decided to take the Stockholm offer. When Coster's letter arrived two days
later, she did not change her mind.

Debye wrote to Coster,

> Dr. Rasmussen was here already on Monday [27 June]. With his visit he
> pursued the same goal as your letter, he is seeking an assistant for Siegbahn's
> new laboratory. Even though I had not yet received your letter it seemed
> to me that I could count on a similar offer from your side; I gathered this
> from a letter of Fokker's. I regret, actually, that I must write that in the end
> Stockholm won. I would have preferred that it be Groningen, but I let myself
> be persuaded by the assistant himself [Meitner], who thinks he will be able
> to accomplish more in Stockholm. . . . Of course I still let him know this
> morning what was in your letter. I knew in advance that nothing could be
> changed, but on the other hand I believed its contents would have a good
> effect on his spirits. I was not mistaken; even under the most ordinary
> circumstances appreciation is something that has excellent effects, and in this

case its effect was even better, as you can imagine. . . . What a pleasure it is for me to see what a couple of splendid Dutch fellows like you and Fokker can do![88]

Satisfied that Meitner was in good hands, Coster and Fokker notified the various donors that their contributions would not be needed after all.[89] Several commented that the Swedish offer was a good one, enabling her to be taken directly into a new working group.[90] Meitner, meanwhile, visited the Swedish consulate, then quietly arranged for her lawyer to transfer her bank account and ship her books and furniture after she was gone. Planck, she noted, was "very deeply shaken [*sehr erschüttert*]" by the finality of her plans to leave.[91]

A day or two later, however, Fokker received a somewhat unsettling letter from Bohr: the Swedish offer, it seemed, was not entirely firm.

All formalities regarding her invitation and her permission to [enter] Sweden are . . . not yet in order, and in case unforeseen difficulties should arise, I shall of course let you know. Otherwise Lise Meitner intends to go to Sweden in the beginning of August to stay with private friends until she can start work in Siegbahn's institute in the beginning of the autumn semester.[92]

Difficulties arose almost at once. On 4 July, Carl Bosch informed Meitner by letter that the policy prohibiting scientists from leaving Germany would soon be strictly enforced, in effect sealing the German borders from within. Meitner and Debye agreed: she must leave immediately.[93] Sweden was inaccessible, Holland the only possibility. Urgently, Debye wrote to Coster,

The assistant we talked about, who had made what seemed like a firm decision, sought me out once again. . . . He is now completely convinced (this has happened in the last few days) that he would rather go to Groningen, indeed that this is the only avenue open to him. He intends to keep his agreement with Rasmussen, but that is only in the future; under no circumstances can he start there right away. I believe he is right and therefore I want to ask whether you can still do anything for him. Perhaps I may now have the pleasure of showing you my laboratory. If you come to Berlin may I ask you to be sure to stay with us, and (providing of course that the circumstances are still favorable) if you were to come rather soon—as if you received an SOS—that would give my wife and me even greater pleasure.[94]

While waiting for a reply, Meitner contacted her former assistant, Carl Friedrich von Weizsäcker, to ask his father, Baron Ernst von Weizsäcker, a high official in the Foreign Ministry, about her application for a German passport. The reply was swift and negative: "Foreign Office same opinion as Ministry of the Interior." It was a useless exercise. The elder Weizsäcker in his capacity as Staatssekretär was at that moment signing orders to German embassies abroad, forbidding emigrating Jews from transferring funds out of Germany, an order that greatly increased the difficulty of emigration. And the fact that her case had now come to the attention of two ministries made it even more urgent that she leave.[95]

Debye's SOS went out on Wednesday, 6 July. It did not reach Groningen until the afternoon of Saturday the 9th. Coster understood and telegraphed: "I am coming to look over the assistant and if he suits me I will take him back with me."[96] But Coster could not say when: he had not yet received permission for Meitner to enter Holland. By the time Fokker called the Hague, it was Saturday evening and government officials had left their offices. He did reach the head of the border guard, who promised him an answer Monday morning. On Sunday they could do nothing but wait. On Monday they heard: Meitner would be admitted. Coster immediately set out for Berlin.[97]

In Dahlem, meanwhile, Monday morning came and went without a word from Coster. At noon Debye sent Fokker a frantic telegram: "WITHOUT ANSWER FROM COSTER CLARIFICATION URGENTLY REQUESTED."[98] Fokker telegraphed back: "DIRK WITH YOU THIS EVENING IN BEST CONDITION." "In best condition" was Fokker's attempt to reassure Debye and Meitner. "It is really rotten," he wrote to Kramers later, "that I could not even telegraph that we had official permission for her to enter."[99]

In Berlin, only Debye, Hahn, Max von Laue, and Paul Rosbaud knew of Meitner's plans. The secrecy heightened the tension. On Monday, Meitner told Planck's wife, Marga, "about Bosch's letter but nothing more. The same with Elis[abeth Schiemann]."[100] There were no farewells.

Coster arrived in Berlin late Monday evening, staying the night with Debye and his family. He planned to leave on Wednesday, 13 July, taking Meitner on a lightly traveled train route that crossed the border at the small station of Nieuwe Schans. One of Coster's Groningen neighbors, E. H. Ebels, was an influential local politician from a large farming family near the border. On Monday, before Coster left for Berlin, Ebels had driven him to Nieuwe Schans where they talked to the immigration officials.

Coster hoped that the Dutch border officers, who were on good terms with their German counterparts, would persuade them to let Meitner pass through undisturbed.[101]

On Tuesday, 12 July, Meitner arrived "early in the institute. Hahn tells me what Coster-Debye propose. Meet Coster in the morning with Hahn. Work in the institute until 8 o'clock at night." Coster took care not to be seen in Dahlem that day.[102] Meitner was careful, too. "So as not to arouse suspicion, I spent the last day of my life in Germany in the institute until 8 at night correcting a paper to be published by a young associate. Then I had exactly 1½ hours to pack a few necessary things into two small suitcases."[103]

Otto helped Lise pack. "Hahn very nervous—too many things. At 10:30 Rosbaud comes, we drive to Hahn's."[104] "At Rosbaud's suggestion I called Scherrer so that if I couldn't get into Holland there might be the possibility of Switzerland. Scherrer understood immediately why I called and said he was waiting for my lecture."[105]

Meitner spent the night at Hahn's house.

> We agreed on a code-telegram in which we would be let known whether the journey ended in success or failure. The danger consisted in the SS's repeated passport control of trains crossing the frontier. People trying to leave Germany were always being arrested on the train and brought back.[106] . . . We were shaking with fear whether she would get through or not.[107]

In 1907, Lise Meitner had come to Berlin as an impoverished student. Thirty-one years later, she wrote, "I left Germany forever—with 10 marks in my purse."[108] And one thing more: a diamond ring Otto had inherited from his mother. He gave it to her when they said good-bye. "I wanted her to be provided for in an emergency."[109]

Rosbaud drove her to the train station. At the last minute, overwhelmed by fear, she begged him to turn back.[110] But Coster was waiting in the train; they greeted each other as if by chance. The trip was uneventful. As they neared the Dutch border, Lise became very nervous. To help her feel unobtrusive, Coster asked her to remove the diamond ring and slipped it into his waistcoat pocket.[111] They crossed the border without incident. In her diary Lise wrote, "13 July. Said good-bye early to Hahn. Ring. Met Coster at the station. In Nieuwe Schans the customs officer was informed. 6 P.M. Groningen."[112]

It was over. Lise was out. The danger had been even greater than she realized. Kurt Hess, whose apartment in the institute villa adjoined hers, had sent a handwritten note to the authorities informing them that she was about to flee. "Fortunately there were two decent people in the notorious Sicherheitsdienst [SD, police] who authorized a scientist whose attitude they knew [probably Georg Graue] to check on this. It was only when Otto Hahn reported that Frau Meitner was in Stockholm [more likely Groningen] that the notification went back [through police channels]."[113] Max von Laue remembered, "The shot that was to bring you down in the last minute missed you. You yourself probably did not notice it. The more so because of it, I waited for news that you had arrived safely."[114]

From Groningen, Coster sent the prearranged telegram to Hahn: the "baby" had arrived, all was well. Hahn sent "heartiest congratulations," adding, "I was of course very happy about the news, as we were somewhat worried recently. What will be the little daughter's name?"[115] Coster was deluged with congratulations, including a telegram from Wolfgang Pauli: "You have made yourself as famous for the abduction of Lise Meitner as for [the discovery of] hafnium!"[116]

For the first time in months, Lise was free to think beyond the moment of escape. Relief turned to shock. Completely uprooted, she had been torn from work, friends, income, language. When Fokker came to see her, he had the feeling that "her sense of being inwardly torn apart is much worse than we can imagine. She completely suppresses it; she speaks only about factual matters on the surface, but under the most severe tension."[117] How unbearable, Miep thought, to be forced at age fifty-nine to "leap into the void."[118]

Stateless, without a passport, she did not know where she would live or how she could travel. Except for a few summer clothes in two small suitcases, she had no belongings. And she had no money, none at all. On her first day in Groningen, Coster gave her a small sum, implying that it came from "some impersonal fund or another," but she soon found out that it was in fact collected from Fokker, Coster, and a few other friends. Upset, she insisted on considering the money a loan.[119]

But money was not her greatest worry. Overriding all else was the fear that she might not find a place to work. It was physics that had kept her in Germany all those years, the loss of her position that finally drove her out. Now the Stockholm offer was unaccountably delayed. Shortly after

Meitner arrived in Groningen, Hahn relayed a message to her from "the Swede from Berlin. . . . His proposals for the evaporation possibilities of overheated gases from sealed vessels are, however, not sufficiently thought through and would not have been useful for our work."[120] Manne Siegbahn, who was of German descent (the "Swede from Berlin"), had apparently not yet obtained permission for Meitner to enter Sweden after her escape from Germany ("evaporation from sealed vessels").

From Zurich, Paul Scherrer sent Meitner his first uncoded letter in months. "I have no idea how your situation developed, as no one comes here from Germany anymore. On the telephone I could learn nothing from Dr. Rosbaud; he spoke in the style of the Koran. Pauli and I were always very concerned for you. Naturally, as an Austrian, Pauli is glad to be here."[121]

"There is a possibility of a position in Stockholm, starting in September," Meitner replied, with a composure she did not feel. "I hope it works out. You see, my situation is not very pleasant, but there is no sense dwelling on it, things will develop."[122]

Fokker and Coster could not understand why the Swedish formalities were taking so long. Having canceled their own efforts in the Netherlands, they now felt responsible for securing her position in Stockholm and pressed Siegbahn to move more quickly.[123] Meanwhile, Meitner stayed busy, spending her days in Coster's institute, her evenings with the Coster family and their friends. Her worries did not prevent her from appreciating the children in the household. In her diary she noted, "Coster's youngest son Hermann very intelligent, has absolute pitch," along with, "I am trying to find out how to get to Sweden, ship or plane?"[124]

At last there was news from Sweden: her entrance visa was approved. On 21 July, Meitner left Groningen for Fokker's home in Haarlem to be closer to the consulates in the Hague and the ports of Amsterdam and Rotterdam. But when Fokker telephoned the Swedish legation, he was told that they were authorized to admit her if she had a German passport, or if she had an Austrian passport with an explicit statement that she was allowed to return to Germany at any time.

Fokker was furious. Again he wrote to Siegbahn.

> I told [the consulate] that she only had an Austrian passport *without* that statement, and that she could never get it from the German authorities. . . .
> You will understand that if the Swedish government is not willing to admit

her without restriction, then she will be lost. I take it that you will succeed in making the point quite clear to your Minister of Foreign Affairs or to whomsoever it regards. . . . Our guest is a brave woman; still she lives under a great stress of anxiety and she will only be relieved from it when she will have set foot in Sweden.[125]

Another delay. Meitner visited W. J. de Haas and his family, old friends, in Leiden, then returned to Groningen. Finally on 26 July, she learned that her Swedish visa had been granted, this time without restrictions. To avoid passing through Germany she would fly to Copenhagen, where she would spend a few days with Bohr, his family, and Otto Robert Frisch, then continue by boat and train to her friend Eva von Bahr-Bergius in Sweden. Coster gave her 200 guilders, enough for travel with a little left over.[126] Before she left, Fokker wrote, "The depths lie open. Only they are not always calm enough for one to look within. . . . May the difficult time that awaits you until you are settled not last long."[127]

On 28 July, Lise Meitner "flew with hidden money to Copenhagen . . . fearful the whole time what would happen to me if the airplane should be forced to land in Germany."[128] But the weather was beautiful and all went well. In the Institute for Theoretical Physics, Frisch showed his aunt the new cyclotron under construction, then Niels and Margrethe Bohr welcomed her to Tisvilde, their summer home by the sea. It was midsummer, splendid weather, perfect for swimming with the children and lazing in the sun. Meitner had always considered Bohr's institute a physicist's paradise; Copenhagen was as beautiful as ever. This time nothing was the same. Memories of earlier, happier visits contrasted starkly with the painful reality of the present.

On 1 August, she left for Sweden. Again the trip was beautiful. Eva von Bahr-Bergius was waiting for her at the Göteborg station. Together they continued by train and then steamer to Eva's home in Kungälv, a small town on the west coast where Lise planned to stay until September. Exhausted and depressed, she hoped that a few weeks rest would give her the strength to plan ahead.[129]

She was not entirely certain she wanted to remain in Sweden. The Stockholm offer was the most substantial she had received: nominally for one year only, it seemed likely that it would be extended for several years, until she reached the mandatory Swedish retirement age of sixty-five. All other offers, including the one cobbled together by Coster and Fokker, had been extremely provisional. Scherrer's and Bohr's invitations had been only a pretext for getting her out of Germany; in August she received invitations from the California Institute of Technology and from E. O. Lawrence in Berkeley to give a few lectures in the fall.[130]

Meitner would have preferred England, and indeed several colleagues were trying to find something for her there. C. D. Ellis commended her as "the most distinguished woman scientist to-day . . . but owing to her having a Jew somewhere in the background she seems to be in danger of losing her post. [She was already in Sweden when the letter was written.] She has a really magnificent record of work." Ellis's letter was forwarded to F. A. Lindemann in Oxford, who was notoriously unsympathetic to women. He judged Meitner's plight "contingent rather than actual," and nothing came of it.[131]

But in Cambridge several physicists were more successful. In late August, W. L. Bragg informed Manne Siegbahn that Girton, a women's college, was prepared to offer Meitner room, board, and a stipend for one year, "but we could not promise anything for a longer time." Meitner did not understand why a women's college was involved, or why Bragg did not write to her directly but instead gave Siegbahn the option: "If you do not like the proposal, and prefer to say nothing to Miss Meitner about it, I shall quite understand."[132]

Nevertheless, Cambridge was home of the Cavendish Laboratory and a "powerful source of attraction."[133] Through her nephew, Lise tried to find out what the chances might be for the appointment to be extended beyond one year. J. D. Cockcroft, a man known for his economy of words,[134] would not speculate. "We—[P. I.] Dee, [Norman] Feather and I—*hoped* that after Girton had seen Miss Meitner they would offer her a fellowship [a permanent appointment] but we cannot say what the chances would be."[135] Meitner thought her reputation should have been well enough known by then; unfamiliar with the British negotiating style, she did not know what to make of it. She decided against the Cambridge offer. At her age, Frisch told Cockcroft, she did not wish to postpone her new start in Sweden.[136]

A "new start." The words were mockingly optimistic. Not since her arrival in Berlin thirty-one years before had she felt so alone. Knowing that thousands were exiles like herself only added to her despair. Her gratitude to Dirk Coster was marred by the knowledge that so many others, lacking influential and devoted friends, were unable to escape from Germany. She was an outcast, suspended between a past that was gone and a future that held nothing at all. "One dare not look back," she wrote to Coster from Sweden, "one cannot look forward."[137]

CHAPTER NINE

Exile in Stockholm

*I often appear to myself like a wind-up doll who does certain things
automatically, with a friendly smile, but with no real life in her.*

In the summer of 1938, only a few friends knew that Lise Meitner would
not return to Germany. On 13 July, the day she took the train to Holland,
Otto Hahn told the institute staff that she had left to visit relatives and even
wrote in his own pocket calendar, "Lise goes to Vienna."[1] A few days later
the institute closed for the six-week summer vacation. In August, the
Ministry of Education moved to dismiss her.

> Frau Professor Lise Meitner, formerly an Austrian national, is working as
> a guest [*sic*] at the Kaiser Wilhelm Institute for Chemistry. Insofar as the
> above-named has become a German national through the Anschluss of
> Austria, it must be certified what fraction Jewish blood she possesses. Pre-
> vious stipulation indicated that Frau Meitner has 25 percent [*sic*] Jewish
> blood. . . . I request an expeditious response in this matter.[2]

The order went to Ernst Telschow, general director of the Kaiser-
Wilhelm-Gesellschaft. Unaware that Meitner was gone for good, Telschow turned it over to Hahn.

> Perhaps you can speak about the situation with Professor Meitner, and let
> me know the answer. I believe that it would—now that the ministry has
> become involved in the matter—be expedient if Frau Meitner herself would
> request a leave of absence until the question of her dismissal or retention by
> the KWG is settled. Such a request would be useful in further discussions
> with the ministry.[3]

Telschow, ever the efficient functionary, was following orders from higher
up and making sure Hahn understood that it was to his advantage to do
the same.

Otto sent copies of the letters to Lise. He was exhausted by the events of the past months and fearful of what lay ahead. With Lise gone he had lost his trusted friend, the scientific leader of the uranium project, the person who had shared with him the direction of the institute. Of course, her departure relieved him of his most pressing political problem, but he was still vulnerable: as a non-Nazi his very presence kept the institute—and the KWG—from a complete Gleichschaltung with Nazi doctrine. This fueled the ambitions of those seeking his position, requiring that he weigh every move with the utmost care. "Starting work and so on will surely not be easy," Otto wrote to Lise in August. "Alone I feel quite helpless."[4]

Lise, meanwhile, had arrived in Stockholm. Torn from everything she knew, she was preoccupied with what she had left in Berlin. She had gone, but her life was still there, eerily untouched: everyone she knew, all she had, her lab coat still on a hook in her office, unanswered letters on her desk, experiments in progress. Adding to her loss was the violation of her character. She, who always tried to be upright in every aspect of her professional and private life, had crossed borders illegally, given up her position without proper notification, left without a word to her co-workers or some of her closest friends. She could not begin a new life in Stockholm until she brought some order to the one she left behind.

The institute was scheduled to reopen on 29 August. On the 23d, she wrote to Carl Bosch, president of the KWG, with a formal request for retirement, and she asked Telschow to meet her lawyer to discuss the conditions of her pension. To Otto she wrote of the

> deep divide this brings into both our lives. . . . We do not need to speak of how we feel; surely we both know how it is. Surely no day will pass in which I do not think with gratitude and longing for our being together in friendship, for our work together and about the institute. But I no longer belong there, and when I think back on the last months, it seems to me that my departure might also correspond to the wishes of the Mitarbeiter. There is no sense talking much about it, facts are facts, one cannot change them. . . . Actually I feel the need to personally thank all the Mitarbeiter and the employees of the institute for the many good years of working together. But perhaps you may think it better if you alone tell them. . . . Let us not, however, discuss the fact of my leaving, please. . . . I myself have not yet really comprehended that what I have written here is a reality, but it *is* a reality.[5]

Preoccupied with his own worries, Otto agreed—too readily—that it was "of no use to discuss these facts any longer. How difficult everything also is and will be for me, I do not have to tell you. In the institute I feel completely unsure of myself, and I am afraid of the days ahead." Kurt Hess, the ambitious Nazi on the third floor of the institute, was a constant threat, and Hahn was sure that his assistant, Otto Erbacher, and Meitner's former assistant, Kurt Philipp, both active Nazis, would expect advancement once they learned that Lise's position was vacant. A few days later he wrote, "On Monday I will first tell Erbacher and Philipp and after that I will call together all the members of the institute and tell them about it with a perspective on our 30 years together and your building of the physical section."[6] After the meeting Hahn reported the staff's reaction: Philipp asked if she might come to Berlin to get her things in order and Arnold Flammersfeld, one of Meitner's students, "wondered if [she] hadn't lost [her] nerve somewhat."[7]

Lise was enormously upset.

> If Flammersfeld asks whether I have lost my nerve, it is because he thinks I have abandoned my responsibilities. Why is it not frankly said that the ministry and Dr. Telschow proposed three weeks ago that I take a leave of absence? . . . And Hähnchen, this overly hasty *Totenfeier* (memorial service) for me, before my resignation is even in effect—what could that mean for *those* people who were perhaps somewhat attached to me? They must surely think I evaded my responsibilities if you do not explicitly tell them it was impossible for me to stay. . . . My future is cut off, shall the past also be taken from me? . . . I have done nothing wrong, why should I suddenly be treated like a nonperson, or worse, someone who is buried alive? Everything is hard enough as it is.[8]

Hahn was taken aback.

> Shall I answer you as bitterly as you have written? . . . Have you absolutely *no* idea of the developments of recent weeks? [All summer Hitler had mounted a frenzied propaganda campaign against Czechoslovakia over the question of autonomy for its ethnic Germans; now invasion appeared imminent, and war hysteria served as pretext and cover for increased terror against Jews.] Can you seriously think that anyone considers you a deserter? . . . Believe me, I know it is hard to be of good cheer when everything is so new. But you are perhaps too optimistic about our current good fortune here.[9]

Lise persisted.

> What upset me, and I beg you to understand, is the possibility that the young Mitarbeiter such as Fl[ammersfeld] might lose confidence in me. I have

regarded this work together as the best and most beautiful part of my life, and it hurts me to think these people might now think I left them in the lurch. . . . Perhaps you will understand that I am not really bitter, although I cannot say that I feel good, and I know that you also, unfortunately, do not have it easy. . . . Don't be bitter or angry, we want to help each other, not make things worse for each other than they already are.[10]

Otto had already given considerable thought to Lise's successor. He was interested in the Viennese physicist Josef Mattauch, a well-known mass spectroscopist who was in some difficulty in Vienna because, like Strassmann, he had not joined the requisite party organizations. "With him here," Hahn noted, "it would be easier to request an enlargement of the institute (laboratory, high voltage) than now. On the other hand, there is the danger that our man upstairs [Hess] would *surely* make [political] demands, so that progress might be even more difficult. . . . Erbacher and Philipp would be section leaders without sections. I can hear them already, asking which section they will get. . . . Please tell me what you think."[11]

However painful it may have been for Lise to discuss her own replacement, she was gratified that Otto was still seeking her advice. This was true even for minor matters: "[Karl-Erik] Zimen has just gotten married. Do you think I ought to give him a present, or hasn't he been [with me] long enough?" Her reply: "I think you should give him a small present, it doesn't have to be anything splendid. Just to emphasize the personal relationships a little, which are really not insignificant in your group."[12]

Meitner and Hahn were both physically and psychologically exhausted. "Even though I sleep quite well, I am as tired as if I had just finished not six weeks of vacation but a year of hard work," Hahn wrote in September. Sixteen-year-old Hanno had just been conscripted into the Hitler Youth, and Edith was still very ill in a sanatorium.[13] In Stockholm, Lise, too, was "as tired as if I never had a vacation. Things are not going very well."[14]

In fact, Lise hardly knew what to make of Manne Siegbahn or his institute. Set in the open, hilly region of Frescati, just north of the city, the Nobel Institute for Experimental Physics was in view of the palatial museum and offices of its sponsor organization, the Royal Swedish Academy of Sciences. Inside, the new institute building was spacious and modern, but instead of the expected hum of a busy laboratory, Lise found it empty and still quite disorganized. Siegbahn greeted her when she arrived, but that was all. Her pay, she found, was not an institute salary but a very modest stipend from the Royal Academy, approximately that of a young assistant. To Otto, Lise wrote nothing at first except, "It will not be easy."[15]

Lise had come to Sweden without possessions or money or even a firm idea of what to expect; unable to speak the language, chilly in her summer clothes, she was living on borrowed money in a tiny room in a residential hotel while her bank account was frozen in Berlin and her pension inquiries went unanswered. Her financial dependence distressed her greatly—already in early September she repaid some money to Dirk Coster, over his strenuous protests[16]—and she pressed Otto continuously with details of every sort, about her books and equipment, misplaced papers and household belongings, journal subscriptions and pension arrangements.

> Dear Otto! . . . Frau Prof. Lampa sent me 2 small manuscripts of her husband's about Boltzmann and Mach. They might be in an envelope in my institute or apartment writing desk or in my briefcase. Could you send them to her?[17]

> Dear Otto! So far there has been no mention of something very important to me, namely the question of my scientific "estate." I hope it is self-understood that I will receive my journals, card-files, slides, diagrams of my experimental apparatus, and so on.[18]

> Dear Hähnchen! . . . Please do not send any reprints that are not about radioactivity. . . . For the books of which I have 2 copies, give my Chadwick-Ellis-Rutherford to Flammersfeld, the Kohlrausch handbook to von Droste. . . . In room 9 are book catalogs, which I don't want, but I would like to have the equipment catalogs. . . . Please do not send the [old] bed and rug . . . and give away the plants, perhaps you would like the ficus, you gave it to me. . . . The bicycle in the cellar, perhaps that can go to Planck's granddaughter if she wants it.[19]

> Dear Hähnchen! About my furniture and books, Emma J[acobsson] offered to store all of it in Göteborg. Given my unsettled situation, that seems most sensible, and if I should ever get an apartment again, at least I won't have to pay for storage until then.[20]

But Otto could not simply ship Lise's belongings to Sweden. Typical of the harassment of the time, the Education Ministry stipulated that everything she owned, even her clothes, must remain in Germany.[21] When Hahn appealed, the Reich called for an inventory, in triplicate, of every item in her apartment: "3 glass plates, 1 cigar cutter, 1 KWG medal, 1 Leibniz Medal, 1 Emil Fischer Medal, 1 Stefan Meyer Medal, 2 linen towels, 2 toothbrush holders . . ." Purchases made after January 1933 were listed separately: "1 radio (325 RM), 6 hand towels . . ." The inventory concluded with a comment about "the very modest furnishings of this relatively large apartment."[22]

> Dear Otto! . . . Here everyone is very friendly, but I am always reminded of [the saying]: if one must rely upon the friendliness of people, one must either be very self-confident or have a great sense of humor; the first I never possessed, and the latter is difficult to call forth in my current situation. And then Max jun[ior] [von Laue] writes to me that I am to be envied. People just don't have the imagination to picture the situation that others really are in. . . . Tell a bit what's happening in the institute. . . . What is our good friend upstairs [Hess] doing?[23]

> Dear Lise! . . . Concerning your belongings . . . the three suitcases with clothing are lying there fully packed, but still open. When permission is granted they can be sent right away. . . . Hertz has told me that I should in any case try to get Mattauch for the institute to bring in some fresh blood. . . . If M. were to be "called" as an associate, he could be offered your apartment. . . . Surely this is very bitter for you to read these things. But *I* must think about it, especially since at the moment I myself am as tired and mentally exhausted as never before.[24]

> Dear Otto! . . . Don't be angry, I must write now about my situation, I feel so completely lost and helpless. And these written conversations, they are enough to drive one crazy. Here I am completely alone and can not ask anyone for advice. . . . L[eist, Meitner's lawyer] sends me a form to fill out and sign, whose meaning I do not understand at all, and which gives me the feeling that I am selling myself body and soul. He writes further that my pension is held up because I have taken up residence outside the German Reich without the approval of the KWG. So why can't the KWG give its approval? . . . I am not at all impatient, but it seems everything is going wrong[,] . . . everything at a snail's pace: just to get rid of me! I know, Hähnchen, that you have a lot to think about and that it is hard for you, but . . . after 30 years of work shall I be left without even a few books? . . . I beg you again, please don't be angry. If you only knew how my life now appears.[25]

Beyond her own troubles, Lise was concerned about her family. Walter and Lotte Meitner were struggling financially in England, her sister Gisela and Gisela's husband, Karl Lion, were about to leave Vienna for England, and her sister Auguste (Gusti) and Gusti's husband, Justinian (Jutz) Frisch, were desperately trying to find a place to go; Jutz, a lawyer employed by a publishing firm, had lost his job. At that moment the situation for Jews

in Vienna was more dangerous than elsewhere in the Reich. Adolf Eich-
mann, a lieutenant in the Viennese Gestapo, was pursuing a program of
extortion and terror that would later be a model for the rest of Germany:
Jews were dismissed from jobs and schools, imprisoned, taxed, fined,
robbed, attacked, and, not infrequently, killed. Everyone was trying to get
out. As the refugees grew poorer and their numbers larger, fewer countries
were interested in taking them in.[26] One of the first things Lise did in
Stockholm was to apply for an immigration permit for Gusti and Jutz.

> Dear Otto! . . . I know exactly how burdened you are; that just adds to the
> pressure I feel. Every time I write to you I feel nauseated about myself, but
> what can I do? . . . You are unjust when you say that I am only concerned
> about myself. I have considerable problems with my brother and sisters,
> whom I am trying to help, and I read the newspapers carefully every day and
> know how dreadful things are everywhere in the world. And no day passes
> in which I do not think of you and Edith especially, but it is impossible to
> keep telling you all this. . . .
>
> Yesterday I saw the *Z[eitschrift] f[ür] Physik* with Reddemann's article. I
> am impressed that he was able to write it without a single word of thanks
> to me. [Shortly before Meitner left, her assistant Hermann Reddemann had
> finished building a high-voltage deuteron accelerator, a project Meitner had
> initiated in 1935 and supervised closely until its completion.][27] For months
> he discussed [it] with me almost daily because at first he did not understand
> the basic principles of it. . . . He doesn't mention my name, but feels he has
> the right to thank Bosch for the money—that Bosch gave *me*, at a time that
> Reddemann wasn't even in the institute. I find it quite depressing from a
> human point of view, but perhaps you will again think I am in the wrong.
> Perhaps you can't quite understand how unhappy it makes me to notice that
> you always believe that I am unjust and bitter. . . . If you would think about
> it, it shouldn't be hard to imagine what it means for me to have none of my
> scientific things. For me, that is much harder than anything else. But I am
> really not bitter, it is just that at the moment I find no real meaning in my
> life and I am very alone.[28]

Otto, too, was at the limit of his strength. "In the second half of 1938,
because of Edith's illness and the agitation about Lise, I myself was so
terribly exhausted—psychologically—that during conversations tears
would often come, which embarrassed me greatly."[29] He had little time
for research[30] and was distracted by the fear of war. "In times like these
one's nerves are extremely tense," he wrote at the end of September. "Your
clothes will be sent. . . . The nervousness here is extreme."[31]

Hitler was about to invade Czechoslovakia; all Europe was preparing for war. In a final effort, Britain and France called for negotiations and then, in their meeting with Hitler in Munich on 29–30 September, forced the Prague government to accede to all his demands. Czechoslovakia was dismembered, but the rest of Europe was overcome with relief and even hope: Hitler had promised to go no further. Neville Chamberlain returned to England, the architect of "peace in our time."

> Dear Otto! . . . Hopefully now the world will be somewhat calmer. . . . Please send my dresses and underwear soon. I need some of them urgently. . . . Good night dear Otto, and many thanks for everything.[32]

> Dear Lise! . . . The laboratory is half deserted because many have gone to the reception for Hitler, as one is supposed to do. Yesterday evening around 6 we had a little champagne to celebrate the past critical days. . . . This past week there was in fact practically no work done, the anxiety was dreadful, and now we are, thank God, freed of this terrible pressure.[33]

A vivid description of the preparations for war came from Erwin and Annemarie Schrödinger, who fled Austria and made their way to England in the last days of September. Although Schrödinger was neither Jewish nor politically active, he left Berlin in 1933, going first to Oxford and then in 1936 to Graz, where the Anschluss caught up with him; by no means a dissident, he was dismissed for reasons of "political unreliability," a legacy, undoubtedly, of the disdain for the Nazis he had shown in 1933.[34] From Oxford a few weeks later, Annemarie wrote to Lise,

> Our "trip" was highly dramatic, as it coincided exactly with the height of the political crisis. . . . Completely unaware, we returned [from vacation] to Graz on September 8 and discovered that Erwin had been dismissed. . . . Erwin went to Vienna to talk to the ministry. During their talk, Erwin was asked, "Tell me, do you still have your passport?" Erwin remained completely quiet, but you can imagine that at this instant he realized what must be done. It was clear to him that it was only because of some bureaucratic mistake that we still had our passports. We wasted no time. On September 14 we went to Rome with three small suitcases, just as if we were going to Rome on a pleasure trip. Italy was the only country which did not require a visa, and in Rome other embassies were available. We shall never forget the friendship, admiration, and help we received! Soon we had all the visas we needed and went to Switzerland on the 22d. As the political situation became more and more critical, we wanted to get to England as quickly as possible before the outbreak of war. So on the critical night of September

27 to 28 we traveled from Zurich to England. There was a blackout in Switzerland. On the other side of the Rhine one could see the bright lights of Germany. Trains were overflowing . . . and greatly delayed. In France we passed airports with countless airplanes standing ready. All bridges and tunnels were under military guard. . . . Then in England in Hyde Park bomb shelters were being feverishly prepared, anti-aircraft guns pointed to the sky, everyone had gas masks. . . . In Paddington we heard reports of the Four Power Conference [in Munich]. That day 80,000 children were evacuated! On the 28th we reached Oxford.[35]

Schrödinger eventually accepted an appointment at Dublin's Institute for Advanced Studies, which had recently been established by the Irish prime minister, Eamon de Valera.

Meitner's anxiety about war was set aside for a time, but her financial worries were constant. She could not understand why her pension inquiries were going unanswered—colleagues who had been dismissed earlier had received their pensions without question.[36] At one point Hahn hinted that the delay might be connected to her illegal departure. "There may be some difficulty because a personal discussion about you was planned with the highest heaven [oberste Himmel, Heinrich Himmler]. When they now hear that you are not in Germany, they might perhaps be upset."[37] In fact, a massive confiscation process was under way, and neither the private sector nor government agencies were obliged to honor contractual agreements with Jews.[38]

It took months before Meitner understood that she would not receive her pension. She also came to understand, but with far more difficulty, that it was impossible for her to sustain even minimal ties to her former Mitarbeiter. For five years she and her co-workers had managed to set politics aside and work amiably together, although nearly all were party members; now Meitner was surprised and pained to find herself cut off from almost everyone except Hahn and Strassmann. Only one of her former Mitarbeiter, Arnold Flammersfeld, wrote to her; most, according to Hahn's reports, had not even asked about her. When Hahn suggested that Meitner congratulate Kurt Philipp on his promotion to professor she responded angrily, "I will not write to Philipp. When a man who has worked for and with me for 15 years—whom I helped whenever he needed it to the best of my ability and knowledge—does not feel the need to react to the end of those times with a single line, then I really cannot congratulate him on his professorship as if nothing has happened."[39]

Deeply hurt, Lise would feel estranged at times from even her closest friends: "It is always a little difficult with [Elisabeth Schiemann]. A few days ago I got a letter from Gertrud [Elisabeth's sister] that was so warm and friendly and natural . . . but Elisabeth never wants to know how I *really* am, rather she wants to be *assured* that everything is going *well.* Of course, that is not a lack of friendship on her part but a general reluctance to face reality."[40] Lise knew that Elisabeth was outspokenly opposed to the oppression of Jews and not without her own difficulties as a woman in academia under the Nazi regime. But their friendship suffered. Later Elisabeth would recall that Lise's letters were "wounded in tone, often bitter, indicating how hard it was for her to adjust to her new situation."[41]

At the core of Lise's unhappiness was her dissatisfaction at work. "The Siegbahn institute is unimaginably empty," she wrote Otto in October. "A very fine building, in which a cyclotron and large x-ray and spectroscopic instruments are being prepared, but with hardly a *thought* for experimental work. There are no pumps, no rheostats, no capacitors, no ammeters— nothing to do experiments with, and in the entire large building four young physicists and a very hierarchical work organization."[42] And in that organization, Meitner seemed to have no place. Neither asked to join Siegbahn's group nor given the resources to form her own, she had laboratory space but no collaborators, equipment, or technical support, not even her own set of keys to the workshops and laboratories.[43] Later it would be evident that Siegbahn never considered her a member of his staff—in his annual reports he would always list her as "apart from the institute's own personnel"[44]—but at first Meitner did not understand why she was so coolly received or why her position was neither defined nor discussed.

In her own institute Meitner had enjoyed the status of professor, an excellent salary, students and assistants at her command, technical help, collaborators of her choosing. Success had not made her humble; her years at a major institute in a large scientific community had not prepared her for the smaller scale of Swedish science. She knew Manne Siegbahn was a major figure in Swedish physics, a superb instrumentalist who had been awarded the 1924 Nobel Prize in physics for his precision x-ray spectroscopy. She may have thought he had far greater resources at his disposal; it is doubtful if she appreciated his career-long struggle for funding, or how hard he had fought to establish his new institute. When Siegbahn turned to nuclear research in 1937, his highest priority was the construction of a cyclotron and other particle accelerators. Although Meitner certainly did not disagree with this, and in fact had just completed her own accelerator in Dahlem, Siegbahn may well have regarded her as old-fashioned; eight years older than he, she had come to nuclear physics much earlier and had made important discoveries with simple equipment. He had always tied his experiments to the advancement of his instruments; she had looked for problems where theory and experiment progress together. She assumed he would be glad to have her; he may have thought she would be content with laboratory space and nothing more.[45]

When Bohr visited Stockholm in October, Meitner welcomed the chance to talk to him. "Of course, Bohr's visit was very comforting," she told Otto, "and possibly it was also useful, as he spoke to Manne S. But S. is very uninterested in nuclear physics, and I am very uncertain whether he wants to have independent people around him. Bohr advised me to be patient, what else could he do? And I can do nothing else but live my life as it actually is. Have my clothes been sent?"[46]

Lise's clothing finally arrived in mid-October, subject to a substantial customs levy.[47]

> If I extrapolate to the value of my furniture and books, then it will be a sum I simply cannot pay. My salary is such that I can pay for my room, food, and small daily expenses like bus fare, postage, etc., only by being very thrifty. Stockholm is very expensive, and I dare not think of what might happen if I should become sick. . . . Could you send me my fever thermometer? I miss all those little things, and I don't want to buy too much. To get phanodorm [sleeping pills], for instance, one must have a prescription.[48]

Lise was quite aware of the burden she was imposing on Otto—"I am always sorry to be such a bother," she would write—but she could not help it.

> I would be grateful if you would have my books packed separately and *insured*. It is hard for someone else to carry out a move to another country, especially as each of you has so much to do. And it is hard for me, too, since of course things happen quite differently than if I would do it. Please understand, it is not ingratitude. It is just inevitable. Just try for once to imagine Edith or Elisabeth in my situation, who are such individuals and are of the opinion that there is only one right way to do things. People (including me) are so strange, they have certain small wishes, even when their head is full of troubles. I hope, Hähnchen, you understand that what I have just

written is supposed to be an apology for the special wishes I sometimes have.[49]

She had still another.

> Please Hähnchen, one more thing I ask of you. I hope that no one will think of giving any special meaning to my birthday this year [Meitner's 60th]. . . . It would not give me the least pleasure if people here would take notice of it; on the contrary, it would *only* be unpleasant and would make me unhappy.[50]

In October, Meitner was asked to lecture in Göteborg on the neutron irradiation experiments with uranium and thorium. She asked Hahn to bring her up to date.

> 23.X.38. Dear Otto! . . . We reported several methods for separating the 2.5h [ekaPt] from the 66h [ekaIr]. . . . The heating method must be right or isn't it? What do you and Strassmann think of it, and what can you say with certainty? And do you believe in the 3.5h substance?[51]

Meitner's references to ekaPt concerned some doubts Hahn had expressed about its separation by sublimation from ekaIr. The 3.5-hour substance was a different and, as it turned out, a far more interesting matter. This was the fairly intense new beta activity that Irène Curie and Pavel Savitch had found in late 1937, when they irradiated uranium with neutrons and searched the entire mixture without separating the "transuranes," as was always done in Berlin. Curie and Savitch first supposed that the new activity was thorium, but after Meitner informed them in January 1938 that she and Strassmann had searched for thorium but could not find it, Curie and Savitch noted that the 3.5-hour activity followed a lanthanum carrier and attributed it to actinium. After failing to separate it from lanthanum, however, they proposed that it was a transurane with lanthanumlike chemical properties, a hypothesis, they admitted, that was difficult to interpret.[52] The Berlin group, thinking that the French had been led astray by some contamination, dubbed the 3.5-hour substance Curiosum and did not pursue it. But they could not ignore it entirely. Curie and Savitch maintained its existence, and it was still very much on Meitner's mind.

When she wrote to Hahn on 23 October, Meitner had not yet seen Curie and Savitch's latest publication in *Journal de Physique et le Radium*. In an attempt to separate the 3.5-hour activity from its lanthanum carrier, the Paris team found that it followed lanthanum rather than actinium but could be separated to some extent from the lanthanum fraction: "Taken together, the properties of the 3.5-hour substance are those of lanthanum, from which it cannot be separated except by fractionation"; for such a substance to be a transurane "would imply some singular irregularities in chemical properties."[53] None of this was new, and Hahn simply glanced at the paper, assuming that the French were, as usual, muddling things with their second-rate chemistry. Offhandedly, he passed the journal to Strassmann. But in this paper Curie and Savitch described their experiments in detail for the first time and gave decay curves for the 3.5-hour substance. Reading it, Strassmann realized that the French really did have something new, that "the experimental conditions in Paris and Berlin were comparable, and the observed effect was too large to be attributed to an 'infection' (contamination due to unclean procedures)." Suspecting that the 3.5-hour activity might contain radium, Strassmann decided to repeat the experiment using his own, cleaner method for the separation of radium.[54]

25.X.38. Dear Lise! Despite dreadful weariness I want to answer your letter of the 23d, at least briefly. . . . Toward the end of last week a new paper by Curie and Savitch about the 3.5-hour substance appeared. . . . Curves are also shown. We are now working on reproducing them and we do now believe in their existence. According to Curie's results we have found the substance, perhaps even better than Curie and Savitch. . . . Curie makes remarkable claims about its properties. The properties do in fact appear to be remarkable. The substance is certainly not identical to the 2.5-hour substance. (Perhaps a Ra isotope has something to do with it. I tell you this only with great caution and in confidence!) . . . A great pity that you are not here with us to clear up the exciting Curie activity.[55]

28.X.38. Dear Otto! . . . Today I obtained the Curie paper and find many statements hard to understand. She gives the intensity of the 3.5-hour [relative] to the 16-minute [ekaRe] as 1:2. That makes it a very intense substance. . . . The very hard beta radiation is also remarkable for the long half-life. Why didn't we find this substance in UX [^{234}Th, used as a carrier for the suspected thorium] when we repeated the Curie experiment [in January 1938], if it does not precipitate with H_2S in 2N [N denotes *normality*, a unit of concentration for solutions] or even in ½N HCl? . . . Please write about Edith, and answer my questions about the 3.5-hour substance.[56]

Meitner could not understand how she and Strassmann could have missed such an intense activity. But in January 1938, when they searched the filtrate—the solution remaining after the transuranes precipitated in acidic H_2S—they were looking only for thorium. Strassmann now thought that the activity might be a mixture of radium and its decay products.[57]

> 30.X.38 Dear Lise! . . . With thorium (UX) [as carrier] we could not find it—it isn't thorium, of course. At most thorium is gradually produced from it. According to Curie it is as capable of being enhanced [yield increased by using slowed neutrons] as ekaRe (16 min). . . . If anything more definite turns up next week, I'll write again.[58]

> 1.XI.38 Dear Otto. . . . How are your experiments going? [In Göteborg] I will definitely say only that I. Curie has found those substances and you are trying to verify them. But I would like to know for myself how your experiments are going. . . . I hope, Hähnchen, that after 30 years of work together and friendship in the institute, that at least the possibility remains that you tell me as much as you can about what is happening there.[59]

> 2.XI.38 Dear Lise! . . . I am certainly not keeping secrets from you about work or institute matters! We really could not (or cannot) say anything definite about the 3.5-hour substance in 2 weeks, despite working day and night, when Curie has been at it for 1½ years. We are now *almost* convinced that we are dealing with several—2 or 3—radium isotopes, which decay to actinium, etc. You know from the thorium work that one cannot be really sure until many trials are done. Perhaps we will feel sure enough by Sunday so that on Monday [7 November] . . . we can write a letter to *Naturwissenschaften*. Because the finding—radium probably by way of an alpha-emitting thorium, perhaps also capable of being enhanced—is really so interesting and improbable that we would like to publish it before Curie gets to it, and before she hears of it from anyone else. That was the reason I wrote that you should please not say . . . anything about it.[60]

Hahn and Strassmann had found several radium isomers and their actinium decay products. The formation of radium, Hahn was suggesting, was the result of a two-step process:

$$^{238}U + {}^1n \rightarrow \alpha + {}^{235'}Th \rightarrow {}^{231}Ra + \alpha$$

If true, it was highly unusual, since (n,α) reactions with slow neutrons had never before been observed, and a neutron-induced *double* alpha emission seemed unimaginable.

Otto continued,

> Naturally we would like it very much if you would think about the situation, how an alpha transformation can come about, probably also with slow neutrons, and at the same time produce several [radium] isomers. As you can

imagine we are making many trials, so that we can rule out errors. On the other hand, it's hot: [Leslie] Cook [a former student of Hahn's], for instance, wrote from Cambridge that [Norman] Feather, [Egon] Bretscher, an Indian, and he are working with strong neutron sources on uranium reactions or will do so.[61]

4.XI.38 Dear Otto! . . . I am extremely eager to think over how Ra or Ac isotopes could be produced if you would only write more factual details. I would definitely say nothing to anyone. Why is it that you think that there are several substances, did you obtain several half-lives? Why do you think it can be enhanced? Did you get considerably more with slow neutrons? And how strong is the activity with equivalent radiation . . . compared to the 16-minute [ekaRe]? Why do you think there are several isomers? Are more than two substances observed? Please be nice and answer these questions, even if it's not so definite.[62]

Meitner was astonished: several half-lives? several isomers? she asked again and again. She could scarcely believe it. An entirely new set of isomers, with uranium already involved in more reactions than they could explain? With her questions about the relative intensity of the 3.5-hour substance to the 16-minute ekaRe, she was trying to determine whether the starting nucleus was a uranium isotope other than ^{238}U. In 1937, she had measured the absolute yield of ekaRe and had concluded that ^{238}U was the nucleus of origin for all the transuranes. On 5 November, Hahn replied with more details: reaction conditions for the new radium isomers were essentially the same as for the transuranes, indicating that the effective nucleus was, once again, ^{238}U.[63]

Hahn and Strassmann submitted their results to *Naturwissenschaften* on 8 November; as always in their joint publications, Hahn wrote the report. He listed three radium and three actinium isomers and noted in passing the resemblance to the thorium experiments reported by the Berlin team earlier in 1938. In both cases it appeared that the process that produced radium was the neutron-induced expulsion of an alpha particle—with one major difference: for thorium only fast neutrons were effective, while the new uranium process was favored by slow neutrons. "Here, surely for the first time," Hahn emphasized, "is a case of α-*particles being split off by slow neutrons*."[64]

The seventh of November was Lise Meitner's sixtieth birthday. In normal times there would have been letters and telegrams, a special issue of *Naturwissenschaften* perhaps, a festive dinner with friends and colleagues.

Now Lise was sure that any celebration would only remind her of what she had lost.

But Otto Robert Frisch, not wishing his aunt to spend her birthday alone in a hotel room, wrote to Oskar and Gerda Klein, who gathered some old friends and new acquaintances for an afternoon tea in her honor. It turned out to be a pleasant afternoon. Meitner had met Oskar Klein in Copenhagen in the 1920s and had come to know him well when she tested and verified the scattering formula he had developed with Yoshio Nishina in 1928. An erudite, humorous man, Klein grew up in Stockholm the son of a liberal, scholarly rabbi, married a Dane, and became one of Sweden's few quantum physicists and then one of its youngest professors when he was appointed to Stockholm University in 1931 at the age of thirty-seven.[65] As a modern theoretical physicist, however, he was not a member of the inner circles of Swedish physics, and he was genuinely glad to have Meitner in Stockholm.

That afternoon, the Kleins invited Gudmund Borelius and his wife, Magnhild, friends of Meitner's since their first meeting in Siegbahn's institute in Lund in 1921. Borelius, a solid-state physicist, was professor at the Royal Institute of Technology a kilometer or so down the hill from Siegbahn's Nobel Institute. There were some new faces as well. Meitner struck up a conversation with Lilli and Hans Eppstein, a pianist and musicologist who had recently emigrated from Germany. Despite an age difference of thirty years, they would quickly become very close friends.[66]

And her birthday was not forgotten elsewhere. Otto sent an album with photographs, "to evoke memories of our 30-year path through life together, memories that come clearly to me again and again in the institute each day."[67] Max von Laue wished her "beautiful results, as so often happened here, right up to the end."[68] From Princeton, Einstein congratulated her for leaving "the dear and grateful Fatherland. . . . I myself am glad to be here and would be truly happy if the human suffering and vileness were not so depressing."[69] Friedrich Hund wrote, as did Hans Geiger: "On some of your work belongs a little star." From the Cavendish laboratory in Cambridge came a joint letter from Bragg, Cockcroft, Dee, Feather, and Aston; a notice appeared in *Nature*.[70] Werner Heisenberg expressed gratitude for Meitner's work, which "so greatly enriched our science and thereby our entire lives," but he ended on a strange note, thanking her "for having done this work in Germany, and for all you have done for German physics."[71] By "German physics" Heisenberg did not, of course, mean the "Aryan physics" of Lenard and Stark (although in other contexts the terms were used synonymously). Nevertheless, the nationalistic tone of Heisenberg's birthday greeting was somewhat peculiar, given the circumstances, and stood out unpleasantly from the rest.

Arnold Berliner was still in Berlin. The former editor of *Naturwissenschaften* had aged considerably since he was dismissed; now he kept to his apartment, as much an exile from his former way of life as Meitner was from hers. "Although I can no longer distinguish myself from other well-wishers by a special issue for the seventh of November," he wrote, "I have of course done so in my thoughts. . . . Now it makes me happy to think that I was able to be at your service now and then and that I did it with pleasure. . . . I would do it again and with the same disposition, if only . . . yes, if only."[72]

By 1938, the condition of Jews in Germany had become much worse. Step by step the Reich was bringing them under its complete control: liquidating Jewish businesses, rescinding the licenses of Jewish physicians and lawyers, requiring Jews to register their property, pay special taxes, assume the middle names Sara or Israel, carry special identity cards at all times. As more and more Jews tried to leave, the gates to other countries began to swing shut. Switzerland, for example, had a long-standing nonvisa agreement with Germany, but in the summer of 1938, alarmed by the prospect of a Jewish deluge, they demanded some form of travel identification. The Reich obliged by stamping the passports of all German Jews with a large red "*J*," which immediately restricted their entry into Switzerland and later served to identify Jewish refugees in France, Belgium, and other countries subsequently occupied by Germany.[73]

Until then, nothing in the history of German anti-Semitism matched the government-incited pogrom of 9–10 November. Throughout Germany that night mobs attacked Jews, destroyed their homes and property, burned synagogues to the ground—a night of destruction so great it became known as *Kristallnacht* for the shattered glass that filled the streets. In the days that followed, the Nazi regime barred Jews from public places, expelled children from schools, and fined the already impoverished Jewish community one billion marks for the damage of Kristallnacht. Most terrifying was the incarceration of thirty thousand Jewish men in concentration camps throughout Germany who could be released only if they emigrated immediately.[74]

On Tuesday, 8 November, Hahn was in Vienna to give a lecture and also to talk to Josef Mattauch about coming to Dahlem as Meitner's successor. On the evening of 9 November, he had dinner in his hotel with the last of Lise's relatives in Vienna, Gusti and Jutz Frisch and Gisela and Karl Lion. Unaware of any disturbances, he had a "fine evening"; the next morning in the Physics Institute there still seemed nothing amiss. That afternoon, however, Gusti called: Jutz had been arrested.[75] After promising Gusti he would speak to Otto Robert, Hahn took the night train back to Berlin.

Knowing only what she read in the newspapers, Lise arrived in Copenhagen on 10 November. She planned to stay a week,[76] and although the trip had been a struggle to arrange—her invalid passport made it difficult to get a visa—she very much needed to leave Stockholm, if only briefly, for the friendship and stimulation of Bohr's institute. A visit with Niels and Margrethe Bohr always did her good; although she had not yet learned of Jutz's arrest, she also wanted to talk to Otto Robert about his parents' situation. In addition, Bohr had invited Hahn to come to Copenhagen, ostensibly to give a talk, and she was eagerly looking forward to seeing him.[77]

When Hahn's train pulled into the station at 6:48 on the morning of 13 November, Lise was there, four months to the day since they last saw each other in Berlin. They had breakfast in Hahn's hotel and talked for hours, went to Sunday dinner at the Bohrs', and returned to the Bohr home, the splendid Carlsberg Residence of Honor,[78] for a gathering that evening. Mindful of the German image abroad, Hahn was pleased by the "very nice reception." Mostly it was a day of intense talk. "Long conversations with Lise, Bohr, etc.," Hahn noted in his pocket calendar. The next morning, after breakfast with Niels and Margrethe Bohr, Lise and Otto Robert took Hahn to the train and saw him off on the 11:13 to Berlin.[79]

Meitner and Hahn kept their meeting an absolute secret outside Copenhagen: they did not mention it in their letters, and Hahn did not speak of it, not even to Strassmann, when he got back to Dahlem. The secret was buried, so much so that years later, when the fear was gone and the need for subterfuge was over, Hahn still said nothing about his meeting with Lise, although in his memoirs he did recall that he went to Copenhagen and that he talked to Frisch, and he always remembered that Bohr was skeptical, indeed "quite unhappy" about the new radium isomers.[80] There can be no doubt, however, that the person who was most skeptical, the one who commanded Hahn's closest attention, would have been Lise Meitner. Face to face, in the strongest possible terms, she must have told him that for all her trust in his and Strassmann's chemical expertise, their new results were a physicist's nightmare, a catalog of everything wrong with the transuranes, and more. We know that it was she who urged Hahn to rigorously reexamine their findings. This was the message Hahn took back to Berlin.

The radium results presented several severe problems in interpretation. As a group, they added to the sheer number of species—sixteen at latest count—that originated with ^{238}U; the multiple isomerism and especially the *inherited* multiple isomerism of radium and its decay products were as inexplicable for radium as it had been for the transuranes. But the most serious difficulty with radium was the mechanism of its formation. For uranium (element 92) to be transformed into radium (element 88), a uranium nucleus would have to expel two alpha particles (helium nuclei); this could not happen, according to theory, unless the neutron that initiated the reaction was highly energetic.[81] Meitner was keenly aware of the parameters for such reactions. Earlier in 1938, in her study of the fast-neutron reaction with thorium, she had concluded that an (n,α) reaction was theoretically possible but only because the neutrons were fast, with just enough energy, according to her calculations, to chip an alpha particle from the thorium nucleus. Had the neutrons been any slower, they would simply have been captured instead.[82] It was at just this point that the radium isomers defied explanation: Hahn and Strassmann had found, as had Curie and Savitch before them, that the process leading to radium was significantly *enhanced* when slow neutrons were used. Something was very wrong. Meitner urged Hahn to test for radium again, more rigorously than before.

Hahn did not tell Strassmann he had seen Meitner in Copenhagen, but Strassmann understood that the directive had come from her. He thought she had written Hahn a letter; he still regarded Meitner as the intellectual leader of their team. Later Strassmann remembered, "In any case (according to a statement from O. Hahn) she urgently requested that these experiments be scrutinized very carefully and intensively one more time. . . . Fortunately L. Meitner's opinion and judgment carried so much weight with us in Berlin that the necessary control experiments were immediately undertaken."[83] These were the experiments that led directly to the discovery of nuclear fission.[84]

Stockholm is dark in November, the days brief and gray. Meitner adapted to the gloomy climate far more readily than to the atmosphere in Siegbahn's institute. "When it snows it is much better," she told Otto;[85] at work there were no comparable signs of improvement. Collegial rapport was not extended to her; equipment and materials were restricted. Having to ask for even the smallest item made her feel permanently unwelcome. She began to blame herself for coming empty-handed to Stockholm. "It would be so much nicer," she told Otto, "if, instead of a request, I could come to Siegbahn with an offering, so to speak, even if it were only a drawing for apparatus."[86] The offerings she did bring to Sweden—experience, talent, ideas—did not seem to interest him.

It was not her greatest worry. After Kristallnacht, Jutz Frisch was imprisoned in Vienna and then shipped to Dachau. Emigration was the only way to get him out, and Lise was desperately trying to obtain a Swedish visa for him and Gusti. "One always thinks life in this world cannot get much harder," she wrote to Laue on her return from Copenhagen, "but one is mistaken."[87]

"I can hardly tell you," Otto wrote, "how depressed I also am by Jutz's situation. And perhaps mostly because of *your* fear and worry for him." Otto had talked to Carl Friedrich von Weizsäcker in the hope that his father, State Secretary Baron Ernst von Weizsäcker, might help, but "the old v. W. is on vacation. Anyway, it is unlikely that he would have intervened or would have been able to intervene." Hahn was able to write more openly than usual, as Bohr's associate Ebbe Rasmussen was visiting Berlin and would mail the letter to Meitner from Copenhagen. "We are all of the opinion that with these mass arrests it would be impossible to have the cruelty and harshness that was the case earlier, when [political] undesirables came to the c[oncentration] camps. But naturally this is small comfort."[88] For some reason Hahn believed that large numbers of Jews would be treated better than smaller groups of Communists or pacifists.

Meitner wrote to Hahn in early December,

There is little or nothing to tell about me. With Gusti everything is unchanged [i.e., Jutz was still in Dachau]. And besides that? I often appear to myself like a wind-up doll who does certain things automatically, with a friendly smile, but with no real life in her. From that you can tell how useful my work is. And yet in the end I am grateful for it, because it forces me to collect my thoughts, which is not always easy.[89]

She began building some apparatus and continued to ply Hahn with questions about the radium isomers.[90]

The Nobel ceremonies take place each year in Stockholm on 10 December, the anniversary of Alfred Nobel's death. In 1938, the physics prize went to Enrico Fermi for his neutron irradiation of the elements, for the new radioactive substances it produced, and for discovering the effects of slow neutrons. In his Nobel lecture, Fermi spoke confidently of the first two transuranium elements, 93 and 94, which the Rome group had named ausonium and hesperium, and of the others up to 96, which had been studied extensively in Berlin.[91] After the festivities, Fermi and his family would leave for New York, never to return to Italy. Although heaped with honors, they too were refugees. Perhaps that was why Laura Fermi noticed how worried and tired Lise Meitner looked, with the "tense expression that all refugees had in common."[92]

CHAPTER TEN

The Discovery of Nuclear Fission

What you wrote about our uranium and thorium studies being a necessary
preparation for your and Strassmann's beautiful end-results is quite right. . . .
And perhaps it would be nice if you could somehow express this quite clearly
in your next paper.

At the very moment that Enrico Fermi was speaking to a festive Nobel audience, the work he had begun in Rome was reaching a dramatic and unexpected climax in Berlin. Otto Hahn and Fritz Strassmann were racing to verify their earlier findings; by early December they thoroughly characterized what appeared to be three radium and three actinium species, determined their half-lives, and eliminated neighboring elements from consideration. As before, they assumed that activities following a barium carrier would be radium, and those following lanthanum and zirconium, respectively, would be actinium and thorium. Then, spurred by the objections Lise Meitner had voiced in Copenhagen, they began a quite different test: an attempt to verify the presence of radium by partially separating it from its barium carrier.[1]

Meitner and Hahn continued to correspond about the radium findings.[2] Her own progress was far from satisfactory. "My so-called work—if I think about it, I can only think and say, Oh God! I have gradually put together a single needle electrometer, a few usable counters, an amplifier, a very poor automatic counter, absolutely no help. Most of the time my head is so full of other things I don't care about the rest."[3]

Neither could concentrate fully on work. After Kristallnacht, Germany was volatile and dangerous; Otto's letters were more cryptic than before. "Unfortunately, there is still nothing new to report about Jutz [Frisch, still in Dachau]," Otto reported in December. "Here quite a few have come back [from concentration camps]. One hears that others have the same fate [as Jutz], e.g., our argon and krypton supplier. He wanted to go [emigrate], but has technical expertise [the same prohibition that prompted Meitner to flee]. However our former colleague above, Richard, later in Munich, is fine at the moment in Switzerland." Actually, Richard Willstätter was still in Munich and not fine at all: The day after Kristallnacht he had narrowly escaped being sent to Dachau, and after that he desperately tried to get out of Germany.[4] Hahn concluded with a message from Arnold Berliner, who hoped Meitner would be awarded "the prize of the noble physicist."[5] Such circumlocutions were necessary: Nobel Prizes had been forbidden in Germany after Carl von Ossietzky, a brilliant political journalist and pacifist who had been in a concentration camp for several years, was awarded the Nobel Peace Prize in 1936.[6]

Otto himself was occasionally mistaken for a "non-Aryan," the result, he would later joke, of his three "Jewish" traits: born in Frankfurt, named Hahn, worked in Dahlem.[7] In December 1938, however, it was not amusing to find his name listed in the defamatory traveling exhibit "The Eternal Jew"; not knowing if it was a mistake or a threat, he placated nervous KWG adminstrators with a new set of affidavits attesting to his "Aryan" heritage.[8] Hahn thought it was due to his resignation from the University of Berlin in 1933, but rumors persisted that he indeed had a distant Jewish forebear or two. True or not, it must have caused him some anxiety.[9]

Lise's finances gave Otto constant trouble. With great effort he managed to get her frozen bank account transferred to Vienna so Gusti Frisch could use it,[10] but he had no success with Lise's pension. After many attempts, Hahn went to the finance bureau in December and was told to fill out a new application and start over. "After I had one of my small temper fits, things went a little better. Believe me, these obstruction tactics affect me too, especially since you must surely think we have done nothing."[11] But there was nothing that could be done. The paperwork was a charade, agreements were not honored, and Meitner's pension never came through. The Reich intended to rid Germany of Jews—but not before it systematically stole all they had or left behind. In Munich, Willstätter obtained emigration documents only after he had signed over to the Reich his home and everything in it, including his books and papers; as he crossed the border to Switzerland, his last few marks went to an "export fee" on his suit, soap, and toothbrush.[12]

In Dahlem that December, Hahn worked when he could and Strass-mann worked constantly, assisted by Clara Lieber, an American chemist, and a laboratory assistant, Irmgard Bohne. Just before Christmas vacation, Hahn wrote to Meitner with startling news.

> 19.12.38 Monday eve. in the lab. Dear Lise! . . . It is now just 11 P.M.; at 11:45 Strassmann is coming back so that I can eventually go home. Actually there is something about the "radium isotopes" that is so remarkable that for now we are telling only you. The half-lives of the three isotopes have been determined quite exactly, they can be separated from *all* elements except barium, all reactions are consistent [with radium]. Only one is not—unless there are very unusual coincidences: the fractionation doesn't work. Our Ra isotopes act like *Ba*.[13]

This was the final test for radium, designed to separate it partially from its barium carrier by the method of fractional crystallization ("fraction-ation") first devised by Marie Curie. Starting with a solution of the pre-sumed barium-radium mixture, Hahn and Strassmann added bromide to the solution in four steps; with each step, a fraction of the barium (and radium) would precipitate as crystals of barium bromide. Because radium was known to coprecipitate preferentially with barium bromide—that is, the proportion of radium that precipitated was larger than the proportion of radium in solution—the first barium bromide fraction was expected to be richer in radium than those that followed. To their surprise, Hahn and Strassmann measured no difference at all: the "radium" activities were evenly distributed among the successive barium bromide fractions. Think-ing that something might have gone wrong with their procedure, they ran control experiments using known radium isotopes.[14]

> Now last week I fractionated ThX [^{224}Ra] on the first floor; it went exactly as it should. Then on Saturday Strassmann and I fractionated our "Ra" isotopes with MsTh$_1$ [^{228}Ra] as indicator. The mesothorium became en-riched [in the barium bromide fractionation], our Ra did not. It could still be an extremely strange coincidence. But we are coming steadily closer to the frightful conclusion: our Ra isotopes do not act like Ra but like Ba. . . . All other elements, transuranes, U, Th, Ac, Pa, Pb, Bi, Po are out of the question. I have agreed with Strassmann that for now we shall tell only *you*. Perhaps you can come up with some sort of fantastic explanation. We know ourselves that it *can't* actually burst apart into Ba. Now we want to test whether the Ac-isotopes derived from the "Ra" behave not like Ac but La. All very complicated experiments! But we must clear it up.
>
> Now Christmas vacation begins, and tomorrow is the usual Christmas party. You can imagine how much I'm looking forward to it, after such a long time without you. Before the institute closes we do want to write something

for *Naturwissenschaften* about the so-called Ra-isotopes, because we have very nice [decay] curves.

So please think about whether there is any possibility—perhaps a Ba-isotope with much higher atomic weight than 137? If there is anything you could propose that you could publish, then it would still in a way be work by the three of us![15]

Hahn was not suggesting that they publish together. He and Meitner both knew that it was politically impossible. But the tone and detail of Hahn's letter clearly show that Meitner was still very much a member of their team and that the barium finding was an integral part and a natural consequence of their four years of work together. Indeed, in the five months since Meitner left Berlin, the pattern of their work had not changed much. As before, nearly all the experiments were done in her former section on the ground floor of the institute, using the neutron sources, paraffin blocks, lead vessels, counters, and amplifiers that she had assembled and built.[16] Although Meitner was no longer present to perform the irradiations and physical measurements, she followed the experiments and provided intellectual direction through her correspondence with Hahn and their meeting in Copenhagen. And it was Meitner's objections that had driven them to make this final test for the "radium" that now appeared to be barium, a most unexpected result.

And now, as so often before, Hahn and Strassmann were expecting Meitner to interpret the findings and place them in their physics context. Without her, Hahn was somewhat adrift. The idea of uranium "bursting" crossed his mind, but he was thinking of nuclear mass rather than atomic number: he did not yet realize that uranium had split in two.[17] It was not solicitude for Meitner that kept Hahn and Strassmann from talking to other physicists; as a team they had been so close, so familiar with each other's work and thought processes, that she was still, in every essential way, one of them.[18] "How beautiful and exciting it would be just now if we could have worked together as before," Hahn wrote on 21 December. "We cannot suppress our results, even if they are perhaps physically absurd. You see, you will do a good deed if you can find a way out of this."[19]

Anxious to publish quickly, Hahn and Strassmann did not wait for Meitner's reply before submitting their report to *Naturwissenschaften* on 22 December. The article, written entirely by Hahn,[20] displays the duality that characterized the uranium investigation: the radiochemical data were

firm, the physical meaning so strange that Hahn wavered between confidence and disbelief. He changed the original title of the report, "Concerning the Determination and Relationships of the Radium Isotopes
Derived from the Neutron-Irradiation of Uranium," by substituting "al-
kaline-earth" for "radium," a generalization that permitted him to distance himself from radium without fully committing himself to barium.[21]
Indeed, the bulk of the article was a detailed description of the "radium"
isomers, with the barium experiments mentioned only at the end and then
"hesitantly [zögernd], due to their peculiar results."[22] As so often in the
uranium investigation, the chemistry inspired confidence, the physics,
doubt: "As chemists . . . we should substitute the symbols Ba, La, Ce for
Ra, Ac, Th. As 'nuclear chemists' fairly close to physics we cannot yet bring
ourselves to take this step which contradicts all previous experience in
nuclear physics. There could still perhaps be a series of unusual coincidences that have given us deceptive results."[23]

Hahn's letter of 19 December arrived in Stockholm on the 21st, Meitner's first news of the barium finding. She answered instantly.

> Your radium results are very startling. A reaction with slow neutrons that
> supposedly leads to barium! By the way, are you quite sure that the radium
> isotopes come before actinium? . . . And what about the resulting thorium
> isotopes? From lanthanum one must get cerium. At the moment the as-
> sumption of such a thoroughgoing breakup [weitgehenden Zerplatzens] seems
> very difficult to me, but in nuclear physics we have experienced so many
> surprises, that one cannot unconditionally say: it is impossible.[24]

Imagine what that letter meant to Hahn! In November, Meitner had
vehemently objected to the radium isomers; now she was surprised but not
opposed, ready to consider the barium an expansion rather than a contradiction of previous experience. Unlike Hahn, Meitner understood that
existing nuclear theory applied only to minor disruptions of the nucleus:
intuitively, she was ready for its massive disintegration. Her response
arrived in Berlin on 23 December, surely the best Christmas present he
received.[25]

Years later, Hahn was known to say that if Meitner had stayed in Berlin,
she might have talked him out of the discovery, might even have "forbidden" him to make it.[26] Many such reports are second- and thirdhand
and require examination of the personal and political agendas of their
sources, but they are consistent with Hahn's later refusal to credit Meitner

with any but a negative role in the discovery. Her letter of 21 December indicates quite the opposite. On the day she first learned of barium, she instantly responded: it is not impossible. And at the time, Hahn must have found her response most reassuring, for it was only *after* he received it that he began to recognize the significance of the barium finding.

Lise left Stockholm on Friday, 23 December, to spend the holidays with Eva von Bahr-Bergius and her family in Kungälv, on the west coast of Sweden. Otto Robert Frisch was invited, too, and came up from Copenhagen. Christmas in Sweden presented some unfamiliar traditions: Lise and Otto Robert would long remember Christmas dinner with their hosts "where we both struggled mightily with the traditional Swedish *lutfisk*."[27] Always close, aunt and nephew shared their worry about Otto Robert's parents. But their thoughts were never far from physics.

On Frisch's first morning in Kungälv—probably the day before Christmas—he found his aunt mulling over Hahn's letter of 19 December.[28] She had nothing else to go on. Hahn had mailed a copy of the *Naturwissenschaften* manuscript to her Stockholm address, and she would not see it for a week. But she had Otto Robert to talk to, a great advantage over her isolated existence in Stockholm.

In his memoirs Frisch recalled,

When I came out of my hotel room after my first night in Kungälv, I found Lise Meitner studying a letter from Hahn and obviously very puzzled by it. I wanted to discuss with her a new experiment that I was planning, but she wouldn't listen; I had to read that letter. Its content was indeed so startling that I was at first inclined to be skeptical. Hahn and Strassmann had found that those three substances were not radium . . . [but] barium.

The suggestion that they might after all have made a mistake was waved aside by Lise Meitner; Hahn was too good a chemist for that, she assured me. . . . We walked up and down in the snow, I on skis and she on foot (she said and proved that she could get along just as fast that way), and gradually the idea took shape that this was no chipping or cracking of the nucleus but rather a process to be explained by Bohr's idea that the nucleus is like a liquid drop; such a drop might elongate and divide itself. . . . We knew that there were strong forces that would resist such a process, just as the surface tension of an ordinary liquid drop resists its division into two smaller ones. But nuclei differed from ordinary drops in one important way: they were electrically charged, and this was known to diminish the effect of the surface tension.

At this point we both sat down on a tree trunk, and started to calculate on scraps of paper. The charge of a uranium nucleus, we found, was indeed large enough to destroy the effect of surface tension almost completely; so the uranium nucleus might indeed be a very wobbly, unstable drop, ready to divide itself at the slightest provocation (such as the impact of a neutron).

But there was another problem. When the two drops separated they would be driven apart by their mutual electric repulsion and would acquire a very large energy, about 200 MeV in all; where could that energy come from? Fortunately Lise Meitner remembered how to compute the masses of nuclei from the so-called packing fraction formula, and in that way she worked out that the two nuclei formed by the division of a uranium nucleus would be lighter than the original uranium nucleus by about one-fifth the mass of a proton. Now whenever mass disappears energy is created, according to Einstein's formula $E = mc^2$, and one-fifth of a proton mass was just equivalent to 200 MeV. So here was the source for that energy; it all fitted![29]

Frisch's story is too well told to be omitted from an account of Meitner's life. It conveys the excitement and delight of a truly new idea, the first recognition that a nucleus can split, and the first understanding of how and why it does.

At its heart was a nucleus with the properties of a liquid drop, a model that began with George Gamow in December 1928.[30] Earlier that year Gamow had explained alpha decay as a quantum mechanical tunnel effect, a highly successful theory that convinced physicists that only small particles could escape the nucleus. The liquid drop nucleus was based on a completely different premise. Instead of assigning individual nuclear particles to quantized energy states and shells, Gamow treated them all as essentially equivalent, drawn together by mutual attraction and surface tension, like molecules in a classic liquid drop. The model was fruitful. Among others, Heisenberg applied it to his theory of neutron-proton exchange forces, and his assistant C. F. von Weizsäcker used it as the basis for a semiempirical formula for nuclear mass defects ("packing fractions") that agreed well with experiment. In many respects the theory corresponded to Meitner's experimental interests, and she followed it closely. She knew Gamow quite well, saw Heisenberg often, and thought highly of Weizsäcker, who was her "house theoretician" (*Haustheoretiker*) for a time in 1936.[31] By then the liquid drop model had proven itself for nuclear stability and mass-energy relationships across the periodic table.

But the model did little for nuclear reactions until Bohr and a young associate, Fritz Kalckar, placed it in a different context. Like Gamow, they began with a nucleus that was liquidlike in its internal cohesion and surface tension; when struck by an energetic particle (such as a neutron or alpha particle), nucleus and projectile merged to form an excited semistable compound nucleus that oscillated and vibrated as a whole, like a liquid drop, until the reaction went to completion. Only two reaction outcomes were considered: energy emission and particle expulsion ("evaporation")—and the expelled particles, it was assumed, would be small. Still under the influence of Gamow's alpha decay theory, physicists missed the full implications of the liquid drop: the idea of a massive split occurred to no one.

First outlined by Bohr in late 1935, the compound nucleus promised an understanding of nuclear processes and emission spectra. Meitner, struggling to find a theoretical interpretation for the neutron-induced reactions of uranium and thorium, referred briefly but hopefully to the work of Bohr and Kalckar soon after it was published in 1937.[32] Otto Frisch was in Copenhagen, directly engaged with the compound nucleus. With others, he measured the absorption of slow neutrons in various elements, participated in theoretical discussions, and made excellent sketches of the compound nucleus to illustrate Bohr's lectures and articles.

When aunt and nephew came together in Kungälv that Christmas, so did the two facets of the liquid-drop nucleus. Once Meitner assured Frisch that barium was a fact, both began thinking of uranium as a nuclear droplet. Frisch, visually gifted, may have been the first to picture its surface oscillations, and both envisioned an elongated nucleus in a dumbbell shape. Together they estimated the surprisingly small surface tension of the uranium nucleus: Frisch calculated the energy needed to drive the two fission fragments apart; Meitner, the mass-defect formula in her head, estimated the energy released when uranium cleaved into two smaller nuclei. It was beautiful. Everything fit.

But still it was based on only one piece of experimental data: barium. Where, Meitner wanted to know, was the second fission fragment? It was obvious that if uranium (with 92 protons) split in two, and one piece was barium (56), then the other would have to be krypton (36). In addition, she suspected a connection between barium and the transuranes: they were formed under the same reaction conditions and exhibited what appeared to be multiple inherited isomerism and an extended sequence of beta decays. Could it be that the transuranes were not higher elements at all but smaller nuclei like barium, which were formed when uranium split? For

Meitner, this was cause for great anxiety. After four years, it appeared that the transuranes might be terribly wrong.

Hahn, meanwhile, was thinking along the same lines. After he received Meitner's letter of 21 December, he too suspected that the transuranes might be lighter elements. A few days later he telephoned Paul Rosbaud, editor of *Naturwissenschaften*, to add the new idea to the page proofs.[33] To Lise he wrote,

> 28.XII.38 Dear Colleague! I want to quickly write a few more things about my Ba fantasies, etc. Perhaps Otto Robert is with you in Kungälv and you can discuss it a bit. . . . Would it be possible that the uranium-239 [^{238}U plus a neutron] breaks up into a Ba and a Ma? [Masurium, the unconfirmed element 43, was subsequently named technetium (Tc).] A Ba 138 and a Ma 101 would give 239. The exact mass number is not important. It could also be 136 + 103 or something like that. Of course, the atomic numbers don't add up. Some neutrons would have to change into protons so that the charges would work out. Is that energetically possible? I do not know; I only know that our radium has the characteristics of Ba; that our Ac does not have the correct characteristics of the true element 89. Everything else is not proved. . . . If there's something to it, then the transuranes, including "ausonium" and "hesperium" [93 and 94] would die. I don't know if that would make me very sad or not.[34]

Hahn had correctly guessed that uranium splits, but he again made the mistake—surprising for a nuclear chemist—of considering mass rather than atomic charge and thus failed to recognize that the number of protons in the two fission fragments must add up to 92. Unfortunately for Hahn and Strassmann, this error in print led others, including Einstein, to believe they did not understand their own work.[35]

> 29.12.38 Dear Otto! Many thanks for your letter of the 28th. . . . Your Ra-Ba results are very exciting. Otto R. and I have racked our brains; unfortunately the manuscript has not yet been forwarded to me, but I have just sent for it and hope to get it tomorrow. Then we can think about it better. Concerning the transuranes, it seems to me that [our chemical] proof that 23-minute U and 40-second U are uranium isotopes must necessarily lead to transuranes. Or do you think those reactions did not confirm the presence of uranium isotopes? Also, what is going on with the so-called actinium [isotopes]? Can they be separated from lanthanum or not?[36]

Meitner was asking quite a lot, and telling very little. She knew Hahn could not publicly acknowledge anything she wrote to him, and she did not want him or the institute physicists to learn too much about what she and Frisch were thinking.

On New Year's Eve, Lise wrote again.

> 1.I.1939, 12:30 A.M. I begin the year with a letter to you. May it be a good year for us all. . . . We have read and considered your paper very carefully; *perhaps* it is energetically possible for such a heavy nucleus to break up. However, your hypothesis that Ba and Mo would result is impossible for several reasons.

In his tiny, cramped script Hahn had actually written Ma and not Mo, but Meitner would have objected anyway: the atomic numbers were wrong in either case. She did not explain, however. She was terribly anxious about the "transuranes," almost certain by then that they were not elements beyond uranium but the fragments into which uranium had split.

> You understand, of course, that the question of the correctness of the transuranes has a very personal aspect for me. If all the work of the last 3 years [*sic:* 4][37] has been incorrect, it cannot be determined from just *one* side. I shared the responsibility for it and must therefore find some way to participate in the retraction. A joint retraction is surely not feasible [for political reasons], so we should consider *simultaneous* statements, one from the two of you and one from me (the latter perhaps in *Nature*). If the transuranes disappear, you are in a much better position than I, since you have discovered it yourselves, while I have only three years [*sic*] of work to refute—not a very good recommendation for my new beginning.

Otto Robert added,

> If your new findings are really true, it is of the greatest interest, and I am eager for further results. Here we had beautiful snowy weather and I did some skiing.

Meitner again:

> The question of isomerism also becomes very doubtful should the transuranes disappear.

And on a separate piece of paper:

> I have an idea that perhaps only the 23-minute U leads to an ekaRe [element 93] (which may be alpha radiating). If this turns out to be true—I have certain theoretical reasons for thinking it is—this could be a starting point for my retraction. (One must be careful . . . that my retraction is not entirely negative.)[38]

Meitner's desire for a simultaneous public retraction of the "trans-uranes" was an effort at damage control. It never took place, in part be-cause Hahn and Strassmann believed in the "transuranes" long after Meitner gave them up, in part because it would have been as politically impossible for Hahn and Strassmann to coordinate a retraction with her as to actually include her in their publications. In any case, the "trans-uranes" were gone. To Meitner it came as a staggering double blow: to be excluded from a sensational discovery, and then to have four years' work proved wrong!

The injustice was bitter. Her contributions to the discovery formed a continuum from her first work in Berlin to the finding of barium and beyond; at the end only her physical presence was missing. Meitner and Hahn's own experience shows that physical presence is not required at all times for every member of a team: in 1917 and 1918, Hahn was in the army, Meitner did nearly all the work, and both, without question, were credited with the discovery of protactinium.[39] Later Strassmann expressed it clearly.

> What difference does it make that Lise Meitner did not *directly* partici-pate in the "discovery"?? Her initiative was the beginning of the joint work with Hahn—*4 years later she belonged to our team*—and she was bound to us intellectually from Sweden [through] the correspondence Hahn–Meitner. . . . [She] was the intellectual leader of our team, and therefore she belonged to us—even if she was not present for the "discovery of fission."[40]

Meitner knew there was nothing she could do, that it had nothing to do with science and everything to do with race, that the same racial policies that had forced her out of Germany made it impossible for Hahn and Strassmann to include her in the barium finding, made it difficult for them even to admit their continued collaboration with a "non-Aryan" in exile. The timing was bad. Unsettled in Sweden, she feared for her reputation, worrying that those who did not know her work or consider the political situation might think she had contributed nothing to the discovery—or worse, impeded it. She never imagined that Hahn himself would soon suppress and deny not only their ongoing collaboration but the value of nearly everything she had done before as well.

Vacation ended on New Year's Day. Otto Robert and Lise parted, arranging to telephone in a few days, after Frisch talked to Bohr and Meitner had a chance to study the literature. Back in Stockholm, Meitner plunged into an intense review of the chemistry of the "transuranes." A day later she was convinced: the chemistry of the "transuranes" resembled the platinum metals, but they could be the *light* platinum metals—Ma (Tc), Ru, Rh, Pd—that lie just above Re, Os, Ir, and Pt in the periodic table. (See Appendix fig. 2.) If true, these light nuclei gave credence to nuclear splitting, eliminated the need for isomerism, and accounted for the long sequence of beta decays. The fact that she could use the false transuranes to substantiate splitting and explain findings that had plagued the uranium investigation began to console Meitner. Her next letter to Hahn was a little more optimistic.

> 3.I.39 Dear Otto! I am now almost *certain* that the two of you really do have a splitting to Ba and I find that to be a truly beautiful result, for which I most heartily congratulate you and Strassmann. . . . As I already wrote to you, I have several reasons for thinking that the 23-minute uranium most probably really is uranium; the two other series [the former processes 1 and 2] probably consist of the light platinum metals. I was always disturbed by those long decay series with the relatively low atomic weight of 239; with this completely different sort of process it is quite possible. In any case, both of you now have a beautiful, wide field of work ahead of you. And believe me, even though I stand here with very empty hands, I am nevertheless happy for these wondrous findings.[41]

In her New Year's letter to Hahn, she had been unable to contain her disappointment. In turn, Otto wrote

> [I am] terribly upset to see you so depressed about our experiments and my vague suspicions about the transuranes. . . . Neither we nor any other physicist could think of elements with low atomic weight. . . . Must Herr Fermi, must we, be ashamed of this, and say everything we did in the last three [*sic*] years was wrong? [He thought she was angry with him for publishing the barium finding without her.] You write: "That cannot be determined by one side only." What else could I have done? In the institute no one knows anything about these newest things. I might perhaps have wanted to discuss things with [C. F. von] Weizsäcker, [Siegfried] Flügge or [Arnold] Flammersfeld. . . . I told the details only to you, so that you can form an opinion and perhaps publish something about it. . . . But it would have been wrong for us not to publish our results as quickly as possible. . . . You write: "The question of isomerism also becomes doubtful if the transuranes disappear." Why? The isomeric Ra-Ba isotopes are definite, why not the others too? . . . Believe me, it would have been preferable for me if we could still work together and discuss things as we did before![42]

4.I.39 Dear Otto! I don't really know how to answer your last letter. Not for a second did I think anything else but that your Ba-Ra results must be published as quickly as possible. I *only* spoke of refuting the transuranes, for which I felt responsible with you. . . . It hurts that misunderstandings can come so easily. Couldn't you trust my friendship a little more?[43]

Frisch echoed Meitner's thoughts.

4.I.39 Dear Prof. Hahn, You have misunderstood our letter somewhat. Not for a moment did L[ise] mean to reproach you for your forthcoming publication. All she meant was that if the results of your joint work (transuranes, isomerism) prove to be false, then the recall of *these* results should not come from you and Strassmann alone, or else people will say that three did nonsense and now that one is gone the two others made it right. Naturally that is not a recommendation for the third. . . . We are of course very enthusiastic about your and Strassmann's new results; surely it is understandable that L. was a little sad not to be a part of it, but this feeling of dejection was suppressed a day later by joy for the beautiful discovery.[44]

It had taken several days for Lise to realize what Otto Robert instantly knew: their theoretical interpretation was also a beautiful discovery, embodying and extending existing nuclear theory, making sense of results that had never been clear before. And as Meitner suspected, the 23-minute ^{239}U was real. Of all the many uranium products, this had always been the most straightforward: formed by an entirely normal resonance neutron capture, with a half-life long enough to be certain of uranium, it necessarily decayed to the first true element 93. The other "transuranes" were false, to be sure, but in their place she and Frisch had discovered something far more satisfying and right.

From Copenhagen, Otto Robert described Bohr's enthusiastic reaction, opening with the affectionate Viennese diminutive:

Dear Tanterl, I was able to speak with Bohr only today [3 January] about the splitting of uranium. The conversation lasted only five minutes as Bohr agreed with us immediately about everything. He just couldn't imagine why he hadn't thought of this before, as it is such a direct consequence of the current concept of nuclear structure. He agreed with us completely that this splitting of a heavy nucleus into two big pieces is practically a classical phenomenon, which does not occur at all below a certain energy, but goes readily above it. (That is also consistent with the very great stability of normal uranium and the very large instability of the [not much higher energy] compound nucleus [of uranium and a neutron].) Bohr still wants to think quantitatively about it this evening and tomorrow we will talk about it again.[45]

Later Frisch dramatized and condensed this conversation with Bohr: "I had hardly begun to tell him, when he struck his forehead with his hand and exclaimed: 'Oh, what idiots we all have been! Oh, but this is wonderful! This is just as it must be! Have you and Lise Meitner written a paper about it?' I said we hadn't yet, but would at once, and Bohr promised not to talk about it until the paper was out."[46]

To speed publication, Meitner and Frisch decided to submit a note rather than a full article to *Nature*. Within the allotted space of one page, they would define and describe the splitting of a nucleus for the first time and give it theoretical justification. In addition, Meitner wanted to add experimental weight to their argument by showing that if the "transuranes" were fragments, the notion of multiple isomerism was unnecessary and the long sequence of beta decays made sense.

Outlining her argument to Frisch shortly after her return to Stockholm, Meitner reasoned that uranium nuclei could divide with varying proportions of neutrons and protons to form a variety of different fragments. The barium, she noted, could be Ba (56 protons) and Kr (36), while the "transuranes" might be part of a decay sequence beginning with Zr (40) and Te (52), or perhaps Sr (38) and Xe (54), or other combinations. In addition, she noted, apparent multiple isomerism could now be explained as several different isotopes of each light element, for example, ^{141}Ba with ^{98}Kr or ^{143}Ba with ^{96}Kr. Furthermore, since the neutron-to-proton ratio is higher in uranium than in lighter elements, the fragments would tend to be heavy with neutrons compared to their normal isotopes, necessitating an extended series of beta decays to reduce the neutron excess. Meitner, who had always been disturbed by the long sequence of beta decays and never satisfied with multiple isomerism, was relieved, finally, to see the transuranes go.[47]

The process was still unclear to Hahn. On 7 January, he wrote to Lise that he and Strassmann had again verified Ba but found no chemical evidence for the sequence Ma(Tc)-Ru-Rh-Pd he had proposed before. "Therefore it seems almost certain that the transuranes stay, and the sum of atomic weights was an unusual coincidence."[48]

Meitner delightedly evaluated the information for Otto Robert.

The more I think of our transuranes, the more likely it seems to me that the scheme I described in my earlier letter to you is correct. . . . Hahn wrote today that they actually believe in the transuranes: their properties do not correspond to the lower homologues. . . . [But] Hahn and Strassmann have just tried another Ba-Ra separation, with the same result, so I do not doubt that [uranium] splits into two nuclei. . . . For now I do not want to tell Hahn about my . . . hypothesis, because if it is right and he verifies it experimentally, then for political reasons he cannot refer to a written communication from me. When it is published, however, he can cite it. . . . Naturally I am really curious about the theoretical explanation Bohr and you have come up with for splitting. If the result is good, I would like to carefully bring up the above explanation for the transuranes, if you agree. Normally, I am really not so concerned about publishing—on the contrary. But in my current bad situation I must, unfortunately, think of such things, to show people that I am not completely dimwitted.[49]

Having labored for years over the false transuranes, Meitner now had the pleasure of mining them for chemical evidence for the splitting of uranium. Frisch, with no such emotional investment in them, was quicker to see the physical implications of nuclear disintegration; moreover, in Bohr's institute he was surrounded by superb physicists and ample equipment, including a high-tension apparatus and a cyclotron capable of generating a strong source of neutrons. On 8 January, he included in a letter to Meitner a first draft of their *Nature* note.[50]

What do you think about trying to find the "recoil" nuclei [fragments] with a proportional amplifier? All such nuclei ought to have about 100 MeV kinetic energy, that would give an incredible number of ions, which one should see even with the uranium alpha-radiation background (one surely must not cover it, I will try to estimate their range). . . . On Friday [6 January] . . . Bohr discussed it at length with me again. He had me show him how I arrived at my estimate of surface tension, and agreed completely with it; he himself had thought of the electrical term but not that it made such a difference. . . . The next morning I began this draft and was able to bring two pages to Bohr at the train station (at 10:29) where he pocketed them; he had no time to read it. . . . Yesterday, when Hahn and Strassmann's paper appeared here, I discussed the whole thing a little, mostly with [George] Placzek who was skeptical, but then he always is. . . . When [George de] Hevesy saw the paper he immediately said that [Irène] Curie had told him already last fall that she found very light elements from uranium, but she obviously did not trust herself to publish it. Well, she already has the Nobel prize, she can be satisfied.[51]

With his "recoil" experiment Frisch was looking for the first physical evidence for nuclear splitting. Expecting the fragments to be violently expelled from the irradiated uranium sample, he prepared to detect the huge number of ions they would create in a proportional amplifier. After Meitner reminded him about the strong natural alpha emission of uranium, Frisch biased the counter to measure only the huge bursts of ionization from the relatively massive fragments. For several days Meitner and Frisch discussed the experiment by letter and telephone, assessing the energy, range, and magnitude of the induced ionization.[52] It would have been to their advantage to submit their theoretical note first. Hahn and Strassmann's *Naturwissenschaften* article had appeared on 6 January and was attracting considerable attention. But at that moment Meitner and Frisch had based their theoretical interpretation on a single piece of experimental evidence—barium—and they very much wanted to have physical evidence to support it.

Hahn and Strassmann, meanwhile, had new results.

> 10. January 39 . . . Our Ra from thorium also appears to be Ba! An experiment we did today indicates that absolutely. Tomorrow we will do another. . . . Our theoretical people have begun to discuss the Ba thing and think about it. They know nothing about the thorium-Ba; we (Strassmann and I) will say nothing about it for now. If you and Otto Robert are going to write something, do it soon. . . . You could also include thorium. Perhaps still lower elements could be burst with higher energy neutrons. Why Ba shows up primarily, I don't know.[53]

On 13 January, Frisch performed the recoil experiment and detected the fragments immediately.[54] He delayed their joint paper for another three days while he refined his data and wrote a separate note describing his experiment.[55] On 16 January, he finally mailed both papers to *Nature*,[56] where their joint paper appeared on 11 February and his recoil paper appeared on 18 February. While he was writing the papers, Frisch asked William A. Arnold, an American biologist working with George de Hevesy in Bohr's institute, what name biologists use for the process of cell division. "Binary fission," Arnold replied. Frisch visualized the droplike nucleus dividing like a living cell. In his paper with Meitner he proposed that the process be named "fission."[57]

In their paper the two physicists likened nuclear fission to the "essentially classical" division of a liquid drop and estimated that the 200 MeV of energy released was "available from the difference in packing fraction between uranium and the elements in the middle of the periodic system." They expected each fission fragment to initiate a chain of disintegrations as its "high neutron/proton ratio . . . readjusted iself by beta decay to the lower value suitable for lighter elements." They predicted, "If one of the parts is an isotope of barium, then the other will be krypton (Z = 92 − 56), which might decay through rubidium, strontium and yttrium to zirconium." They thought it "rather probable" that all the "transuranes" were light elements, possibly "masurium decaying through ruthenium, rhodium, palladium and silver into cadmium" and that it "might not be necessary to assume nuclear isomerism." They took care to cite Hahn and Strassmann only for results already in print. For the fission of thorium, they noted that in 1938 the Berlin team had obtained isotopes from the neutron irradiation of thorium similar to those from the fission of uranium, "suggesting a fission of thorium which is like that of uranium." (They could not report that Hahn and Strassmann had already confirmed the fission of thorium—nor did they wish to omit thorium entirely, since they had thought of the possibility themselves.)[58] The paper concluded with the one aspect of the earlier work Meitner still believed to be true: "The body with 24-minute half-life . . . is probably really ^{239}U, and goes over into an eka-rhenium [element 93] which appears inactive but may decay slowly, probably with emission of alpha particles. (From inspection of the natural radioactive elements ^{239}U cannot be expected to give more than one or two beta decays; the long chain of observed decays has always puzzled us.) The formation of this body is a typical resonance process."[59]

It was an excellent piece of work, brief yet elegant in its theoretical discussion. Although it could not compensate Meitner for being excluded from the barium finding, it tied her to it, permitting her to be the first to bury inherited isomerism and the false transuranes and to point to the one species that was still valid, the precursor to the first true element 93.

Then there was good news from Vienna. On 14 January, Lise and Otto Robert learned that Gusti and Jutz Frisch had a Swedish visa, Jutz would soon be released from Dachau, and they would come to Stockholm.[60]

Relieved of her worries about her family and finished with her work with Otto Robert, her situation in Stockholm suddenly seemed bleaker than before. She had ideas for follow-up experiments but no material or equipment with which to do them. And she could not penetrate Siegbahn's coldness and lack of collegiality. When Hahn sent Siegbahn a counter he wanted, at Meitner's request, she wrote to Hahn, "Once again he said not a word to me, although I asked him about it. It is all very depressing, but I see no possibility of changing it."[61]

Frisch had detected the large ionization pulses from fission fragments by connecting a linear amplifier to an ionization chamber that was lined with uranium and irradiated with neutrons. Such physical evidence for fission was dramatic, but the quantity was far too small for analysis. Meitner was interested in collecting a larger sample of the fission products, determining half-lives and some chemical characteristics, comparing the former transuranes with fission products, and possibly identifying the first true transuranium element. The method she proposed was a typical recoil separation: "If a metal plate is placed close to a uranium layer bombarded with neutrons, one would expect an active deposit of the light atoms emitted in the 'fission' of uranium to form on the plate,"[62] while elements 93 and above would remain embedded in the uranium. Recoil, which Meitner first used thirty years before, was still a practical separation technique.

But the experiment required counters, supplies, and a strong neutron source, none of which was available to her in Stockholm. She thought she might be able to measure half-lives of the fission fragments with a basic counter, and perhaps perform some simple chemical separations. Frisch was ready to provide long-distance assistance—irradiating uranium with neutrons, collecting the fission products on a plate, and sending it via three-hour air express to Stockholm.[63]

As Meitner and Frisch planned the new experiments, they were unaware of the excitement that fission was causing elsewhere. When Bohr left Copenhagen for New York on 7 January 1939, he promised not to mention Meitner and Frisch's work until after their paper was submitted for publication: American scientists would not receive the 6 January *Naturwissenschaften* for another two weeks.[64] On board the *Drottningholm*, a blackboard in his stateroom, Bohr went over and over the mechanism of fission with another physicist, Léon Rosenfeld, until at last Bohr was satisfied that he understood it. Bohr then spent a few days in New York, while Rosenfeld attended a seminar in Princeton where he immediately told everyone about fission—Bohr had forgotten to tell him to keep it to himself. When Bohr heard, he tried to protect Meitner and Frisch's priority, but as Frisch later related, it was almost too late. "[A] fantastic race was already going on in a number of American laboratories, where the news had spread from Princeton, to perform the same easy experiments I had already made to detect the fission fragments."[65]

The secret was out, and Bohr waited impatiently for news from Frisch that he had submitted his and Meitner's theoretical note. Unaware of the competition, Frisch did his experiment first, took his time with the *Nature* note, and did not send the news that Bohr was anxiously awaiting until 22 January—and then he sent it surface mail to New York. By 20 January, Hahn and Strassmann's *Naturwissenschaften* paper reached American shores, and there was no holding back. On 26 January, at a meeting of physicists in Washington, D.C., Bohr announced Hahn and Strassmann's finding and did his best to preserve the priority for Meitner and Frisch's interpretation. He still did not know of Frisch's recoil experiment.

So intense was the interest that "[a] group of local physicists, who . . . only heard of the matter at the meeting, rushed to their laboratory and worked without pause for two days in order to be able to announce at the meeting that they too had seen the fission fragments. . . . Rosenfeld described to me [Frisch] the scene he and Bohr witnessed: a physicist simultaneously . . . recording fragments . . . and phoning to an anxious newspaperman: 'Now, there's another one.'"[66] Others got there sooner: word from Princeton reached Columbia University, where Herbert Anderson had already detected fission fragments on 25 January.

Not knowing when Meitner and Frisch had submitted their paper, Bohr sent Frisch "telegram after telegram, asking for further information and suggesting further experiments . . . but we [Frisch] had no idea of the reasons which could prompt Bohr to such unusual impatience."[67] To his aunt, Frisch wrote, "I had to send [Bohr] a report by telegram, which cost the institute 40 Kroner. When he finally got my letter [on 2 February] with the copies of both manuscripts he sent me a congratulatory telegram, probably at the same time as yours."[68] Bohr was evidently immensely relieved: the priority for the fission interpretation did remain with Meitner and Frisch, and the first physical detection of fission was Frisch's.

A letter of Frisch's to Meitner, written 6 February 1939, provides a sampling of the variety of physical experiments spawned by the discovery of fission.

In one telegram Bohr asked if the splitting occurs immediately after neutron capture or if beta emission occurs first. Perhaps he was thinking that the nucleus emits a beta particle first and thus becomes even more unstable (it seems to me you thought of that too, in Kungälv). I then did a few experiments with a source on the edge of a rotating disk and can say that the average delay in fission must always be smaller than a twentieth of a second. . . . I built a second amplifier for the high tension, where we shall soon have neutrons and then want to test for the splitting of Bi, Pb, Tl, Hg, Au, etc. We shall also try to initiate fission with Li gamma radiation, as you proposed. As soon as we have neutrons I will do irradiations for you. It would really be very nice to find the "transuranes" in the recoil; evidently Hahn still believes in them, but they seem terribly unlikely to me.[69]

None of the experiments was very difficult, but in Stockholm, Meitner lacked the equipment to do anything. "If only you could see my daily work!" she complained to Hahn. "I have *absolutely no* help, I put counters together, evacuate them, test them for leaks, in short I am doing the work that Herr Dorffel [a technician in Dahlem] did, only very poorly."[70] In frustration, she watched as others completed experiments she had thought of weeks before. By the end of January, Meitner was so discouraged that she inquired again about a position in Cambridge. Sir William Bragg invited her to visit in March, but with her passport problems she could not travel that soon.[71] She wrote to Hahn, "I sleep very poorly and people say I am thin as a line, but I feel quite healthy, even though my head is full of troubles."[72]

Hahn, meanwhile, was also under considerable stress. In the hardening political climate, his position seemed vulnerable. Kurt Hess, the Nazi on the top floor of the institute, was always a threat, and now that the 6 January *Naturwissenschaften* article was out, physicists in his institute—nearly all of them party members—were complaining loudly because he had not shared the barium result with them before publication.[73]

Scientifically, too, Hahn was floundering. Without direction from physics, he lacked confidence in the barium finding, so that he and Strassmann spent the first weeks of January 1939 verifying it over and over again. They also tested for a few of the light platinum elements in the transurane precipitate, primarily because Meitner insisted this was the case, but found nothing. Without firm chemical evidence to the contrary, Hahn maintained that the transuranes were still valid. In part, it may have been reluctance to abandon the results that had led to the barium finding. "There is so much good work in our uranium and thorium research that had it not been for that experience, Strassmann and I would surely not be moving ahead quickly now."[74]

In the first weeks of January, Meitner deliberately told Hahn very little. Only after she and Frisch submitted their articles to *Nature* did she tell him more—and then, cautiously.

> 18.I.39 Dear Otto! . . . If I was somewhat secretive about this, it was because you really are not in a position to cite it before publication, and it includes a number of contentions which could be proved experimentally. And it will appear at the earliest in 3 weeks. Theoretically, the most amusing thing is that although two or more α-decays are energetically impossible, a split into two lighter nuclei is, in contrast, energetically possible—because of the deep valley in the mass defect curve between Z = 40 and Z = 60—and is also understandable on the basis of the liquid drop model for the nucleus. I beg you, however, *not* to tell anyone else about this. I am quite convinced that the transurane series are those of light nuclei, except for the resonance process [that forms ^{239}U]. . . . Is it possible that ekaIr is palladium and ekaPt is silver? . . . When we have received an acceptance from *Nature*, I will send you a copy of the manuscript—then you can refer to it as if it were already in print. What you wrote about our uranium and thorium studies being a necessary preparation for your and Strassmann's beautiful end-results is quite right and I also thought it over in the same way. And perhaps it would be nice if you could somehow express this quite clearly in your next paper, *not* to justify our earlier work—we don't need that—but to indicate that without the earlier results and the development of the chemical and physical techniques it would not have been possible to clarify the Curie-Savitch observations so quickly. But if you don't feel like doing that, it's all right with me.[75]

Meitner's letter, it seems, finally prompted Hahn to consider the fission fragments in terms of atomic number rather than mass and to recognize, finally, that when barium is one fission fragment, krypton must be the other. Her letter arrived in Dahlem on 20 January; on the 23d, Strassmann tested for and found radioactive strontium and yttrium. It is uncertain whether they regarded these as indicators for krypton (by way of the beta decay sequence $_{36}$Kr $\rightarrow$ $_{37}$Rb $\rightarrow$ $_{38}$Sr $\rightarrow$ $_{39}$Y) or if they were merely testing for strontium as a possible alternative for barium, also a group II (alkaline earth) element.[76]

On 24 January, Hahn received copies of the Meitner-Frisch *Nature* manuscripts.

> 24.I.39 Dear Lise. . . . It is wonderful how quickly you and Otto Robert thought of the physical experiments and carried them out, so that some of our laborious chemical trials would have been completely unnecessary. . . .

We also have confirmed that our radiums are Ba from thorium as well, despite frightfully low activities, but not so elegantly as with uranium. Now after your experiments, this confirmation would not have been so necessary. . . . We do not believe in your view of the "transuranes" [being false]; in the absence of something better we think they are still, or perhaps again, transuranes. Their characteristics do not agree with [the sequence] Ma to Ag. But more experiments are necessary there. On the other hand, we have also thought of krypton, etc., as the second product of splitting. Until now we have not been able to confirm the presence of krypton or rubidium. But perhaps strontium and yttrium. . . . How shall we handle this [in print]? I think that we will briefly describe our results and say *that the two of you also suggested this possibility.* Others have made the same supposition. . . . The physicists here naturally have also thought about the difference between the atomic numbers 92–56, after the difference in atomic weights didn't work out.[77]

Hahn was impressed enough by Frisch's physical experiment to consider some of his own trials useless, but his only reaction to the theoretical arguments of the Meitner-Frisch paper was to insist that krypton had occurred to many others, including himself and Strassmann, before they learned of it from her. Perhaps he was irritated with Meitner for not confiding in him sooner; undoubtedly he was nervous about the grumbling within his institute. But whatever the reason, he was ungenerous, possibly even dishonest, about the krypton. For he and Strassmann had not looked for strontium until 23 January—after they received Meitner's letter of 18 January, which pointed to the importance of atomic numbers—and they did not begin their search for krypton or rubidium until the evening of January 24—*after* copies of the Meitner-Frisch manuscripts arrived in Dahlem.[78]

Dismayed, Meitner thought Hahn sounded angry.[79] But she knew how tired he was, and thinking he was just feeling a bit sorry for himself, she responded with an especially affectionate letter.

25.I.39 Dear Hähnchen!. . . Your laborious experiments are by no means "unnecessary." Without your beautiful result—Ba instead of Ra—we would *never* have had anything to consider, and you can hardly imagine what it meant to me for once to do something that really seemed like scientific thinking for a few days. Naturally you should publish your Sr-Y results, they are just beautiful, just cite that we have stated it in our *Nature* note on the basis of our simple considerations. . . . The recoil experiments confirm only the fact of splitting, but *not* into *what* it splits. That, after all, can only be determined by chemistry, and therefore every one of your experiments is extremely important.[80]

After Meitner had suggested in her 18 January letter that one of the "transuranes" might actually be silver, Strassmann had tested for it on 20 January and found none.[81] This bolstered Hahn's faith in the transuranes, he wrote Lise on the 25th, although

> the puzzling similarity [to Ba] in excitation and enhancement factors seems to indicate the opposite. So do your arguments, which we don't understand in detail. . . . Of course we do not doubt it, but since we cannot show your manuscript without your permission, we cannot ask any physicist. Please write when we can show your manuscript. V[on] Droste and Reddemann have or had the idea of looking for fast beta radiation (or something like that) as a consequence of splitting using Droste's apparatus and Reddemann's [neutron] source. I am sure this would take a year at Droste's pace, but perhaps they won't even begin once they know of Otto Robert's result. . . . Last not least: In addition to the Sr (+ Y) (no Ba!) which we determined a few days ago we found the corresponding rubidium, as a short-lived substance, twice when delivering the hypothetical krypton into dilute acid. . . . Please treat th[is] as confidential (except for Otto Robert). Now we want to write our work up quickly. . . . For reasons of expediency—our chemists in the institute!—it might be advisable if in your paper you cited the chemical references for the transurane papers in addition to those in Zs. für Physik.[82]

> 26.I.39 Dear Otto! . . . Your results are really wonderful! . . . The fact that Kr-Rb-Sr + Y + Zr should be present, we deduced from the constancy of nuclear charge (56 + 36) and we mentioned only that one or another of the Ba-La series could be Sr or Y. . . . O.R. telephoned Monday evening, told me that the page proofs were there and that he would send them off right away; I did not see them at all. Therefore the literature citations cannot, unfortunately, be changed. Unhappily I overlooked the fact that only our joint papers from Zs. f. Physik are cited (O.R. did the literature, as the paper is written in English) and not our first papers from Naturwiss[enschaften] or those in Chem[ische] Ber[ichte]. I am really very sorry, but I hope that no one will think it was on purpose, especially since your and Strassmann's really wonderful results are stressed as the basis—as a matter of course—in both notes. I think that everything you have done since then is absolutely marvelous. Really half the periodic system is included in the splitting of uranium, and in these few months you have earned several "first" prizes. . . . You can certainly show our notice to anyone, as I assume that it will appear quickly.[83]

On 28 January, Hahn and Strassmann submitted their newest findings to Naturwissenschaften.[84] Hahn was exhausted. After mailing a copy of the manuscript to Meitner, he prepared to leave for two weeks in the Italian Alps.

Hahn and Strassmann devoted the first part of their paper to their barium proof, the second part to other fission products, including the unsuccessful search for light elements among the transuranes, the discovery of strontium and yttrium, and their early evidence for krypton. Throughout his narrative Hahn gave not the slightest indication that the idea for several of the experiments had come from Meitner. When he finally mentioned the Meitner-Frisch paper, it was almost as a postscript: "As we were writing up our last experiments [the search for krypton] we received manuscripts of two communications which will appear in *Nature*, kindly sent to us by their authors, Lise Meitner and O. R. Frisch, and O. R. Frisch. Meitner and Frisch discuss the splitting of uranium and thorium nuclei into two large fragments of approximately equal size, e.g., barium and krypton, and base the possibility of such an occurrence on Bohr's new liquid drop model for nuclei."

By placing this paragraph directly after the experimental discussion, Hahn—intentionally or not—characterized the Meitner-Frisch work as a fairly meaningless description of known results. Moreover, the Meitner-Frisch manuscript did not, in fact, arrive as Hahn and Strassmann were writing up their krypton results but before that experiment began—and may have been the impetus for it. And Hahn said nothing to indicate that Meitner prompted the search for light elements and anticipated the discovery of strontium and yttrium.

Already depressed about her situation in Sweden, the paper plunged Meitner into despair. It seemed that nothing she had done before or after fission would be recognized. She feared her reputation was irreparably damaged and she would never improve her status in Stockholm.

5.II.39 Dear Otto! . . . Your results present a wonderful closed chain of results, and it is marvelous what you have accomplished in these few short weeks. Unfortunately I fear from the manner in which you brought up our note that you are personally angered by the lapse in our literature citations. I really am terribly sorry. I had hoped a little that our note would have given you some pleasure also, and it would have been so nice for me if you had just written that we—independently of your wonderful findings—had come upon the necessity for the existence of the Kr-Rb-Sr series.[85] You yourself wrote in your first answer to our manuscript (24.I.1939), "I believe we will briefly describe our experiments with Sr and Y and say that you also stated this supposition." . . . With me things are not good at all. I have a place to work here but no position that gives me the least right to anything. Try for a moment to imagine how it would be if instead of your own beautiful institute, you had *one* work room in a *strange* institute, without the *slightest* help, without any rights, and with Siegbahn's attitude: he only loves big machines, is very sure of himself and self-confident, and probably does not want anyone independent around him.[86] And I with my inner insecurity and self-consciousness, that I have to do all the little jobs that I haven't done for 20 years. Of course it is my fault; I should have prepared my departure much better and much earlier; should have had at least a drawing for the most important apparatus, etc. Siegbahn said to me once, Debye wrote nothing about co-workers or assistants (I asked Debye to do that several times) and he has little room. To me that doesn't seem true, the institute looks empty, there are few people around. But the important thing is that I came with such empty hands. Now Siegbahn will gradually believe—especially after your beautiful results—that I never did anything and that you also did all the physics in Dahlem. I am gradually losing all my courage. Forgive this unhappy letter. I never wrote how bad it really is.[87] Sometimes I do not know what to do with my life. Probably there are many people who have emigrated who feel as I do, but still it is very hard.[88]

She wrote the next day to her brother Walter,

Unfortunately I did everything wrong. And now I have no self-confidence, and when I once thought I did things well, now I don't trust myself. The Swedes are so superficial; I don't fit in here at all, and although I try not to show it, my inner insecurity is painful and prevents me from thinking calmly. Hahn has just published absolutely wonderful things based on our work together—uranium and thorium nuclei split into lighter nuclei such as barium and lanthanum, krypton, strontium, etc. And much as these results make me happy for Hahn, both personally and scientifically, many people here must think I contributed absolutely nothing to it—and now I am so discouraged; although I believe I used to do good work, now I have lost my self-confidence.[89]

7.II.39 Dear Lise, I just received your letter of February 5, and I want to answer it right away. You cannot imagine how your situation affects me, and how glad I would be if I could help you. With small things I tried to do so when I could. With large things I could not. And now, completely unintentionally, I have done something wrong in the way that I have cited you. I only know that at first my citation was even briefer, then I added something about the Bohr liquid-drop model to make it a little longer. Now I see that instead of [saying] 2 pieces, I should have mentioned Sr and Y by name. Nevertheless I cannot imagine how that can make a difference. Strassmann and I already had thought of Sr (and therefore naturally of Kr as the source), then this came from all sides. . . . We refused all discussions; I refused

Wefelmeyer [*sic:* Wefelmeier] three times despite Weizsäcker's recommendation. Therefore we could not write anything other than what we did. I fear that to some extent it is also held against me that we strictly refused to tell anyone about our findings. I heard something like that: if we had told others a little more, then Droste—he now says—could have done some experiments with Reddemann earlier, and he would automatically—he now says—have gotten Frisch's results. Of course I assign no value to such things, but I do not want to confess to these gentlemen that you were the only one who learned of everything immediately. . . . Believe me, at times I am somewhat ashamed that we got such clear results so quickly while you sit in Stockholm in an empty institute. But with the others at our heels, you understand that we wanted to move quickly. Once during the Droste commotion, Philipp said to me directly that perhaps we should have discussed our results in the institute first and not published right away; I had to answer him that with our previous experience with Irène [Curie] we did not want to be too late again. He agreed with that.

I don't understand how you can believe that Siegbahn thinks that Strassmann and I also did the physics. In all our work we absolutely never touched upon physics, instead we only did chemical separations over and over again. We know our limits and of course we also know that in this particular case it was useful to do only chemistry. The question of [who will get] your apartment is not settled. . . . In that regard the uranium work [fission discovery] is for me a heaven-sent gift. Namely I was fearful sometimes that Dr. K. would first [take over] your apartment, then eventually parts of the institute.[90]

Hahn's letter is testimony to the self-deception brought on by fear: fear of his collaborators, of the opportunistic "Dr. K." and others, of the ambitious Kurt Hess, of anyone and everyone who was poised to take advantage of his political vulnerability. A year before, just after the Anschluss, when Lise's presence seemed to endanger the institute, he had instinctively distanced himself from her. Now that she was gone, he did not want to admit that they had continued their collaboration. If in December he still thought of a publication by Meitner almost as "work by the three of us," it was because he still needed her active contribution. By February, he could barely bring himself to cite her name. In those six weeks, he had come to think of the barium finding as an isolated discovery that relied only on the chemical separations he and Strassmann had done in December. Politically, it was much safer that way. Fission was a "gift sent from heaven" to protect him and his institute, a gift with no strings attached, no ties to the past, owing nothing to physics or Meitner: "In all this work we absolutely never touched upon physics." Hahn was divorcing fission from physics—and himself from Meitner. It was an exceedingly expedient thing to do.

Meitner did not comprehend all of this right away; she, too, was concerned for Hahn's well-being and the integrity of the institute they had shared. And she was preoccupied with her own professional difficulties in Stockholm.

> 12.II.39 Dear Otto! . . . I came here with completely wrong information. I have gradually gotten it out of Eva v. B[ahr-Bergius] just how it went. Siegbahn actually did not want to have me. At the time he said he had no money, he could only give me a place to work, nothing more. Then Eva wrote to [C. W.] Oseen [an Uppsala theoretical physicist] and he said that the Nobel Foundation could grant some money. But no one thought of the fact that I cannot work without help, particularly not in an institute where nothing is available for my field. And when I noticed that here I was quite despondent, and of course that is not the right situation for asserting oneself. In Dahlem, if I needed something from a shop, I talked it over with Gille, he did the mechanical work and also the necessary drawings. Here I am told every time: please make a drawing. So I do the work that Dorffel did, and actually I don't know how. And Siegbahn, who is a wonderful draftsman and technically very gifted, now will gradually believe that I can do nothing. Perhaps he is right. I am often so terribly despondent. Of course I am at fault, I should not have left the way I did, should have taken along many more drawings of the important apparatus . . . and above all I should have discussed the pertinent working conditions before I came.[91]

And to Eva von Bahr-Bergius,

> All these things weigh on me so, that I am losing all my self-confidence. I am making a real effort to hold on to my courage and I tell myself again and again that until now I have done very respectable physics. But under the current conditions I won't be able to do anything sensible and the fear of such an empty life never leaves me.[92]

Lise Meitner had come to Sweden with nothing but her scientific reputation. Now, seeing how little that meant in exile, she was reliving the terrifying insecurity of her first months in Berlin: again she was a stranger in a foreign country, unwelcome in a male profession, living poorly on bread and black coffee in a tiny room—and devastated to find that thirty years in physics had not kept it from happening again.[93] In the winter of 1939, her unhappiness deepened into despair. She blamed herself for her decision to come to Sweden, for her inability to cope with Manne Sieg-bahn, for failing to prepare herself for emigration. She lost hope of ever improving her situation in Stockholm: she feared that her professional life was over, her scientific past destroyed. In her despondency, Meitner exaggerated the importance of recognition from Germany and greatly underestimated the impact of her theoretical note with Frisch. The term "fission" that they proposed was instantly accepted as the name of the new process,[94] and over the next several months their fission interpretation would receive wide attention as Bohr used it as a starting point for further theory. Bohr also relied on physical measurements done by Meitner in 1937 to deduce that ^{235}U and not ^{238}U undergoes fission with thermal neutrons.

But her conditions in Siegbahn's institute did not improve. In Sweden there was no general sympathy for refugees from Nazi Germany: the country was small, with a weak economy and no immigrant tradition, and its academic culture had always been firmly pro-German, a tradition that would not change appreciably until the middle of the war when it became obvious that Germany would not win.[95] Members of Siegbahn's group saw Meitner as an outsider, withdrawn and depressed;[96] they did not understand the displacement and anxiety common to all refugees, or the terrible worries about friends and relatives, or the exceptional isolation of a woman who had single-mindedly devoted her life to her work.

Although Lise Meitner would adjust, work, and live in Sweden for another twenty years, her forced emigration, her exclusion from fission, and her poor relationship with Siegbahn would have lasting effects. Forced out of one country, held at arm's length in the other, she would not fully establish herself in physics again. And she would never be at home.

Lise Meitner, about 1900, age twenty-two. (Courtesy Churchill College Archives Centre, Cambridge)

Hedwig Skovran Meitner, Lise's mother. (Courtesy Churchill College Archives Centre, Cambridge)

Philipp Meitner, Lise's father. (Courtesy Churchill College Archives Centre, Cambridge)

Ludwig Boltzmann, 1898, shortly before he became Lise's teacher. (Courtesy Bildarchiv der Österreischischen Nationalbibliothek)

Stefan Meyer as a young man, about 1900. (Courtesy Deutsches Museum, Munich)

Max Planck, as he appeared around 1900. (Courtesy Deutsches Museum, Munich)

Emil Fischer, the great organic chemist, about 1900.
He reluctantly allowed Meitner to work in his institute
in 1907. (Courtesy Deutsches Museum, Munich)

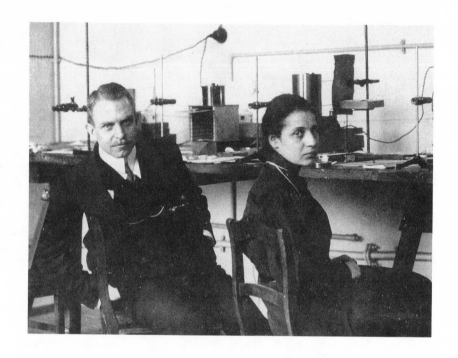

Meitner and Otto Hahn, in their laboratory in Fischer's institute, about 1910. (Courtesy Archiv zur Geschichte der Max-Planck-Gesellschaft, Berlin)

Hahn, Meitner, and Emma (or Grete) Planck, about 1910. The Planck sisters were identical twins; usually only one would be in a photograph, no doubt because the other was taking it. (Courtesy Churchill College Archives Centre, Cambridge)

Grete (or Emma) Planck, Meitner, and Elisabeth Schiemann, about 1913. (Courtesy Churchill College Archives Centre, Cambridge)

The small instrument is the simple beta spectrometer first used by Meitner, Hahn, and Otto von Baeyer in 1910. Meitner used the larger instrument for her studies of beta-gamma spectra in the 1920s. (Courtesy Deutsches Museum, Munich)

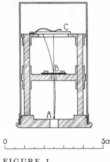

FIGURE 1

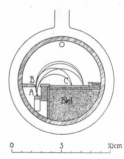

FIGURE 2

In the small instrument, figure 1, sample **A** emits beta particles (negative electrons) that travel upward through slit **B** and are recorded as a dark line on photographic plate **C**. In a magnetic field perpendicular to the plane of the diagram, the electrons are deflected into a circular path, the deflection being greater for less energetic electrons. Meitner, Hahn, and von Baeyer observed discrete lines on the photographic plate, evidence of monoenergetic electron groups whose energy they determined from the position of the lines. In the larger instrument, figure 2, the orientation of the photographic plate **C** has been changed to improve the resolution of the electron lines. (Courtesy Deutsches Museum, Munich)

Meitner with Eva von Bahr-Bergius, Kaiser Wilhelm Institute for Chemistry, about 1920. The women met before World War I, when Eva was a student in Berlin; later, Eva would be Lise's closest friend in Sweden. (Courtesy Churchill College Archives Centre, Cambridge)

Colloquium with Niels Bohr in Berlin, 1920. From left: Otto Stern, Wilhelm Lenz, James Franck, Rudolf Ladenburg, Paul Knipping, Bohr, E. Wagner, Otto von Baeyer, Hahn, George de Hevesy, Meitner, Wilhelm Westphal, Hans Geiger, Gustav Hertz (with pipe), Peter Pringsheim. (Courtesy Archiv zur Geschichte der Max-Planck-Gesellschaft, Berlin)

Miep and Dirk Coster in Groningen on their silver wedding anniversary, 1944. (Courtesy Ada Klokke-Coster, Epse)

Kaiser Wilhelm Institute for Chemistry, about 1930; view from Thielallee. The smaller building at the left is the institute villa, where Meitner lived in an apartment during this period. (Courtesy Churchill College Archives Centre, Cambridge)

Lise Meitner, about 1930.
(© Atelier Lotte Meitner-Graf, Courtesy
Archiv zur Geschichte der Max-Planck-
Gesellschaft, Berlin)

In the laboratory, about 1930.
(Courtesy Archiv zur Geschichte der
Max-Planck-Gesellschaft, Berlin)

Solvay Congress, Brussels, 1933, which was attended by the leading atomic and nuclear physicists of the time. Seated, from left: Erwin Schrödinger, Irène Joliot-Curie, Niels Bohr, Abram Joffé, Marie Curie, Paul Langevin, Owen Richardson, Ernest Rutherford, Théophile de Donder, Maurice de Broglie, Louis de Broglie, Lise Meitner, James Chadwick. Standing: E. Henriot, Francis Perrin, Frédéric Joliot, Werner Heisenberg, Hendrik A. Kramers, E. Stahel, Enrico Fermi, Ernest Walton, Paul Dirac, Peter Debye, Nevill Mott, B. Cabrera, George Gamow, Walther Bothe, Patrick M. S. Blackett, M. Rosenblum, J. Herrera, E. Bauer, Wolfgang Pauli, M. Cosyns, J. Verschaffelt, E. Herzen, John D. Cockcroft, Charles D. Ellis, Rudolf Peierls, Auguste Piccard, Ernest O. Lawrence, Léon Rosenfeld. (International Institute of Physics and Chemistry, Courtesy American Institute of Physics, Emilio Segrè Visual Archives)

Otto Robert Frisch, age 29, shortly before emigrating from Germany in 1933. (Courtesy Ulla Frisch)

Fritz Strassmann in 1936, age 34. (© Hanne Zapp-Berghäuser, Courtesy Irmgard Strassmann)

Max von Laue in his automobile, mid-1930s. (Courtesy Archiv zur Geschichte der Max-Planck-Gesellschaft, Berlin)

*Meitner and Hahn, about 1935. (Courtesy Archiv zur
Geschichte der Max-Planck-Gesellschaft, Berlin)*

ARBEITSTISCH VON OTTO HAHN

Meitner's physical apparatus, used by the Berlin team from 1934 to 1938 for the work that led to the discovery of nuclear fission. Erroneously displayed in the Deutsches Museum, Munich, as the "Worktable of Otto Hahn." (Courtesy Deutsches Museum, Munich)

    Arranged on a wooden table like the one in Meitner's laboratory, the display includes a neutron source, uranium, and a paraffin cylinder (right rear), detectors, amplifiers, and counters (center), a bank of batteries (below), and lead housings (front) for shielding the counters and handling the radioactive specimens. The display is a composite: in practice, it was necessary to carry out the irradiations, the measurements, and the chemical separations (symbolized by the flask at right) in three separate rooms.

*Kaiser Wilhelm Institute for Chemistry and the institute villa
(right) after air raids, February 1944. (Courtesy Archiv zur
Geschichte der Max-Planck-Gesellschaft, Berlin)*

Meitner and President Harry S. Truman, 9 February 1946, Washington, D.C. Meitner was honored as "Woman of the Year" by the National Women's Press Club. (Courtesy Churchill College Archives Centre, Cambridge)

Meitner lecturing in Bonn after receiving the Max Planck Medal of the German Physics Society, 23 September 1949. (Courtesy Theodore Von Laue)

Elisabeth and Gertrude Schiemann with Meitner,
about 1950. (Courtesy Archiv zur Geschichte der
Max-Planck-Gesellschaft, Berlin)

Meitner at Bryn Mawr College, 1959. (Courtesy American Institute of Physics, Niels Bohr Library)

Hahn, Werner Heisenberg, Meitner, and Max Born at the Lindau conference of Nobel Laureates, 1962. (© Franz Thorbeck, Courtesy Archiv zur Geschichte der Max-Planck-Gesellschaft, Berlin)

The 1966 Enrico Fermi Prize, presented to Lise Meitner in October 1966 in Cambridge by Glenn T. Seaborg, chairman of the United States Atomic Energy Commission. Otto Frisch is at Meitner's right. (Courtesy Max Perutz)

Priorities

But it is so easy for misunderstandings to crop up with these things.

While Lise Meitner languished in Stockholm, physicists elsewhere were rushing ahead. The pace was hectic, the experiments often simple and quick. By the end of February 1939, dozens of physicists had confirmed the fission process. Otto Robert Frisch had been the first to detect the huge pulses of ionization from fission fragments, on 13 January, but his *Nature* note did not appear until mid-February, by which time Frédéric Joliot had published a similar experiment in *Comptes Rendus*[1] and physicists all over the United States and Europe had made similar observations.[2] Joliot collected the fission fragments, as did Edwin McMillan in Berkeley a little later,[3] but neither attempted to identify the collected material with the former "transuranes." Lise Meitner would do that experiment in Copenhagen at the end of February.

Meanwhile, several other investigators found that the ionization pulses fell into two distinct energy groups, an indication that uranium splits into two fragments of different size; the almost continuous variation within each group indicated that splitting can occur in many different ways.[4] Ionization chamber experiments by Frisch, Joliot, and others showed that fission in thorium occurs only with fast neutrons while uranium favors slow neutrons, in agreement with the reaction conditions reported by Meitner, Hahn, and Strassmann in their investigations several years before.

In February, Joliot irradiated uranium in a cloud chamber, photographing the thick tracks of fission fragments that started at a common point and diverged in opposite directions.[5] D. R. Corson and R. L. Thornton in Berkeley, noting that the fission fragments were undeflected by collisions with lighter nuclei in the cloud chamber, estimated the mass of the fission fragments to be at least 75.[6]

Prior to the discovery of barium, Philip Abelson in Berkeley had been studying the characteristic x-radiation of the "transuranes"; associated with the 66-hour "ekaIr" (presumably element 95), he detected a relatively soft radiation that he assumed was its L x-rays. After learning of Hahn and Strassmann's result, however, he realized that these were in fact K x-rays from iodine; subsequent chemical tests showed that the 66-hour "ekaIr" was an isotope of tellurium (element 52), and its daughter, the 2.5-hour "ekaPt," was indeed iodine (53), the source of the observed K x-rays.[7] Abelson's finding, verified by Norman Feather and Egon Bretscher in Cambridge,[8] was an independent chemical confirmation of fission and the first identification of "transuranes" with isotopes of lighter elements.[9] With that, the entire chemical foundation of the uranium investigation crumbled, since all "transuranes," including "ekaIr" and "ekaPt," had been thought to resemble platinum in chemical behavior, yet tellurium and iodine are not platinumlike at all.[10]

But the earlier physical measurements held, and they were essential to a deeper understanding of the fission process. In two short notes written in Princeton at the end of January and early February 1939, Bohr outlined a theoretical approach that tied Meitner, Hahn, and Strassmann's 1937 findings to the expected behavior of compound nuclei. The experimental picture was complex: fission in uranium occurs slightly with fast neutrons, readily with thermal neutrons, and not at all with moderate-energy neutrons, which, Meitner had found, undergo a typical resonance capture by ^{238}U to form ^{239}U.[11] Bohr's first note, submitted a few days after his arrival in the United States, was written primarily to defend Meitner and Frisch's priority against the rush of American results; in his second note, submitted to *Physical Review* on 7 February, he again made a point of emphasizing Frisch's "direct proof" of fission and Meitner and Frisch's "ingenious suggestions" that fission be explained in terms of the compound nucleus. According to compound nucleus theory, Bohr noted, a nucleus that captures resonance neutrons of moderate energy requires faster neutrons for other reactions, such as fission, to occur. The "peculiar effect" that uranium undergoes fission with both fast and thermal neutrons seemed to indicate two independent fission processes, one for the abundant (99.3%) ^{238}U isotope and another for the very scarce (0.7%) ^{235}U. For Bohr, the resonance process was the key: if ^{238}U is not split by moderate-energy neutrons, he reasoned, it surely will not be split by thermal neutrons—so the enhanced fission seen for thermal neutrons must, therefore, come from ^{235}U.[12] In Princeton over the next several months, Bohr and John A. Wheeler developed a quantitative theory for the mechanism of fission that explained known fission phenomena and established criteria for other fissile nuclei.[13] Crucial to the theory, however, was Bohr's proposal that fission in natural uranium is due primarily to ^{235}U. That would be verified experimentally in February 1940, when Alfred O. Nier in Minnesota separated a tiny amount of ^{235}U in his mass spectrometer and mailed it to Columbia University where John R. Dunning, Aristide von Grosse, and Eugene T. Booth determined that it was indeed fissile with slow neutrons.[14]

Meanwhile, new experiments raised the prospect that fission might be used to generate immense quantities of energy. In March, Hans von Halban, Joliot, and Lew Kowarski in Paris, followed immediately by Herbert Anderson, Fermi, and H. B. Hanstein at Columbia, reported that several free neutrons are released for each uranium that splits. They recognized that under suitable conditions such secondary neutrons might in turn induce fission in other nuclei, setting in motion a chain reaction capable of sustaining itself until the uranium was entirely consumed.[15]

The possibility of a chain reaction aroused fascination and fear. Frisch reported,

> [My] immediate answer was that in that case no uranium ore deposits could exist: they would have been blown up long ago by the explosive multiplication of neutrons in them! But I quickly saw that my argument was too naive; ores contained lots of other elements which might swallow up the neutrons; and the seams were perhaps thin, and then most of the neutrons would escape. With that, the exciting vision arose that by assembling enough pure uranium (with appropriate care!) one might start a controlled chain reaction and liberate nuclear energy on a scale that mattered. Of course, the spectre of the bomb was there as well.[16]

This all took place before the end of March 1939; more would follow in the months ahead. The field was wide open, the experiments fast and exciting.[17] As physicists rushed into print, some neglected to fully cite the work of others; some results appeared in newspapers and magazines weeks before publication in scientific journals. The work was surest and quickest where both talent and neutron sources were strong: in Copenhagen, Paris, Cambridge, Berkeley, New York, and Princeton.

Otto Hahn felt left out. Having convinced himself that the discovery of barium owed little to Meitner and nothing to physics, he was irked to find fission so thoroughly dominated by physics—and by physicists with no special connection to him. The local physicists did credible work but had difficulty keeping up. Although many resented the fact that Hahn had not informed them earlier—and continued to complain for decades to come[18]—none had been close enough to the uranium project when Meitner was still in Dahlem to take the initiative for the physics after she left. Meitner's former assistants Hermann Reddemann and Gottfried von Droste had a good neutron source and ample electronic equipment but took several months to publish their measurements of secondary neutrons;[19] von Droste and Siegfried Flügge, the institute's theoretical physicist, quickly arrived at some of the same theoretical conclusions as Meitner and Frisch, but their work did not appear until March. And nearly a month after the publication of barium, Wilfrid Wefelmeier and Carl Friedrich von Weizsäcker submitted two related theoretical articles in which they still discussed the isomerism of the transuranes.[20]

Throughout the spring of 1939, Hahn worried constantly about his and Strassmann's priority for fission; sensitive to every inadequate reference from abroad, apprehensive about the grumbling within his institute, he feared that his "gift from heaven" might slip away.[21] In part his anxiety befitted his situation: he was a non-Nazi under duress in Germany, a German becoming estranged from the international community. But much of it was his own doing. To bolster his professional standing, he had claimed fission for chemistry alone, distanced himself from Meitner, and been ungenerous to Curie and Savitch—only to find himself out of the fission mainstream. In the first few months after the discovery, when Meitner and Frisch were highly visible among physicists and the French were ignoring Hahn and Strassmann as best they could, Hahn was exceedingly unhappy.

3.III.39 Dear Lise! . . . I am quite convinced that you and Otto Robert want to be as objective as possible in your publications. There can't be any doubt of that. There may, perhaps, be some doubt whether your manner of setting things forth was very adept. Because gradually I must say [Otto] Erbacher and [Kurt] Philipp are right when they say that the priority for splitting uranium is gradually slipping away from Strassmann and me.[22]

Hahn angrily listed several reports from America and elsewhere that attributed the observation of barium to Hahn and Strassmann and the interpretation to Meitner and Frisch (true enough), but Hahn did not see it that way: lacking an understanding of physics, especially theoretical physics, he believed that his mention of splitting constituted an interpretation on a par with Meitner and Frisch's. Hahn was especially livid about the French scientists, who occasionally omitted him and Strassmann entirely while diligently citing Joliot, Curie and Savitch, and Meitner and Frisch (revenge, perhaps, for Hahn's stingy references to the work of Curie and Savitch). What bothered Hahn, of course, was the politics: "These things are widely discussed during the institute coffee hour, and most think the work by Strassmann and me has not been treated justly.... The fanatics among our people are making it somewhat political, and I regret that especially."[23]

In his *Nature* note of 20 January, written to emphasize Meitner and Frisch's priority, Bohr had mentioned Hahn and Strassmann's finding in the opening sentences but cited only the publications of Meitner and Frisch. Hahn was annoyed by this, and he misunderstood the title of Meitner and Frisch's note. He complained to Meitner,

Your title was somewhat disturbing. "A New Nuclear Reaction, etc." ... Our people say: Str[assmann] and I already mentioned [fission] in January against the opinions of every physicist; even a possible second fragment (masurium) is already in it. In this regard Philipp is somewhat angry with me and says it serves us right. Because we said nothing about our results before the article was out; if we had, Flügge and the others could have obtained physical verification as quickly as several others did in a few days. (I personally am not convinced of that. But of course I dare not say that I always kept you completely current and not the institute. That would be greatly held against me. When Strassmann was asked if we sent you the manuscript he had the good sense to deny it indignantly.) ... The only one who observes these things absolutely calmly is Strassmann himself who, one must say, comes away the worst in every sense. He says: every thinking person who reads our paper cannot doubt that we observed the fission of uranium and also stated that it was so. Perhaps it was wrong that in our conscientiousness we wrote it so cautiously. But we had to assert something quite alone, contrary to every physical authority. ... If we were as careless as Irène Curie, we would have published the Ba in November. Even you could hardly make up your mind to believe the results.[24]

It is hard to know what Hahn means here: Meitner never doubted the barium finding. What is clear, however, is that he wanted full recognition not just for barium but for fission as a whole. And yet, in his 6 January article he had weakened their claim by hedging—"as nuclear chemists close to physics we cannot yet bring ourselves to take such a drastic step"—and then by making the truly fundamental error of using atomic mass rather than charge to propose masurium (technetium) as the second fission fragment. This led others to believe that he and Strassmann were unaware of the meaning of barium until Meitner explained it.[25] And despite Hahn's denials, this was largely true. For it was not until after Meitner assured him that a large-scale nuclear disintegration was possible that he added the idea of splitting (and the unfortunate masurium) to the barium article. Hahn had apparently forgotten all this.

> Strassmann and I know that we were as sure as possible of the splitting of uranium, without leaning on any other observation or hypothesis. After we knew at the beginning of January that the transuranes were not masurium, etc., the search for the other fragments was self-understood.[26]

Here again Hahn's memory was selective. He and Strassmann did not search for strontium and yttrium until after they received Meitner's letter of 18 January or for krypton until after they had read the Meitner-Frisch manuscript on 24 January.[27]

> If [credit for fission] turns out otherwise, as it now appears, there is nothing we can do. We both (you and I) already experienced this with thorium; I myself with radioactive recoil.[28] . . . I have not the slightest doubt that you [and Otto Robert] and also Bohr are loyal. If others are not, we will do our best to straighten things out objectively. We will not stoop to the methods of the French, that would be just too shabby.[29]

One is struck by Hahn's exaggerated defensiveness. After all, no one could deny that they had found barium, no matter how hesitantly they may have announced it, and no one could dispute that they were the first to publish the idea, however poorly articulated, that uranium had split. The danger that Hahn and Strassmann might lose their place in the history of the discovery was nil. Why, then, was he so upset?

Hahn's problem was that he wanted more from fission than a place in its history: it was to be his salvation, the heaven-sent gift that would render him politically and professionally invulnerable. For this, he had to transform the discovery into something that was his and his alone, the product of his institute, his Mitarbeiter, his discipline—unencumbered by ties to other scientists and other disciplines.

What made this difficult was that the discovery was so inherently interdisciplinary. Everyone in the field knew that the investigation was begun by physicists and, for all its reliance on analytical and radiochemical data, driven by physics at every step; that fission was being interpreted, verified, and extended by physicists into a fertile new field of nuclear physics to which chemists, including Hahn and Strassmann, were contributing relatively little. It was obvious, too, that others had contributed mightily and come very close, that Curie and Savitch's work had pointed Hahn and Strassmann in the right direction, that Meitner's absence from Dahlem was a political artifact that kept her from sharing in a discovery that in normal times would have been regarded as the culminating achievement of an interdisciplinary team.

Hahn's response was to define the discovery as the three weeks' work he and Strassmann had done in Dahlem in December. Although at Christmas 1938 he still regarded Meitner as a partner and the discovery as "a kind of work by the three of us," by February 1939 he was sure he and Strassmann had "absolutely never touched upon physics, instead we only did chemical separations over and over again," and in March he believed the discovery was "contrary to every physical authority." The evolution was inexorable: physics—and Meitner—were absent, irrelevant, opposed, obstructive. For a radiochemist in the shadow of physics, the success of his discipline must have been as appealing as his disengagement from Meitner was politically useful. He had nothing to lose, except for one thing: by cutting himself off from physics, Hahn risked being submerged by the flood of new physical discoveries. His hypersensitivity in the spring of 1939 was an expression of his fear that physicists might be ignoring him just as he was trying to cut himself off from them.

He need not have worried. In Germany there was an early and pronounced interest in powerful new explosives. Although Hahn did not initiate contact with the military, he did not avoid it when it came.[30] By summer the threat to his institute was over: he was safe, his position secure.

As he recalled after the war, "The splitting of uranium saved that whole situation."[31]

Lise, meanwhile, tried to soothe Otto's nerves and correct his more obvious errors.

> 6.III.39 Dear Otto! . . . I think . . . that you or your gentlemen have misunderstood some things. [Our] title, "A New Type of Nuclear Reaction," referred *as a matter of course* to your and Strassmann's findings, and we only tried to show that one can explain this "new nuclear reaction" *on the basis of the purely classical liquid drop model,* and that it must be accompanied by energies of the order of magnitude of 200 MeV. That is exactly what Bohr meant when he says, word for word, "that the authors (i.e., we) propose an *interpretation* of the remarkable findings of Hahn and Strassmann." . . . He discusses in conceptual detail only the theoretical aspect, and takes your findings as so certain, that he bases his discussion on them.[32]

The next day, 7 March, Lise wrote to Otto again, a sixtieth birthday letter with no unpleasant discussions about credit.

> You gave yourself the most beautiful possible birthday present—the wonderful discovery by Hahn and Strassmann. . . . Tonight Otto Robert and I will drink a glass of wine to you. And perhaps in 10 years there will be a bottle of wine for us all together—that will have to be a big one.[33]

Lise had been in Copenhagen since mid-February. With Otto Robert she carried out the experiment she had earlier proposed for collecting fission fragments and testing them for the presence of the former transuranes. The Bohr institute's new high-voltage apparatus served as a powerful neutron source;[34] a thin layer of uranium hydroxide was irradiated with neutrons, and the fission fragments were collected on the surface of water a millimeter away. Only light fission fragments had the energy to free themselves from the uranium and reach the water; heavy nuclei, including any true transuranic elements, were expected to remain embedded in the uranium. After applying the standard "transurane" hydrogen sulfide precipitation to the collected activities, Meitner and Frisch obtained decay curves in perfect agreement with an old curve for "transuranes." They concluded that "the 'transuranium nuclei' originate by fission of the uranium nucleus. . . . So it appears that the 'transuranium' periods, too, will have to be ascribed to elements considerably smaller than uranium."[35] In other words, the hydrogen sulfide precipitate contained nothing but fission fragments; for elements that were truly transuranic, one would have to look elsewhere.

The experiment had been done early in March, but Lise waited until after Otto's birthday to tell him.

10.III.39 Dear Otto! . . . It really is the case that the two long transurane series must be lighter elements. Obviously the only process that produces a heavy element is the resonance process. Actually I am not surprised at our result, since after your discovery of the splitting of the uranium nucleus I no longer believed in the transuranes, as I wrote to you many times. . . . We had this result several days ago, but since I thought you might feel sad at first, I did not want to write to you about it just before your birthday. But you and Strassmann could not have made your beautiful discovery if we had not done the earlier uranium work. . . . Our evidence is very simple and, it seems to me, unequivocal. We collected the recoil nuclei and I did the usual hydrogen sulfide separations. And in the hydrogen sulfide precipitate we were able to find the 16 min, 59 min, 5.7 h, 66 h, + 2.5 h substances. . . . It seems to me that there is, therefore, no doubt that the transuranes are lighter nuclei. Exactly which ones they are you will have to find out; we are not going to try.[36]

13.III.1939 Dear Lise, This time it is we who heartily congratulate you upon your exciting result with the recoil particles and the transuranes. We cannot find any holes in your interpretation and from your findings we must indeed declare that the transuranes are dead. For us—Strassmann and me—this result was, however, completely incomprehensible until now because we absolutely could not say what the transuranes might be. . . . In any case you [and Frisch] were the first to achieve a completely clear result for the physicists based on experiment and not on vague suppositions.[37]

Actually, Abelson had been the first to identify a "transurane" as a lighter element and, Meitner surely must have thought to herself, the theoretical arguments amounted to more than "vague suppositions."

In a similar experiment, Meitner collected the fission fragments of thorium and found, in addition to barium and lanthanum, activities with the chemical properties of "transuranes." In previous thorium studies, no such search had been made, since neutron irradiation of thorium had not been expected to yield transuranic elements. The fact that essentially the same elements were produced in the fission of both uranium and thorium provided additional proof that nuclei can split in so many ways that at least some fission products from two different nuclei can be the same.[38]

Before leaving Copenhagen at the end of March, Meitner wrote to Bohr, who was still in Princeton, to tell him her results and thank him for the "beautiful, productive time" in his institute. "Not only did everyone from [George de] Hevesy to the youngest Mitarbeiter strive to help me in the friendliest way with necessary materials and working space. The entire scientific attitude, the natural way of discussing problems with me in which I could ask and be asked anything, is wonderfully beneficial. Of course I owe my thanks for this beautiful atmosphere above all to you."[39]

Then Meitner returned to Siegbahn's institute and an atmosphere that was anything but friendly. While she was away, Gusti and Jutz Frisch had come to Stockholm. Lise was planning to share an apartment with them, but they had already received their furniture and books from Vienna while her belongings, after months of bureaucratic wrangling, were still impounded in Dahlem. She appealed to Otto.

> Perhaps you cannot imagine what it means for a person my age to live for 9 *months* in a little hotel room, with none of the comforts of home, with no scientific material and with the fear that no one has the time to move my situation forward. My sister and brother-in-law have rented an apartment beginning 1 May and are counting on me, but I don't have my things and they have barely enough for themselves.[40]

As Otto described it, there was no end to the harassment.

> You must think we haven't done a thing, but the opposite is true. After the official expert examined your things, your silver, etc., we thought it would go quickly. Then came a new order that all books, etc., must be examined. That couldn't be done without a list of every book and publisher. . . . After many telephone calls to the Literature Office, they turned us over to the superior authority: the Ministry of Propaganda. . . . Tomorrow two men will finally come to inspect all your books.[41]

In his next letter Hahn included a government memo: "As required by the Reich Literature Department the library in the rooms of Frau Lise Sara Meitner has been inspected today. Part was designated forbidden/undesirable or of importance to the state. The Customs Bureau of the National Finance Office is required to secure the excluded books and deliver them to the Reich Literature Department."[42]

"Your books were inspected," Otto explained, "and some were removed. I had already taken away some Thomas Mann, Werfel, Zweig, etc., but I couldn't clear everything out, that would have been too obvious. Therefore now, e.g., the Lily Braun, Gorki, the rest of the Thomas Mann, etc., were taken. A pity, but nothing can be done. . . . Everything has become unbelievably strict. . . . Only one or two settings of silver may be sent out! [Everything] was packed, then came the new order and all the utensils had to be unpacked because of the silver, etc. How things stand right now, I don't know."[43] This letter, more open than usual, was mailed from Paris by Clara Lieber.

Lise's fixation on her belongings went beyond practical necessity. Her few pieces of silver came from her mother's home; her books were collected over a lifetime. It was also a matter of self-respect. One wanted to keep *something:* the Nazis should not have it all. For help, Lise had a new lawyer in Berlin, whose stationery bore the name Dr. F. W. Israel Arnold and the notice, "Licensed for the legal representation of Jews only."[44] (As of 1 January 1939, German Jews had been required to add to their own names the middle names Sara or Israel; Jewish lawyers had lost their licenses some time before.) She also called a Swedish lawyer. She wrote to Otto, "On the telephone [he] offered to lend me a bed and linens. Thus after more than 30 years' work, I have gotten to the point where a total stranger must lend me a bed."[45]

Otto answered,

> Truly it is a scandal. I cannot tell you in a letter all that we have done in the last weeks and months. I just get *too* angry. . . . Your case is different somehow, perhaps because you are out. Something does not fit the usual pattern, and over and over again someone or other pronounces himself not responsible. You must have patience for a few days. . . . Surely this will not make you feel better, but believe me, the scandal with shipping your things has affected my nerves more than all the transuranes and the other not so pleasant things taken together.[46]

Paul Rosbaud assured Lise that Otto was doing all he could: "Your furniture will be sent, but he is completely at the mercy or lack of mercy of the official in charge. When I was sending furniture to my wife [in England] I practically had a nervous breakdown, and I remember several telephone calls during which I was seized by an insane desire to kill."[47]

What was left of Lise's household belongings and books arrived in Stockholm in May. The Swedish mover who delivered the shipment told Lise he had never seen anything like it—furniture in splinters, the bed broken, books with torn pages thrown randomly in with the furniture, the china cabinet filled with shattered dishes and glassware that had never been taken out and packed.[48] Lise must have wept bitterly at this message from Germany. To Otto, she wrote only, "In two weeks it won't seem important, but right now it makes me feel pretty low."[49] She never mentioned it to him again but a few weeks later reported that her health was poor. "In the last week I collapsed three times, but recovered quickly. Probably I am too thin. I weighed myself yesterday, 47 kg [103 lb] with clothing, a bit low." Edith Hahn, home with Otto and feeling much better, begged Lise to "eat tirelessly" to get back to her previous weight.[50]

Despite her troubles, Lise knew that in Germany she would have been far worse off. Between 1933 and 1938, one-third of all German Jews had emigrated, most in a fairly orderly fashion; in the ten months between Kristallnacht and the beginning of World War II, another third got out, any way they could. As a group, those left behind were older and weaker, were displaced from their former livelihoods, had had their property confiscated, and were cut off from normal social contact with the German population. By the beginning of 1939, Jews were forbidden to use hospitals, schools, and universities, forbidden to practice licensed occupations, and required to carry identifying papers and to assume the name Sara or Israel; by the end of the year, there were food rations, forced work, punitive taxes, and restrictions on housing, transportation, and communication. With no aspect of their existence exempt from coercion, Jews were a captive population whose only hope was that things would not get worse.

Otto did not write to Lise about such things. It was Max von Laue who kept Lise informed about events in Germany. "Please do not feel inhibited about writing here!" he assured her in the winter of 1939. "An absolutely black sheep need not fear becoming blacker. For me it is at least a small substitute for our talks in the KWI."[51]

Laue and Meitner were particularly concerned about Arnold Berliner. The seventy-six-year-old former editor of *Naturwissenschaften* was in poor health, confined to his apartment, isolated from nearly all his former colleagues and friends. In January Laue wrote, "Yesterday I spoke to our old friend Berthold Charlottenburger. . . . Do you understand this letter? When I use a pseudonym, you must emit a positron from the first letter, so that the letter declines one place in the periodic system [i.e., 'Berthold Charlottenburger' was Arnold Berliner.]"[52]

February: "Here things are getting really interesting. Do you remember from your school days the story of how to catch a crocodile? . . . Because the animal can bite through any rope, one captures it using a number of thin threads that together have the strength of a rope. But when the beast bites, the threads go between his teeth so that he can do nothing with them. When you read about intellectuals, remember this! I visit Berliner once a week. He speaks of many things: for me it is always a lesson in general education. I get more from these visits than I did before from his presence at lunch, because now I have him to myself, and if he needs to look something up he can go to the bookcase right away."[53]

March: "Your . . . letter came today, poorly sealed. I sent it on to Arnold." Laue ignored the censors. "In the university one of the organizations hung a large, large poster, exhorting every student to do his duty to make sure that 'distinguished specialists in every field should leave the university.' Really, those were the words!"[54]

April: "Our friend on Kielganstrasse [Berliner] is not very well; he has been complaining for weeks about his heart. . . . He reads a great deal of Goethe and the Epistle to the Romans, is intellectually as alert as ever, but extremely sad."[55]

In 1937, Laue had sent his son Theodor to the United States so that he would not be forced to fight for Hitler.[56] But in 1939 when Theo decided not to return, Laue was torn.

> Theo writes that he wishes to remain permanently in the USA. It is to his credit that he would rather be on his own there, with no support from home, than to come back here to the present situation. . . . [But] I would advise him not to do anything that would make his relationship to Germany impossible in the long run; he should, for instance, report for military training if he is called up. I fear that he might get homesick sometime, and then he would have a bad time if even a short visit home would be forbidden. . . . At the moment life is easy for no one.[57]

That was certainly true for Otto Hahn. In his priority disputes with foreigners, he could assume the role of embattled German, but with other Germans, he was vulnerable. When Ida Noddack entered the fray in March 1939, he was especially upset. He wrote Lise,

> Another not very pleasant situation is a letter to *Naturwissenschaften* by Ida Noddack, in which she accuses me of not citing her, after she predicted already in 1934 that uranium splits into lighter nuclei. I'm sure you re-

member her article. . . . The letter—to *Naturwissenschaften*—is unusually unfriendly. . . . She mocks what Strassmann and I have done: we always retracted and changed, etc., etc.; "One need only read all the articles one after the other to see how we have contradicted ourselves," etc. With all this one is supposed to have one's head clear for work.[58]

For someone who did not engage in name-calling, Lise's response was unusually strong: "I am so sorry that you now have this ugly difficulty with Frau Ida. I have always known that she is a disagreeable thing [the German *eine unangenehme Ursche* is more pungent]. I remember the article itself only dimly, evidence of how meaningless it was. Where did it appear?"[59]

Noddack's article, written for *Angewandte Chemie* in September 1934, was not meaningless, but it had never been taken seriously. In a criticism of Fermi's first tentative proposal of element 93, Noddack had objected to his chemistry and suggested that nuclei could break apart into several sizable fragments.[60] As it turned out, she was right, but in 1934, no one, including Noddack herself, pursued it experimentally. Her suggestion was briefly discussed in Rome and then dismissed; in Berlin it was hardly considered. Members of Fermi's group later remembered a "slight bias" against her; with Hahn and Meitner there was active disdain.[61] Noddack's idea went nowhere. It influenced neither the direction of the uranium investigation nor the recognition of fission when it finally came.

Now Noddack was publicly attacking Hahn with unconventional hostility. Furious, Hahn and Strassmann prepared a sharp response, but the editor of *Naturwissenschaften*, Paul Rosbaud, withheld it for a possible second round, noting instead, "The gentlemen Otto Hahn and Fritz Strassmann inform us that they have neither the time nor the desire to answer. . . . They want their colleagues to be the judges of the correctness of Frau Ida Noddack's demands and the manner of her presentation."[62]

With that the controversy died, at least on the pages of *Naturwissenschaften*. "I did not suppress it," Rosbaud wrote to Meitner. "Now she has made a fool of herself, and that is what I wanted to accomplish." Meitner agreed. "Nothing could better illustrate her unscientific small-mindedness and envy than her own words. She really has made a great fool of herself."[63]

Strong words, especially from Meitner, who tended to be quite circumspect. The animosity apparently grew out of a long-standing scientific controversy that began in 1925 when Ida Tacke, together with her future husband, Walter Noddack, and their colleague Otto Berg, reported the discovery of the two missing higher homologues of manganese: element 43, which they named masurium, and element 75, rhenium. Their initial identification of both elements was based in part on x-ray spectroscopic data that were difficult to interpret, but after they isolated enough rhenium to determine its chemical and physical properties, the existence of Re was confirmed. Their evidence for masurium, however, was considered inconclusive; the Noddacks nevertheless insisted they had found it. Meitner and Hahn were highly critical, judging the masurium affair to be long on self-promotion and short on scientific integrity. In Dahlem, it was said, the Noddacks were taboo.[64]

In Palermo in 1937, Emilio Segrè and Carlo Perrier analyzed molybdenum targets that had been bombarded with deuterons in the Berkeley cyclotron and found several radioactive isotopes of element 43. Because the Noddacks had always described masurium as present in natural ores, with the implication that it was quite stable, their claim was seriously eroded. The discovery was credited to Segrè and Perrier, and their name for the element, technetium, was adopted after World War II. Although the Noddacks' discovery of rhenium was a significant achievement that garnered them several Nobel nominations,[65] the masurium controversy seems to have diminished their reputation overall and contributed to the dismissal of Ida Noddack's ideas in 1934.[66] In 1939, Hahn viewed her demands as yet another attempt at self-promotion, this time at his expense.

Still, the degree of acrimony is evidence of something more visceral. Years later, several of the Noddacks' associates claimed that their enthusiasm for national socialism was never very great, while Emilio Segrè, just as firmly, believed that it was.[67] Such retrospective judgments are difficult to evaluate, and Meitner and Hahn's correspondence offers only a hint: "She doesn't have many friends here among us," Otto remarked in April 1939, "but obviously some in other circles!" That was all he would commit to paper; Rosbaud, however, regarded the Noddacks as pro-Nazi opportunists.[68] Ida Noddack's political connections could not have been extremely strong: her employment status, like that of most professional women, remained poor.[69] Hahn and Rosbaud treated her more as a nuisance than a threat, and she had little to say after that.

But Noddack's attack and the other priority disputes in 1939 permanently shaped and hardened Hahn's view. He was adamant: the discovery was his and Strassmann's only. To Ida Noddack he would not yield so much as a footnote. As to Curie and Savitch? They did not know what they were doing. And Lise, once his best friend and closest colleague? From that time forward, Hahn would insist they were friends while doing all he could to destroy the memory of their scientific ties. By the end of the war, Hahn would imply that fission could not be discovered as long as Meitner was in Dahlem.[70] Near the end of his life he was known to say that Lise might have forbidden him to make the discovery.[71]

In 1939, Lise understood only that Otto was struggling to secure his position. She completely concurred in this: she wanted him to retain control of the institute they once shared. In her 18 March article with Frisch she made a point of citing Hahn and Strassmann repeatedly. Hahn was grateful but incensed that in the same issue of *Nature*, Halban, Joliot, and Kowarski failed to cite them at all. "In France Joliot is regarded as the discoverer of fission," he complained angrily. "From this one sees how systematic suppression and false citations have their intended effect."[72] Lise sympathized, but she was no longer entirely in Otto's camp.

> The French are crazy with their noncitations. Now again the paper by Joliot and Halban! But I almost got a laugh out of it, because right after it was the paper of ours [Meitner and Frisch] that begins with your names.[73] But it is so easy for misunderstandings to crop up with these things. You even doubted our loyalty, and in contrast, I heard in Copenhagen that perhaps in your first Ba paper you should have cited Curie and Savitch more explicitly, as they actually did say they had a substance with the properties of lanthanum.[74]

In May 1939, when the British physicist Norman Feather surveyed several dozen fission reports in a review article for *Nature*, he discussed Curie and Savitch's contributions in some detail.[75] Upset, Hahn protested that Feather made it appear that Curie and Savitch believed their 3.5-hour substance to be lanthanum while Strassmann and he merely verified their finding and identified barium.[76] Lise agreed that Feather had made a mistake, but she was growing tired of the endless quarrels.

> Feather is not quite fair to you, but you once again are not quite fair to Curie-Savitch. . . . In one of their *C[omptes] R[endus]* articles they emphasized strongly that their 3.5h substance had very remarkable chemical properties and emphasized the similarity to lanthanum. The fact that they tried to place it among the transuranes doesn't change their experimental findings. And these findings led you to begin your experiments. And again you have not stated that quite clearly. . . . One must not take people's words so literally.

Curie obviously saw that something remarkable was going on, even if she did not think of fission. In November [1938] Hevesy heard her say in a lecture that the entire periodic system arises from U + n bombardment. . . . If one were to take your words out of context, your January 6 *Naturwissenschaften* article closes with these words, "As nuclear chemists who are close to physics . . . we cannot commit ourselves to take this drastic step, etc." and "It could still be a series of unusual coincidences that have given us false indications." . . . It always makes trouble to take people's words too literally. . . . If I may give you some advice, I think you should not get involved in discussions of priority. . . . The English don't like that sort of thing.[77]

A few weeks later, after Hahn wrote an article with several incomplete citations,[78] Lise noted tartly,

Feather could perhaps hold it against you that you did not mention his chemical enrichment of the 2.5 hour substance with AgI, and the same is true of [Philip] Abelson's articles . . . for which you cite only his first short note. . . . Also, I did not quite understand why you attributed the theoretical explanation of the fission process to Bohr's (3-year old) liquid-drop model and not to [Frisch's and my] work.[79] . . . About Feather, we found it surprising that you demanded a correction from him about an incorrect sentence and at almost the same time failed to cite his chemical proof of I (and indirectly Te). Doesn't that show that you're a little angry and aren't seeing things so clearly anymore?[80]

Lise and Otto had met only twice since she left Berlin: in November 1938 in Copenhagen and again in April 1939 when Hahn lectured in Stockholm. On the Swedish trip Otto was accompanied by a colleague with political credentials, and although he had made time for Lise, he had been cautious. In June, Hahn was in London, again with a chaperone. "Since this morning I am here," he wrote Lise. "Together with a party member. Therefore I cannot yet tell when I shall have the free time to visit your brother. But I shall do it on any account. . . . My colleague is very nice, but I better am a little careful." Hahn's "colleague," a chemist named Wever, later submitted a secret report citing his "disgust" at meeting German refugees in England.[81]

Lise told her brother Walter to expect Otto's visit. "He will not be completely free because of a colleague. And perhaps inwardly he is also not completely free."[82]

Despite everything, Lise and Otto were still friends, and he still turned to her with questions of physics. Several times he asked why so many different fission fragments were obtained from uranium, and how uranium and thorium could have fission fragments in common.[83] "Bohr will perhaps think I'm a cretin, but even after 2 of his long explanations I again don't understand it."[84] Patiently Lise explained, "There are many possible ways for fission to occur, and the evaporation of neutrons increases the number of possibilities even more."[85]

One of the most confusing findings of the uranium investigation had been the fact that slow and fast neutrons seemed to produce the same "transurane" activities, misleading physicists, including Meitner, to conclude that both slow and fast processes originated with the same isotope, ^{238}U. After the discovery of fission, physicists realized from the profusion of fission fragments that the fission of different nuclei would yield some fragments that were the same. Because the "transurane" precipitation always selected out the same few species from the wide range of different fission products, the products of the slow-neutron irradiation of ^{235}U and the fast-neutron irradiation of ^{238}U appeared to be identical.

Bohr had published a succinct explanation already in February,[86] but Hahn found it difficult to grasp. Meitner wrote again.

> Now I want to try once again to set forth what Bohr meant when he said that the same products can be produced in the fission of ^{239}U and ^{236}U as well as ^{233}Th. For these nuclei there are very many fission possibilities, because in the atomic weight range from about 80–150 all processes are almost equally possible. . . . In this way, despite different atomic weights and different nuclear charges, uranium and thorium can yield a series of identical isotopes . . . as well as isotopes that are not identical.[87]

Hahn was also interested in the conditions for a fission chain reaction. By then most scientists had concluded that a controlled chain reaction would be difficult to achieve, an explosive device even more so. Because fission in natural uranium was due primarily to the reaction of thermal neutrons with ^{235}U, a fairly slow, controlled chain reaction might succeed—if a critical mass of uranium could be assembled and if neutrons within the uranium could be sufficiently slowed, or moderated, to promote the fission of ^{235}U. But a rapid, uncontrolled chain reaction—a bomb—seemed impossible, at least in natural uranium, because ^{235}U was too sparse and thermal neutrons too slow, and because ^{238}U captured neutrons of moderate energy and extinguished the chain reaction. As Lise explained to Otto in July 1939, "Naturally, as soon as a chain reaction produces a temperature corresponding to 25 ± 10 [electron] volts, the resonance capture takes over and the chain is broken. I am surprised that [Siegfried] Flügge didn't see that, since he has done a calculation which in itself is very nice."[88] (Flügge had just published an article in *Naturwissenschaften* discussing the huge energy potential of a fissionable uranium "machine.")[89]

"We don't understand it," Hahn persisted, "neither Flügge, nor Strassmann, nor I. . . . Flügge thinks the explosion will take place long before the resonance energy is reached."[90]

"Bohr's opinion," Meitner replied, "is that an explosion of uranium can certainly be initiated at first, but with rising temperatures, when the 'thermal' neutrons also become more energetic, the capture cross section [of neutrons by ^{235}U] falls rapidly, and when the resonance energy is reached, resonance capture [by ^{238}U] predominates over everything else. Joliot and his co-workers demonstrated experimentally that even at normal temperatures the resonance process extinguishes 16% of all fission, and this fraction will rise rapidly with higher temperatures. Therefore a uranium-H_2-(Cd) mixture would certainly be very explosive, but not nearly so much as without the resonance process."[91]

Hahn's questions may not have been entirely academic. Beginning in April 1939, he had been invited to secret meetings with military officials and civilian scientists exploring the military potential of fission. By then there was no doubt that Europe was heading for war. In March, Germany took the rest of Czechoslovakia, causing England to unilaterally guarantee the security of Poland and the Soviet Union to seek an agreement with Germany. Meanwhile the Spanish Republic succumbed to Franco, Mussolini invaded Albania, Hitler clamored for the return of Danzig and prepared to invade Poland.

But to the relief of most scientists, a fission bomb seemed unlikely—unless ^{235}U could be separated in quantity from natural uranium, an isotopic separation that seemed virtually impossible. Nevertheless, scientists were afraid to completely rule it out. No one could be sure their experiments were correct or their theories complete. As German scientists, including Hahn, attended secret government meetings,[92] refugee physicists in the United States and Britain, impelled by fear of Hitler's Germany, warned of the potential for "extremely powerful bombs of a new type."[93] On both sides attempts were made, unsuccessfully at first, to suppress the open publication of fission-related scientific reports.

Meitner was more anxious than ever to leave Stockholm. She had given up hope of improving her working conditions in Sweden or—what amounted to the same—developing a collegial relationship with Manne Siegbahn.

In July 1939 she went to Cambridge, with "an especially friendly invitation" from W. L. Bragg.[94] There J. D. Cockcroft and Bragg offered her a position at the Cavendish Laboratory and a three-year contract with Girton College,[95] a decided improvement over their offer the year before. On 3 August she "practically accepted"[96] but then decided to delay her move to Cambridge until the summer of 1940. It was a major mistake. "If I had a talent for regret," she would write later—and in this case she certainly did—"I would not be able to get over the fact that I returned to Stockholm from Cambridge."[97] She had promised to take on a young Dutch physicist in Stockholm that winter, and she knew it would take months to get the necessary permits for emigration to England. Very likely she was reluctant to go through another unplanned and semilegal emigration.

"It would be nicer for you if you could go to C[ambridge] already in the fall," Hahn advised, thinking it best she move before war began.[98] In Germany summer vacations had been canceled, a prelude to mobilization. "It is completely impossible now to think about or do anything scientific. The political situation is too interesting for that," Hahn wrote on 24 August, the day Germany signed a nonaggression pact with the Soviet Union.[99] The pact sealed the fate of Poland. "At the moment the world really looks menacing," Lise agreed. "One cannot make plans more than two days ahead."[100]

Hitler invaded Poland on 1 September 1939. Two days later, Britain and France declared war on Germany.

CHAPTER TWELVE

Again, World War

I will have nothing to do with a bomb!

Although expected, the war came as a shock. Civilians were somber, displaying little of the reckless patriotism of 1914. "We lived through all this 25 years ago," Edith Hahn told Lise. "We know what it means."[1] Otto was gloomy, too. "One cannot tell how long this will last, and what will then become of Germany."[2] Poland was crushed by the end of September, Britain and France having done nothing to help. Then the war stalled for months, a "phony war," potential rather than actual.

Meitner's Berlin friends suffered from little at first except some consumer shortages and a feeling of isolation. Lise began sending small packages to Otto and Edith: soap, cigars, coffee. When Otto's *Nature* and *Comptes Rendus* stopped coming, he asked Lise to inform him of new developments in fission;[3] Max von Laue designated Lise go-between should he be unable to reach his son Theodor in Princeton. Although the mail between Germany and Sweden remained regular, they were all aware of wartime censorship. Planck became "Uncle Max senior"; Laue "Uncle Max junior" or "Theo's father";[4] Arnold Berliner, who lived on Kielganstrasse in the Charlottenburg district of Berlin, was variously "Arnold," "Charlottenburger," or "Dr. Kielgan"; the hated Kurt Hess was given the endearment "Kürtchen."

The most notable feature of that first winter of war, however, was how little there was of it. Berlin braced for attack, but no enemy planes appeared in the blacked-out skies. According to Laue, the family air raid shelter was so comfortable "we shall soon declare it our parlor," while the blackout had "the pleasant consequence that one can easily see stars, unusual for a large city. One must certainly be careful, though, when walking. Many men smoke while walking, as a burning cigar is the only source of light that does not have to be extinguished." An avid motorist, Laue's only complaint was that he could not use his car. "Even I am forced to go into town on the S-train. . . . Charlottenburger is now still more isolated, as it is hard for me to get to him."[5]

The war defeated Lise's plans for moving to Cambridge. "Practically the whole of the nuclear physics school is now doing government work in other places," J. D. Cockcroft wrote in October. "In view of this I think you ought to think over very carefully the question as to whether you still wish to come to Cambridge." If this was a signal for Lise to bow out gracefully, she ignored it. "I always come to the conclusion that it is next to hopeless to do any useful work in Stockholm. So if there is a possibility of work in Cambridge at all . . . I wish to come." But the Cambridge physicists were working on radar and other military problems, and nuclear physics seemed less relevant to the war effort. W. L. Bragg tried to sugarcoat their decision: "One of the great attractions of your coming here was that the members of our research team would have the benefit of collaborating with you. Since they are all away, I had much rather that you should come here in happier times." Mail between England and Sweden was sporadic, and Lise did not get Bragg's letter until early December.[6] Having told Manne Siegbahn of her intention to leave, she was now more unwelcome than before.

And the young Dutch assistant for whom she had delayed her move to Cambridge did not come to Stockholm after all.

> Dear Otto! . . . Perhaps in principle [Siegbahn] does not dislike that it worked out this way, he would much rather use the institute money for mechanics and big machines than for an academic person and scientific problems. In the whole big institute, only 5 academic people are employed, and they too work almost entirely on problems of apparatus. Scientifically I am completely isolated, for months I speak with no one about physics, sit alone in my room and try to keep myself busy. You cannot call it "work."[7]

Lise would always be a stranger in Siegbahn's institute. In Berlin, she had molded her professional and personal life into a satisfying whole, with friendship and collegiality an integral part of doing physics; in Stockholm, her ideas were of no interest and her experience was not sought after, so that she found it necessary to keep her "natural and unfettered personality under control."[8] Never had she worked with so little human contact or in such inhospitable surroundings. Plagued by her old feelings of insecurity, she struggled to carry on.

When she went to Copenhagen her confidence returned, only to evaporate the instant she returned to Stockholm. On one such trip in November 1939, Meitner prepared a detailed publication of the experiments she and Frisch had done the previous March;[9] she discussed the work at length with Bohr and, she wrote Hahn, "for once really heard some physics again, which is always really nice." In Stockholm there was always some unpleasantness. "While I was with Niels [Bohr] I [heard] that [Svante] Arrhenius's youngest daughter was in [Siegbahn's] institute asking about me, about working with me. When I returned, I heard that she was working with S. S. himself said not a word to me about it."[10]

In Siegbahn's institute, she wrote Otto a few weeks later, her work was "as meaningless as ever. Just to keep busy, I did some experiments with the rare earths. I am not very interested in it, but one must do something. All day long I sit alone in my workroom, and for days I do not speak a word. In this way one has plenty of time to think over the world situation, a good method for staying slender."[11]

Meitner's work was not meaningless, but it was restricted by the material and equipment available. In the fall of 1939, she measured the cross section for the thermal neutron capture of thorium, lead, and ^{238}U using radium (α)-beryllium as a neutron source—Siegbahn's cyclotron was not yet ready—and the rare earth element dysprosium as a neutron detector.[12] In her subsequent studies of the neutron-induced activities of dysprosium and other rare earth ores, Meitner showed that the activities are unique for each element and directly proportional to the quantities present, even for exceedingly small amounts.[13] This followed an earlier discovery by George de Hevesy and Hilde Levi in Copenhagen, who were already using neutron activation to analyze rare earth mixtures, a method that was later developed into the nondestructive assay technique known as neutron activation analysis.[14]

But, as Meitner told Hahn, none of this interested her much. She longed to extract one great success from her earlier work with uranium: the discovery of the first true transuranic element. She knew precisely where to look, for in Berlin in 1936 she and Hahn had shown that the slow-neutron irradiation of uranium produces a beta-emitting uranium isotope of half-life 23 minutes and in 1937 she, together with Hahn and Strassmann, had proved that this was ^{239}U, formed by the resonance capture of slow neutrons by ^{238}U.[15] It was obvious that a beta-emitting ^{239}U would decay to the next higher element, but the Berlin team never found even a trace of element 93. Preoccupied with the false transuranes, they never really searched for it, and their neutron sources were too weak.

By the late 1930s, however, most major laboratories had given up their old radium-beryllium neutron sources in favor of the intense neutron beams generated by high-tension apparatus or cyclotrons (in Dahlem, Meitner had built a high-tension apparatus for this purpose which was finished just before she left in 1938). In 1939, when Edwin McMillan used the Berkeley cyclotron to irradiate uranium with neutrons, he detected a new beta activity of half-life 2.3 days associated with the 23-minute ^{239}U.[16] With him was Emilio Segrè, a former member of Fermi's group who had made his way to Berkeley following his dismissal (for racial reasons) from Palermo in 1938.[17] As the co-discoverer of element 43, Segrè was familiar with the manganese group and the chemistry expected for ekarhenium, but the chemistry of the 2.3-day activity, he found, was more like a rare earth.[18] Since the rare earth elements come just after barium in the periodic sequence, he concluded that the new activity was a fission fragment—and missed the discovery of element 93.

Ironically, Segrè's error mirrored the earlier misconceptions that had delayed the discovery of fission: where fission fragments were once thought to be transuranes, Segrè now mistook a true transuranic for a fission fragment. Both errors were based on the same faulty assumption: that transuranics would be homologous to third row transition elements and that element 93 in particular would resemble rhenium. No one ever gave much thought to the proposal made by Bohr in 1922 that the elements beyond uranium might be higher homologues of the rare earths.[19]

Meitner, however, had a physical basis for doubting that the 2.3-day activity was a fission fragment. She noted that it remained embedded in uranium, even though McMillan and Segrè had deliberately used a very thin layer of uranium to permit light, energetic fission fragments to recoil and escape. The idea of nuclear recoil had been with Meitner since she and Hahn first explored it in 1909. She was sure the 2.3-day activity was no fission fragment. It had to be element 93.[20]

But she could not prove it without repeating the experiment, and for this she needed an intense source of neutrons. Through the winter of 1940, Meitner waited for Siegbahn's cyclotron to become operational. In April she finally gave up and went to Copenhagen to use the cyclotron in Bohr's institute. She arrived late in the afternoon of 8 April. At dawn the next morning, Copenhagen awoke to the sound of German planes overhead. A declared neutral, Denmark had signed a ten-year nonaggression pact with Germany just the year before; it was utterly without the means to fight. (George Placzek, one of Frisch's Copenhagen friends, once cracked, "Why should Hitler occupy Denmark? He can just telephone, can't he?")[21] Denmark surrendered before breakfast.

There was every reason to believe the Germans would pay special attention to the Institute for Theoretical Physics. Occupation of Denmark gave Germany not just butter, bacon, and access to Norway but its first cyclotron (its second, the cyclotron in Joliot's laboratory in Paris, was in German hands two months later). Of more immediate concern was the fact that Bohr was "non-Aryan" (his mother was Jewish) and, as everyone knew, his institute had been a haven for displaced Jewish physicists since 1933. As Bohr was out of town the morning of 9 April, George de Hevesy (also of Jewish origin) immediately began destroying correspondence and other records of Bohr's efforts to help refugee scientists.[22] When Bohr arrived later that day, he remembered that Max von Laue had left several medals in Copenhagen, including his large gold Nobel medal and his Max Planck medal, to keep them from being confiscated by the German authorities. As Hevesy wrote to Laue after the war,

> After the occupation of Copenhagen, your medals were Bohr's first concern. He was not interested in your medals but in your person. I proposed that we bury the medals, but since your name was engraved on them this did not satisfy Bohr. Dissolving the medals was the only way to make them disappear. I spent the entire first day of occupation with this not very easy task [this because gold is exceedingly unreactive and difficult to dissolve]. . . . [Later] the Nazis occupied Bohr's institute and searched everything very carefully, especially the vault where your medals had been stored. If they had found your medals in their original state, you would probably have landed in prison and would surely have wished you had never received them.[23]

Laue's medals quietly waited out the war in a solution of aqua regia.

Meitner stayed in Copenhagen three weeks.[24] Apart from the scare of German occupation, she had no trouble. In a show of benevolence toward fellow "Aryans" who had surrendered with virtually no resistance, Germany granted Denmark a semblance of self-rule. For a time Jews in Denmark were not disturbed.

On her return to Stockholm, Lise informed Otto Robert, who had left Copenhagen for Birmingham in the summer of 1939. As required by the British censor, she wrote in English.

> I had come to C[openhagen] just 12 hours before the great event took place, when nobody suspected in the least that such a thing should happen. . . . We were awakened by the noise of plenty of aeroplanes at about a quarter to six in the morning and there was nothing to do but to wait what might happen next. The central post office, the offices of newspapers, radio-station and police-stations were occupied almost immediately, but you saw but very few soldiers in the street and they all—mirabile dictu—were speaking Danish. They did not interfere officially with anything. [Of her own reaction, Lise said nothing.] As long as I was there, no special difficulties arose in the common life. The scientific work was going on as usual, only all kinds of meetings (including "doctor-disputations") have been forbidden. . . . Of course Niels and Margrethe were very unhappy about the events but he does not have in view to give up his work although he got three offers from other places in the course of the first two days. . . . I cannot drop this matter without mentioning how wonderful Niels was during the whole time.[25]

Before she left, Bohr asked Lise to send a telegram from Stockholm to a British friend, physicist Owen Richardson, assuring him that the Bohrs were all right. When it arrived, her telegram seemed somewhat cryptic: "Met Niels and Margrethe recently both well but unhappy about events please inform Cockcroft and Maud Ray Kent   Meitner"

The telegram was discussed at length by a committee of physicists who were exploring the military feasibility of a fission chain reaction in uranium. As they could make no sense of the words "Maud Ray Kent," they looked for a coded message about German research on a secret weapon. John Cockcroft, who was working on radar, was intrigued by the word "Ray," while Otto Robert Frisch and Rudolf Peierls, who had recently set forth the theoretical conditions for constructing a fission bomb from ^{235}U, tried the anagrams "radium taken" and "U and D may react." None of this seemed quite right, yet no one thought to ask Meitner for further explanation. The fission committee named itself the Maud Committee or (more impressively) M.A.U.D., but the mystery was not solved until after the war:

Maud Ray, a former governess to the Bohr children, lived in Kent; her street address had been dropped from the telegram.[26]

The occupation of Denmark cut off Lise's search for element 93. In Berkeley just seven weeks later, Edwin McMillan and Philip Abelson identified the 2.3-day activity as a decay product of ^{239}U, a transuranic element they later named neptunium.[27] Bitterly disappointed, Meitner wrote to Hahn, "I never believed that this [2.3-day] activity was a rare earth, because it was too unlikely that a fission product would not fly out by recoil, and I intended to prove it [when I was] with Niels. Unfortunately nothing came of it, but the fact remains that the first true transuranic (the unavoidable decay product of uranium-239) was found by us [in Berlin]."[28] While it is true that the Berlin team had characterized ^{239}U and deduced the existence of its decay product, they never actually found element 93: Meitner could certainly claim the parent, but she had no rights to its offspring. That was terribly difficult for her to accept, more so as McMillan and Abelson's neptunium, a beta emitter, was precursor to yet another transuranic, element 94. Of the many heartaches Meitner suffered after leaving Berlin, her failure to find element 93 grieved her most. It would remain a "crêve coeur," as she put it, the rest of her life.[29]

She tried not to dwell on it. Personal troubles paled against the terrible events of 1940: Denmark's surrender, the fall of Norway, Belgium, and Holland, the collapse of France, the bombing of Britain. From Germany she received urgent pleas for help. Deportations had begun. Emigration to England and France ended with the outbreak of war; other avenues of escape were severed or clogged. Prior to 1939, the American quota for German immigration had not been filled; now there was a waiting list several years long. In desperation, Jews tried Sweden, South America, Russia, even China. No country would accept impoverished immigrants without affidavits of financial support. Relief organizations were drained, and individuals, often recent émigrés themselves, were generally too poor to help.

In December 1939, Max Born wrote to Meitner about the plight of the Breslau physicist Hedwig Kohn. One of the first women to undergo Habilitation in Germany, Kohn had been at the University of Breslau until she was dismissed in 1934; for a time she worked as an industry consultant, but by 1939 she was completely unemployed and faced deportation to Poland.[30] As Kohn's quota number for U.S. immigration was very high,

Born hoped Meitner could get her into Sweden on a temporary basis until she was admitted to the United States.[31]

Meitner's income was too low for her to qualify as a sponsor, but she wrote scores of letters to Jewish agencies, women's organizations, and individuals, finally obtaining £50 from a women's group in London and a pledge of $300 from a physicist in Washington, D.C.[32] That was not enough for the Stockholm Jewish community to underwrite Kohn's stay.[33] In April 1940, however, Kohn learned that the United States had eased the admission of scientists on nonquota visas, so that her stay in Sweden might not be long. But she begged Meitner, "Since I last wrote to you and my helpers in the U.S. the conditions here have become such that it is no longer a question of my desire to be employed as a physicist again, but an urgent necessity that requires the greatest haste. If by early June I do not definitely know where I can go, help may be too late."[34]

Kohn arrived in Stockholm in June and with Meitner's help obtained her American visa in a few weeks. From Sweden, Kohn went by ship to Leningrad, crossed Russia to Vladivostok, waited there for a Japanese transit visa, and then sailed to San Francisco, a journey of six months. Exhausted and ill, she was cared for by James Franck in Chicago, then took a faculty position at Sweetbriar College and then the Women's College of the University of North Carolina. (By that time a few American physicists had taken government and military jobs, creating some openings for refugee scientists.) With a fairly secure position and an annual salary of $1,000, she tried to get a visa for her brother, whose letters from Breslau, she told Meitner, were "increasingly sad and increasingly urgent."[35]

The news from Germany was ominous. Having lost all means of normal employment, Jews were subject to forced labor, at minimal wages reduced by punitive regulations and special income taxes. Food rations were cut to the point of hunger, even for children. Economically helpless, Jews were further isolated by a deliberate policy of social segregation: forced into concentrated housing, forbidden to associate with Germans, forbidden to own radios or use public transportation, forbidden even to use the telephone. Already branded with revealing middle names and stamped identification papers, Jews were made still more vulnerable beginning in 1941, when all Jews from the age of six were required to wear large yellow Stars of David on their outer clothing whenever they appeared in public.[36]

This was only a prelude. In ever-larger numbers, Jews were shipped to the labor camps of occupied Poland. Gradually it became known that the deportees' mortality rate was enormously high. Reports trickled back. Deportation was more than slave labor: it was a sentence of death.

In May 1941, Meitner learned from Hahn that Stefan Meyer was in great difficulty in Austria.[37] Meyer, one of the leaders in radioactivity since its inception, had been director of the Radium Institute in Vienna since it was built in 1910. After the Anschluss he knew he would be dismissed; hoping that a nonconfrontational stance would protect him and his wife, Emilie, who were both part-Jewish, he retired from his position in April 1938 and withdrew from all academies. By 1939, however, when the situation in Vienna deteriorated and it appeared they might be deported, he and Emilie retreated to their vacation house in the mountain town of Bad Ischl, where they were protected for a time by a sympathetic town official. The official died in 1940, however, and then they were in great danger, particularly since the government had begun to treat so-called *Mischlinge*—"mixed-race" persons—as Jews.[38]

Lise appealed to Hans Pettersson, her friend at the Oceanographic Institute in Göteborg whom she had first met in Vienna when he worked with Meyer in the 1920s. An immigration permit required five to ten thousand Swedish crowns as security; Siegbahn promised to collect part of it, perhaps Pettersson could collect some in Göteborg? They had to work quickly, Meitner told him, "so that M[eyer] can state that he is coming here when the question of being transported to Poland becomes acute. . . . It is very urgent. M. will hardly let himself be taken to Poland: I fear he would prefer a voluntary death, as so many have done. Laue wrote to me recently that the wife of Geh[eimrat Heinrich] Rubens and the widow of the . . . x-ray physicist Ernst Wagner have committed suicide. I hope with all my heart that one can spare the Meyer couple a similar fate."[39]

Meitner and Pettersson tried every possible avenue of escape for their friend. In December 1941, Meitner wrote to some American colleagues who were considering inviting Meyer to the United States; knowing that American universities were not free of anti-Semitism, she took care to point out that Meyer and his wife were "Germans of Lutheran religion."[40] Her letter, written one day after the Japanese attack on Pearl Harbor, was of no use: German immigration to the United States was cut off the moment America entered the war.

Meyer optimistically began learning Swedish, but because he was beyond the mandatory retirement age of sixty-five, no university or institute in Sweden could offer him anything but a temporary position. This did not satisfy the German authorities; early in 1942 Lise learned that Jews were prohibited from leaving the Reich without the guarantee of a permanent position abroad.[41] Even then, she told Pettersson, Meyer and his wife might not get out—Jews were forbidden to travel by train.[42] It was evident that Germany was no longer encouraging Jewish emigration but thwarting it. The "final solution of the Jewish question" was under way. Mass killings had begun in Poland and Latvia and the Ukraine; gassing facilities had been installed and tested in Auschwitz, and additional death camps were being built. On 20 January 1942 at a meeting in Wannsee, in western Berlin, Nazi officials decided on deportation procedures and killing methods and ordered the pace to be accelerated.[43]

After two years and hundreds of letters, Meitner and Pettersson succeeded in negotiating an industrial position for Meyer and collecting 2,200 kroner for a Swedish immigration permit and living expenses.[44] By then, however, it was 1943 and too late: Meyer and his wife could not get out. "He is in a very insecure situation," Pettersson wrote to Lise. "At any moment the local authorities could yield to outside pressure and refuse to help him."[45] There was nothing more that Pettersson or Meitner could do.

These were terrible years, when Germany was dominant across the face of Europe and annihilating Jews without interference. Lise knew that she was exceedingly fortunate to be out of Germany; fortunate, too, that she had not settled in Holland or Denmark. But she could not relish her safety amid pervasive suffering. And at work she was miserable, unable to adjust to her dependence and poor working conditions, hurt by the coldness and lack of collegiality in Siegbahn's institute. Feeling isolated from physics and deprived of meaningful work, she could regard Sweden as a haven but never a home.

Meaningful or not, work was all she had. During the war Meitner remeasured the resonance energy of neutron capture in thorium, investigated in detail the radioactivity of ^{46}Sc, and studied the interaction of secondary and primary electrons in beta spectra.[46] But she took little satisfaction in it. It was just "bits and pieces," she told Eva von Bahr-Bergius, "always the same unsatisfying muddling." To get anything from a workshop took "weeks of petitioning"; she had virtually no help, supplies, or equipment.[47] For the thorium study, she "begged here and there" for indium and rhodium foils; for the [46]Sc experiments, George de Hevesy prepared a strongly activated scandium sample in Copenhagen and Bohr brought it with him when he visited Stockholm in 1941.[48] Not only were materials and equipment in short supply but scientific journals were irregular and colloquia so infrequent that at times Lise felt as though her "brain was drying up."[49]

The construction of Siegbahn's cyclotron took longer than expected, and when it was finally ready, Meitner seldom got to use it.[50] "The cyclotron has been in operation now for a few weeks," she wrote to Hahn early in 1942, "but things are irradiated only for Manne. He is working with a biologist on radioactive phosphorus, and the cyclotron is not generating neutrons. Instead deuterons are used for direct activation . . . so there is no room in the cyclotron for anything else. I have no idea how long things will go on this way."[51] Two years later, in 1944, her complaint was the same: "My work is going infinitely slowly; for more than two months it has been impossible to get anything irradiated in the cyclotron—it is used almost exclusively for biologists and for Manne's son [Kai] whose work is urgent because his doctoral exams are coming up. . . . Manne is . . . interested only in his son's research, which is overrated in every way."[52]

Meitner was hardly presenting a balanced picture: she knew that cyclotrons were commonly shared with biological and medical researchers—in Copenhagen it was no different—and she almost certainly underestimated Kai Siegbahn, who won the Nobel Prize in physics in 1981. But she was offended by the unabashed nepotism—when her nephew Otto Robert first applied for a job in Germany in 1927, she had been reluctant even to recommend him[53]—and it galled her that with Siegbahn she was never a colleague, not even when his son began work in beta spectroscopy, a field in which she had been a pioneer.[54] When Meitner became friendly with some of the younger scientists, Siegbahn tried to discourage it. "He doesn't want me and the young people to talk; they do it anyway and come to me for advice." Every aspect of the institute was under Siegbahn's control. "One can't really work if one can't make needed changes during the course of the experiment. . . . Recently I asked Manne if I could get something irradiated, whereupon he pulled out his notebook and said, 'Which day? All the work must go through me, you know.'"[55] Siegbahn required deference, something Meitner was quite unable to give.

Siegbahn's authoritarian bent was well known: already years before Svante Arrhenius had referred to the "small popes" of Swedish science, "especially Siegbahn."[56] Meitner's friends sympathized. "My only consolation," Pettersson wrote during the war, "is that Siegbahn is of German descent, and the prefect psychosis from which he suffers is imported material."[57] A Danish physicist in the institute wryly suggested that a fitting punishment for Siegbahn would be to "split him in two, and force one half to work for the other half."[58]

Meitner's greatest complaint was that she had no assistant. "Siegbahn refused all my requests for one," she told Pettersson. "In 1939 a Dr. Cohn-Peters wanted to work for me without pay to get into nuclear physics—I knew him through [Oskar] Klein—but Siegbahn would not allow it because C. P. is German."[59] The shortage of young scientists was not Siegbahn's fault, and there may have been employment restrictions on immigrants such as Cohn-Peters, but Siegbahn's lack of civility always seemed to add insult to injury. In 1939, he took on Arrhenius's daughter without telling Meitner; again in 1943, when Pettersson sent her a promising young physicist, Siegbahn took him instead. "[The young physicist] did come to see me, and made a good impression, but he seemed very uncertain, and asked right away whether my field could lead to a job in industry later. . . . Siegbahn, without saying a syllable to me beforehand, offered him a position. . . . If someone wants to work in industry, he would not want to antagonize Siegbahn."[60]

Of those who did antagonize Siegbahn, Meitner was permanently on the list. It is unclear if it began with bad personal chemistry—Meitner was a forceful woman, forthright and not particularly diplomatic—or irreconcilable scientific disparity or Siegbahn's resentment over being forced to take into his institute a person he did not choose, whose prominence in nuclear physics exceeded his, at the formative stage of a program he had struggled for years to establish. Siegbahn's associates remember him with fondness as an inspiring scientist and generous director who took care of his own, qualities not inconsistent with his ambition to direct the slim resources of Swedish science toward his own scientific goals.[61]

But his conduct toward Meitner was from the outset so negative that one suspects it had less to do with Meitner herself than with her ties to others, most probably Oskar Klein and Hans Pettersson. Klein, a theoretician, was not close to Siegbahn, and Pettersson had clashed with

Siegbahn several times, most recently in 1936 when Pettersson publicly challenged him and C. W. Oseen, both members of the Nobel physics committee, for their practice of withholding prizes and channeling the unspent funds into their own research.[62] Siegbahn had lost that battle; when Meitner arrived two years later, he may have seen her primarily as a member of the opposition.[63] Her subsequent complaints and the sympathy she engendered must have made her presence in his institute all the more intolerable.

Other than sympathizing, however, Lise's friends could do little to help. Siegbahn might have been arrogant and high-handed, but he was a driving force in Swedish physics: Nobel laureate, director of a major institute, well known in government and industry, an active member of the Royal Swedish Academy of Sciences, influential in the selection of Nobel awards. Meitner was of no real consequence to him, and their poor relationship would prove detrimental only to her.

Nevertheless, Lise's capacity for making and keeping friends did not diminish. Eva von Bahr-Bergius remained her closest friend in Sweden; they discussed religion, philosophy, and world events in their letters, visited when they could, and generally served as each other's confidantes. For Christmas, Lise usually joined Gudmund and Magnhild Borelius and their family and friends, and each summer she vacationed near the Borelius's country place in northern Sweden. She liked Oskar and Gerda Klein's lively family and somewhat disorganized household, and she participated regularly in Klein's theoretical physics seminars at his institute at Stockholm University. In Siegbahn's institute she developed a special rapport with Sigvard Eklund, a young physicist whose research touched on hers, and she became close friends with Lilli Eppstein, the refugee pianist she had met at the Kleins' on her sixtieth birthday. She saw other colleagues and friends less often but maintained an active correspondence: in Uppsala, The Svedberg, the eminent physical chemist; in Göteborg, Hans Pettersson and also Emma and Malte Jacobsson, the friends who had first invited her to Sweden and fed her so amply in the summer of 1919. In most respects Meitner's Sweden was neither inhospitable nor cold. But her friends could not completely fill the void. Her separation from physics gnawed at her. She regarded her life as empty, "so empty that I truly cannot say one word about it."[64] Her image of herself as "a wind-up doll . . . with no real life in her"[65] persisted; she was a stranger to herself, someone she no longer knew or liked. "If I were to tell you 'the truth about Lise Meitner,'" she told Otto, "you would surely not enjoy it."[66] Unable to work as before, she felt herself an exile, homeless.

For thirty years her center had been the physics community in Berlin. Now her name seemed to vanish from the German record, as if written in smoke. From the start, her 1939 theoretical paper with Frisch was largely ignored in Germany[67]—in a 1941 article on fission theory, her once-close associate Carl Friedrich von Weizsäcker never mentioned it[68]—and when her earlier papers were cited, her name was omitted. Lise took this as a final blow, the theft of the one thing she thought she could not lose—her scientific past. Bitterly she wrote to Hahn in 1943,

> The references of [Fritz] Houtermans in Phys[ikalische] Ber[ichte] struck me particularly; his selection principle for citing [my work] is somewhat astonishing. Uranium 239, thorium 233, all things [you and I did without] Strassmann, bear his name as the second author, and [not mine]. It is really wonderful what people can do. If one just lives long enough and is shaken up thoroughly enough—and there is no lack of that nowadays—one can learn a great deal. But it reminds me of the story of the peasant who wants his cow to become accustomed to not eating, and who tells his friends that just as the beast finally learned so well how not to eat, it died.[69]

On the overwhelming scale of the Holocaust, the failure to cite the intellectual contributions of Jews may be minor, but it is not trivial. The attempt to expunge Jews from Germany's collective memory was part of the genocide: ruin, expel, and kill Jews—and then obliterate their record so that nothing, not even the tiniest bit of their former existence, would survive. Primo Levi, a survivor, wrote, "The entire history of the brief 'millennial Reich' can be reread as a war against memory, an Orwellian falsification of memory, falsification of reality, negation of reality."[70] The distortions and omissions that were intentionally introduced into the scientific literature are but a minuscule part of that war against memory— and even these have never been corrected.

Lise tried to maintain perspective. "I don't take this all very tragically, because I no longer take my life very seriously," she wrote to Eva. "One sees far too much suffering to be overly concerned about oneself."[71] She was appalled to learn in 1942 that Richard Willstätter died quite alone in Switzerland;[72] her heart "simply stood still" when she heard that mentally ill people were being put to death in Germany.[73] The war depressed and distracted her: air raids were battering Berlin, relatives in England were living under a hail of bombs, the Coster family was nearly starving in Holland. She worried constantly that her relationship to old friends in Germany might never again be the same: they did not seem to comprehend that "every good German must wish that Germany not be victorious."[74] Yet she could not view the war as a strategic series of victories or defeats. Having seen casualties firsthand in 1915, she found it "just as terrible when a German ship is sunk with people on it as when an English ship goes down. And I always get an inner fright when I realize that even here people forget completely about other human beings, and view these sad necessities more with a feeling of success than sorrow."[75]

Lise's lifeline was her correspondence with friends outside. From Dublin, Erwin Schrödinger wrote enthusiastically of new work, exhilarated by the "paradise" he and Annemarie had found in Ireland.[76] Max and Hedi Born were quite content in Edinburgh. In addition to their own children, they were looking after two young women, daughters of the physicist Heinrich Rausch von Traubenberg and his Jewish wife, Maria, who had remained in Germany. For five years Lise was the intermediary between the Traubenberg parents and daughters, transcribing German letters into English and vice versa, as required by the censors of both countries. In 1944, the mother was suddenly arrested and the father, in his terrible anxiety that she would be sent to a concentration camp, collapsed and died of a heart attack. Lise relayed the news to Edinburgh. Heinrich's death precipitated just what he had feared: Maria, now without protection, was deported to Theresienstadt. She survived and joined her daughters in England after the war.[77]

Contact with occupied Holland was difficult, but Lise occasionally managed to send food to the Costers; she was distressed by their enormous gratitude, an indication of their great need.[78] For a time the occupation of Denmark was more benevolent. In September 1943, however, Danish Jews were suddenly threatened with immediate deportation to concentration camps; with the help of a network of Danish citizens, Bohr and his family along with thousands of others crossed the narrow Øresund by night in small boats and reached safety in Sweden. For the first time, Sweden came out from its cover of neutrality: radio broadcasts assured refugees that their boats would be welcome in Swedish waters, and officials met the refugees onshore and prepared their entry. Each refugee was given a subsistence allowance, housing, and, when possible, work, without distinction for an individual's former pay or status in Denmark. In a decent and egalitarian way, the Danish refugees were treated as welcome long-term guests.[79]

Niels Bohr soon left for England, but Margrethe and several other family members stayed behind in Stockholm until the war was over.[80] Their son Erik completed his chemical engineering degree in Siegbahn's institute.[81] Lise, who had always been fascinated by Margrethe's "uncommon mixture of cleverness, warmth and charm," was glad to have her near.[82]

It was harder to sustain real contact with friends in Berlin. Lise's relationship with Otto had become somewhat fragile. Their concern for each other was deep, but scientific quarrels, politics, and the individual difficulties of their separate lives had come between them. When Otto wrote—less often than Lise liked—his tone was curiously flat and his subjects narrowly personal: his health, Edith's well-being, Hanno's leaves from military service. Both avoided anything political: the institute or colleagues, the fate of Jewish friends, the war.

Lise's friendship with Elisabeth Schiemann also suffered. Elisabeth's letters were stilted and cautious; the contact was broken. Nevertheless, Lise was aware of Elisabeth's "bitter troubles" under the Third Reich: she was dismissed from the University of Berlin in 1940 and was without a paid position for several years. And so Lise persisted, keeping the tone of her letters to Elisabeth very light. But the closeness was gone. "It takes half or even a whole lifetime to make a few friends," Lise lamented, "and then one loses them in the blink of an eye."[83]

If politics estranged some friends, it drew others closer. One was Paul Rosbaud, whom Lise had known even in Berlin as a highly political person with unusual connections: it had been Rosbaud who tried to get her an illegal passport in June 1938, who helped her pack, drove her to the train, and saw to it that she left for Holland in July. Later that summer he had brought his wife, Hilde, who was Jewish, and their daughter, Angela, to England for safety, but he returned to Berlin to be near the regime he hated, eager to fight the Nazis from within.[84] Lise was the principal intermediary between Paul and Hilde, relaying their messages throughout the war.

By training an engineer, Rosbaud was a scientific consultant for Springer Verlag's technical publications, which included scientific books and the journal *Naturwissenschaften*. Gregarious and charming, he traveled freely in Germany and the occupied countries, a familiar presence among people of all political stripes in universities, industry, and the military. His publishing connections were the perfect cover for espionage.[85] Although it is unlikely that Lise knew many details, she did know that Rosbaud intervened on behalf of Jewish friends, contacted Jewish prisoners in Norway and Poland, and on occasion smuggled food into concentration camps.[86] She regarded him as a hero and must have known he was a spy. In essence he told her so in 1944, when he wrote that Hilde and Angela would be "proud of the father's successes, even though they are now in another field. . . . My work now has a different character, but it gives me a great deal of pleasure."[87]

Of all the Berlin friends, Max von Laue remained the closest. Later Lise would describe him as her only truly faithful friend, the one friend she could always rely on.[88] They wrote to each other often, nearly every week. Their letters were formal in some respects—Meitner and Laue never adopted the familiar "Du"—but intellectually intimate, trusting, and open. In Berlin they had met each week in the KWI for Chemistry; their correspondence was a continuation of their talks. Frequently letters arrived poorly sealed, but Laue was not intimidated by the censors. His correspondence with Lise was a necessity: for friendship, for a window to the reality outside Germany, for a means of bearing witness to the events within.

As intermediary between the Laues and their son Theo in Princeton, Lise regularly transcribed their family news. When the Laues' daughter Hilde took a job as laboratory assistant in the KWI for Chemistry, she inherited Lise's lab coat and described the institute for Lise in a way that Hahn would not: personalities, politics, even the fact that Hahn's deputy, Otto Erbacher, read the party newspaper *Völkische Beobachter* "from morning to night" while neglecting his research.[89]

Laue's letters to Lise were an ongoing memoir, sent out page by page for information and safekeeping. When the Prussian Academy of Sciences and the Viennese Academy underwent Gleichschaltung in 1938, Laue sent Lise a list of the dismissed members.[90] He was one of the few to publicly counter scientific revisionism: "Herr Lenz took my breath away with his demonstration that the theory of relativity originated with [French physicist] H. Poincaré. You will read it in the next issue of *Naturwissenschaften*.

Hopefully soon afterwards, my reply. Here in this country there certainly are great politicians."[91]

While theoretical physicists like Wilhelm Lenz were trying to sanitize relativity by eliminating Einstein entirely, C. F. von Weizsäcker took a subtler tack. His stance, he told Laue, would be to say that most of special relativity was already present in the work of Lorentz and Poincaré and that Einstein did not properly recognize them. Einstein may have discovered relativity, Weizsäcker decided, but it would have been discovered without him. Germany did indeed breed politicians, of every stripe.[92]

In almost every letter Laue sent news of Arnold Berliner. The former editor of *Naturwissenschaften* was ill, nearly blind, and, except for a few friends, quite alone. "Dr. Kielgan [Berliner] . . . suffers from troubles that afflict him more than the rest of us," Laue wrote in March 1940, as the Reich cut housing, transportation, and food rations for Jews. Berliner retreated into his apartment and lived almost like a hermit, in part because theater, restaurants, and concerts were forbidden to Jews, in part "so as not to have to wear the Jewish star, without which no non-Aryan was allowed to be seen on the street."[93] He was, as Laue later described him, "an absolutely cultured person," a physicist with an overview of all science, a freethinker steeped in the history of the papacy, a man immersed in classical literature, music, and art. His apartment was filled with works of science, literature, and cultural history. Among his treasured belongings were rare manuscripts, modern paintings, and a bust by Rodin of his friend Gustav Mahler.[94] Berliner was the author of a successful physics textbook, now banned; he was working on the sixth edition in the hope of better times to come. But he found his forced isolation almost impossible to bear. He wrote to Lise in 1940,

One is pulled so far off balance spiritually that the body also cannot endure it for long. The loneliness and loss of employment burdens my existence most severely. I read as much as my eyes permit, often, in fact, more. But the joy I once had from it, when reading was a counter-weight to professional work, that is gone. And then: a life without music! . . . A purely scientific experience, a performance of *Figaro* under Mahler, a lecture by Herr von Harnack. . . . Thank God we experienced these! But what we are living through now breeds . . . only the fear of fear itself. I have only now truly learned to understand that saying. My connection to physics is close to zero; luckily friend Max keeps me informed of this and that. He is touchingly loyal and good, he visits every week for an hour or two.[95]

His relationship with Laue reminded Berliner of his friendship with Mahler. "With Mahler I learned what a great musician is," he told Lise, "now I experience a great physicist. You have already had this good fortune with the other great Max [Planck]."[96]

The year 1942 was marked by mass deportations of Jews. Berliner was ordered to vacate his apartment by the end of March. On 22 March, Laue informed Lise, "About our . . . 80-year-old, here on Kielganstrasse, things look bad for him. No one knows what will happen to him next week. He is tired of living, but then speaks in a very animated manner about everything possible as soon as he is distracted from his personal situation."[97]

Arnold Berliner took poison that night, probably hydrocyanic acid. Laue informed Lise of Berliner's death, then succumbed to nervous vomiting. Later Laue learned of Berliner's last hours.

Kielgan . . . spoke with his housekeeper until 11:30 that night, distributed books, cleared up other things, which she thought was related to his move to another—not yet found—apartment. The only unusual thing was that he adamantly refused . . . to have a warm evening meal. Kielgan then did not go to bed, and the next day, when his housekeeper came from her job in the long-distance telephone office, she found him, cold and stiff in his armchair, sitting up. . . . I miss Kielgan very much. He was one of those rare people with whom one could talk about things other than the commonplace, and in whom one could confide completely. Since you left, he was really the last person with whom I could do that. . . . His last words, as I left him at noon on the 22d, were that I must not be sad. He surely would have said the same to you.[98]

Arnold Berliner was cremated and his ashes interred in a cemetery in Berlin-Westend. A few friends came: Laue, Hahn, Rosbaud. His books and valuables were immediately taken by the police.[99] The journal he had founded and edited for twenty-three years took no notice of his death.[100]

The deportations continued. In June 1942 Laue wrote to Meitner, "Once again a colleague (this time the author of a paper in *Zeitschrift für Physikalische Chemie* 62 [454]: 1908) must leave Berlin. . . . One becomes accustomed to the abnormal."[101] The deported colleague was A. Byk. In September: "The wife of the discoverer of canal rays [Eugen Goldstein] does not live in Berlin any longer. . . . About Prof. Traube . . . no one knows exactly where he is." In October, "The former professor of chemistry, Traube, is dead."[102]

By 1943, Berlin's Jews were gone and Laue's letters focused on the destruction of the city. After a Brahms concert under Furtwängler: "As the concert hall stands completely intact in the midst of ruins, and the music was so wonderful, we had the feeling we were in another world."[103] He wrote of Paul Rosbaud's house, gutted by fire; of Max Planck's house, damaged and then completely destroyed in February 1944. The same air raids severely damaged the KWI for Chemistry. "I saw your former rooms," Laue wrote to Meitner. "Splinters! By the director's room . . . great heaps of ashes."[104] Details of the location and intensity of air strikes could have been useful to the enemy: on occasion the censor cut Laue's letters to shreds.[105] To detect interception, Laue and Meitner numbered their letters in sequence.

Laue was more cautious when writing about scientists and their projects. As head of a theoretical section of the Max Planck (formerly Kaiser Wilhelm) Institute for Physics, he knew that part of the institute had been requisitioned for a highly secret project: the military development of nuclear fission. This took place in 1939 just after the war began; Peter Debye, the institute director, was ordered to give up his Dutch citizenship or his job. Debye eventually left for America, never to return, but other scientists, including Werner Heisenberg, Carl Friedrich von Weizsäcker, and Otto Hahn, were recruited. Within a few months their so-called uranium club—an uninspired cover name!—outlined the basis for further research. The work was distributed among several laboratories; Heisenberg remained at his post in Leipzig, but his theoretical expertise dominated the fission project from the start.[106] None of this was explicit in Laue's letters, but Meitner knew soon enough. Word spread as scientists attended meetings and talked to colleagues in Germany, Copenhagen, and Stockholm.

During the first winter of the war, therefore, it appeared that Germany's top scientists were actively pursuing a nuclear weapon while the British were barely considering it and the Americans were blithely publishing relevant new results. Among the Allies, some scientists—émigrés, mostly—were alarmed; most outspoken was Leo Szilard, who urged that fission results be kept secret and persuaded Einstein to warn President Franklin D. Roosevelt of the danger in August 1939.[107]

But most physicists took comfort in Bohr's contention, buttressed by the Bohr-Wheeler theory, that the dominant fission process in uranium, fission by thermal neutrons, was due to ^{235}U. The scarcity of ^{235}U (0.7% of natural uranium) meant that a self-sustaining chain reaction would require several tons of natural uranium, an assembly whose very size would make the reaction slow and relatively weak—useful for a reactor, perhaps, but by no means explosive enough (and far too unwieldy) for a bomb.

Early in 1940, however, Otto Robert Frisch and Rudolf Peierls, an émigré theoretical physicist who was also working in Birmingham, began thinking about the behavior of pure ^{235}U and estimated that only a few pounds of it would suffice for a devastating nuclear explosion. In a three-page report, Frisch and Peierls set forth their theoretical reasoning for the size of a bomb, suggested methods for separating ^{235}U and for detonating the bomb, estimated its military destructiveness, and warned of its radiation danger—an altogether remarkable feat, considering that many of the essential measurements had not yet been made. They communicated their findings in secret to the British authorities. The fact that their estimate for the critical mass of ^{235}U was quite low made it appear that an atomic bomb was possible after all—if Bohr and Wheeler were right about ^{235}U, and if ^{235}U could be separated in quantity from ^{238}U. The latter was a very big *if* indeed, but the suspicion that the Germans had already begun isotope separation made it imperative to try.[108]

At the University of Minnesota that spring, Alfred O. Nier used a mass spectrograph to separate a minute quantity of ^{235}U; at Columbia University, Eugene Booth, John Dunning, and Aristide von Grosse showed that it was indeed fissile by thermal neutrons. The report was published in *Physical Review* in May 1940 for all to read;[109] Lise Meitner called it to Max von Laue's attention in June.[110] At its heels came Lise's crêve coeur, McMillan and Abelson's identification of element 93, the beta emitter that decayed to the still undetected element 94.[111] (The two transuranics, later named neptunium and plutonium, respectively, were both of mass 239.) To those concerned with weapons, these findings were strongly linked, for the ^{235}U confirmation was crucial to the Bohr-Wheeler theory, and the theory in turn pointed to fissile alternatives to ^{235}U.[112] According to the Bohr-Wheeler theory, ^{235}U's susceptibility to fission was due to two factors: its relatively low neutron-proton ratio (which made it less stable than ^{238}U) and its odd-even neutron-proton count (which led to a destabilizing release of energy on neutron capture). Exactly the same could be said for plutonium-239, with the added advantage that ^{239}Pu, unlike ^{235}U, would be separable from uranium by ordinary chemical reactions and might thus be extracted in sufficient quantity for an atomic bomb.

Long before the first plutonium nuclei were actually detected (in Berkeley in the winter of 1941),[113] its military potential was recognized on both sides of the Atlantic: by British scientists, who angrily protested McMillan and Abelson's open report of element 93; by the Princeton physicist Louis A. Turner, who thought about using ^{238}U to breed plutonium but decided not to put his ideas into print;[114] by C. F. von Weizsäcker, who mulled it over while riding the Berlin subway on his way to Dahlem in the summer of 1940.[115] By the end of the year fission had disappeared from American journals, a clear signal that the United States had finally entered the race for an atomic bomb.

In Stockholm, Lise read between the lines, paying particular attention to the whereabouts and activities of Werner Heisenberg. Years before, when Lise knew him, he had not been a Nazi; as a theoretical physicist, he had in fact been attacked as a "white Jew" for his failure to renounce the "spirit of Einstein."[116] And yet, disturbing character hints came to mind: Heisenberg's visit to Max Born in Cambridge in 1934 with a government invitation for him to return to Germany—without his wife and children;[117] Heisenberg's tactlessly nationalistic greeting to Lise on her sixtieth birthday;[118] Heisenberg's utter devotion to his homeland, widely discussed by American physicists after he visited the United States in the summer of 1939 and refused even to consider several exceedingly attractive offers to stay.[119] Unlike Rosbaud and Laue, who had stayed to fight the regime they hated, Heisenberg had made it clear that he could not live away from the Germany he loved. What was he doing now? Working on a nuclear weapon? Or withholding his talent somehow, masking resistance with a show of nationalism?

Like many others, Lise reserved judgment until she learned of Heisenberg and Weizsäcker's visit to occupied Copenhagen in September 1941. The two were scheduled to speak at a lecture series on astrophysics, an event unilaterally arranged by the local propaganda office for German culture and designated a "German" conference.[120] Rejecting the implication that Denmark had become part of Germany, the scientists in Bohr's institute boycotted the conference[121] but welcomed Heisenberg and Weizsäcker nevertheless as colleagues and old friends. It soon became obvious, however, that Heisenberg's attitude was "quite peculiar," so much so that Margrethe Bohr would later describe it as a "hostile visit."[122] As a young Danish physicist, Christian Møller, told Lise a few months later, Heisenberg came to Copenhagen "to stage a German physics congress, and was absolutely unable to understand that this was not fair. He was entirely imbued with the wish-dream [*Wunschtraum*] of German victory and had developed a theory of higher-level people and serf-people to be ruled by Germany; in this connection he considered the occupation of Denmark and Norway to be 'regrettable' [presumably because Danes and Norwegians could be classified as "Aryans"]."[123] Møller's report and Meitner's later account of it were seconded by others. Stefan Rozental, a Polish émigré in Bohr's institute, heard Heisenberg speak "with great confidence of the progress of the German offensive in Russia"—Stalingrad was yet to come—and stress "how important it was that Germany should win the war."[124] The Danish scientists concluded that Heisenberg's chief purpose in Copenhagen was to induce or coerce Danish cultural collaboration and to establish his own political reliability with the Nazi regime.

Under the circumstances, Bohr was wary when Heisenberg approached him for some private conversation. When Heisenberg began talking about the wartime use of nuclear energy, Bohr thought the worst—that the Germans attached great military importance to atomic energy, that they were making progress, that Heisenberg was betraying their once-close friendship by probing him for Allied secrets.[125] For Bohr, the other Danish physicists, and the Allied scientists who were soon informed, the question of Heisenberg's integrity was decided.[126]

Following Christian Møller's visit to Stockholm, Lise felt it necessary to warn Laue that Heisenberg and Weizsäcker were not to be trusted. Unable to write plainly, she said only,

> Dr. Møller was here for a few days. . . . He told me a lot about Niels and the institute . . . and about the visit of Werner and Carl Friedrich. . . . In addition to other peculiar things, C. F. seems to have curious thought processes: he appears to believe in certain "constellations" [astrological signs], but I beg you to keep this confidential. I was quite sad to hear all this, as at one time I thought highly of both of them. It was a mistake.[127]

Laue already knew.

> I have often wondered about the inner attitude of Werner and Carl Friedrich, but I believe I understand their psychology. Many people, especially younger ones, cannot bring themselves to come to terms with the

302 AGAIN, WORLD WAR
302 AGAIN, WORLD WAR

great irrationality of the present, and so in their imagination they construct castles in the air. It is an enormous task they have undertaken, to find some good aspects in things they can do nothing about. The ones named are not alone in this.[128]

After the war, Heisenberg repeatedly tried to mend his damaged reputation. Although he never explained his "peculiar" attitude, he and his supporters offered several versions of his meeting with Bohr. J. Hans D. Jensen, another German physicist visiting Copenhagen at the time, suggested that the "high priest of German theoretical physics" (Heisenberg) had gone to "the Pope" (Bohr) seeking "absolution" for his work with fission. Heisenberg himself claimed otherwise: troubled by the "spectre of the atomic bomb" and afraid that the Allies would develop their own atomic bomb and use it on Germany, he had intended to propose to Bohr an international boycott of work on nuclear weapons.[129] It appears that after the war, at least, Heisenberg was indeed seeking absolution, if not for the "sin" of nuclear weapons research (he always insisted his research was directed "only" toward a reactor), then for his compromised behavior during the Third Reich. Over and over again, he, and later his wife, portrayed his decision to remain in Germany as an "inner exile," describing his work on the fission project as important for keeping young scientists out of the military and providing an "island of stability" for the preservation of German science. The notion was exceedingly self-serving. Unlike true "inner exiles" who were deeply opposed to the Nazis and in some instances suffered greatly for it, Heisenberg was a favored member of the establishment who had nothing to lose from the continued success of the Nazi regime; his identification with Germany was so reflexive that he appears to have welcomed Nazi military victory as a means of imposing German "culture" on others. His biographer notes that while Heisenberg found it impossible to sever his attachment to Germany under the Nazis, he would have been ready to emigrate immediately in the event of a Russian takeover.[130]

After the war, Heisenberg intimated that he had been on moral high ground, that he had sought international cooperation while Allied scientists had not, that he and the other German scientists had worked only on a reactor while the Allies had built and used atomic bombs.[131] With this, however, he stumbled. Whether through insensitivity, arrogance, or (as Lise Meitner believed) intellectual dishonesty,[132] he did not seem to recognize that Germany's crimes had made international cooperation impossible and moral parity nonexistent.

The Danes, it should be noted, were not alone in their perception of Heisenberg's attitude. Early in the war, Paul Rosbaud had listened to Heisenberg talk about the necessity for German victory and the ensuing "purification" and had judged Heisenberg's views to be "childish, I almost would say immoral."[133] When Heisenberg visited the Netherlands in 1943, he told the Dutch physicist Hendrik B. G. Casimir that it had "always been the mission of Germany to defend the West and its culture against the onslaught of eastern hordes . . . and so, perhaps, a Europe under German leadership might be the lesser evil." Talk of "Asiatic hordes" was a staple of West European war propaganda, and Heisenberg's statements might have been camouflage, but it seemed clear to his colleagues that he relished the role of German goodwill ambassador, that he traveled widely and often—an unmistakable stamp of political approval—and that he seemed unperturbed by the genocidal proclivities of German "leadership."[134] In 1943 when Dirk Coster appealed for help in getting Samuel Goudsmit's parents released from a Dutch concentration camp, Heisenberg responded vaguely that he would be "very sorry if, for reasons unknown to me, difficulties were to arise for his parents in Holland." The elderly Goudsmits had already been deported to Auschwitz and killed.[135] To Coster and others, it seemed that Heisenberg had blinded himself to everything but physics and the German tribe. His and his supporters' later attempts to change that impression never dispelled it.[136]

After Heisenberg's Copenhagen visit, Allied scientists followed his activities closely.[137] In early 1942, the momentum of the war began to turn; the German army was stopped at Moscow and Leningrad, and America entered the war. No longer confident of speedy victory with the weapons at hand, the German military reevaluated the fission project. When Laue told Meitner in June 1942 that Heisenberg would leave Leipzig to head the Max Planck Institute for Physics in Dahlem,[138] she gathered that fission research had become more important. "Heisenberg has been appointed director of the Max Planck Institute," she wrote to Max Born in Edinburgh. "I think that to be quite an interesting fact."[139] Earlier in 1942 German physicists had made a concerted effort to convince authorities to increase the status and funding of the nuclear fission project;

Heisenberg's appointment was indeed an indication of their success, although the scale of their project never approached that of the Allies'.

The tone of Meitner's note to Born immediately raises the question, was this mere collegial chitchat? Or was she acting as a conduit for information, transmitting scientific intelligence to the Allies?

To Lise's friends in Sweden, the very question is both preposterous and abhorrent. They believe she could never have been an informant: it would have violated everything they know of her character and personality. Lise was politically aware, they say, but never an activist; more important, she was by nature open, incapable of deceit. Above all, they insist, Lise would never have put friendship or her profession to political or military use. She would have regarded this as a betrayal, not only of friends but of her most strongly held values.[140]

And yet, Meitner's innate openness may have made it impossible for her *not* to pass on what she knew to people she trusted. From Laue, and perhaps others, she learned bits and pieces of the fission project: Heisenberg's assignment, the comings and goings of other scientists, the level of activity, the fact that the project was important enough—yet still sufficiently small—to be moved to southern Germany when Berlin became too dangerous in 1944. It is unlikely that Laue's information was ever very technical—he was not a member of the uranium club—but it may have helped the Allies form a general picture of the scope and progress of the German fission project. Although Laue's information was peripheral, it was reliable—and Lise knew he expected her to pass it on. And she did, in letters and private talks with trusted colleagues: to Max Born, as we have seen; to Otto Robert Frisch when he was still in England; to Bohr after 1943 through Danish scientists with ties to British intelligence; and very possibly through Margrethe Bohr who remained in Stockholm until the end of the war. Lise was not apolitical and she was not naive: she knew that both sides were engaged in the military development of fission, she made her disclosures in one direction only, and she understood that by not doing so she would benefit only Germany. As to risk, she may have told herself that Laue's information, taken alone, was of no great value. Laue may have thought the same. After all, every one of their letters was read and censored.

With Paul Rosbaud, the situation was different: he was a spy who gathered scientific intelligence of great importance. From Rosbaud the

British learned of German weapons and military installations, the progress of the fission project, the heavy water plant in occupied Norway, and the rocket installation in Peenemünde—data of such astounding quantity and quality that for a time the incredulous British suspected him of being a double agent. It is unlikely that Rosbaud sent much vital data through Meitner: he had other, more direct routes to his British contacts, and the work was dangerous: Stockholm was a prime listening post during the war, teeming with agents of every nationality. Nevertheless, it is known that Rosbaud used book codes (a natural choice for a publisher's agent) and it is true that he regularly sent scientific books to Lise;[141] Rosbaud's biographer believes that Lise on occasion passed coded books to British agents or possibly to Njål Hole, a young Norwegian physicist in Siegbahn's institute who reported through the Norwegian delegation to British intelligence. Lise's friend Sigvard Eklund, a contemporary and friend of Hole's, does not believe that she or Hole was an informant or agent,[142] and indeed it may have been more important to Rosbaud that Lise's function as intermediary to his wife and daughter not be jeopardized. Unsurprisingly, there is no written corroboration: except for occasional veiled references to a mutual friend in a prison or concentration camp, there is no obvious clandestine undercurrent to their wartime correspondence, and after the war Meitner expressed only gratitude to Rosbaud for the books he had "so kindly selected."[143]

Passing on information was one thing, active participation in the war effort quite another. In 1943, Meitner was asked to join a group of British scientists, including Otto Robert Frisch, who were bound for Los Alamos to work on the atomic bomb. The offer promised intriguing physics, valued colleagues, and escape from Sweden.[144] Yet she refused. Hers was not an intellectual decision, carefully considered and weighed. It is unlikely, for example, that she knew enough about the German fission project to be reassured by its lack of progress, or that she was told of the successful self-sustaining reaction Fermi and his colleagues had achieved in Chicago in December 1942. Rather, it appears that she recoiled from putting her life's work to military use. Her refusal arose from a deep revulsion: *"I will have nothing to do with a bomb!"*[145]

Among scientists worldwide, almost none refused outright to work on weapons during the war; among the Allies, nearly all scientists, especially refugees, were driven by fear of the Nazi regime, the huge early successes of the German military, and the presumed superiority of German science. They had no moral compunctions about working on the bomb: they saw it as a race with Hitler that they had to win. Only after Germany was defeated without the bomb, and it was used on Japan over the protests of many leading scientists, did they understand that they worked on something over which they had no control. Perhaps Lise already understood that, after her experience in World War I. Hahn later suggested that Meitner could not bear to work on a weapon that might be used against the country that had been her home and people who were still her friends.[146] This may be part of the truth, but not all. Meitner wanted no part of deaths anywhere: she could not commit herself and her physics—the two were not distinct—to a weapon of war. She had seen the casualties firsthand in 1915–1916; she had heard the screams. She could not do it. Her decision was instantaneous and absolute: there was no discussion. She would not work on the bomb.

Lise did not join the Allies, yet she no longer shared the outlook of her German friends. When Hahn visited Stockholm in the autumn of 1943, Lise tried to convince him that even decent Germans had, with their passivity, contributed to the terrible misfortune Germany had brought on the entire world. Otto found Lise's position unjust, as did Max von Laue who visited Lise in 1944. Only Max Planck agreed with her that Germany's defeat was necessary and that all Germans, himself and other anti-Nazis included, shared responsibility for the crimes of the Third Reich. "Terrible things must happen to us," he told Lise in the summer of 1943. "We have done the most horrible things." Lise's love for Planck was unqualified. "He used the words 'we' and 'us.' And yet this 85-year-old man was more courageous in his resistance than all the others."[147] It did not matter to Lise that Planck, like nearly everyone else, had acquiesced and compromised to survive; the resistance she admired was deep within, his courage an expression of his moral strength.[148]

Lise waited for the end of the war in anxious isolation. Her situation in Siegbahn's institute remained poor. At times she was exhausted by the demands of her correspondence, particularly the twice-written relayed messages, some in English, which she did not find easy.[149] She worried constantly—about Stefan Meyer, about friends in Berlin, about the Costers going hungry in occupied Holland. Hanno Hahn, her godchild, was severely wounded in action in 1944 and lost an arm. The tension was unrelenting. Her weight dropped to ninety pounds; at times her health faltered.

After parts of Dahlem were heavily damaged by air raids in 1944, several scientific institutes, including Max von Laue's and Otto Hahn's, were relocated, with their personnel, to the relative safety of southern Germany. Relieved that these friends were no longer in Berlin, Lise was nevertheless disturbed to learn that Heisenberg's fission project was also relocated, even though it had not been damaged.

Lise's contact with Max Planck and his wife became intermittent. She knew that their house in Berlin was destroyed early in 1944 and that their visits to friends and relatives in other cities seemed to coincide with heavy air attacks. Then to her horror, Lise learned that Planck's oldest son, Erwin, had been seized and tortured to death by the Gestapo for his part in Count von Stauffenberg's bungled attempt to assassinate Hitler.[150]

Late in 1944, the Royal Swedish Academy of Sciences awarded the Nobel Prize in chemistry to Otto Hahn. Because Germans were forbidden to accept Nobel awards, Hahn was notified secretly and no public announcement was made.[151] Nevertheless, word spread among Swedish scientists, and Lise feared for Otto's safety. Otto hastily wrote that he could not possibly accept, and Max von Laue stressed that even a mention of the prize might be disastrous for Otto and them all. "Please do not speak of this," Lise begged Hans Pettersson. "It could be dangerous for Hahn."[152]

The fear of a German atomic bomb was never far from Lise's thoughts. In June 1944, shortly after D-Day, Lise wrote to Eva von Bahr-Bergius,

I am certainly glad that the invasion has finally begun, but I fear what the Germans will do when their situation appears hopeless. I am very anxious to know if the [German] robot-bombs have something to do with uranium bombs. I don't actually think so, but I do know that the institute that was working on those things [Heisenberg's] was transferred to southern Germany, even though it was not hit by bombs at all. That makes me wonder, and I am disturbed to think that it may be possible to make uranium bombs after all.[153]

Lise knew little about the German fission project—only that they had been unable to separate ^{235}U—but she was afraid they might have found another path to a bomb.[154] With this fear she waited out the final months of the war. The brutal fighting went on. As the Allied armies closed in, the utter horror and unimagined scale of the Holocaust were revealed.

Germany's defeat was inevitable, yet Lise was deeply pessimistic about the future. She wrote to Eva in March 1945,

> The letters from German friends sound very depressed, yet I do not think they comprehend just what sort of fate has befallen Germany through their passivity. And they understand even less that they share reponsibility for the horrible crimes Germany has committed. These thoughts make me terribly unhappy. How shall the world trust a new Germany when its best and intellectually most prominent people do not have the insight to understand this and do not have a burning desire to make whatever amends are possible? Not only must they feel this strongly, but at the proper time they must state this openly. But I fear they are far from it. For this reason I do not believe that for the most part they had a strong inner resistance.[155]

It was difficult to imagine a lasting peace among the ruins of Europe.

CHAPTER THIRTEEN

War Against Memory

That is indeed the misfortune of Germany, that you have all lost your standards of justice and fairness.

The war in Europe ended on 8 May 1945, and the numbing search for survivors began. Lise Meitner lost contact with Otto Hahn and Max von Laue when their region of southern Germany fell to the Allies; in her thoughts she composed letter after letter to them, worried as always that they would no longer understand each other.[1] She did not know if Max Planck and his wife, Marga, were alive, and despite appeals to the Russian legation in Stockholm, she could not find out if Paul Rosbaud and Elisabeth Schiemann had survived the fighting in Berlin. In late June, when an American on his way to Germany offered to hand-carry a letter to Hahn, Lise finally set down on paper the thoughts she had rehearsed so often in her mind.

Your last letter is dated March 25, so you can imagine how anxious I am to receive news from all of you. I followed the English war reports very carefully and think that the region that you and Laue are in was occupied without fighting. So I hope that you personally did not suffer. Naturally it will now be very hard for you, but that would be unavoidable. On the other hand, I am very worried about the Plancks, there was bitter fighting near them. Do you know anything about them and the Berlin friends? This letter will be taken by an American; he is coming soon, and so I am writing in a great hurry, even though I have so much to say that weighs on my heart. Please keep this in mind, and read this with the certainty of my unbreakable friendship. I have written many letters to you in my thoughts in the last few months, because it was clear to me that even people like you and Laue have not comprehended the reality of the situation. I noticed this when Laue wrote to me at the time of Wettstein's death, that his death was a loss in the general sense because

309

W. could have been very useful with his diplomatic expertise at the end of the war. Now how could such a man, who never spoke against the crimes of the last years, be useful for Germany? That is indeed the misfortune of Germany, that you have all lost your standards of justice and fairness. You yourself told me in March 1938 that [KWG official Heinrich] Hörlein told you that horrible things would be done to the Jews. So he knew of all the crimes that were planned and later carried out, but despite that he was a Party member and you—also in spite of everything—considered him to be a very decent person and you permitted him to influence your relationship with your best friend.

You all worked for Nazi Germany and you did not even try passive resistance. Granted, to absolve your consciences you helped some oppressed person here and there, but millions of innocent people were murdered and there was no protest. I must write this to you, as so much depends upon your understanding of what you have permitted to take place. Here in neutral Sweden, long before the end of the war, there was discussion of what should be done with German scholars when the war is over. What then must the English and the Americans be thinking? I and many others are of the opinion that one path for you would be to deliver an open statement that you are aware that through your passivity you share responsibility for what has happened, and that you have the need to work for whatever can be done to make amends. But many think it is too late for that. These people say that you first betrayed your friends, then your men and your children in that you let them give their lives in a criminal war, and finally you betrayed Germany itself, because even when the war was completely hopeless, you never once spoke out against the meaningless destruction of Germany. That sounds pitiless, but nevertheless I believe that the reason that I write this to you is true friendship. You really cannot expect that the rest of the world feels sympathy for Germany. In the last few days one has heard of the unbelievably gruesome things in the concentration camps; it overwhelms everything one previously feared. When I heard on English radio a very detailed report by the English and Americans about Belsen and Buchenwald, I began to cry out loud and lay awake all night. And if you had seen those people who were brought here from the camps. One should take a man like Heisenberg and millions like him, and force them to look at these camps and the martyred people. The way he turned up in Denmark in 1941 is unforgettable.

Perhaps you will remember that while I was still in Germany (and now I know that it was not only stupid but very wrong that I did not leave at once) I often said to you: as long as only we have the sleepless nights and not you, things will not get better in Germany. But you had no sleepless nights, you did not want to see, it was too uncomfortable. I could give you many large and small examples. I beg you to believe me that everything I write here is an attempt to help you.[2]

Lise wrote this letter in the early summer of 1945. Just seven years before, at that same time of year, she had been desperately trying to find a way out of Berlin. In those seven years, Lise had never been an immigrant to Sweden so much as a refugee from Germany: a refugee whose reference point was still German, whose friends were German, whose strongest memories and deepest longings were for the life and work that had been hers in Berlin. During the war she did not know if it would ever be possible to return, but she herself did not rule it out. Only after the fighting was over did she begin to understand that things would never be the same. "Concerning Germany," she wrote to Dirk Coster, "I appear to myself like a mother who sees clearly that her most beloved child has gone hopelessly astray."[3] The metaphor was not especially apt—in many ways she was more the abused child than the anguished parent—but emotionally it was right. For all her anger and grief, she could not cut her ties to Germany or her German friends. Her letter to Otto shows how involved she still was, with him and with Germany.

Later, Otto told Lise that he never received this letter, and Lise, perhaps disbelieving, wrote with quotation marks, "A letter that did not reach him" at the top of her copy.[4] But it was true. The American who promised to deliver it (and may have suggested she write it) was Morris (Moe) Berg, a bookish Jewish baseball player who had retired as catcher for the Boston Red Sox in 1942 and become an intelligence agent for the American secret service, the OSS. In 1944, Berg was in Zurich, where Paul Scherrer, Meitner's friend at the Swiss Federal Institute of Technology and a strong anti-Nazi, supplied him with whatever information he had about the location and activities of German atomic scientists. Berg came to Stockholm in June 1945 to probe Meitner for what she knew about German fission research, to make sure she kept quiet about it in public, and—always a bedrock of American anxiety—to discover if she or other scientists were likely to go to the Soviet Union. As an introduction he brought a chatty letter from Scherrer, offering her a temporary position in his institute. Although the offer was genuine enough, it was designed primarily to keep Meitner away from Russia, something Berg surely did not tell her at the time.[5] He took her reply, a pleased but cautious "maybe,"[6] back to Scherrer, but he never, it seems, had any intention of delivering her letter to Hahn. The record shows that he read it and turned it over to the OSS.[7]

Holland was not liberated until the final surrender in May 1945. For five years Germans had eaten Dutch food and exploited Dutch labor, killed Holland's Jews, and brought severe deprivation to nearly everyone else. Dirk Coster first contacted Lise in June 1945. He wrote in English, "not being able to use the language of our criminal oppressors," and with a shaky hand; the years of severe malnutrition and cold had worsened his multiple sclerosis. "But," he continued, "you can help me by answering me in German so that I again become accustomed to the idea that it is still used by decent and kindhearted people also. . . . For three days stiff fighting took place in Groningen. . . . All our automobiles are stolen."[8] The German occupation had created the "gravest difficulties" for the education of the Costers' four children—their oldest son, Hans, a university student, had been in prison for a time—but they were grateful to be alive. Coster wrote again a few weeks later.

All is quite well with all of us, though from time to time we had a narrow escape. I hope you will not think that we did any important so-called illegal work; but when something came in our direction we tried to do our duty and did not evade. . . . Here in the north we never had real hunger as in the big cities in the west. Agriculture is much developed here and though all good things were stolen by the German bandits, there still remained enough potatoes, e.g., to save us from starvation. But in the . . . west with their dense population it was most horrible and you may believe all the terrible stories told. . . . For us it was not to bear: we knew our friends and relatives were dying from hunger and the German bandits made it impossible for us to give any help. At any rate they tried to make it impossible but nevertheless we succeeded from time to time to deceive them and to send food to some friends ([Hendrik] Kramers, e.g.) and to the many relatives I have. But it took all our energy at day and at night you could not sleep at the idea that at some 100 km distance all those terrible things happened. . . . The Germans have left a total vacuum here, stolen the locomotives and railway wagons, the machinery of many factories, blown up the bridges, the harbour installations, and many factories. . . . I think the German people will rouse your pity. But I hope you will not forget that the people as a whole (with the exception of some few as Hahn and von Laue) is guilty of the horrible crimes committed in the name of the Reich. Yesterday I spoke with a man from the American scientific mission [possibly Samuel Goudsmit][9] who had been in Germany. The German scientists, he said, are eager to work under our direction, they will do all we command. But we are not interested: they have prostituted science on behalf of the most terrible criminals who ever had power. We will help them to restore education, they need physicians and the like, but as far as pure science this does not (concern) us. Don't you think he was right? But Otto Hahn and von Laue! I hope they will leave Germany in the near future. . . . Heisenberg was in a scientific committee during war time to study the possibility of using nuclear fission for war purposes. Do you understand the mentality of the majority of the German scientists?[10]

Stefan Meyer and his wife had come through the war in Bad Ischl. "It is really a miracle to be alive," he wrote to Lise. "My wife and my mother-in-law were on the list to be deported; until the end of the war the Nazis had arrest orders for my daughter and me; my brother, a professor in Prague, and his son and a large number of relatives were murdered in concentration camps. One dares not think about it. Everything we owned in Vienna was seized and confiscated, but we still had the house in Ischl and that saved us." Meyer returned to Vienna and the Radium Institute in 1945.[11]

July passed without a word from Otto Hahn or Max von Laue. In August Lise left Stockholm for the peaceful countryside near Leksand, Dalecarlia, where she stayed in a small hotel not far from the country cottage of Gudmund and Magnhild Borelius. No one disturbed the long summer evenings by turning on the radio, and therefore Lise did not learn what had happened on 6 August until early the next day, when a long-distance telephone call jarred the morning quiet. Alarmed at first, she was momentarily relieved. It was not a relative or friend but a reporter from the *Expressen*, a Stockholm newspaper. Then his words burned through: "The first uranium bomb has been used over Hiroshima, said to be the equivalent of 20,000 tons of ordinary explosives."[12]

Stunned, Lise escaped to open air. For five hours she walked, on country roads and footpaths, through woods and fields, alone. Later that day, perhaps to calm herself, she recorded in detail the route she had taken, but not whether she wept or screamed or kept silent. When she returned to her hotel, she found a stack of telephone messages and a reporter and photographer from a newspaper in Leksand. Hastily she wrote a brief article[13] as calls from the Swedish press and international news agencies poured in. "A lot of nonsense will surely be printed," she noted in her diary that night. "Everyone I talked to understood nothing about it."[14]

Lise spent the next day trying to fend off the press and correct their more blatant mistakes. Both efforts were futile; the press went after her with an intensity that bordered on assault. Reporters spent the night in a nearby boardinghouse and camped on her doorstep. When she told them she could tell them nothing about the bomb, they invented interviews many columns long, complete with "stupid, tactless statements" she never made.[15] "I was too awkward to get rid of them," she told Hilde Rosbaud. "They followed me when I tried to escape, and neglecting all my protests took the silliest snapshots."[16] In one Swedish newspaper photo, Lise is standing near a goat outside her hotel; in another she is "discussing atomic fission" with a local woman in peasant dress.[17] Under the banner headline "FLEEING JEWESS," an article described how Lise escaped from Hitler with the secret of the bomb and gave it to the Allies.[18]

Besieged with requests, Lise agreed to a radio interview with Eleanor Roosevelt in New York on the night of 9 August. That day the second atomic bomb destroyed Nagasaki. The interview went badly; speaking from a radio studio in Leksand, Lise was exhausted and tense. "Nothing was tested. I did not know how loud I should speak, or how close to the microphone to be. Mrs. R. began. I understood only half of it and lost my composure when I answered. Much too near the microphone, and too soft. It was quite idiotic. . . . I was inwardly ashamed of my helplessness." The next evening she went to Oskar and Gerda Kleins' for dinner and listened to a rebroadcast: "Horrible."[19]

But a few days later, Lise agreed to another interview with a different radio station in New York; she wanted to counter press reports that Hahn was a Nazi. This time she was better prepared and there was a surprise: Lise's sister Frida, who lived in New York, was in the studio.[20] Hearing her sister's voice for the first time in many years, she was overcome with an intense longing for her family.

Lise's friends had never seen her so distraught. She knew none of the intricacies of bomb physics, none of the tricks the bomb builders had used to multiply neutrons or the specifics of bomb assembly or detonation; she did not know how the Americans had managed to separate the ^{235}U they used over Hiroshima, or the details of the reactor that bred the plutonium that destroyed Nagasaki. She knew only that these things had been done and that she had been present at the beginning. She had split uranium nuclei for four years without knowing it, recognized fission, explained it, and calculated the energy released; she had anticipated the creation of plutonium by showing how ^{239}U was formed and that it decayed to element 93, the precursor to plutonium. She knew the physics but could not comprehend what it had come to. Physics would never be the same; the world would never be the same.

Lise was the unwilling center of a media circus that lasted for weeks. As the press avidly sought a human face for the atomic bomb, she was the celebrity of choice: the "fleeing Jewess," the woman scientist who had snatched the secret of the bomb from Hitler and delivered it to the Allies. This was fiction, of course, a fable whose origins can be traced to an article written in 1940 for the *Saturday Evening Post* by William L. Laurence, later an esteemed science writer for the *New York Times*. In a mix of gee-whiz science and docudrama, Laurence described Meitner staring sadly out the train window on her way from Berlin to Stockholm as the idea of fission began forming in her mind and told how she telegraphed the idea to her "friend" Otto Frisch in Copenhagen who then cabled his "father-in-law" Niels Bohr, who passed it on to scientists in America.[21] After the bombs, when emotions were running high and facts were in short supply, the 1940 story was resurrected, repeated, and embellished until Meitner was nothing less than the "Jewish mother of the bomb"—wrong on every count— and all aspects of her private and professional life, even her looks, clothes, and personal habits, were front-page news.[22]

Many years before, Max Planck told Lise how upset he was whenever he inadvertently gave out wrong information, whether in scientific publications or casual conversation, since despite all efforts at retraction, he knew that at least some untruth would always persist.[23] This certainly applied to the story of Lise Meitner and the bomb: some of it was roughly true—she did flee Germany, she and Frisch did tell Bohr, Bohr did bring the news of fission to America—but the rest was wrong—Hahn and Strassmann were usually left out (or in some versions made into Nazis), the chronology of the discovery was wrong, it was not a secret at the time, Meitner was never involved with making a bomb. But as Planck would have predicted, the story was too dramatic and far too appealing ever to completely disappear.[24]

Lise was stung by the disregard for the truth, embarrassed among her colleagues, and unhappy to find herself embraced by the Jewish community, for which she felt no special affiliation. "I feel like an imposter," she confided to Frida, "when American Jews (or from my point of view one should say Jewish Americans) praise me especially, because I am of Jewish descent. I am not Jewish by belief, know nothing of the history of Judaism, and do not feel closer to Jews than to other people. And just now, when one wishes so strongly that all racial prejudice be eliminated from the world, isn't it unfortunate if Jews themselves document such racial prejudice?"[25]

In the limelight she felt her loneliness more acutely than before. Known everywhere, she belonged nowhere: isolated from physics, far from family, a "Jew" who was not Jewish, an Austrian exiled from Germany to Sweden, a person with no real home. "I live here among people some of whom are very friendly to me, but they do not speak my language in the deeper sense of the word. I always feel homeless."[26]

At the end of August 1945, Hilde Rosbaud wrote that Paul was alive.[27] From Berlin, Paul himself wrote euphorically to Lise.

> I have experienced the greatest triumph of my life: I exist, and those who wished to exterminate us all have vanished forever. . . . The greatest happiness for me, of course, is to be in contact with my wife and daughter again. . . . I can hardly express how thankful I am for every line you wrote to me these last years. For me every letter was a message from another and better world, and you helped me to hold on and never to lose courage.[28]

In December an exhausted and starved Paul Rosbaud was smuggled out of Berlin by the British and joined his family in London.[29]

Otto Hahn and Max von Laue were already in England, but not voluntarily. They and eight other scientists had been taken into custody by *Alsos*, an American intelligence mission charged with finding and disabling the German fission project. The *Alsos* scientific group, led by the Dutch-born physicist Samuel Goudsmit, entered southern Germany with the first Allied troops in late 1944 and quickly learned that the German project posed no immediate threat—no bomb, not even a critical reactor. By April 1945, the *Alsos* group had located and dismantled an unfinished reactor in Haigerloch, tracked down Heisenberg in Bavaria, and found Hahn at his institute in Tailfingen and Max von Laue and Carl Friedrich von Weizsäcker in nearby Hechingen. Altogether, ten scientists were rounded up and taken to Farm Hall, a comfortable country estate near Cambridge that was wired throughout with hidden microphones in preparation for the distinguished prisoners. The reasons for their "detainment" (Hahn liked that term) were never fully specified. Originally they were seized to stop their work and to keep them from the Russians and also, given the left-wing reputation of Frédéric Joliot, away from the French who were about to occupy southern Germany. The Farm Hall "guests" (the British officer in charge enjoyed that term) would be released only at the end of 1945, when conditions in Germany were better and they were considered unlikely to succumb to Soviet enticements on their return.[30]

In September 1945, Lise learned that Hahn was in England and also that Edith Hahn was well, Hanno married, and the Plancks safe. "Naturally I don't know if you are allowed to write to me; it would make me very happy," she wrote, worried that Otto's long silence was due to the critical letter she sent in June. "I hope you understood that in spite of some harsh-sounding things I said about the German situation that it was intended in the sincerest friendship for you. . . . I have heard nothing from the Plancks since February, but I have heard that they are now in the American zone and so I hope that things are not so hard for them now. It is true, isn't it, that their son was tortured to death by the Nazis? That poor old man, that he had to live to experience that!"[31] Erwin Planck, a former undersecretary in the government, had been in close communication with some of the leading plotters in the abortive attempt to assassinate Hitler in July 1944. He was arrested and tortured, and although at one point his family was told to expect his release, he was hung, in an exceptionally cruel way, in the Plötzensee prison in Berlin on 23 February 1945.[32] With Erwin's death, Planck had lost all four children from his first marriage.

Hearing from Lise made Otto "especially happy." He had not received her June letter. (Moe Berg was annoyed that she found out. "It's fun to be double-crossed like this," he complained to his commanding officer. "Why didn't the London people censor Meitner's latter letter to Hahn if they saw fit to deliver it?")[33] Hahn explained,

> We were really completely shut off from the outside world. I found out that Hanno got married only long afterwards. How you know of it is a puzzle to me. Personally we are fine, that is eating, drinking, living space. Nevertheless we are still carefully guarded and naturally we are not permitted to leave the house and the part of the garden to which we are assigned. . . . I heard from Edith twice since April. I am very glad that Hanno and Ilse [Hanno's wife] are living with her, she couldn't make it alone. If only I could send them some of our good food. In the last few months Hanno has lost 30 pounds. He is by no means completely healthy and under the current conditions cannot get enough to eat. In the French zone it is different from the Anglo-Saxon zones which do not requisition food from our land, which has been bled dry.[34]

Otto certainly knew that Germany had stripped every country it occupied of food, labor, and resources; his suggestion that Germany was now unjustly being "bled dry" did not escape Lise.[35] About the atomic bombs, Hahn had little to say except that the Farm Hall scientists had all been "tremendously surprised."

In fact, they had been incredulous, then shocked and chagrined. At first, early on the evening of 6 August, Heisenberg was sure the announcement was false, just American propaganda. (Those who first read the secretly recorded Farm Hall conversations took Heisenberg's utter disbelief as evidence of his arrogance—and how far behind the German fission project actually was.) "If the Americans really have a uranium bomb," Hahn remarked, "then you're all second-raters. Poor old Heisenberg."[36] Later that evening, when the scientists realized the report was genuine, their reactions varied greatly. Walther Gerlach, a physicist who had enjoyed the title Plenipotentiary for Nuclear Physics under Reichmarshal Hermann Göring (Bevollmächtigter des Reichsmarschalls für Kernphysik), acted like a "defeated general," went to his room, and sobbed. Max von Laue expressed relief that he had never been part of the fission project, Hahn that he had not constructed a bomb.[37] Others, especially the younger physicists, were very upset. As British officers listened on hidden microphones, recriminations flew: they had been hampered by the lack of cooperation and leadership, by insufficient government support, by the heavy Allied bombing that disrupted their research.

The next morning (as Lise Meitner was wandering the Swedish countryside) the Farm Hall scientists learned from the newspapers that the Allies had separated ^{235}U on a scale they had never attempted; three days later, after the plutonium Nagasaki bomb, they realized that the Allies had produced element 94 in quantity, which required a working reactor. They themselves had never come close to separating ^{235}U, never isolated elements 93 or 94 in quantity, never determined their chemical properties or tested them for fissibility, never even brought their "uranium machine" (reactor) to a self-sustaining chain reaction, something Enrico Fermi and his group had achieved in Chicago in 1942.

As Germans, they were defeated and tainted; now they were failures as scientists as well. Within hours an explanation began to take form. On Hiroshima night, Carl Friedrich von Weizsäcker suggested, "I believe we didn't do it because all the physicists didn't want to do it, on principle. If we had all wanted Germany to win the war we would have succeeded." "I don't believe that," Hahn responded, "but I'm thankful we didn't succeed."[38] Out of earshot one of the younger physicists, Erich Bagge, objected angrily, "I think it is absurd for Weizsäcker to say he didn't want the thing to succeed. That may be so in his case, but not for all of us."[39]

Absurd as Weizsäcker's suggestion may have been, it was also exceedingly attractive. By implying *could have* along with *would not*, it turned German scientific shortcomings into evidence of moral scruples: now the Americans were the mass murderers and the German slate, at least with respect to fission, was wiped clean.

The whiff of moral superiority infuriated those who first read the Farm Hall transcripts. Just before the *Alsos* mission entered Germany, Samuel Goudsmit had gone to the ruins of his boyhood home in The Hague and learned how his parents had been taken away to their deaths. He could not comprehend how a man like Heisenberg could have stayed in Germany and worked on fission, and he did not believe that Heisenberg and the others would not have created a weapon if they could. Goudsmit was convinced that the German scientists had tried hard, known little, and accomplished less—he attributed it to the control of science and repression of intellect by the Nazis—and that anything they said to the contrary was a self-serving attempt to explain away their lack of scientific progress and absence of moral scruples.[40]

In a protracted public debate, Heisenberg later insisted that he had been scientifically successful and ethically uncompromised, a stance later historians have termed a "postwar apologia."[41] Heisenberg's wartime dilemmas might have been better understood outside Germany if he had been more candid after the war. His American biographer notes that the effort by Heisenberg and his supporters "to distance themselves from the Nazi regime while at the same time claiming they had done great but harmless work under it" was a position almost impossible to maintain.[42]

Such efforts were endemic in postwar Germany. So deliberately and so pervasively did Germans turn away from their past that they soon believed that they were responsible for nothing and that they were themselves victims, of the Nazis and the war. Such willed amnesia may have helped Germans come together for the rebuilding process, but outside Germany others could not go on without trying to understand, or at least examine, the mentality and behavior that had so horrified the world. The fact that nearly all Germans, even "good" Germans, had no desire to do the same created an enormous psychological gulf between them and non-Germans, and eventually between those Germans who experienced the Third Reich and the generation that followed.[43] In June 1945, when Lise urged Otto and his colleagues to openly admit their passivity and examine their responsibility, she was asking them to take the first steps to narrow that gulf. Otto Hahn's generation assumed the right to forgive themselves and suppress the memory of what had been done to others, by them and in their name. They wished only to forget and ignore, and the gulf remained open.

The controversy over the German fission project is a prime example. After the deaths of hundreds of thousands of civilians in Hiroshima and Nagasaki, many Allied scientists experienced great moral anguish over the fact that they had been willing to build weapons without ever being in a position to control their use. They were appalled to see no evidence of comparable soul-searching among their German counterparts, only smug self-congratulation for not having built an atomic bomb and an apparent lack of candor about how hard they tried. As a result, the controversy continues to this day, flaring instantly with each new set of facts or interpretation.[44] In Germany, especially, the legend of heroic German scientists who refused to build a bomb persists;[45] in that vein one American writer has recently suggested that Heisenberg's steep learning curve after Hiroshima indicates that he understood a great deal and even concealed what he knew during the war to avoid building a bomb for Hitler.[46] Others see the truth as closer to the exact reverse: to them, Heisenberg's initial disbelief and wrong guesses in Farm Hall show that he understood relatively little of reactor construction and had no real idea of bomb physics until he gleaned key parameters from the newspapers.[47] Still others have analyzed the German fission project in detail and found no evidence that Heisenberg deliberately hindered the German fission project or indeed that he had much control over it; these historians regard the legend of heroic resistance as an outgrowth of the "old apologetic story" that was born in Farm Hall immediately after the war.[48]

The Farm Hall transcripts, recently made public (presumably in full) after nearly fifty years, do not shed much new light on the tangled issues of competence and intent. But the conversations among the scientists, recorded over an eight-month period, do reveal their intense nationalism, thoroughly permeated with the standards of the Third Reich. The men appear at once naive and callous, devoid of historical sensibility, unable to see the connection between Germany's aggression and its ultimate defeat. In the transcripts Weizsäcker complains righteously about English air raids killing German women and children, as if Germany had never bombed Britain;[49] Hahn admits he had hoped for Germany's defeat but never imagined such "terrible tragedy" for his country;[50] Heisenberg suspects the Allies were fearful of being attacked by the Russians "just as we were attacked [*sic*]";[51] Gerlach expresses his disapproval of the Nazis but wishes nevertheless that the fission project had succeeded and Germany had won the war.[52] After listening for months, the British officer in charge, Major T. H. Rittner, commented on his guests' "inborn conceit"[53] and their general attitude, which "seem[ed] to be that the German war was a misfortune forced on the Germans by the malignancy of the Western Powers, who should by now have forgotten that it had taken place (the guests seem to have done so) and that the United Nations should all be largely concerned to set Germany on her feet again."[54]

The scientists all knew of murders and atrocities, but again the record shows reactions that are curiously flat. Paul Harteck thinks Goudsmit is unlikely to help get them out of Farm Hall because "Goudsmit can't quite get away from the fact that we killed his parents";[55] Heisenberg tells how he had been asked to help five Polish and Belgian scientists who "were murdered by our people"; Karl Wirtz recognizes that Germans did things "unique in the world," murdering Jews in Poland and murdering Polish high school girls to "wipe out the intelligentsia."[56] Yet no one disagrees when Erich Bagge likens their internment to German atrocities: "If Hitler ordered a few atrocities in concentration camps during the last few years, one can always say that these were done under the stress of war but now we have peace . . . and they can't do the same things to us now."[57]

During World War I, Otto Hahn worked with poison gases, with the result (he later wrote) that "our minds were so numbed that we no longer had any scruples about the whole thing."[58] In their recorded conversations, the Farm Hall scientists appear similarly numbed. They knew their country had initiated world war and committed unimaginable crimes, yet they felt nothing: no grief, no responsibility, no shame.

Several scientists expressed horror over Hiroshima, but only Laue seemed to understand that Germany itself had unleashed the fear and hate that produced the bomb. "The . . . émigrés' passionate hatred of Hitler was . . . the thing that set it all in motion," he wrote to his son Theo, after learning that German immigrants in England had been the first to bring atomic energy to the attention of British military authorities.[59]

Others were unable to make such connections; they were thinking along different lines. In the early part of the war, when the plutonium alternative was first considered, Weizsäcker, the son of a diplomat, had the idea of going to Hitler and telling him about atomic bombs, in the expectation that Hitler would be sensible and stop the war before things got out of hand.[60] Weizsäcker never did talk to Hitler, but he still thought of himself as someone close to power, and even at the end of the war, he still thought Hitler's Germany would have a salvageable moral legacy. In Farm Hall, one day after Hiroshima, he announced, "History will record that the Americans and the English made a bomb, and that at the same time the Germans, under the Hitler regime, produced a workable machine [critical reactor]. In other words, the peaceful development of the uranium machine was made in Germany under the Hitler regime, whereas the Americans and the English developed this ghastly weapon of war."[61]

Weizsäcker was mistaken about history's eventual judgment of Hitler's Germany, and his statement was factually false: German scientists tried but never achieved a critical reactor, and they were well aware that once they had a reactor it would not only supply energy to further the war effort but could be used to breed element 94 (plutonium) for a weapon. In a brief memorandum written on 8 August and intended as a press release, the Farm Hall scientists closely followed Weizsäcker's line, refuting British press reports that they had worked on a bomb, stressing how close they had come to a critical reactor, saying little of their efforts to separate ^{235}U and nothing about the plutonium alternative.[62] The younger scientists refused to go along at first, presumably because much of their research was omitted, but Heisenberg urged them to sign, and solidarity prevailed. They still thought their reactor expertise was more advanced than the Allies; it was their one bargaining chip, and Heisenberg saw the "possibility of making money."[63] The next day they learned that the Allies were way ahead, having produced and separated enough plutonium for the bomb over Nagasaki.

With its description of Germans engaged in peaceful atomic research in time of war, the 8 August memorandum marks the beginning of the "myth of the German atomic bomb."[64] Although Max von Laue had not participated in the German fission project, he believed the others, agreed with the memorandum, and signed it.[65] Many years later, in a letter to Paul Rosbaud, Laue repudiated it as a *Lesart*—literally a "reading," or version of the facts that was not actually true. "The *Lesart* was developed that the German atomic physicists really had not wanted the atomic bomb, either because it was impossible to achieve it during the expected duration of the war or because they simply did not want to have it at all. The leader in these discussions was Weizsäcker. I did not hear the mention of any ethical point of view. Heisenberg was mostly silent."[66]

The memorandum of 8 August was not the only Lesart to come out of Farm Hall. After Hiroshima, Otto Hahn was distressed to see Lise Meitner's name in all the newspapers—a typical article on 7 August, headlined "A Jewess Found the Clue," attributed the discovery of fission to her and never mentioned him.[67] "In part untrue reports about the discovery," he noted gloomily. "Especially at first Lise Meitner played a big role in it; I myself was not mentioned."[68] The press inaccuracies were annoying because Hahn and the entire Farm Hall group were anxious to claim that pure research, including the discovery of fission, was in the German domain. An addendum to the joint memorandum of 8 August pointedly states that "Professor Meitner had left Berlin a half year prior to the discovery and did not participate in the discovery."[69] On the same day Hahn expanded this into a second memorandum that was as much a Lesart as the first.

> As long as Prof. Meitner was in Germany the fission of uranium was out of the question. It was considered impossible. Based on extensive chemical investigations of the chemical elements which resulted from irradiating uranium with neutrons, Hahn and Strassmann were forced to assume by the end of 1938 that in these processes uranium splits into two pieces, of which one piece, the chemical element barium, was determined with certainty.
>
> The production of barium from uranium was communicated to Prof. Meitner in Stockholm in a number of letters even before publication in Germany. With her nephew Dr. O. R. Frisch she explained the experimental findings of Hahn and Strassmann, the "nuclear fission" which had previously been thought impossible.[70]

About Meitner's contributions before leaving Berlin, Hahn said not a word, except to repeat that fission was considered impossible while she was there. (In fact, the idea of nuclei splitting was not considered at all.) He did not actually say, but strongly implied, that Meitner had done nothing in Berlin except prevent fission from being discovered sooner.

This was the summer of 1945. It could have been a time for Hahn to reflect on the past and set the record straight: a time to remember Meitner's intellectual leadership of the Berlin team, the injustice of her forced emigration, her crucial criticism in Copenhagen, the fact that her participation in the uranium project, scarcely diminished by her absence, continued until the discovery of barium and beyond. Instead, Hahn was restating the version of the discovery he first expressed in February 1939: that he and Strassmann "never touched upon physics, but only did chemical separations over and over again."[71] In 1939, Hahn made it clear that fear for his professional survival drove him to pull the discovery out of its physics context and to distance himself from Lise Meitner. But in 1945 the war was over and the Third Reich gone. Why was he still pushing her away?

Hahn was calling on fission to serve again—not just himself this time, but his defeated country. He was famous, and fission was more sensational than ever. He would use the importance of the discovery and his personal prestige to call attention to Germany's misery and rebuild German science. He saw no purpose in looking back to the injustices of the Third Reich; he felt no personal necessity to make amends. He wanted the discovery to be his, alone. And Germany's.

In the years ahead, Hahn would become a major public figure in Germany. Tirelessly he would repeat his version of the discovery, in countless interviews, articles, reminiscences, and memoirs, scarcely reviewing the scientific literature and never, apparently, checking his "memory" against correspondence or personal diaries. In time, his Lesart would assume a life of its own. Followers and associates—none close to the discovery, and none with firsthand knowledge—would echo his contention that physics and Meitner had nothing to do with the discovery of fission. Fission, they would insist, had been discovered by the chemists in spite of the physicist.[72]

Among the most ardent supporters of Hahn's Lesart would be several members of the Farm Hall fraternity: Erich Bagge, Walther Gerlach, Heisenberg, and Weizsäcker. Farm Hall had forged them into an exceedingly loyal group; they would speak with one voice about fission and its development. Their message was simple: the discovery of fission belonged to chemistry, to Hahn, and to Germany. In creating a scientific hero for the new Germany, they elevated Hahn and set Strassmann to one side. A detailed, open examination of the science was never on their agenda: it would have exposed the interdisciplinarity of the discovery and the injustice, fear, and dishonesty of that time. Most especially, Lise Meitner was not to be part of the history they were constructing. With respect to her, as with so much else, they abandoned whatever standards of justice and fairness they may once have had.

Thus the falsification of memory and reality[73] that began under the Nazis persisted long after the Third Reich was gone. Strassmann, Laue, and Meitner occasionally spoke out, but in Germany it was extremely difficult to counter the phenomenon of Otto Hahn. Elsewhere it was not much easier, in part because the 1944 Nobel Prize for the discovery was awarded to Hahn alone. Announced just after the war—the timing was perfect for Hahn's agenda—the Nobel Prize would stamp Otto Hahn with the scientific community's most prestigious seal of approval and thereby cement his Lesart into the historical record.

CHAPTER FOURTEEN

Suppressing the Past

*If the best Germans do not understand now what has happened and what must
never happen again, who should instruct young people that the path that was
tried was tragic for Germany and the world?*

Autumn is Nobel season in Sweden. Speculation begins in October, builds
to a flurry of November announcements, and ends in a glittering royal
ceremony each 10 December, the anniversary of the death of Alfred Nobel.
In 1945, rumors floated for weeks that Lise Meitner would share in one
or another of the prizes. On 16 November, the Royal Academy of Sciences
announced its Nobel decisions: the 1944 chemistry prize to Otto Hahn and
the 1945 physics prize to Wolfgang Pauli.

At Farm Hall the German scientists celebrated, raising their glasses to
Otto Hahn.[1] But in Sweden, Lise's friends were furious.[2] They viewed her
exclusion as neither omission nor oversight but deliberate personal re-
jection, the work of Manne Siegbahn. From Göteborg, Hans Pettersson
wrote angrily, "[We] are indignant about the one-sidedness of the distribution of the Nobel Prize. We are certainly glad that Hahn got the
chemistry prize, but by all rights the physics prize should have gone to you.
I personally am quite sure that this would have been the case had not
Sweden's most practiced *krokbensläggare* [one who pulls others down, i.e.,
Siegbahn] been against it, for dark reasons of prestige."[3] Borelius and Klein,
certain that Siegbahn had kept Meitner from the physics prize, regarded
her as a victim of "royal Swedish jealousy [*kungliga Svenska avundsjuka*]."[4]

The injustice was apparent beyond Lise's immediate circle. Birgit
Broomé Aminoff, a scientist herself and the wife of a prominent mineralogist on the board of the Nobel Foundation, wrote,

> Long before the release of nuclear energy had been realized on a practical
> scale, it seemed to me that Professor Meitner had reached a status equivalent

to that of many Nobel Prize recipients. It must therefore have been very bitter that for completely unrelated reasons you were forced to leave the laboratory where the now-rewarded discovery was so close, and thereby lost the possibility to complete a work which promised to be the natural climax of a long and devoted career as a scientist.[5]

Lise responded,

Surely Hahn fully deserved the Nobel Prize in chemistry. There is really no doubt about it. But I believe that Frisch and I contributed something not insignificant to the clarification of the process of uranium fission—how it originates and that it produces so much energy, and that was something very remote from Hahn. For this reason I found it a bit unjust that in the newspapers I was called a *Mitarbeiterin* of Hahn's in the same sense that Strassmann was. Your letter, therefore, was a double present; a warm, understanding word can mean so much. A thousand thanks for it.[6]

The press coverage had added insult to injury. Just three months before, Swedish newspapers could hardly get enough of Meitner; now if she was mentioned at all, it was as Hahn's *medarbetare*[7]—a term that in Swedish, German (Mitarbeiter), or English (co-worker) means a junior associate, a subordinate.[8]

"The article in *D[agens] N[yheter]*, [a leading Stockholm newspaper] was almost insulting," Lise wrote to Eva von Bahr-Bergius. "Really, I was never just a Mitarbeiter of Hahn's; I am, after all, a physicist, and haven't I done a few quite decent physical things? I know something about the particulars of the deciding Nobel session and hope to be able to tell you about it sometime in person."[9]

Details of the physics sessions may never be fully known, but from contemporary accounts it is evident that the chemistry deliberations were stormy. After the war the Nobel chemistry committee voted to reconsider its 1944 award to Hahn—an unprecedented move, and evidence that the original decision was flawed. Oskar Klein, newly elected to the Royal Academy,[10] provided Bohr with the details. In September 1945, Klein wrote that Bohr would surely be "pleased to hear" that the chemistry committee and *klass* (section) had just voted to postpone distribution of the 1944 prize for a year, to assess the contributions that others had made to the discovery of fission.[11] According to Klein, The Svedberg, a member of the chemistry committee, and Arne Westgren, an academy member, "strongly pointed out" that the foundation for the committee's 1944 decision had changed with new information from America and France regarding Meitner and Frisch's contributions and also the new importance of Meitner and Hahn's earlier discovery of ^{239}U, precursor to plutonium. But later, in sessions of the academy as a whole, there was resistance to amending an earlier decision. It was argued that Westgren and Svedberg had reversed their 1944 position (which they admitted) and that the field was too complex to evaluate fully; Klein suspected an important factor was the academy's desire to avoid the appearance of yielding to American influence. In the final plenary session in November 1945, a slim majority voted to leave the award unchanged, to Hahn alone.[12] The vote was procedural rather than substantive: Strassmann's role was not considered, nor was the effect of Meitner's forced separation from the Berlin team, nor was the fact that her physical measurements in Berlin permitted Bohr to deduce, shortly after the fission discovery, that ^{235}U undergoes fission with thermal neutrons. Quite possibly the decision to stay with Hahn was also constrained by the Nobel three-person rule: if Meitner joined Hahn, it would be difficult to exclude Strassmann and Frisch.[13]

The discussions that took place in 1945 illustrate the complexity of evaluating an interdisciplinary discovery, given the statutory requirements of the Nobel awards, the jurisdictional division between chemistry and physics, and the various precedents that had grown over time.[14] In contrast, the 1944 decision appears to be impulsive and uninformed. One wonders about the haste to award the prize to Hahn at a time when it could not be announced, or even mentioned, without endangering him. One wonders also how the chemists on the Nobel committee could have missed Strassmann's contributions—it appears they mistakenly regarded him as a latecomer to the investigation[15]—and why they ignored Meitner's role, given that The Svedberg had already nominated Hahn *and* Meitner for the chemistry prize in 1939.[16]

In 1944, it seems, Hahn was simply the sentimental favorite: his time had come.[17] Because the committee worded its citation—for his achievements in radiochemistry and the discovery of fission—to recognize all of Hahn's scientific career and not just the fission discovery, Meitner and others could not help but agree that he "fully deserved" it, but the chemistry committee, not recognizing that fission would be the overwhelming focus of the award, completely failed to examine the extent to which analytical chemistry and physics had contributed to it.

In 1944, there may also have been a political edge: at a time when their old cultural ties to Germany were crumbling, Swedes could still take pride in honoring a man like Otto Hahn.[18] And they valued him personally, as a colleague and friend who had been Doktorvater to Swedish scientists and a welcome visitor to Swedish universities and institutes; in 1943, he had been elected a foreign member of the Royal Academy of Sciences and had come to Sweden to lecture on fission in Stockholm and Göteborg.[19] The 1944 Nobel decision followed soon after. Perhaps it is not surprising that the chemistry committee's view of fission coincided so exactly with Otto Hahn's. Only when the attention of the world was riveted on fission in 1945 did they have second thoughts and begin to seriously evaluate the science.

Much later, some of Hahn's associates contended that the dispute over his award in 1945 was purely political, motivated by animosity toward Germany and revulsion against the use of nuclear weapons.[20] There is no evidence for this in Klein's reports to Bohr. On the contrary, it appears that the 1944 decision was swayed by personality and politics, while the 1945 vote to postpone was a belated effort to properly sort out the science. Underlying their change of heart, Klein believed, was "the chemists' desire to erase a possible bad impression in the choice of Hahn."[21]

Klein was dismayed that the academy elected to stay with Hahn, taking it as a major setback for Meitner. He thought it unlikely that another chemistry prize would go to fission and regarded Siegbahn's influence in the physics committee and klass as a "dangerous obstacle" to a prize in physics.[22] Nevertheless, Klein hoped Bohr would submit a nomination: "I regard it as a fairness, a needed rehabilitation after the underestimation—in any case on the part of the Swedish physicists—that she has had to put up with (and perhaps is still exposed to) and to which Hahn's references to her and Frisch's achievements undoubtedly contributed, while she has always been fair to Hahn in all her publications."[23] Bohr was of the opinion that the 1944 chemistry prize for fission should "in no way" prevent Meitner and Frisch from receiving a physics prize. He nominated Meitner and Frisch for physics in 1946 and chemistry in 1947 and 1948, to no avail.[24] Lise's friends were convinced she would have shared in one or another of the Nobel Prizes had she emigrated to any other place but Sweden.[25]

After the furor over the 1945 prizes subsided, Meitner's friends were determined to find a position for her outside of Siegbahn's institute. Gudmund Borelius was proposing a new facility for nuclear physics at the

Royal Institute of Technology; he wanted Meitner to join him there, possibly in a joint appointment with Klein at Stockholm University.[26] This was not merely an act of friendship but an effort to mobilize resources for the development of nuclear energy, a field that had scarcely been touched in Sweden during the war, not least because of the reluctance to include Meitner in the military/scientific establishment. As she was above the mandatory Swedish retirement age of sixty-five, however, such an appointment would require special approval from parliament, but the government was newly supportive of nuclear research and there was considerable sympathy for Meitner in government and academic circles.

Lise meanwhile was preparing for her first trip to the United States. She wanted to spend time with her sisters, Lola in Washington and Frida in New York, and to see Otto Robert, who was still in Los Alamos, and the many friends who had scattered all over. Lola's husband, Rudolf Allers, a professor of psychology at the Catholic University of America in Washington, D.C., arranged for a visiting professorship for Lise for the spring semester of 1946.

As Lise left for the United States, Otto Robert had some last-minute advice.

> As regards publicity, you had better cultivate a philosophical attitude. Half of what is printed in American newspapers is false anyway, so why worry? The best scheme is to be nice to newspaper reporters, then they will at least write nice things about you. . . . And remember, in this country it is regarded as perfectly natural to have one's breakfast habits, one's favorite color in stockings, and one's opinion of Beethoven and Mickey Mouse discussed in the newspapers, so you need not feel self-conscious about it. The idea that people have a private life is a stuffy and un-American notion.[27]

Thus fortified, Lise arrived in New York on 25 January 1946. *Time* reported that when the "pioneer contributor to the atomic bomb" stepped from the airplane and saw the swarm of reporters and photographers below, she went back in to collect herself. Once on the ground, she was given a "push and pull welcome" from newsmen, murmured "I'm so awfully tired," and rushed into the arms of her relatives—Frida and Lola, their husbands, Leo and Rudi, and, to her surprise, Otto Robert, who had traveled by train two days and nights from Los Alamos. With her family again, she was "utterly happy" for the first time in years.[28]

Lise was swept into a whirl of activity that continued unabated until she left five months later. It began in New York in February when dozens of acquaintances and old friends welcomed her to a meeting of the American Physics Society; in Washington a few days later, more friends and colleagues gathered at a reception in her honor at the Catholic University. On 9 February at a banquet given by the Women's National Press Club, she was designated "Woman of the Year"; seated next to President Harry Truman, she accepted an inscribed silver bowl with a smile, a bow, and no speech—"painless," she told Otto Robert later—and enjoyed the political repartee between reporters and the attending "bigwigs." The president gave the impression of a "jovial, laughing youth," an American trait, she supposed. They talked about the bomb, both expressing the hope that it would never be used again. Then Truman gave a speech that began with a few jokes—another Americanism, she told Eva—and ended with a very serious appeal to Americans to give food to starving Europe.[29] The chill of Lise's seven years in Sweden was beginning to thaw.

At the Catholic University, Lise was scheduled to teach a course in nuclear physics and conduct a weekly seminar. At first she worried about her spoken English and also about her ability to understand American English, but the four hundred people who crowded into her first lecture "did not seem unhappy," and more than one hundred continued to attend her subsequent lectures, fifteen in her seminar. She was in constant demand. She talked to high school winners of the Science Talent Search, gave scientific seminars at universities, addressed women's colleges and organizations, attended countless meetings, and accepted honorary degrees. Everywhere she left an indelible impression: a small figure, a gentle voice, and an unmistakable air of authority when the subject was physics.

Early on Lise held a press conference, hoping to be spared private interviews with their "idiotic questions."[30] It didn't help. The press pursued her every step; her face was everywhere, from the *New York Times* to the center of a crossword puzzle in the *San Francisco News* to article after article in newspapers and magazines.[31] She was even the subject of a sonnet—quite a good one—published in *American Scholar*.[32] "I see you hate publicity as before," James Franck wrote during her stay. "It is a sign of respect, a childishness of this young nation."[33] But it was too much. Strangers stopped her on the street, waitresses and taxi drivers wanted her autograph, and mail flooded in, 300 letters by the end of March, 500 by May: requests from publishers, magazines, schools, Girl Scout troops, individuals with odd causes, organizations of every sort. Lise felt obliged to answer even those for whom she could do nothing. "I [am asked] to help keep south Tyrol in Austria, to join a Board of Trustees for aviation, to help 5,000 Jewish children emigrate to Palestine, to get radioactive substances from the bomb for a man with cancer, etc. Among others I received two hate letters for giving lectures at the Catholic University, accusing me of selling myself body and soul to these criminals."[34]

Hollywood was interested, of course. Lise was shown a script for a Metro-Goldwyn-Mayer film, *The Beginning of the End*, which she dismissed as "nonsense from the first word to the last. It is based on the stupid newspaper story that I left Germany with the bomb in my purse, that Himmler's people came to the Dahlem institute to inform me of my dismissal and more along the same lines. I answered that it was against my innermost convictions to be shown in a film, and pointed out the errors in their story."[35] When MGM countered with a much higher offer, Lise threatened to sue.[36] "I would rather walk naked down Broadway!" she told Otto Robert.[37] Such scruples—old-fashioned and perhaps a bit snobbish—were not to her immediate advantage. With a sizable sum, she could have left Sweden and worked anywhere, but careless public exposure—perhaps especially the American brand of exposure—was more than she could bear.

For the most part, however, Lise's American visit was an exhilarating return to life, a reunion with family, friends, and physics that helped assuage the isolation of the years in Stockholm. When she gave seminars in Princeton, she spent evenings with Rudolf Ladenburg and his family, talked for hours with Einstein and Hermann Weyl, discussed physics with Hylleraas, Yang, and Lee. In New York, she lectured at Columbia, discussed experiments relevant to beta decay theory with I. I. Rabi, and visited Selma Freud, her student friend from Vienna. She spoke at Harvard and MIT, visited Goucher and Pembroke, received an honorary degree from Brown, addressed professional women, the American Association of University Women, and congresswomen in Washington. Visiting Hedwig Kohn and Hertha Sponer in Durham, North Carolina, she took time out to enjoy the dogwood and spiraea in bloom. Everywhere she saw that former émigrés had found places for themselves in America in ways not available to her in Sweden.

By 1946, peace had already given way to tension between East and West. Many scientists who had worked on the atomic bomb were urgently concerned with its political implications: the balance between international cooperation and national security, the consequences of scientific secrecy, the dangers of a nuclear arms race. Many of these problems had been foreseen before Hiroshima. In the spring of 1945, when it was evident that the war in Europe would end without an atomic bomb, a committee of Chicago scientists chaired by James Franck had urged the government against military use of the bomb; in 1946, the Chicago committee introduced its bimonthly *Bulletin of the Atomic Scientists* as a forum for public debate. In Washington, Lise spent an evening with James Chadwick, who had served for three years as head of the British mission to the American bomb project. She had hoped to discuss these issues with him but was inhibited by unspoken constraints: "Chadwick was very nice, although several problems were anxiously avoided. I would have liked to know how he stood on the question of the bomb, and the wishes of the Scientists' Committee, but I felt that the manner in which he spoke of the work on the bomb was intended to cut off any questions from me."[38]

A more amusing encounter took place a few days later when Lise met Leslie Groves, director of the Manhattan Project. Drily she recorded their conversation: "Cocktail party, General Groves and wife. Groves told me that when he saw the first pile in Chicago, he understood everything in half an hour, and was able to give 6 or 7 good suggestions. He: Bohr was of no help at all. I: Nevertheless, he is the greatest living physicist. He: All theoretical physicists are prima donnas. (He praised Oppenheimer highly)."[39]

In May, Lise visited Otto Stern at the Carnegie-Mellon Institute in Pittsburgh, went to Purdue University for an honorary degree, and then to the University of Illinois. Early in June she arrived in Chicago for an extended visit with James Franck and his daughters and their families. Ingrid Franck had died in 1942, after being ill for years with multiple sclerosis. Now he and Hertha Sponer planned to marry, and Lise was glad to see them "happy as children" together.[40] She found Franck "as lovable as ever, with an admirable, wise understanding for many things that scarcely concerned him before."[41] Franck believed America and the immigration experience had taught him tolerance, something he supposed Lise had learned long before in Vienna.[42]

Traveling again, Lise spoke to Smith College graduates about women's education in her youth, then returned to Chicago, where she met Fermi, Edward Teller, Victor Weisskopf, and Leo Szilard at a meeting of the

American Physical Society. The meeting itself was clouded by "oppressive secrecy," so that Lise felt "more like a member of a secret society that excludes the public than a participant at a scientific meeting."[43] Whenever the subject was nuclear physics or reactor construction, "all papers had been censored by the military authorities and stopped just at the point where you could hope to learn something new. On [Eugene] Wigner's script one could see that large parts had been clipped, whereat the audience broke out into a roar of laughter. I could not help feeling rather unhappy, it was like a caricature of a scientific meeting, one did not dare ask a question."[44]

After a final week in New York with Frida and Leo, Lise boarded the *Queen Mary* for England on 8 July. The crossing gave her a few quiet days to rest and think. She was conscious of how good it had been to be with her family, how stimulating to be part of the physics community again.[45] By the end of July, Lise was in Cambridge for an entirely unfettered nuclear physics meeting and a reunion with Erwin and Annemarie Schrödinger, Wolfgang Pauli, and Max Born.

In London the British were belatedly commemorating the 300th anniversary of Newton's birth; the only German they invited was Max Planck.[46] Since Lise had seen him in 1943, he had suffered beyond all measure: his house and possessions destroyed, the final terrible battles of the war, his son tortured and put to death. Lise grieved to see him now, frail and forgetful. But when she was alone with him, "his human and personal qualities were wonderful as ever."[47] It was their last meeting.

Max von Laue was also in London that summer, the only German invited to an international crystallography conference. He had been looking forward with great anticipation to seeing Lise again, and when they met he greeted her with unqualified joy. "What a reunion that was!" he wrote to Theo in America. "Almost as if I had found you there."[48] Physically Laue had hardly changed, but Lise thought he was under the utmost tension, on the verge of weeping or a fit of rage. In his presence Lise was torn between affection and hurt; the tension she sensed in Laue reflected her own. It was painful to see how much their lives had diverged, more painful still to see that Laue took for granted what she had not yet fully accepted: the permanence of her exile. Coming between them were the separate experiences of many terrible years, Lise still struggling to comprehend things that Laue had already set aside.

He sees Germany's misery, but for the most part he sees too little what else is happening in the world. . . . We discussed how poor England has become, and that the disaster that nazism brought to Germany did not happen by chance, but was the consequence of an ideology that developed for over a hundred years, and therefore it is understandable that the English and the Americans do not want these traditions to continue.[49]

Where Lise was preoccupied with the question, "How and why did it happen in Germany?" Laue sought a return to normalcy. When she expressed outrage that Heisenberg and others had contributed to the German war effort, Laue countered that English, American, and French scientists had done the same for the Allies. He did admit to an ethical distinction, but he believed outsiders had no right to judge: "I am not sure that all those who assign blame would have acted differently if they happened to have been born in Germany." It should be forbidden, he thought, to speak of the past in terms of accusations against individuals, because the occupation forces could never justly assess the individual guilt of millions. He advocated punishment for major Nazi criminals, general amnesty for everyone else.[50]

Lise had heard much the same from Otto Hahn. On his return from Farm Hall in January 1946, he had accepted the presidency of the Kaiser-Wilhelm-Gesellschaft, newly relocated to Göttingen in the British occupation zone. The Germany he came back to was miserable, cold, and hungry, and his letters to Lise were a litany of hardships: shortages of food, coffee, cocoa, cigars; problems with British and American officials; demeaning travel restrictions; apartments requisitioned by occupation forces; the deaths of Hans Geiger, Otto von Baeyer, and others hastened by poor food and difficult living conditions. Lise sympathized but then sharply reminded him of the suffering and death that Germany had inflicted on others. "The question, 'Why go on?' is one I asked myself often enough, especially in the early years in Stockholm. And for millions of people with the same question, a horrible death (gas, etc.) cut off any answer. I cannot ever forget it."[51] Otto's nationalistic self-absorption worried Lise deeply. "How can Germany regain the world's trust if the best Germans have already forgotten what happened?"[52]

Even before the end of the war, Lise had urged her friends, as representatives of Germany's "best," to openly acknowledge their nation's crimes, admit the contributions of their own passivity, and disavow the

Germany of the Third Reich. Max Planck had done so already in 1943, when he had told her in Stockholm, "We have done the most horrible things; terrible things must happen to us." Planck recognized individual responsibility and collective guilt, but among Lise's friends, almost no one else did.

To Hahn, Lise stressed the practical benefits of such penance, for without it the world could not trust Germany or know which Germans to trust. But for herself she was seeking something deeper: horror, rage, anguish—an emotional reaction to match her own. Samuel Goudsmit, who had been unable to save his parents in Holland, remarked at the time, "I was gripped by that shattering emotion all of us have felt who have lost family and relatives and friends at the hands of the murderous Nazis—a terrible feeling of guilt."[53] Was it only Jews and other victims of the Nazis who were stricken with grief and guilt, while Germans professed innocence or ignorance or—remarkably—their own victimization? Did Germans feel nothing, did they have nothing whatever to say? Was it, in the end as in the beginning, always to be a "Jewish problem" and never a "German problem"? If so, Lise knew that her old friendships were mostly memory and nothing remained of the Germany she had once cherished.

Lise's experiences in America had clarified her perspective. The person who most closely shared her views was James Franck; he, too, had no hesitation about sending food packages to Germany, but he found the process of renewing old ties painful and subject to misunderstanding. Others judged Germany far more harshly. Einstein turned his back on Germany forever, vehemently dismissing reports of postwar difficulties as a shopworn "campaign of tears [*Tränencampagne*]." The Germans would slaughter again, he insisted: "I see among Germans not a trace of guilt or remorse."[54] Otto Stern's distrust nearly matched Einstein's; even James Chadwick held what Lise thought was the "very sharp" opinion that even the most decent Germans shared responsibility for the crimes of the Third Reich.[55]

In the United States, Lise found that most Americans simply lumped all Germans together. When she began her talks with a plea for aid to starving Germany, the response was invariably, "Have you forgotten what the Germans did?"[56] And when she tried to refute the widespread press reports that Hahn had been a Nazi, she realized that few seemed to think it made much difference. Theodore von Laue, fresh from Princeton with a doctorate in history (and a slight Americanization of his name), agreed. "At present the distinction between Nazis and German nationalists of the old national-liberal school, great as it was in Germany, means little to the world at large. . . . And there is no assurance that from the civilized nationalism of our German friends (who are a small minority) there will not arise a more brutal version of it among the masses. I fully share the world's suspicion." He, too, had noticed that his father and Hahn seldom looked beyond the hardships of occupation to the underlying causes of Germany's ruin. "Now they probably find the distance between [A]llied mentality and their own attitude still more impassable than that between Nazis and themselves. What tragedy, on top of losing once more all their material security."[57]

In England, Lise heard the same. British scientists who had visited Germany after the war reported that their German colleagues, almost without exception, knew the facts but had no real understanding of what happened and were, moreover, full of self-pity. Lise looked at the bombed buildings and streets in London, saw how much was ruined and poor, and never heard an English person complain.[58]

In London, she was asked by the British commissioner for science in Germany to convince Laue that changing the name of the Kaiser-Wilhelm-Gesellschaft was essential for a symbolic break with the militarism of the past. Planck understood; Laue resisted. And in Göttingen, Hahn objected bitterly: "The fact that the name is so repellent to the U.S. does not, in my opinion, speak for a great deal of generosity. . . . If now everyone who had held a position in Germany since '33, and everything that has happened be damned, then one must wonder why the outside powers did not break off relations after '33." Hahn's argument was that responsibility for nazism was so widespread that no single group, certainly not Germans, could be held accountable. In this version of history, atrocious behavior was not confined to Germans or Nazis but was a general attribute of all nations and people. "I would almost doubt that the behavior of the occupying forces is so much nobler than that of the Germans in the occupying countries. . . . Naturally one must not generalize, and certainly we did not do it differently in Poland or Russia, but the Allies claim to have fought for humanity and justice, and today the war is over. Why then do the war methods continue?"[59]

Lise was aghast.

The fact that the Allies consider the era of Kaiser Wilhelm to be unfortunate and for that reason want the name of the KWG changed, is considered everywhere to be so self-evident that no one can understand the resistance to it. The idea that Germans are the select people and have the right to subjugate "inferior" people by any means was constantly repeated by historians, philosophers, and politicians, and finally the Nazis tried to carry it out. [Here Lise cited a passage from Fichte's *Reden an die deutsche Nation.*] What the best people among the English and the Americans wish is that the best Germans understand that this unfortunate tradition, which has brought the whole world and Germany itself the greatest misfortune, must finally be broken. And a small token of this insight is to change the name of the KWG. What meaning is there in a name, when one is concerned with the existence of Germany, and with it, Europe?

That a thoughtful German could seriously say, "I would almost doubt that the behavior of the occupying powers today is so much nobler than that of the Germans in some of the occupied countries." In Poland 2 million people were killed, not in war, but methodically killed. [The current estimate is over 3 million.] When [pastor Martin] Niemöller gave a speech declaring that he shares the guilt, he said that 6 million Jews were murdered, and that in Dachau 283,765 people were put to death between 1933 and 1945. In England a prominent physicist who values German science and culture very highly told me that German professors were sent to Poland with instructions to destroy all books of Polish history, and that this assignment was carried out with German expertise and thoroughness. The "serf-people" [*Helotenvolk*] of Poland were not to retain even the memory of their independent existence. I write all this in honest love for Germany. If the best Germans do not understand now what has happened and what must never happen again, who should instruct young people that the path that was tried was tragic for Germany and the world? In the reports of the Nürnberg trials it was said that every time visual evidence of the horrors of the concentration camps was shown, [former Reichsbank Pres. Hjalmar] Schacht looked away. As if by doing that it didn't exist. The enormous problems that the Nazis have created do not permit one to look away. One must not forget this with daily concerns. I certainly do not doubt that the occupation of Germany brings with it various unnecessary hardships, how could the Americans or the English be perfect? Please send me Hanno's current address; I would like to send him a package. Unfortunately almost all food here is still rationed. . . . Do you think his little boy could use dry milk?[60]

In September the British decided that the Kaiser-Wilhelm-Gesellschaft would be dissolved and succeeded by the newly formed Max-Planck-Gesellschaft. (In the American and French zones, the corresponding reorganization took another two years.) Otto accepted that (he had no choice) but responded to the other issues Lise raised in a most superficial manner.

> When my nephew was on a Japanese ship from Ceylon to Japan he met an Englishman who seemed very nice. After a while the Englishman gave what was intended as a serious compliment: "You are such a nice pleasant person—you could be English." Such things probably happen all over. I don't know if our nation is worse than any other. [Hahn was making no distinction between simple-minded chauvinism and genocide.] I hope that I'm not irritating you with this controversy. Don't be angry. But believe me, we also want to know what happened and how all of this could have taken place. . . . Do you really believe that the majority of Germans knew of the horrors of the concentration camps or the gas chambers?

Edith entered a plea for truce.

> Again you are fighting with each other. That is such a pity, as basically we all agree, don't we? But we simply can't continue, and you must understand that one nervous breakdown in 1938 was enough for me, and I simply can't go on. We don't care what Fichte said, there are people like that here and there, and there are even quite decent Nazis, only I think very stupid, just as there are indecent anti-Nazis. And horrible to say, there will always be wars, and we can only hope that we won't live to see the next one. It would be better if you wrote about yourself and your family, don't be angry, really. Your Edith.[61]

By then it was November 1946, and the Hahns would soon arrive in Stockholm for the Nobel ceremonies. Edith had written that they were looking forward to "perhaps the only days we will spend in a free country for the rest of our lives . . . and to the shops, and good food, and no political arguments, Lise, we are so tired of them, and tired and worn out ourselves. It was 12 years, of which you experienced only the 5 easiest ones."[62]

Lise stopped arguing. It had been futile anyway, and she did not want to spoil the Nobel occasion for Otto and Edith. November brought one more disappointment: the 1946 Nobel Prize in physics was awarded to Percy Bridgman. On the day of the announcement, Lise sent Otto a card. "The chance that I might become your Nobel colleague is finally settled. If you are interested, I could tell you something about it."[63] Otto did not respond.

"The Hahns are coming early in December for the Nobel ceremonies," Lise wrote to Margrethe Bohr. "Of course I am looking forward to it, but it will not be easy."[64] Lise had a new evening dress made to replace the one black velvet dress she had worn for eight years. "Like it or not, I must attend the Nobel banquet, which I've never done before. But if I don't go this time, when the Hahns are being honored, I fear it might be misunderstood."[65] To Otto Robert she confided, "Mentally it will be a bit of an *Eiertanz* [like walking on eggs]."[66]

It began the moment Otto and Edith stepped from the train on 4 December. On hand were Lise, a Swedish official, friends, and journalists. *Dagens Nyheter* carried a front-page photo of Hahn's arrival, noting the presence of Lise Meitner, Hahn's "former pupil." On the front page of *Stockholms Tidningen*, Hahn was shown open-mouthed, eating grapes; "on hand was Prof. Lise Meitner, Prof. Hahn's world-famous pupil, whom he heartily embraced when she came to meet him." *Svenska Dagbladet* reported that Hahn spoke at length about Strassmann, about the bombing of his institute in Dahlem and the loss of his correspondence with Rutherford, about Russia and his hopes for the peaceful use of atomic energy.[67] Hahn himself had not mentioned Lise's name.

That afternoon, as Lise shared a moment with the Hahns in their suite in the Grand Hotel, a *Svenska Dagbladet* reporter came to interview Edith. "Do you help your husband with his work?" "No, I am only a lay person. It is Lise who was my husband's right hand in the institute for many years." Here the reporter noted that Professor Meitner "politely but unmistakably" cut in with the remark that Frau Hahn was still very weak from the hardships in Germany. Whereupon the "splitter of atoms" asked his wife, "Won't you have another cup of coffee, Mammi?"[68] The interview was over.

Otto and Edith's presence stirred old loyalties, bitter disputes, the unresolved tension of their recent correspondence. At dinner that evening with the Hahns and several friends, Lise's Eiertanz faltered briefly. "Agitated discussions with Lise," Otto noted in his diary. "But these subside. Lise does not mean it badly. But the others agree, in Sweden we should forget politics for once."[69]

The problem was that Hahn himself had no intention of forgetting politics. He had come to Sweden to use every iota of his Nobel publicity as a platform to plead for Germany. Lise's preparations were more personal. She hovered over Otto and Edith, sent chocolates to their hotel room, took them sightseeing around Stockholm and shopping on the elegant Kungsgatan. They needed everything: warm clothes and shoes, formal evening wear for the Nobel ceremonies, housewares, gifts, and above all, food. They looked forward to every meal and prepared dozens of food parcels to send back to Germany.[70] Nearly every day Lise arranged small dinners for them at her home or larger gatherings with Swedish friends and others who had come to town: Wolfgang Pauli was there, George de Hevesy and his wife came for a visit, and Otto Stern, always one of Lise's favorite people, had come from Pittsburgh to collect his 1943 Nobel Prize in physics.[71]

Meanwhile in press conferences, interviews, and radio broadcasts, Otto single-mindedly campaigned for Germany. Unhappy Germany! Oppressed by the Nazis first and now the Allies, struggling to make a future for its unfortunate youth! Lise expected that, but she was taken aback by Otto's metamorphosis from apolitical scientist to zealous propagandist. In disbelief she heard him say he was glad Germany had not built the bomb and caused the needless deaths of thousands. When he claimed that the Allied occupation of Germany was as unjust and cruel as Germany's former occupation of Poland and Russia, Meitner, Stern, and Pauli objected angrily. Hahn was undeterred.

Several days went by before Lise fully understood that she no longer had a place in Otto's life, or even his memory. When asked by the press to reflect on his life's work, he did not speak of their thirty years together. When he talked about fission, he did not mention her contributions. Not once, in any of his public statements, did he so much as speak her name. Lise's friends were shocked. They wondered at her restraint,[72] but there was nothing she could do: she could not beg for fairness, force him to remember what he had willed away, breathe life into a friendship that had become a hollow shell.

The Nobel ceremonies took place, as always, on 10 December. By tradition, the king of Sweden dispenses the awards and a great banquet follows, at which the science laureates usually say a few words and the literature laureate delivers the main speech of the evening. In 1946, the person with the most to say was Otto Hahn. Speaking for Germany, "probably the most unfortunate country in the world," he wanted his international audience to know that it was "really not true that all Germans and especially German scientists subscribed to the Hitler regime with flying colors." He asked that German youth not be judged harshly, for they had "no chance to form their own opinion, no free press, no foreign radio broadcasts, and they could not get to know foreign countries." In fact, most Germans deserved sympathy, for "not many people outside Germany know the extent of the oppression which most of us experienced for the last 10 or 12 years."[73]

Hahn's speech was well received, "going straight to the heart," according to one newspaper.[74] Certainly Hahn was pleased; it established him as a leading voice for German science, and for Germany as a whole.

In his Nobel lecture three days later, he could not ignore Meitner entirely, but by describing the discovery exclusively in terms of radiochemistry he gave no sense of their ongoing collaboration and emphasized instead that the discovery had been made in opposition to the experience of nuclear physics.[75] In her diary the next day Meitner noted, "His lecture: showed him his letter of 21 December 1938."[76] In 1938 Hahn had written, "How beautiful and exciting it would be just now if we could have done our work together as before!" and "You see, you'll be doing a good deed if you can find a way out of this."[77]

Hahn's behavior in Sweden hurt Meitner personally, damaged her professionally, and contributed to her ongoing isolation in Sweden. There is no indication that he ever understood, or cared. In a personal autobiography many years later, he scrambled the facts, meaning, and emotional context of their discussions in Stockholm, retaining only the bitter aftertaste.

On his first night in Stockholm, according to Hahn,

> I had quite an unhappy conversation with Lise Meitner, who said I should not have sent her out of Germany when I did. The discord probably stemmed from a certain disappointment that I alone was awarded the prize. I did not talk to Lise Meitner about that, however, but a number of her friends alluded to it in a rather unfriendly manner. But I was really not at fault; I had only been looking after the welfare of my respected colleague when I prepared her emigration. And after all, the prize was awarded to me just for work I did alone or with my colleague Fritz Strassmann, and for her achievements Lise Meitner received many honorary doctorates in the USA and once even was "Woman of the Year."[78]

Because so much of this passage is patently untrue, it reveals more than Hahn may have realized. Never did Lise believe she had left Germany too soon—not at the time in 1938, and certainly not with hindsight[79]—and Hahn neither "sent her out" (an absurd notion) nor "prepared her emigration." Nor had he been concerned for her welfare only. As is evident from his impulsive visit to Heinrich Hörlein in March 1938, Hahn's instinct had been to protect himself and his institute, not her: the truth was not that she believed she had left too early but that he believed she had stayed too long.[80] Was his faulty memory the guilty expression of having *wished* her away? Clearly Hahn was misleading the reader by presenting himself as her savior and Meitner as a disappointed also-ran who begrudged him his Nobel Prize.

Hahn wrote the passage late in life, after decades of unparalleled public acclaim. One wonders: Why did he feel the need to make himself look good and Meitner bad, when he could easily have said nothing at all? Why would he portray Meitner as ungrateful, unreasonable, and unjustified in her claim for scientific recognition when his own prestige was not threatened? Why was his memory so faulty and, with respect to Meitner, so consistently self-serving?

Hahn was not resting comfortably on his laurels. At the center of his discomfort, always, was Lise. With her, his character was always somewhat in question; with her, he could never be completely sure his version of the discovery would hold. And so, in what he knew would be his final memoir, written when he was in his eighties, he asserted his Lesart one last time: he distorted the facts of Lise's emigration to avoid acknowledging that she would have shared in everything had she not been forced to flee; he denied that she took part in the discovery because he could not admit he had excluded her to protect himself and then lied about it ever since; he twisted the issue of the Nobel Prize, because he would not concede that it was wrong for him to be rewarded for work all three had done. The last sentence of Hahn's passage is oddly defensive, a bald assertion that he alone deserved the prize and Lise should have been satisfied with whatever else came along. Lest he be considered ungenerous or averse to giving credit where due, Hahn pointedly let his readers know that he gave part of his Nobel award—"a substantial sum"—to Fritz Strassmann. Again, this was more self-serving than accurate. The amount was less than 10 percent. Strassmann's wife called it a "tip" (*Trinkgeld*), and Strassmann refused to use it.[81] Here, too, Hahn leaves the distinct impression that his conscience was not clear. For all his fame, he was not at peace.

In 1946, Hahn departed Sweden far more content than his later re-membrances would indicate. A charming person, accustomed to being liked by everyone he met, he had relished the warmth of his reception in Stockholm. "We had a wonderful time in Sweden!" he gushed in a card to Lise on their way home. "A thousand greetings!"[82] Later, still glowing, he wrote from Göttingen to thank her again "for all your friendship for Edith and me. The Christmas candles you gave me and the Laues were on the little tree that Edith and I enjoyed on Christmas eve." He appreciated their "productive discussions" with Stern and Pauli: "We differ only in our opinion of Germany." And he was still savoring his contact with royalty. "I was somewhat shocked by the rapid decline in Planck's intellectual strength. He has aged greatly in the last 6 months, when I compare him, for instance, to the King of Sweden."[83] It cannot be said that Otto Hahn was a sensitive man.

After the Hahns left, Lise could barely summon the strength to do anything at all. Slowly she unburdened herself in a series of letters. Her first was a very tired Christmas letter to Eva von Bahr-Bergius.

> I found it quite painful that in his interviews [Hahn] did not say one word about me, to say nothing of our thirty years of work together. His motivation is somewhat complicated. He is convinced that the Germans are being treated unjustly, the more so in that he simply suppresses the past. Therefore while he was here his only thoughts were to speak for Germany. As for me, I am part of the suppressed past, the more so as I wrote him before he came here and tried—in the friendliest possible way—to let him know that decent Germans can help Germany only if they are objective. In a letter to me he wrote, for instance, that the Americans in Germany are now doing the same as the Germans used to do in the occupied countries. I answered him, that he could not have meant that seriously, that although undoubtedly every occupation is unpleasant and injustices do occur, he cannot disregard the fact that the Germans killed millions of innocent people. But he did not respond to this and when he was here he asked me not to discuss politics with him. . . . Hahn himself surely did not want to work on a bomb, but in one interview he made the statement, "He was glad the Germans were not burdened with the responsibility for the bomb and the resulting meaningless deaths of thousands of people in Hiroshima." . . . What he ought to have said was that he was glad about it because Germany has done so many other things that were much worse, but he was unable to say that. Perhaps it is asking too much, I don't know myself. I have struggled all these years not to become bitter or distrustful, one must take people as they are. But joy in living suffers with this attempt.[84]

To her sister Lola, Lise wrote,

The Hahns enjoyed all the material comforts and the very attentive way that
they were celebrated here. . . . For the first few days they simply could not
get enough to eat. . . . Hahn came here completely permeated with the idea
that Germans are being treated unjustly, especially by the Americans, which
from a certain point of view is understandable. Less understandable to me,
he absolutely suppresses the Nazi crimes, and is thereby led to very wrong
convictions. He does have the faults of his virtues. Probably one cannot be
such a charming person and also very deep.[85]

And finally in January 1947, Lise wrote in detail to James Franck.

Hahn said that the Allies are doing in Germany what the Germans did in
Poland and Russia . . . at which point [Otto] Stern became very upset. He
absolutely did not respond to our objections; he suppresses the past with all
his might, even though he always truly hated and despised the Nazis. As one
of his main motives is to gain international respect for Germany once again,
and since he does not have a very strong character nor is he a very thoughtful
person, he deceives himself about the facts, or belittles their importance. . . .
Stern told me that . . . a legend is now being propagated, similar to the
Dolchstosslegende, that the Germans did not produce the bomb for motives
of purest decency.

After World War I, the legend of the Dolchstoss (dagger stab) held that
the German army had been stabbed in the back by Jews and Communists
at home. Although not entirely analogous, the new legend also rationalized
defeat with a fiction—this time, the superior morality of Germans who
refused to build an atomic bomb. Like the first, it would prove exceedingly
difficult to dislodge.

Naturally I told Hahn that he would not help Germany by making incorrect
assertions, that the Allies know precisely how surprisingly little the Germans
knew of the necessary conditions for an explosive chain reaction. . . . For a
moment it seemed to make sense to him, but he did not pursue it, and all
his subsequent interviews sounded the same. Just forget the past and stress
the injustice happening now in Germany. As I am part of the suppressed past,
Hahn never, in any of his interviews about his life work, mentioned our long
years of work together, nor did he even mention my name. I received a series
of indignant questions about the origins of Hahn's behavior, from Eva von
Bahr-Bergius among others. It is clear to me that Hahn was hardly aware
of his lack of friendship, and he wrote to me after his return thanking me
completely naively for my "great friendship." . . . Perhaps our generation

is too old to see things clearly and no longer has the strength to battle the prevailing spiritual disorders that go back more than a hundred years and happened to find an especially gruesome expression in nazism.[86]

Lise was suddenly very tired.

CHAPTER FIFTEEN

No Return

When I wrote that I was not sure of the trust of the Mitarbeiter, in part because
I am Austrian, and in part because of my Jewish origins, Hahn said nothing
and Laue wrote that it was a pity, as they had no objections to Austrians.

After Otto's visit to Stockholm, Lise was at the end of her emotional and physical strength and spent the first two months of 1947 slowly recovering.[1] It was not a good time to be so exhausted, as she was just then in the process of moving out of Siegbahn's institute and into her own laboratory at the Royal Institute of Technology (KTH). Gudmund Borelius had long wanted to bring nuclear physics, and Lise Meitner, to the KTH; after the atomic bomb, the political climate for such a proposal was very favorable. Concerned that Sweden had no capability in a field that might be vital for defense and energy, the government hastily appointed a blue-ribbon scientific panel in the fall of 1945 for the purpose of devising a national policy for atomic energy; the Atomic Committee's first report was issued a few months later and unanimously approved by the Riksdag (parliament) in the spring of 1946. It was evident to all that the long-range development of weapons and civilian nuclear reactors would first require funding for basic research.[2]

The beginnings were modest, however. There had been essentially no fission research in Sweden, due in part to the lack of support for Meitner's work in Siegbahn's institute.[3] Pending the construction of a new laboratory for atomic research at the KTH, Meitner's facility was temporarily housed in the Academy of Engineering Sciences (IVA) research station, a small building on Drottning Kristinas Väg not far from the KTH. In it she had three small rooms, two assistants, technical help, and an equipment budget: a laboratory of her own. Occupying adjoining rooms was Sigvard Eklund, a friend from Siegbahn's institute, a "very good younger physicist" and a "fair and pleasant person."[4] Together they planned a Van de Graaff generator and a small high-voltage apparatus.[5] Meitner was to have the title of "research professor" (that is, without teaching duties) and a professorial salary, a welcome improvement after nine very lean years on a minimal stipend. Although her appointment still required special parliamentary approval, that was expected in the spring 1947 session of the Riksdag. She was supported by the minister of education, Tage Erlander, a Social Democrat, who had a background in science, a physicist wife, and an interest in wartime physics developments; he was also the minister responsible for the Atomic Committee.[6] Meitner's assignment, with Eklund, was to create a nuclear physics section for the KTH. It was the sort of position she had hoped for when she first came to Sweden, and it permitted her, once again, to think she might really be of use. At her age she regarded it as "a wonder."[7]

But for a time Lise was too tired for optimism. She was grateful for the new position but worried, suddenly, whether she was up to it—not the science but everything else.[8] Despite the friendship and support of Borelius and Oskar Klein, she "could not get over the feeling that [she didn't] fit in, with them or with Sweden"; after her visit to America she was even more conscious of how alone she was in Stockholm.[9] The years in Siegbahn's institute had left their mark. As she told James Franck, "That little bit of self-confidence that the Berlin colleagues, above all you and Planck, worked so hard to instill in me, the Swedes have driven out."[10] Given that Lise had good friends in Sweden and opportunities for work, the statement seems overly negative, but it expressed her struggle to maintain her self-assurance in a place that gave her no sense of home.[11]

She realized that she herself had not adjusted very well. "The main reason for my dissatisfaction can be found within me, therefore it is in that sense my fault. I absolutely do not fit in here."[12] When she tried to analyze why that was so, however, she tended to blame the Swedish "mentality," in sweeping generalities that were not always fair or accurate. Sometimes she ascribed it to gender discrimination: "Here, just being a woman is half a crime and . . . having one's own opinion is completely forbidden."[13] Although she recognized that social equality for women was more pronounced in Sweden than in many other countries, she believed that women were still excluded from higher positions.[14] Often she complained about a "general lack of collegiality, by no means just against me, that I would not have thought possible in scientific circles."[15] Having lived in cities with populations comparable to that of all Sweden, she thought the country's small size might explain the "sharpened competition" and "general lack of courtesy."[16] The language was a problem: "to really learn a completely strange language in addition to scientific work would require more effort than I am able to muster," and for this reason she expected to be always "half-excluded from things."[17] Compared to German, Swedish is by no means a completely strange tongue, but Lise felt foreign, not just in speech but thought processes, with all but a few close friends.[18] She did not mention anti-Semitism as a factor, but it could only have added to her outsider status in a homogeneous society. She repeatedly described herself as homeless, "never really an equal . . . always inwardly alone." The condition of being without a homeland, she told Laue, was something she wished he would "never experience, or even understand."[19]

Had she been younger when she immigrated, had she learned to speak the language well, had she not had the debilitating years with Siegbahn, she might perhaps have liked Sweden well enough. Apart from her good friends in Sweden—the Kleins, Lilli Eppstein, Borelius—she had lost nearly all the connections, large and small, that had given meaning to her life and work in Germany.[20] She blamed Sweden for a certain coldness of spirit; it is impossible to know if she would have been much happier somewhere else. After eight years she had yet to set firm roots in Sweden, and so she clung to past ties, however frayed.

In 1947 that meant sending food and other necessities to German friends: to the Plancks; to the Laues and their daughter, Hilde, and her husband and child; to the Hahns and to Hanno and his wife and son.[21] For Otto's birthday she prepared a special supplement of sugar, cigars, and cigarettes; to others she sent clothing, shoes, household items, anything she could spare. She sent packages to people she hardly knew, including one of Boltzmann's relatives whose husband "probably let his coat blow the wind [i.e., was a Nazi]. But when it comes to sending packages," Lise shrugged, "it makes no difference to me. If people could only look ahead to better times!"[22]

But when former Mitarbeiter asked Meitner for help with their "denazification" proceedings, her reaction was far more reserved. "Denazification," as mandated by the Allied powers and administered by German courts, required all Germans to declare their previous affiliation with the

Nazi party or related organizations; civil servants, including teachers in schools and universities, were closely scrutinized. A person could be demoted, dismissed, or sent to prison, but individuals were allowed to present affidavits from well-known people in their defense.[23]

Nearly all of Meitner's former Mitarbeiter had been party members, and very few had contacted her after she left. When her former assistant Gottfried von Droste wrote to her in February 1947, for the first time since she left Berlin, he allowed that he had a "somewhat bad conscience" for his long silence but quickly got to the point. His membership in the party had been a mistake—"I never subscribed to the party program, never let myself be pulled into party duties and never participated in party events"— but now his livelihood was threatened, and he needed her to confirm "what a poor Nazi I was. . . . You would be one of the best to give an opinion of me."[24] Meitner hardly knew how to react. She remembered that von Droste had worn an SA brown shirt to the institute for years, and she was quite sure that another of her students, Herbert Hupfeld, had recruited him into the party early on.[25]

A similar request came from Hermann Fahlenbrach, the young Assistent who had brought charges against her in 1934. Now with "three little children, aged 9 to 3 years," he was threatened with "loss of employment and bread, perhaps even arrest!" He regretted his "wretched accusations," which, he wanted her to understand, were "not political or anti-Semitic . . . but based on an immature character . . . [that] could, as a man, subordinate itself professionally to a woman only with inner resistance. Very respected Frau Professor Meitner, if you can answer me in my need quite soon, I would be exceptionally grateful."[26]

Of the two letters, Meitner judged Fahlenbrach's "the more decent"; von Droste's, "lies from beginning to end." Nevertheless, because she regarded denazification as "sheer madness,"[27] leading only to "bitterness and lying," and because she did not want to see young people who had grown up in an "atmosphere of madness" permanently deprived of their professions,[28] she carefully vouched for their behavior, without mentioning their character.

"I knew very little about the details of your [party] membership and position," Meitner answered von Droste, "so in that sense I can state . . . that you did not propagate Nazi ideas or express them by your manner."[29] For Fahlenbrach, she even summoned a glimmer of warmth. "The motives of personal relationships, good and bad, are usually much more compli-cated than they seem. . . . I am certainly willing to believe that anti-Semitism was not the driving force for you. You did apply . . . for a position as assistant in my section in May 1934."[30]

For Lise, the entire exchange produced a flashback to the Hitler years, the "emotional equivalent of physical nausea." She was surprised that Otto wrote a positive letter on von Droste's behalf: "In the interests of Ger-many, I hope there are better representatives of the 35- to 50-year-old generation than these former *Mitarbeiter* of ours."[31]

A few months later, during the investigation of Kurt Hess, the Nazi who denounced her in 1938 and tried to sabotage her escape, the prosecutor sent Meitner a set of detailed questions. She refused to answer.[32] Her opinion of denazification coincided with that of Paul Rosbaud, who con-sidered it dishonest and dangerous. Upset that well-placed Nazis and their sympathizers, among them Walter and Ida Noddack, were being reha-bilitated and restored to their former positions—"they all seem to be in high spirits"—he called the exercise "renazification" and sent Allied friends lengthy character descriptions, good and bad, of people he knew.[33] Meitner, however, would not make negative statements for the record. She refused to testify against Hess, probably the most malevolent Nazi she had known personally, and she "strictly declined" to provide the Allies with her opinion of Ernst Telschow, the general secretary of the Kaiser-Wilhelm-Gesellschaft, although privately she noted that he had been an "enthu-siastic Nazi" who was "humanly untrustworthy" for the position.[34] Most likely she was repulsed by the entire process, with its questionnaires and testimonials that were reminders of the notorious *Fragebogen*, defamations and denouncements of the Nazi period.

In April 1947, Lise met Elisabeth Schiemann in London. Elisabeth had been reinstated in the university position she had lost in 1940 and was in England for several months at the invitation of British geneticists to survey the literature for a new edition of her book.[35] It had been nine years since they had last seen each other, and their meeting was awkward. Over lunch, when Meitner mentioned Hahn's "admitted anti-Semitism," Schiemann remarked that "there was a great deal of anti-Semitism in England." At one point Elisabeth asked Lise if she was thinking of returning to Austria. "When I told her that I couldn't *return* to Austria since I never held a position there, only in Germany, she then said in a roundabout way that the Austrians could surely find something for me. . . . Did she mean to say that I am not wanted in Germany? . . . Despite much reflection, I do not understand how she really feels about me."[36]

Lise knew of Elisabeth's difficult years under the Nazis; she may have known that at one time Elisabeth and her sister Gertrud concealed a Jewish friend, the pianist Andrea Wolffenstein, in their apartment for months, at great risk to themselves. (When Wolffenstein had to move, Fritz and Maria Strassmann hid her in their apartment for several months, endangering their own lives and that of their 3-year-old son. Wolffenstein survived,[37] and in 1986 Strassmann was honored at Yad Vashem, the Israeli Holocaust memorial.)[38] Lise did not doubt Elisabeth's essential decency, but she was hurt that Elisabeth seemed not to care if she came back or not, and she was distressed to see that Elisabeth's politics were as nationalistic as before, without much concern or understanding for anything but Germany. Where Elisabeth regarded nazism as an aberration in an otherwise superior German culture, Lise saw it as the end result of tainted German ideology; Lise judged Elisabeth unpleasantly strident and politically naive, whereas Elisabeth thought that Lise had become bitter and unforgiving.[39] They remained friends, but mostly because they had known each other such a long time. As with Otto Hahn, the friendship was damaged.

Lise was not ready even to visit Germany. When Max von Laue invited her to attend a physics conference in the summer of 1947, she hesitated, then refused, in part because she still had no passport, more because of what she called "a spiritual problem." Unlike her friends, she told Laue, she did not look at world problems "*only* from the German point of view," and therefore she was afraid she might jeopardize her friendships with political arguments. Her recent experiences with the Hahns and Elisabeth worried her. "I am ready to consider any honest viewpoint and if I cannot share it, that has nothing to do with my friendship. But how is it on the other side? I have written very frankly, dear Laue, please try to understand me correctly."[40] To some extent her misgivings applied to Laue as well, whom she considered "better qualified for thoughtful reflection than Hahn . . . but his fundamental attitude is not very different from Hahn's."[41]

In November 1947, Lise traveled through Germany by train on her way to Paris, where she participated in a French commemoration of the tenth anniversary of Rutherford's death. Secure in her compartment, Lise began to replace the imagined with the real: here, great heaps of debris where towns once stood; there, people at local stations in warm coats and good shoes who did not look underfed.[42]

Just before leaving for Paris, Lise received an unexpected offer from Fritz Strassmann. For two years Strassmann had been engaged in reorganizing and rebuilding the Kaiser Wilhelm Institute for Chemistry at its new permanent site in Mainz, on the grounds of a new university that was also under construction. (In 1949, the Kaiser-Wilhelm-Gesellschaft [KWG] would be replaced by the Max-Planck-Gesellschaft [MPG] and the institutes renamed accordingly.)[43] Strassmann had a dual appointment as professor of chemistry at the university and head of the chemistry section in the institute. Josef Mattauch, the Austrian physicist who succeeded Meitner in 1939, had originally been appointed institute director, but Mattauch was ill with tuberculosis and unable to work, and now Strassmann was asking Meitner to return as head of the physics section—her old position—and director of the institute. As president of the KWG, Otto Hahn had agreed. "[He] was as convinced as I am that this would be best solution for the institute," Strassmann wrote, "but he did not think you would even consider such a proposal. Since I am an optimist, I will ask you anyway." He listed the positives: he and Lise would work well together, the completed institute promised to be better than the one in Dahlem, and among the Mitarbeiter there would probably be no one who would not work enthusiastically under her direction. Admittedly, conditions in postwar Germany were difficult, but he hoped she would give the offer serious consideration.[44]

Lise did think it over for several weeks and then plainly expressed her concerns to Strassmann.

> Quite frankly, if the proposal had come from anyone but you, I could only have answered "No," even though I have never stopped longing for my old work situation. But what remains of this, and what is going on in the heads of the younger generation? . . . A mutual human understanding is indispensable for true cooperation. I do not doubt you, but that alone is not sufficient.[45]

"For now I answered somewhat evasively," Lise confided to Eva von Bahr-Bergius, "but I personally believe that I cannot live in Germany. From all I see in letters from my German friends, and other things I hear about Germany, the Germans still do not comprehend what has happened, and they have completely forgotten all the horrors that did not personally happen to them. I think I would not be able to breathe in such an atmosphere."[46]

In October 1947, Max Planck died, at the age of eighty-nine. Lise grieved. He had been a central figure in her life for forty years, and she had loved him, as she did no one else, with absolute trust in his wisdom and character.[47] The commemoration was to take place in Göttingen on 23 April 1948, his ninetieth birthday, for which a celebration had already been planned.[48] It gave her the impetus she needed to visit Germany, and an excuse to delay the Mainz decision until she could talk to Strassmann in person.

She approached Germany warily. She preferred not to accept lodging from the KWG, she told Hahn, because she did not wish to be "a guest, so to speak, of [KWG General Secretary Ernst] Telschow . . . who, among other things, fully accepted the Hitlerism-decreed inferiority of women and put it into practice."[49] In Göttingen, Lise had no sense of home-coming. This was where James Franck had been for many years, and Max Born; as Lise walked the familiar streets of the undamaged old town, Germans were going about their lives as if nothing had happened—as if no one were missing—while she was thinking only about those who were not there; she could not look at the faces of strangers on the street or former associates without wondering where they had been all those years and what they had done. Being among Germans aroused half-buried memories of the Hitler period, and she was upset to see that an ambitious party member like Telschow was being reinstated into a high position, with Hahn's wholehearted support.[50]

Returning to Stockholm, she focused her anger on an article by Max von Laue that had just appeared in the April 1948 issue of the *Bulletin of the Atomic Scientists*.[51] During the previous two years, Samuel Goudsmit and Werner Heisenberg had been debating the competence of the German atomic scientists and their fission project, Heisenberg insisting that the Germans had been able but unwilling to build a bomb and Goudsmit saying the opposite—positions that one historian has labeled "apologetic" and "polemical."[52] After Hiroshima and Nagasaki, when many Allied atomic scientists were grappling with their consciences, they were incensed that their German counterparts seemed to have no similar qualms about having worked for the Third Reich and were even suggesting that moral scruples had kept them from building a bomb. In 1947, Goudsmit published his account of the *Alsos* mission, including a summary of the transcribed Farm Hall conversations that bolstered his argument; he called the German scientists' claim of peaceful intentions a "brilliant rationalization of their failure" to make a bomb and ridiculed Heisenberg's postwar apologia as "the new theme song of German science."[53]

In his review of *Alsos* for the *Bulletin of the Atomic Scientists*, the physicist Philip Morrison stated flatly that although both Allied and German scientists had worked for the military, it would "never be possible to forgive" those who worked "for the cause of Himmler and Auschwitz, for the burners of books and the takers of hostages"; the fact that some "brave and good men like von Laue" could resist the Nazis proved that it would have been possible for scientists to stay aloof from the German war effort.[54] Laue responded forcefully. He argued that German scientists as a group had not given strong support to Hitler, that Morrison was "keep[ing] alive hate," and that Goudsmit and others who had lost relatives to the Nazis must feel "unutterable pain" at the mention of the "mere word Auschwitz" and were "for that very reason" incapable of unbiased judgment. Laue's implication that Germans could be objective but Jews could not[55] and that Americans in general might give Germans a fair hearing but Jews would not indicates that he had not been completely unaffected by the racism of the Third Reich. It was not his finest moment. Morrison lashed back: "It is not Prof. Goudsmit who cannot be unbiased, not he who most surely should feel an unutterable pain when the word Auschwitz is mentioned, but many a famous German physicist in Göttingen today . . . who could live for a decade in the Third Reich, and never once risk his position of comfort and authority in real opposition to the men who could build that infamous place of death."[56]

The exchange went to the heart of the emotional divide between the Allies and the Germans: on one side, moral outrage without much thought for the future; on the other, self-serving denials with no apparent remorse for the past. Certainly Laue was not unreasonable in asking for collegial understanding, but he stumbled badly in his remarks about Goudsmit, and he was curiously blind to the revulsion and anger, not just among Jews, that Germans and Germany still evoked in the West. That very blindness, Morrison was saying, was evidence of the "tragic failure" of many learned Germans: they had ignored their human obligations and worked to the advantage of the Nazi state, a moral failure they still seemed unable to recognize. And if Laue, whose integrity had been much praised by the Allies, now saw fit to indiscriminately defend his compatriots, he would be lumped in with the rest.

That was what Meitner had been saying all along. She believed that Laue had damaged his credibility, and she disagreed with his premise. "There is too much evidence that many scientists (with or without conviction) went along with Hitler," she wrote heatedly to Otto, citing Einstein's immediate expulsion from scientific academies in 1933, the Göttingen Dozents who went on record against Franck, the very small number of scientists who dared attend the Haber memorial, and more. At Meitner's request, Hahn showed the letter to Laue, who replied that Einstein's political involvement made his expulsion from the Berlin Academy inevitable,[57] a response Meitner regarded as "so naive" that she did not discuss it further.[58]

She concluded in a letter to Hahn,

I think I cannot take the position in Mainz. I have little fear of uncomfortable living conditions, but considerable concerns about the mentality. Except for physics, every time I would have a different opinion from the Mitarbeiter I would surely be met with the words: of course she doesn't understand the German situation, because she is Austrian, or because of her Jewish origins. I stressed these doubts to Strassmann also, and he responded only by repeating how necessary I would be for the institute. Thus he did not dare to refute my doubts. . . . It would be a similar battle to the one I waged—with very little success—in the years '33–'38, and today it is very clear to me that I committed a great moral wrong by not leaving in '33, because staying had the result of supporting Hitlerism. These moral conflicts do not exist today of course, but . . . my personal situation would not be very different now than it was then, I would not have the trust of my Mitarbeiter and therefore would not be truly useful.[59]

Hahn countered,

You speak of the battle you waged. What battle was that? Would you, if you had been in our place, have acted differently from so many of us, namely to make forced concessions and to be inwardly very unhappy about them? . . . Think of the concessions made by Geheimrat Planck, so revered by you and by us all! . . . One cannot do anything to counteract a terror regime. . . . How can one constantly reproach an entire people for their behavior during such times? . . . We all know that Hitler was responsible for the war and the unspeakable misery all over the world, but there must be some sort of world understanding also for the German people.[60]

Hahn's view, that Hitler was responsible but Germans as a whole were not, was exactly the point on which he and Lise would never agree.

Lise cited a recent international genetics conference in Stockholm, from which Germans had been expressly excluded at the request of foreign scientists, especially the Norwegians and the Dutch. "I mention this only to show that you do not always assess foreign opinion correctly, and above all that you don't fully understand that even your devoted personal friends cannot always dismiss everything that happened and cannot grasp that such fair people as, for instance, Laue want to retroactively defend almost everything, even though they were very unhappy about many things at the time."[61]

This was Lise and Otto's last explicitly political argument. Lise never understood the extent to which "good" Germans like Hahn and Laue felt harmed by the Nazis. And Otto never understood that she was reproaching him and others not so much for failing to fight the Nazis as for their attitude afterward: for remembering so little and caring less, for refusing to examine their past, and for reflexively defending German honor, as if the Nazis had not been German and the Germans had not been Nazis.[62] Lise and Otto had been arguing fruitlessly for years, and she was tired of it. What struck her most about Otto's letter was that he said not one word about Mainz.

"They wanted me to go to Mainz," she told James Franck later that summer. "When I wrote that I was not sure of the trust of the Mitarbeiter, in part because I am Austrian, and in part because of my Jewish origins, Hahn said nothing and Laue wrote that it was a pity, as they had no objections to Austrians."[63] Franck had declined an offer from Heidelberg with similar misgivings, convinced that "misunderstandings would arise from differences in experience alone."[64] By the end of summer 1948, the discussion was over.

Still, the Mainz offer meant something to Lise. It was an attempt at restitution and thus eased the tension in her relations with German colleagues and institutions. Gradually she accepted the fact that there were many in Germany who wished her well. In November 1948, she was deluged with seventieth-birthday greetings from former Mitarbeiter, colleagues, and government officials, including Austria's president and the rector of the University of Vienna, and she accepted her election as a foreign member of the Max-Planck-Gesellschaft with genuine pleasure.[65]

From then on she visited Germany often and associated quite freely with Germans at conferences and commemorations. Her political antennae were still sensitive and her memory was sharp, but she allowed herself to soften personally. In Bonn in 1949, former students and colleagues greeted her so warmly that she was surprised and moved; in Basel, she was disheartened by the "idiotic theories" some Germans had developed to avoid responsibility for the Third Reich; at a conference in Como in 1949, she talked to Heisenberg several times, although they both "very carefully" avoided politics.[66] Meitner would eventually resume a reasonably cordial relationship with Heisenberg, but when she met Carl Friedrich von Weizsäcker in 1948 in Göttingen, she found his attitude so strange that she hardly ever spoke to him again. Margrethe Bohr, undoubtedly thinking of Heisenberg and Weizsäcker's "hostile visit" to Copenhagen in 1941, wanted to hear about Meitner's talk with Weizsäcker. "It is a difficult problem with the Germans," Margrethe wrote, "very difficult to come to a deep understanding with them, as they are always first of all sorry for themselves."[67]

There were many émigrés, including Einstein and Otto Stern, who adamantly rejected all overtures from Germans and Germany; Einstein in particular stated his "irrefutable aversion" to taking part in German public life, "simply out of a need for cleanliness."[68] But others, including James Franck and Max Born, were doing the same as Lise: tentatively renewing old contacts, visiting Germany, accepting membership in the societies and academies from which they had been expelled. They believed it was important to help Germany become less isolated, more normal.[69] In 1949, when Meitner (together with Hahn) was awarded the Max Planck Medal of the German Physics Society, she was very pleased, in part because of her "love and reverence" for Planck, in part because she regarded as a "very valuable gift every bond that ties me to the old Germany that I loved very much, the Germany to which I can hardly be grateful enough for the crucial years of my scientific development, for the deep pleasure in scientific work and a very dear circle of friends."[70] Clearly, Lise's renewed ties to Germany went a long way toward making her own life whole again.

Meitner's situation in Sweden was also quite satisfactory. In 1949, she moved out of her temporary quarters in the IVA research station into her own permanent laboratory at the KTH nearby. Contrary to all promises, however, her professorship fell through—the minister of education who was sponsoring her appointment, Tage Erlander, became prime minister a few months later, and the proposal was never even brought to the Rikstag.[71] But Borelius and Oskar Klein managed to fund her position, so that she was, officially, a research scientist "with professor's salary" and without immediate financial worries for the first time since arriving in Sweden.[72] In 1949, she became a Swedish citizen; her petition for dual citizenship with Austria required a special act of the Riksdag that did go through.[73] In 1945, she had been elected a foreign member of the Royal Academy of Sciences; in 1951, her foreign membership was converted into a full membership, permitting her to participate, as do all members, in the Nobel process.[74]

Her scientific work was also making progress. In the spring of 1947, plans were approved for an experimental nuclear reactor, Sweden's first, to be built deep into the granite underlying Drottning Kristinas Väg.[75] Sigvard Eklund was the project director; together with Meitner, they attracted young physicists and engineers interested in basic nuclear research and reactor technology.[76] Among other studies, Meitner measured neutron capture cross sections for several heavy elements, for fast and slow neutrons. The results were of theoretical interest, as it appeared that certain spin selection rules might be responsible for some abnormally small slow-neutron capture cross sections in lead and bismuth; the investigation was also relevant to reactor function, as fission produces numerous new isotopes with a wide range of neutron absorption characteristics.[77]

While still in Siegbahn's institute, Meitner had attempted to relate the energy released by uranium fission to the manner in which the nucleus divides. In her experimental arrangement, a uranium sample was bombarded with fast neutrons in the cyclotron and the fission products were collected by recoil on layers of foil. The experiments were difficult and the results not very conclusive, but the question was of considerable theoretical interest, as the Bohr-Wheeler theory predicted that the highest energy release would be associated with the most symmetrical division of the nucleus.[78]

After the war, Meitner returned to the subject in a number of speculative articles, this time relating the modes of fission to the stability of the product nuclei. Since the early 1920s, she had looked for empirical correlations between nuclear stability and structure: in 1921, she postulated the existence of subnuclear alpha particles to explain patterns of radioactive decay;

in 1926, she suggested that the higher abundance of elements of even atomic number might be due to even-numbered proton complexes in the nucleus; in Leningrad in 1934, she lectured on the role of neutrons in nuclear stability overall; and she had always been interested in relating nuclear masses and binding energies to geological and cosmological questions of element formation and abundance.[79] These speculative surveys, interspersed with her major experimental investigations, typified Meitner's approach to physics. She looked for the point where experiment and theory might advance together, relying on theory to choose and guide her experiments and using her experimental results to resolve some theoretical problems and raise new ones, surveying the field for existing data that might illuminate underlying patterns of structure and behavior.

Others had done that, too. In the early 1930s, Walter Elsasser and Kurt Guggenheimer pointed out that particularly stable and abundant nuclei are associated with certain recurring numbers of protons or neutrons. Tin and lead, for example, with 50 and 82 protons, respectively, are especially abundant and have many stable isotopes; moreover, there exist an unusually large number of nuclides that have 50 neutrons or 82 neutrons. The recurrence of these "magic numbers" (as Eugene Wigner first called them) suggested the existence of a shell structure within the nucleus—analogous to the electronic structure of the atom—with protons and neutrons occupying shells that are most stable when completely filled. After World War II, when accumulated data were released for a large number of new species, the list of magic numbers grew longer (2, 8, 20, 28, 50, 82, 128) and the nuclear shell model was widely discussed. In 1950, Maria Goeppert Mayer in Chicago and J. Hans D. Jensen in Heidelberg independently published a fundamental theory of nuclear shells that accounted for all the magic numbers; they shared the Nobel Prize in physics with Wigner in 1963.[80]

In a note that was written before the theory appeared but published at about the same time, Meitner applied magic numbers to fission; she postulated that even a major nuclear disruption like fission would leave filled nuclear shells intact, with the excess energy being distributed among the "loose" protons and neutrons outside the closed shells. The slow neutron fission of ^{235}U is asymmetric, she reasoned, because its 144 neutrons can divide into filled nuclear shells of 50 and 82 neutrons (with some left over), while ^{209}Bi (which undergoes fission by neutrons or deuterons) has fewer than 132 neutrons and thus splits symmetrically, into two nearly equal nuclei with filled shells of 50 neutrons each (and again some extras). Meitner also noted that bombarding U with very fast particles could disrupt the shells and produce symmetrical fission, an effect that had been experimentally observed.[81] This work was published in April 1950 and expanded the following year; Meitner was pleased with it and gratified that subsequent experiments by others seemed to confirm her thesis.[82] These were among Meitner's last scientific publications,[83] although she continued to work with Eklund on the design and construction of the nuclear reactor.

Early in 1954, at the age of seventy-five, she retired and moved to Eklund's laboratory, where the underground reactor was nearly complete. The reactor reached criticality on 13 July 1954, marking the end of Sweden's so-called heroic era of reactor development and its entry into the atomic age.[84] Sweden embarked on an ambitious program of nuclear reactor construction that ended in the late 1970s. A nuclear weapon was apparently never developed.[85]

In Eklund's institute Meitner had a carefully furnished writing room—"an old-age home," she called it—and no formal responsibilities other than those she chose: attending weekly colloquia and following new developments in physics.[86] In 1960, she retired fully and moved to Cambridge to be closer to Otto Robert Frisch and his family. A year later Sigvard Eklund left Sweden for Vienna to serve as director of the International Atomic Energy Agency (IAEA).[87] With this, Lise Meitner's modest influence on the Swedish physics community was essentially over.

CHAPTER SIXTEEN

Final Journeys

Science . . . teaches people to accept reality, with wonder and admiration,
not to mention the deep joy and awe that the natural order of things brings
to the true scientist.

When the war was over, and it became possible to think beyond the present once again, reminiscences and memoirs began to appear. Among the first was a notice written by Otto Hahn in 1947 for Stefan Meyer's seventy-fifth birthday. "The events of the time did not allow his 70th birthday to be openly mentioned," Hahn began. "At the time he was robbed of his position of director of the Radium Institute in Vienna." Meyer had returned to Vienna and his former position; in 1948, he presented a seventieth-birthday gift to Hahn and Lise Meitner, his "Memories of the Early Days of Radioactivity." Meyer remembered everything: his first meeting with Lise in the old physics building on the Türkenstrasse, her interest in radioactivity, even the bundle of slender metal tubes that she used in 1907 for her studies of the scattering of alpha particles. Meyer began working in radioactivity soon after Marie Curie; he had participated in its entire development. He died in 1949 of a heart attack, in Bad Ischl.[1]

Richard Willstätter's autobiography, written in Switzerland shortly before his death in 1942, was published in 1949. Lise and Otto both agreed that it was a work of art, illuminating their own experiences and bringing to life people they had known. Lise thought back to the time she lectured in Munich in 1925, shortly after Willstätter resigned his professorship to protest the university's anti-Semitic hiring practices. He came to her talk with a bouquet of roses, invited her to dinner, and told her the details behind his resignation. "I had the distinct feeling," Lise remembered, "that he had taken this step . . . in the hope that it would bring people to their senses."[2] Like Lise, Willstätter had experienced a Germany that was "tolerant, liberal, just"; in his autobiography he expressed "consuming grief . . . at the human retrogression" that destroyed the society he had known.[3]

After the war, Meitner was asked repeatedly to contribute to a biography of herself, or to write an autobiography. She always refused, insisting that her scientific papers were available and that biographies of living people were "either insincere or tactless, usually both."[4] Her experiences with the press had made her wary of exposure, and she was unwilling to surrender her privacy to a biographer, but she did preserve a large collection of her personal correspondence and documents. Perhaps she did so merely because she was too much the scientist to discard valuable data; more likely, she wanted to leave a record that there was more to her life than science—and more to her science than appeared in her scientific publications. Meanwhile she began to evaluate the past, in a limited way and on her own terms, most often in the German tradition of commemorating major birthdays and anniversaries.

For the silver jubilee of Dirk Coster's professorship in 1949, Lise looked back to their first meeting in Lund in 1921; her first lecture tour in Holland in 1923, when Coster was her "affectionate impresario"; Coster's visit to Dahlem in 1935, when they heard golden orioles in her garden, birds whose penetrating cry seemed "oppressive and foreboding, a symbol of the conditions of the time." Above all she remembered the summer of 1938 and his "enormous act of friendship," for which she had no words, "only the heartfelt friendship that binds me to you."[5] They never saw each other again. Ill with multiple sclerosis for many years, Dirk Coster died on 12 February 1950, at the age of sixty. Miep Coster died three years later, at age fifty-seven, of a stroke.[6]

On James Franck's seventieth birthday in 1952, Lise went back forty-five years, to her arrival in Berlin. "Almost from the first day, we knew we spoke the same language," she wrote, recalling the many evenings in their home, Ingrid at the piano, James whistling or playing the violin, Brahms *Lieder* perhaps, Lise humming along. "It makes me happy just to remember." Twenty years before, on Franck's fiftieth birthday, she and Gustav and Ellen Hertz had spent a whole day of vacation in the Tyrol trying to create an amusing poem and finally sent a telegram: "*Senem seniles salutant*" (The senile salute the old).[7] Such silly pleasures now seemed inconceivably distant, and innocent, but their friendship was unchanged.

Max von Laue, too, went back to earlier times. "I don't know when we first met," he wrote to Lise after his seventieth birthday in 1949, "but we first got to know each other in 1920. . . . You lived in a boardinghouse in Dahlem and had the major problem of buying a floor mop. After you discussed this problem for quite some time with my wife, I allowed myself to remark that discovering a new chemical element [protactinium] must have been simpler than finding a mop." Laue also remembered darker events: Fritz Haber's "spiritual torment" in April 1933; the suicides of Arnold Berliner and the widow of Heinrich Rubens, when they were about to be deported in 1942.[8] Shortly after Berliner died, Laue wrote an obituary that was not published until 1947. It began, "Five years after the death of its founder, *Naturwissenschaften* is coming around to mentioning it in a commemorative article."[9] In the postwar years Laue suffered from bouts of depression,[10] and in that condition "between mental death and physical death," he thought deeply about the Nazi period and his own actions.

His letter to Meitner for her eightieth birthday, in 1958, was almost confessional. "We all knew that injustice was taking place, but we didn't want to see it, we deceived ourselves. . . . Come the year 1933 I followed a flag that we should have torn down immediately. I did not do so, and now must bear responsibility for it." He was grateful to Lise during those years, he wrote, "for trying to make us understand, for guiding us with remark-able tact. . . . Your goodness, your consideration had their effect. . . . I have made many mistakes, I do know that, but I was prevented from certain things for which I would never have been able to forgive myself."[11]

Many years had passed since his internment at Farm Hall, and Laue no longer defended German scientists as a group. In 1959, in a letter to Paul Rosbaud, he repudiated the idea that "German atomic physicists really had not wanted the atomic bomb." It was a Lesart, Laue said, a "version" developed during Farm Hall discussions that were led by Carl Friedrich von Weizsäcker, in which Heisenberg was "mostly silent" and Laue "did not hear the mention of any ethical points of view"—except, presumably, to rationalize their failure to build a bomb.[12]

Laue's repudiation was a response to Robert Jungk's *Brighter than a Thousand Suns*, a highly readable chronicle of the atomic bomb that portrayed German scientists "obey[ing] the voice of conscience" while their Allied colleagues "concentrated their whole energies" on building bombs.[13] Jungk, an Austrian-Jewish émigré journalist, relied heavily on discussions with Weizsäcker and Heisenberg; years later he conceded that he had not been sufficiently critical because, during the McCarthy era in America, he had been anxious to show that no government, not even a totalitarian Nazi regime, could force scientists to do its bidding.[14] At the time, however, many recognized that the book was unreliable. Meitner and Hahn discussed Jungk's inaccurate description of their research,[15] Laue found so many mistakes he did not finish reading it,[16] and Rosbaud was made "quite sick reading of German 'resistance' to the Nazis";[17] he did not doubt the sincerity of the author but had "every doubt in the sincerity of some of the people he has interviewed. . . . It is full of half truths and many things are completely wrong or distorted—Damn it all!"[18] Overall, these objections had little impact, and Laue asked that his letter to Rosbaud not be made public during his lifetime.

Publicly Meitner and Hahn also avoided controversy. For many years, in articles for each major birthday, they engaged in a ritual exchange of mutual admiration. For Lise's seventieth, in November 1948, Otto reviewed her work in Germany, her "success" in Sweden, her "admirable intellectual activity and physical hardiness"; he used the sanitized phrase "under pressure of external conditions" to explain her emigration but otherwise did not discuss it.[19] Four months later Otto turned seventy; Lise noted his "uncommon intuitive talent" and "almost indefatigable cheer."[20] For Lise's seventy-fifth, Otto's article emphasized her work on the nuclear shell model, her "enviable" intellectual productivity, and her hikes in the Austrian mountains;[21] Meitner responded with a memoir of their Dahlem institute that stressed the "strong feeling of community" and "mutual trust," but she did not mention fission.[22] And so it went to age eighty-five: Lise's career and awards; Otto's musical ability and charm[23]— all kind words, wise sayings, and quotations from Goethe but no serious evaluation of their scientific work and no hint that any disagreement had ever come between them.

The harmony was underscored by the many awards and honors Meitner received in Austria and Germany after the war: the city of Vienna's Prize for Science and Art in 1947; the Max Planck Medal in 1949 (jointly with Hahn); the prestigious Otto Hahn Prize, of which she was the first recipient, in 1955; the Orden pour le mérite, West Germany's highest civic award, in 1957 (the same year as Hahn); the Wilhelm Exner Medal in

Vienna in 1960; the Dorothea Schlözer Medal of Göttingen in 1962; and assorted honorary doctorates and awards from universities and memberships in scientific societies and academies. She accepted them all: from Austria, with nostalgic pleasure; from Germany, with her critical faculties somewhat more intact, aware that those making the awards were motivated by any number of things, from genuine appreciation for her to image repair for themselves. She sometimes shrank from the formalities, and she often said that young people needed the recognition more, but she took the awards as they came. Especially at first, the stipends were important to her,[24] and she welcomed the renewed relationship with Germany. Exactly the opposite of Einstein, she believed it was right to participate in German public life again.

Yet Meitner was never really part of German life again: her role was that of honored guest, visitor from the past. And the awards, for all their number and prestige, were without scientific resonance. She was associated primarily with fission, but only at the margin; her earlier work was often cited, but not its relationship to the discovery. She was neither here nor there: her work was regarded as important, but it was not clear why. It may be that a full evaluation of Meitner's work, especially her contributions to fission, was not possible so soon after the Third Reich: it would have exposed the conditions that had separated her from her laboratory in Berlin, a Pandora's box very few Germans cared to open. And so they sidestepped the issue, honoring Lise the person far more generously than Meitner the scientist. The many awards functioned as Germany's conciliatory gesture for having forced her out, rather than as scientific restitution for what she had lost in the process.

The difficulty, always, was her connection to Otto Hahn. When it came to fission, his was the deciding voice—and he did not believe she was owed any scientific restitution at all. To him it was simple: she had missed the discovery of fission not because she was driven out but because the discovery was chemical and she was a physicist. He never, apparently, reviewed the record to see what she had actually contributed or what he himself might have forgotten or repressed; it was easier to write about Meitner's "success" in Sweden than to consider the injustice of her forced emigration or the difficulty of her exile. After the war, Einstein accused Germans in general of lacking remorse; with Hahn it was a lack of fairness, an unwillingness to look below the surface.

During his lifetime and for some time after, Hahn was accorded the unusual luxury of determining the history of the discovery, unhindered by substantive critical analysis from historians of science or other contemporaries. At the time only a few people saw it differently. Strassmann later wrote, "What did it matter that she did not *directly* participate in the discovery?? . . . Lise Meitner had been the intellectual leader of our team." Laue saw her forced emigration as an "unprecedented tragedy . . . otherwise she would undoubtedly, in one way or another, have participated in the discovery of fission."[25] Others, not quite so close, simply sensed that things were not right. Max Volmer, a professor of physical chemistry in Berlin, and his wife, the physicist Lotte Volmer-Pusch, had known Meitner and Hahn since the 1920s. They questioned the awarding of the Nobel Prize to Hahn alone and were convinced that Hahn was suppressing Meitner's contributions.[26]

Hahn's attitude was typical of the willed historical amnesia of the period; what made the situation unusual was his overwhelming prominence, in Germany and abroad. He was Germany's anointed postwar scientific icon, venerated by scientists and everyone else: the decent German, great scientist, Nobel laureate, president of the Max-Planck-Gesellschaft, discoverer of fission but against the bomb, nationalist but never a Nazi, decorated veteran of World War I, presentable abroad, affable, witty, photogenic—in short, a symbol of all that was good about the old Germany and the new, with no unpleasant reminders of the twelve bad years in between. During his life he was made an honorary member of nearly every scientific society on earth and awarded countless honorary doctorates, medals, keys to cities, and honorary citizenships; his face was on a stamp and his name was on buildings, institutes, schools, libraries, streets, a nuclear-powered ship, a prestigious prize, and an unknown number of baby boys. (And posthumously: trains, a moon crater, coins, an element, an Antarctic island, plaques of the "Washington-slept-here" type, brigades, bridges, plazas, and more.)[27] At the center was nuclear fission, a discovery he increasingly referred to as "mine." Strassmann, making excellent progress in Mainz, was permanently in his shadow.[28] It is no wonder that the work Meitner did independently of Hahn was mostly ignored and her work with him consistently undervalued.

And yet, despite the phenomenal "Otto Hahn Effect," the "Lise Meitner Question" never quite went away. It came through in a certain tension, especially in official matters. A well-documented example involves the dedication of a chemistry building and the naming of a research institute in Berlin in the mid-1950s. By then the old KWI for Chemistry in Dahlem had been restored and was being used by West Berlin's Free University as a chemistry building; the decision was made to name it the Otto-Hahn-Bau (Otto Hahn Building). Just down the street was Max von Laue, the director since 1951 of Haber's former KWI for Physical Chemistry, which had been renamed the Fritz-Haber-Institut and was also part of the Free University. Laue and other university professors were chafing under the restrictions the Allies had imposed on nuclear research in West Berlin; a major ceremony dedicating the Otto-Hahn-Bau and honoring Hahn at the site of the fission discovery seemed an ideal way to call attention to their case. Bronze tablets were ordered, all of official Berlin was invited, and Laue, rather tactlessly, asked Meitner to give the speech of the day. Hahn thought that might be asking too much, given "her many painful memories of being driven out in 1938," especially since "she would otherwise, as far as anyone can tell, have participated in the discovery of fission" (an admission completely at odds with his public statements); he unenthusiastically agreed that she should perhaps be honored in some way, "although not in connection with fission." The ceremony was held in December 1956. Meitner did not attend and was not honored in any way. Bronze tablets bearing Hahn's and Strassmann's names were mounted inside and outside the Otto-Hahn-Bau, but Meitner's name was entirely absent from the building in which she worked for over twenty-five years. The incident has been cited as an example of Robert K. Merton's Matthew effect, in which honors are heaped on an already-famous honoree because doing so benefits the honorers. A corollary, also evident in this case, is that the inordinate attention paid to one person detracts from the other.[29]

The episode left Laue with a bad conscience. The day after the ceremony, he proposed that a new institute for nuclear research, then under construction in Wannsee, on the outskirts of Berlin, be named for Lise Meitner. When the decision came up in 1958, however, her name seemed incongruous: no one could imagine naming a nuclear research facility for anyone but Hahn. But his name alone would not do, as there was already the Otto-Hahn-Bau in Berlin and an Otto-Hahn-Institut in Mainz. Accordingly, the Wannsee institute was named the Hahn-Meitner-Institut für Kernforschung (nuclear research). In the words of a contemporary

German historian of science, "Thus did Hahn profit one last time from Lise Meitner."[30]

But in this instance, one could also say that Meitner profited from Otto Hahn. An institute with her name was a significant honor and a permanent tribute. At the dedication ceremony in March 1959, Mayor Willy Brandt turned to thank her and then, noting that she "had not been spared bitter suffering," thanked her "all the more" for the work she had done in Berlin.[31] It was not scientific restitution, but it ensured that her name at least would not disappear from view.

Thus Meitner maintained a relationship with Germany. If she had not been so isolated in Sweden—if she had established herself with work and friends, perhaps in England or the United States—she might not have remained so bound to German culture and friends. But Germany had given her the best years of her life and then taken them away. She returned again and again, if for no other reason than to salvage what she could.

That need was evident in her friendship with Hahn. He had been her "colleague-brother" for many years, and although the collegiality was long gone, the familial relationship was strong. She was informed of every family event, illness, or emergency: of the assassination attempt on Otto in 1951; of Edith's nervous breakdowns; of the deaths of Hanno and his wife, Ilse, in an automobile accident in 1960, from which Edith did not recover, leaving Otto to care for her and his teenaged grandson, Dietrich.[32] And Lise shared her sorrows with him, including the deaths of sisters, brothers-in-law, and her beloved younger brother, Walter, in 1961. Otto was almost the only person who had known them all; if nothing else, Lise needed him for that. She expressed it quite poignantly in 1951, when she thanked him for writing "warmheartedly and affectionately" soon after Gusti and Jutz Frisch died. "There are no people here who would have known me or my brother and sisters in the early years to whom one could say: Do you remember? or Do you still know? I once read that Chamisso's Peter Schlemihl, the man without a shadow, is supposed to represent a man without a fatherland. I often think of that. But when I read your letter, I had a shadow."[33] It explains, at least in part, Lise's loyalty to Otto, a loyalty that might otherwise seem incomprehensible.

Privately Hahn may have given Meitner a shadow, but his public image hugely overshadowed hers. In Germany, he loomed so large that Meitner was, as a matter of course, almost always considered his Mitarbeiterin, a subordinate rather than an equal. This infuriated her, not only because it was untrue, but because it was so carelessly untrue: hero worship, prejudice, and gender bias wiped out her independent scientific record with a single word. And when officials in the MPG—in effect, her former employers—and physicists such as Heisenberg, who knew her work and had been her colleague, also referred to her as Hahn's Mitarbeiterin, she was especially dismayed and angry.[34] In 1953, she wrote to Hahn,

> Now I want to write something personal, which disturbs me and which I ask you to read with our more than 40-year friendship in mind, and with the desire to understand me. In the report of the MPG there is a reference to a lecture I gave in Berlin (a purely physics lecture) and I am referred to as the "longtime Mitarbeiterin of our president [Hahn]." At the same time I read in an article in *Naturwissenschaftlichen Rundschau* by Heisenberg *about the relationship between physics and chemistry* in the last 75 years, where the only mention of me . . . is as follows: "Hahn's longtime Mitarbeiterin, Frl. Meitner." In 1917 I was officially entrusted by the directors of the KWI for Chemistry to create the physics section, and I headed it for 21 years. Try for once to imagine yourself in my place! What would you say if you were only characterized as the "longtime Mitarbeiter" of me? After the last 15 years, which I wouldn't wish on any good friend, shall my scientific past also be taken from me? Is that fair? And why is it happening?[35]

Hahn never answered those questions, nor did he ever set the record straight. He himself did not refer to her as his Mitarbeiterin, of course, but he did little to dispel the impression that she was just that. In his autobiographies and memoirs, his portrayal of Meitner the person was perfunctory—except for a few coarse anecdotes,[36] she appears more a stick figure than a real person—and his treatment of Meitner the scientist was correspondingly thin. (In fact, Hahn's autobiographies are uncommonly superficial for nearly everyone else, including himself, but even casual readers sense something amiss in his depiction of Meitner.) In all his memoirs, in hundreds of retrospective articles, talks, and interviews, in dozens of languages and countries, Hahn single-mindedly stayed with the course he set in February 1939: fission belonged to chemistry, alone. He would always stress his radiochemistry, usually understate Strassmann's analytical chemistry, and never fail to repeat, mantralike, that physicists had deemed fission impossible—that they had "forbidden" it—and thereby delayed the discovery.[37] Not once, in anything he wrote or said, did he ever mention Meitner's initiative in 1934, her leadership of the

Berlin team, their ongoing collaboration after she left, or their crucial meeting in Copenhagen in November 1938. Doing so would not have diminished his and Strassmann's achievement in the least, but an honest description that included physics and Meitner would have shown that she was unjustly denied her share of the discovery; that he had been afraid and therefore lied to his Mitarbeiter about his continuing contacts with Lise, failed to acknowledge her hidden collaboration, and deceived himself into thinking the discovery had been made without physics or her. And that would have spoiled his political message: that fission was a pure achievement of German science, that German scientists were upright men who had not succumbed to Hitler or tried to turn fission into a weapon of war. (That, too, was mostly false, as was Hahn's claim that his institute never did military work during the war.)[38]

Meitner's reaction was muted. In 1963, however, shortly after the appearance of Hahn's scientific autobiography,[39] she wrote an article for *Naturwissenschaftliche Rundschau*, a nontechnical scientific periodical. Without contradicting Hahn directly, she placed the uranium investigation solidly in its physics context, beginning and ending with references to Fermi, and for the first time stated publicly that she had recruited Hahn for the investigation, that she had been very disturbed by the physical implications of their results, and that their great mistake had been to investigate only the "transurane" precipitate and not the filtrate.[40] In an oral history interview, also in 1963, she emphasized that while she was in Berlin she had "tormented" the chemists to try to get them to examine the filtrate, but they "absolutely didn't want to."[41] Finally, she defended herself and physics: Hahn and Strassmann's "wonderful result . . . did not justify the chemists' often-expressed opinion" that physicists had declared fission impossible and therefore were to blame for delaying the discovery: "No one really thought of fission before its discovery."[42] Her message was clear: in this complex interdisciplinary problem, both disciplines had made mistakes, and both were responsible for the final success. She did not explicitly say, perhaps because it was so painfully obvious, that the obstacles to recognizing the complex foundations of the discovery were not scientific but political—at the time of the discovery and since.

Hahn's version dominated. In 1953, it was enshrined in the Deutsches Museum, a science and technology museum in Munich. There, in the museum's chemistry section, was a display of the apparatus used in the fission discovery: neutron sources, paraffin blocks, lead vessels, Geiger-Müller tubes, power supply, counters, amplifiers—the physical apparatus that Lise Meitner built and assembled in her laboratories in her physics section on the ground floor of the Dahlem institute.[43] Above the table was a large sign: *Arbeitstisch [Worktable] von Otto Hahn.* A smaller sign on the wall mentioned Hahn and Strassmann; Meitner did not appear at all. Some thirty years later, in response to criticism from the public, the museum posted a very small sign to one side, citing Meitner for the first time—as Hahn's Mitarbeiterin. The display, in a well-known museum employing historians of science, was accurate only in reflecting the majority opinion of the history of the discovery.[44]

Over the years, Hahn's unparalleled celebrity status spawned an enormous quantity of derivative material. A chorus of former associates, none close, echoed his contention that fission had nothing to do with physics or Meitner. The chemist Kurt Starke, ignorant of Meitner's advice to Hahn in Copenhagen, wrote that "no hint from [the physicists] had ever induced Hahn and Strassmann to do their crucial fractional crystallizations."[45] (Starke's political acuity can be deduced from his explanation that Meitner's emigration "was forced upon her by the loss of her [Austrian] citizenship.") Karl-Erik Zimen, a chemist and director of the Hahn-Meitner Institute, insisted that physicists, including Meitner, had been convinced of the impossibility of fission, that they delayed the discovery, that fission was "discovered by the chemists in spite of the physicists."[46] As "proof," Zimen suggested that Manne Siegbahn knew Meitner personally and would have nominated her for a Nobel Prize had she deserved it. The statement is disingenuous and mean: Zimen, who was working in Sweden at the end of the war, surely knew that it was precisely the poor relationship between Meitner and Siegbahn that kept her from the prize.[47]

The denial of physics by the chemists was, to some extent, a form of discipline chauvinism. By 1949, it was so blatant that the theoretical physicist Siegfried Flügge thought it necessary to review the physics behind fission in order to "rescue it from oblivion." Flügge had been the Haustheoretiker in the KWI for Chemistry in 1938–1939 and thus, except for Meitner, the physicist closest to the discovery; he regarded theoretical physics as the framework for the entire investigation and, like Strassmann, thought that the physicists' objections to the "radium" isomers were the impetus for Hahn and Strassmann's final experiments.[48] Compared to

Flügge, the theoretical physicist Carl Friedrich von Weizsäcker knew far less about the discovery, but he had already made up his mind. Weizsäcker has repeatedly written and spoken of "Hahn's discovery," always to the exclusion of Meitner and usually Strassmann: he claims that Meitner had no desire to investigate the filtrate because it would have been too much trouble (obviously untrue, as that would have been a job for the chemists); that Hahn "nobly" credited Meitner with a share in the discovery although she had nothing to do with it (both false); that his father, the *Staatssekretär*, somehow assisted Meitner's escape in July 1938 (for which there is no evidence, except to the contrary).[49]

Other physicists gladly repeated and published whatever second- and third-hand tidbits came their way. According to Rudolf Fleischmann, Erich Bagge told him that Hahn told the Farm Hall group in 1945, "If Miss Meitner had still been in the institute in December 1938, she would have talked us out of barium";[50] according to Heisenberg, Hahn would "let slip" over a glass of wine, "I don't know, I'm afraid Lischen might have forbidden me to make the discovery."[51]

To evaluate their statements, one must know who these people were: Fleischmann was a physicist who worked on the German fission project during the war in the National Socialist university of Strassburg; Bagge was the scientist at Farm Hall who regarded his internment as worse than the atrocities in concentration camps; Heisenberg, as we know, had many reasons for wanting fission to be German. These men were not motivated by the desire for historical accuracy. Some of them were true believers, preaching the gospel according to Hahn. Others were propagandists with agendas of their own.

In part this can be seen as an example of people gravitating to a famous man so that his aura would extend to those who agreed with him. Most certainly this was also an exercise in male chauvinism, rendering Meitner as a subordinate, misguided, obstructive woman-who-didn't-know-her-place—and doing it with calculated impunity, aware that others who knew even less would tend to accept the stereotype. It is highly unlikely that Heisenberg would have referred to a male colleague of Meitner's prominence as anyone's Mitarbeiter or would have relished recalling (or imagining)—and then publishing—a chummy scene in which two men disparage a third while reminiscing over a glass or two of Johannesberg Riesling.

But in the hostility to Meitner there is also a pronounced nationalism. After the war, when it became known that Meitner and Frisch had been informed of the barium finding before publication, some physicists reacted as if German physics had been the victim of a scientific Dolchstoss. Over the years the more aggressive Meitner detractors claimed that any number of German and Viennese physicists would have been the first to interpret and confirm fission, if only Meitner and Frisch had not been given advance notice. These men had nothing to say about the fairness of Meitner's (or Frisch's) forced emigration, but they were apparently quite aggrieved that they (or Germany) had not reaped the maximum benefit from it. Although it was too late to accuse Hahn of not playing by the rules, Fleischmann, Zimen, and others counted the days of Meitner's and Frisch's "head start" and virtually disputed their priority.[52] (Occasionally, their complaints reached sympathetic ears. In Sweden, for example, the authors of a recent history of physics garbled Meitner's contributions to fission and incorrectly asserted that publications by other groups preceded Meitner's and Frisch's. The authors are former associates of Siegbahn's.)[53]

To insist that Meitner contributed nothing to the fission discovery, to imply that Meitner and Frisch had been given an unfair advantage—these were ways of denying that she had been treated unjustly and, in a larger sense, of refusing to confront the injustice and crimes of the Nazi period. Rather than acknowledging that Meitner's exclusion from fission was political, Hahn and his hangers-on invented spurious scientific reasons for it. Arrogantly, and with misplaced national pride, they denied the injustice, created new injustice—and implicated themselves.

Meitner herself did not do battle with Otto Hahn. There were other things she wanted to do with her life, and in any case she would have preferred to be known for her earlier work and not exclusively for something that had already killed hundreds of thousands and continued to terrify people the world over.[54] In part for that reason, she was "not particularly disappointed," not even at the time, that she did not share the Nobel Prize with Hahn. And, she told James Franck in 1955, she would not have wanted the prize without Frisch. In any event, the absence of a Nobel Prize was "in no way an open wound."[55]

She was greatly troubled by the perilous state of the world and the degree to which science, especially physics, added to the danger. Nevertheless, she refused to sign appeals that called for nuclear disarmament or a nuclear test ban, calling them useless and, worse, dangerous illusions;[56] instead she thought every effort should go into making war itself obsolete.[57] Her opinion was close to that of Niels Bohr, who believed that the sheer destructiveness of nuclear weapons would make all-out world war impossible. And in this there was some room for optimism: people and their governments would have to choose restraint, openness, and global cooperation.[58] In a speech in Vienna in 1953, Meitner pointed out that although science and technology had been used for war throughout history, the ethical traditions of scientific research and international scientific cooperation offered a model for human betterment and understanding. "Science makes people reach selflessly for truth and objectivity; it teaches people to accept reality, with wonder and admiration, not to mention the deep joy and awe that the natural order of things brings to the true scientist." Her passion for physics was as strong as ever.[59]

But world problems restricted the "unconditional nature" of her love for physics. Sometimes Lise thought back to the years before Hitler, when despite war, inflation, and other troubles, "one could love one's work and not always be tormented by the fear of the ghastly and malevolent things that people might do with beautiful scientific findings."[60] When the Russian *Sputnik* went up in 1957, she thought the satellite was "wonderful in itself," but she worried that it might drive a new wedge into the relationship between East and West;[61] she wrote Lilli Eppstein, "spiritually and morally, we are in no way keeping pace with technical advances."[62] Her heart pounded each day as she read the newspaper.[63]

In these years her closest confidant was James Franck. They saw each other frequently and wrote often, trusting each other's views on people, science, and politics. They spoke the same language, as Lise often said, and laughed at the same things. "Your true friendship is among the most cherished and precious experiences life has given me,"[64] she wrote for his seventy-fifth birthday; it was "never marred by even the slightest disappointment."[65] In 1952, Hertha Sponer, Franck's second wife, spent a sabbatical year in Stockholm, and Lise became close to her as well. Lise enjoyed being with happy couples, and James and Hertha were "so happy with each other that it makes one glad just to see them together."[66] She worried about Franck's health and begged him not to work too hard. "But of course," she wrote when he was eighty-one and she eighty-five, "I still do understand how wonderful it is to be possessed by one's own work."[67]

She traveled as often as her health allowed, often more. A letter written in July 1959 at age eighty is representative: "In January I was in England, mid-March in Göttingen (for Hahn's 80th birthday) and Berlin, and April and May in America [at Bryn Mawr College] where I gave several lectures, and in Washington, Durham [to see the Francks], Chicago (Argonne), and of course New York, where I visited my sisters and their children, friends and laboratories. . . . Now I am getting ready to go to Kitzbühel on Thursday."[68] Although high blood pressure and occasional dizzy spells almost made her cancel her American trip,[69] she delivered three lectures at Bryn Mawr, including a perspective on the status of women in the professions.[70] "As long as I don't read the newspapers, I feel fine," she told Franck.[71] In Kitzbühel she hiked every day, slightly chagrined about having to take a cable car to the top and walking from there instead of hiking all the way up on her own.[72]

She took great pleasure in her visits to Vienna. Returning to Germany at first, especially Dahlem, was painful, "as if I were going through the ruins of my own past." But in Vienna the days were "almost unbelievably beautiful";[73] "I walked around in a dreamlike state; my entire happy youth came alive and all the difficult years . . . retreated into the background."[74] Later her travels in Germany were less traumatic, but Vienna was still the place for unencumbered sentiment. It probably helped that she did not know many people in Vienna and that those she did know were very close friends. When she visited, she stayed with Sigvard Eklund, her colleague from Stockholm who was director of the International Atomic Energy Agency, and his wife, Anna-Greta, a charming, lively couple whose company Lise especially enjoyed.[75]

In 1963, Meitner was asked to give a speech at the Urania Volksbildungsanstalt, a cultural center in Vienna, on the topic "Memories of Fifty Years in Physics." Perhaps because the talk was in Vienna, perhaps because it was given to a general audience with many young people, Lise's portrayal of her life in physics was glowing and youthful, almost innocent; she described the physics she had done and the "great and lovable personalities" she had known as a "magic musical accompaniment to [her] life."[76] When she finished, the enthusiastic crowd nearly mobbed Lise, then eighty-five years old and tinier than ever, in an effort to get closer to her.[77]

She never went long without music. For her seventy-fifth birthday in

1953, Oskar Klein organized friends from all over to buy her a good phonograph and records; Lise's brother Walter came from London, and she woke up on the morning of 7 November to a house filled with flowers and Mozart's clarinet quintet.[78] Whether at home or traveling, she went to concerts: chamber music and symphony, mostly classical, but she also enjoyed Carl Nielsen and other contemporary composers and conscientiously tried to appreciate more modern music. In Bonn she was introduced to Paul Hindemith and his wife and heard one of his quartets performed in concert, which she "just about" understood.[79] In a concert in Vienna in 1963, David and Igor Oistrakh performed Mozart's *Sinfonia concertante* under the direction of Yehudi Menuhin, and then Menuhin played the Beethoven violin concerto—"overwhelmingly beautiful."[80] And in Geneva that year, she heard Rudolf and Peter Serkin play Mozart's Concerto for Two Pianos. "It was so incredibly beautiful that I was under a spell for days."[81]

In 1960, at age eighty-two, Lise moved to Cambridge to be near Otto Robert, who was a professor of physics and a Fellow of Trinity College. They had been close since he was a boy and she had thought him clever and observant; she admired his "excellent combination" of clear thinking and warm heart. In 1951, Otto Robert had married Ulla Blau, an artist who was also of Viennese extraction, and they had two children. Otto Robert helped Lise move out of her Stockholm apartment and find a place in Cambridge; he played the piano for her, talked to her about physics, attended to the practical details that she disliked. He and Ulla made Tante Lise part of their family, looking after her "like loving children."[82] Lise was very grateful.[83] And she was surprised to find that her Swedish experience had been more positive than she realized. In those twenty-two years she had made good friends, and she missed them.[84]

Lise had an affinity for England and the English: she admired their stoic qualities during the war and after,[85] and she appreciated their helpfulness: whenever she stepped from a bus or prepared to cross a street, several people would reach out a hand; when she asked directions to a shop, people would offer to take her even if it was blocks away. And Cambridge was beautiful, its huge beeches and oaks more spectacular than any she had seen elsewhere. Her only complaint, typical for Lise, was the chill: underheated houses, single-paned windows, doors that flew open in a strong wind.[86] She was very thin, about 41 kilograms, or 90 pounds.[87] Food did not interest her; in a restaurant she could become absorbed in conversation and completely forget about lunch.[88]

She kept up with physics to the extent possible, but it was increasingly hard to do. "Physics keeps slipping further away from me," she complained when she was eighty-two. "For those of us who don't work any more, many of the articles seem to be written in a kind of secret code. . . . Even the symbols in the equations are not explained."[89] At eighty-four, she decided that physics had become too abstract. Max Born felt the same: when asked about the future of physics, he felt like "a retired captain of sailing ships [who is asked] to comment on the condition and future of steamboat travel." Hearing this from one of the inventors of quantum mechanics consoled Meitner quite a bit: "If a Max Born says that, what can I expect?"[90] At eighty-five, she was still trying to keep up. She relied on Otto Robert to explain new developments and kept a small notebook with a "Questions for O.R." section that included "spin-spin relaxation effects?" and "nuclear form factor?" along with "highfalutin?" and "juke box?"[91] She regularly read *Physikalische Blätter* and *Nature* for an overview, some history of science and nonscience for pleasure, and she generally had an opinion on the most recent Nobel Prizes.[92]

The severest penalty for growing old was outliving her friends and colleagues. When Enrico Fermi died of cancer in 1954, at the age of fifty-three, everyone was shocked: "He was one of the most gifted physicists of our time; how hard it must be for his wife."[93] Irène Curie died in 1956, at fifty-eight, also of cancer: "Beneath her almost unfriendly exterior, few people recognized her inner warmheartedness."[94] Einstein died in 1955: "For all my great admiration and affection for Einstein, during the Berlin years I often stumbled inwardly over his absolute lack of personal relationships. . . . Only later did I understand that this separation from individuals was necessary for his love and responsibility toward humanity."[95] Lise last met Erwin and Annemarie Schrödinger in the summer of 1959 near Kitzbühel; she saw that Erwin's health was very bad, and she wrote to the Francks "I'm afraid he knows it." He died in January 1961 at the age of seventy-three; Anny lived another four years.[96]

In 1960, Max von Laue died of injuries after an automobile accident, and his wife, Magda, died a year later. "How small the old dear circle has become!" Lise wrote to Franck. "And how much more valuable every dear friend one still has."[97] Her brother Walter, the baby born when she was already thirteen, died in 1961. "He was my friend even as a child . . . and remained so all his life."[98] "I have lost one of my truest and dearest friends."[99] When Bohr died in November 1962, she was shattered. "But," she wrote Lilli Eppstein, "my head and heart are full of wonderful memories of . . . his greatness and harmony in intellect and heart."[100] In Copenhagen the next summer, at a commemoration of the fiftieth anniversary of the Bohr atomic model, Lise felt Bohr's influence so strongly she almost forgot he was not there. As ever, she admired Margrethe's "loving charm and deliberate composure."[101] "Take care of yourselves!" Lise urged James and Hertha. "I don't want to lose any more friends before I myself am on that list."[102]

In May 1964, Lise took an extended trip with the Francks in Germany. Returning home, she described it to James as "a dream come true; every day a joyful holiday, every conversation a mutual understanding—even in the few situations where we might think differently—and always the feeling of immutable reciprocal friendship."[103] In Göttingen a few days later, on 21 May 1964, Franck collapsed from a heart attack and died. Trying to comfort Hertha, Lise managed a few lines in a shaky hand.[104]

At the end of 1964, Otto Robert attended a meeting in the United States, and Lise went along with him to visit friends and family. While there she suffered a heart attack from which she never fully recovered; she had considerable pain in her arms, walked with difficulty, and was very weak.[105] Still she had physics questions when Otto Robert came to see her, went for small walks each day in the garden, and gradually improved.

In 1966, Hahn, Meitner, and Strassmann were jointly awarded the U.S. Atomic Energy Commission's Enrico Fermi Prize "for their independent and collaborative contributions" to the discovery of fission. Although Hahn had proposed Strassmann alone for the prize,[106] the AEC, under the chairmanship of Glenn Seaborg, decided to honor them as a team.[107] In a letter to Hahn, Lise said she was glad for him and Strassmann but that her own feelings were "somewhat mixed"; she did have "a kind of pleasure" in it.[108] Hahn asked Otto Robert, why only "a kind of pleasure?" Did Tante Lise still have "resentments," thinking she had left Germany too soon?[109] No, Otto Robert replied gently, she never thought she had left Germany too soon, she was always grateful to the people who helped her get out. She had mixed feelings because of the bomb.[110]

As neither Hahn nor Meitner were strong enough to travel to Washington, the award ceremony was held in Vienna in September 1966; Hahn and Strassmann attended, but Meitner was too ill, and Otto Robert went in her place. Later Glenn Seaborg came to Cambridge where he presented the award to Lise Meitner at the home of Max Perutz, a Vienna-born molecular biologist, on 23 October 1966.[111] She was not well, but she had prepared a few words of thanks. Seaborg noticed how difficult it was for her to remember what she intended to say.[112]

In 1967, Lise Meitner broke her hip in a fall and suffered several small strokes; it was difficult for her to speak or recall words, but she had no pain and made a partial recovery.[113] She gradually grew weaker and spent the last several months of her life in a nursing home, quite unaware of her surroundings. Otto Robert did not tell her of Otto Hahn's death on 28 July 1968, or of Edith's death two weeks later. Lise Meitner died in her sleep, just after midnight on 27 October 1968, a few days before her ninetieth birthday.[114]

She had asked to be buried in the same country graveyard in Bramley, Hampshire, where Walter had been buried a few years before. Adjoining the grassy graves is the old stone church of St. James, a Church of England parish church, where the burial service was held. Only family members attended; there were no eulogies. The church organist played Bach.[115] Otto Robert selected the headstone and the inscription:

Lise Meitner: a physicist who never lost her humanity.

It was exactly the way she would have wanted to be remembered.

APPENDIX

THE URANIUM SERIES

Radioelement	Corresponding Element	Symbol	Radiation	Half-Life
Uranium I ↓	Uranium	^{238}U	α	4.51×10^9 yr
Uranium X₁ ↓	Thorium	^{234}Th	β	24.1 days
Uranium X₂* ↓	Protactinium	^{234}Pa	β	1.18 min
Uranium II ↓	Uranium	^{234}U	α	2.48×10^5 yr
Ionium ↓	Thorium	^{230}Th	α	8.0×10^4 yr
Radium ↓	Radium	^{226}Ra	α	1.62×10^3 yr
Ra Emanation ↓	Radon	^{222}Rn	α	3.82 days
Radium A 99.98% | 0.02%	Polonium	^{218}Po	α and β	3.05 min
Radium B	Lead	^{214}Pb	β	26.8 min
Astatine-218	Astatine	^{218}At	α	2 sec
Radium C 99.96% | 0.04%	Bismuth	^{214}Bi	β and α	19.7 min
Radium C′	Polonium	^{214}Po	α	1.6×10^{-4} sec
Radium C″	Thallium	^{210}Tl	β	1.32 min
Radium D ↓	Lead	^{210}Pb	β	19.4 yr
Radium E ~100% | 2×10^{-4}%	Bismuth	^{210}Bi	β and α	5.0 days
Radium F	Polonium	^{210}Po	α	138.4 days
Thallium-206	Thallium	^{206}Tl	β	4.20 min
Radium G (End Product)	Lead	^{206}Pb	Stable	—

Fig. 1. Radioactive decay series. From Samuel Glasstone, *Sourcebook on Atomic Energy*, 3d ed., (New York: D. Van Nostrand, 1967), 152–154. Reprinted by permission of Van Nostrand Reinhold Co.

THE THORIUM SERIES

Radioelement	Corresponding Element	Symbol	Radiation	Half-Life
Thorium ↓	Thorium	^{232}Th	α	1.39 × 10^{10} yr
Mesothorium I ↓	Radium	^{228}Ra	β	6.7 yr
Mesothorium II ↓	Actinium	^{228}Ac	β	6.13 hr
Radiothorium ↓	Thorium	^{228}Th	α	1.91 yr
Thorium X ↓	Radium	^{224}Ra	α	3.64 days
Th Emanation ↓	Radon	^{220}Rn	α	52 sec
Thorium A ↓	Polonium	^{216}Po	α	0.16 sec
Thorium B ↓	Lead	^{212}Pb	β	10.6 hr
Thorium C 66.3% | 33.7%	Bismuth	^{212}Bi	β and α	60.5 min
Thorium C′	Polonium	^{212}Po	α	3 × 10^{-7} sec
Thorium C″	Thallium	^{208}Tl	β	3.1 min
Thorium D (End Product)	Lead	^{208}Pb	Stable	—

Fig. 1. *Continued.*

THE ACTINIUM SERIES

Radioelement	Corresponding Element	Symbol	Radiation	Half-Life
Actinouranium	Uranium	^{235}U	α	7.13×10^8 yr
↓ Uranium Y	Thorium	^{231}Th	β	25.6 hr
↓ Protactinium	Protactinium	^{231}Pa	α	3.43×10^4 yr
↓ Actinium 98.8% | 1.2%	Actinium	^{227}Ac	β and α	21.8 yr
Radioactinium	Thorium	^{227}Th	α	18.4 days
Actinium K	Francium	^{223}Fr	β	21 min
↓ Actinium X	Radium	^{223}Ra	α	11.7 days
↓ Ac Emanation	Radon	^{219}Rn	α	3.92 sec
↓ Actinium A $\sim$100% | $\sim$5 × 10^{-40}%	Polonium	^{215}Po	α and β	1.83×10^{-3} sec
Actinium B	Lead	^{211}Pb	β	36.1 min
Astatine-215	Astatine	^{215}At	α	$\sim$10^{-4} sec
↓ Actinium C 99.7% | 0.3%	Bismuth	^{211}Bi	α and β	2.16 min
Actinium C′	Polonium	^{211}Po	α	0.52 sec
Actinium C″	Thallium	^{207}Tl	β	4.8 min
↓ Actinium D (End Product)	Lead	^{207}Pb	Stable	—

Fig. 1. *Continued.*

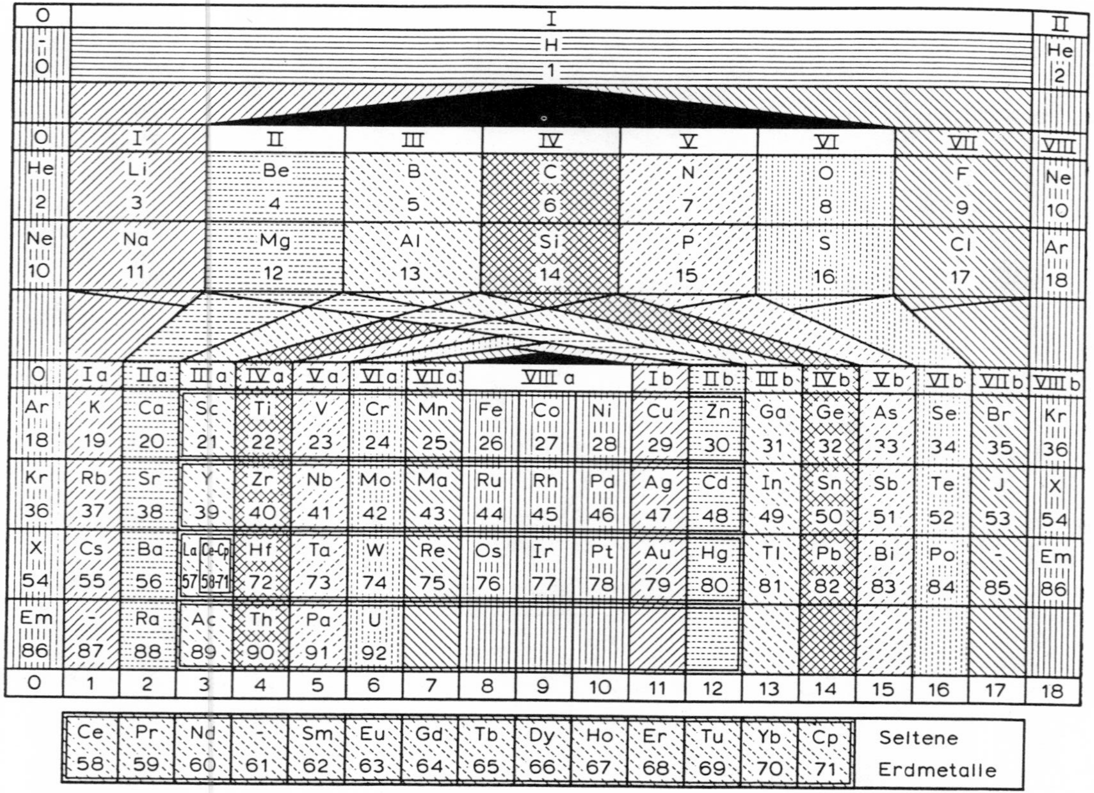

Fig. 2. Representative periodic table of the 1920s and 1930s (Andreas von Antropoff, 1926). Ac (89) to U (92) were classified as transition elements; the rare-earth elements 58–71 were grouped below. From J. W. van Spronsen, *The Periodic System of Chemical Elements* (Amsterdam: Elsevier, 1969), 160.

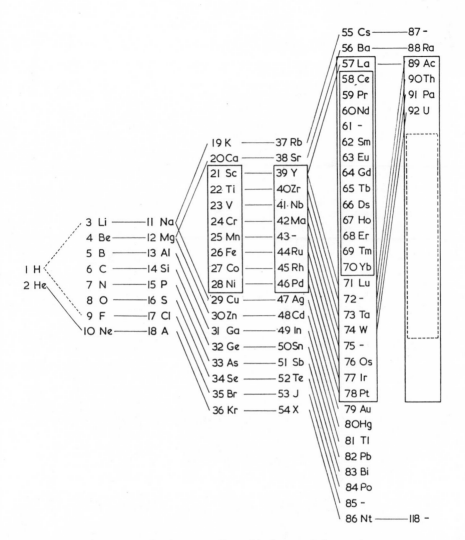

Fig. 3. Periodic system of Bohr, 1922. From his theory of electronic structure, Bohr proposed that a second rare-earth series (dashed lines) would appear somewhere beyond uranium. From Spronsen 1969:156.

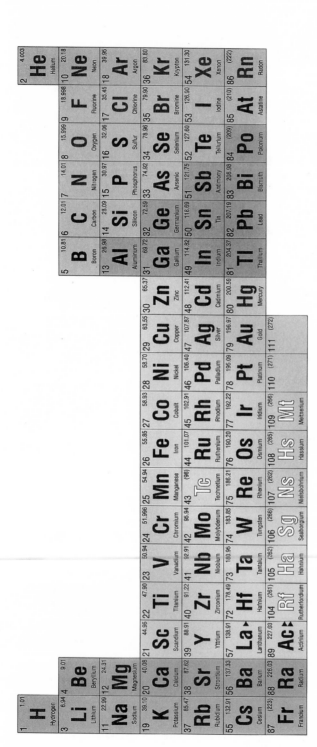

Lanthanide series

Actinide series

Fig. 4. Periodic table, 1995. After the discovery of Np (93) and Pu (94) it was evident that the actinides are homologous to the lanthanides. In 1994, the International Union of Pure and Applied Chemistry (IUPAC) approved the name *meitnerium* for element 109, but several others are still unresolved, including the placement of *hahnium*, and the name *seaborgium* (for Glenn T. Seaborg) for element 106. See *Chemical & Engineering News*, 5 December 1994, pp. 25–29, and *Chemical & Engineering News*, 8 May 1995, p. 7. Source: Lawrence Berkeley National Laboratory, University of California.

ABBREVIATIONS

ARCHIVES AND COLLECTIONS

AIP   American Institute of Physics, New York (later Maryland)
BPC   Bohr Private Correspondence, Niels Bohr Archive, Copenhagen
BSC   Bohr Scientific Correspondence, Niels Bohr Archive, Copenhagen
DM   Deutsches Museum Archives, Munich
EC   Meitner correspondence, private collection of Lilli Eppstein
FP   James Franck Papers, Joseph Regenstein Library, University of Chicago
GP   Samuel A. Goudsmit Papers, American Institute of Physics, New York
LP   Max von Laue Papers, Deutsches Museum Archives, Munich
MB   Adriaan Fokker Papers, Museum Boerhaave, Leiden
MC   Meitner Collection, Churchill College Archives Centre, Cambridge
MPG   Archiv zur Geschichte der Max-Planck-Gesellschaft, Berlin
NBA   Niels Bohr Archive, Niels Bohr Institute, Copenhagen
OHN   Otto Hahn Nachlass, Archiv zur Geschichte der Max-Planck-Gesellschaft, Berlin

JOURNALS AND PERIODICALS

Alman. Österr. Akad. Wiss.—Almanach der Österreichischen der Wissenschaften
Amer. J. Phys.—American Journal of Physics
Angew. Chem.—Angewandte Chemie
Angew. Chem. Intl. Ed. Engl.—Angewandte Chemie, International Edition in English
Ann. Phys.—Annalen der Physik

Ann. Rev. Nucl. Part. Sci.—*Annual Review of Nuclear and Particle Science*
Ark. Mat. Astr. Fys.—*Arkiv för Matematik, Astronomi och Fysik*
Bild der Wiss.—*Bild der Wissenschaft*
Biog. Mem. Fell. Roy. Soc. Lond.—*Biographical Memoirs of Fellows of the Royal Society, London*
Ber. Dt. Chem. Ges.—*Berichte der Deutschen Chemischen Gesellschaft*
Ber. Wissenschaftsgesch.—*Berichte zur Wissenschaftsgeschichte*
Brit. J. Appl. Phys.—*British Journal of Applied Physics*
Bull. Atom. Sci.—*Bulletin of the Atomic Scientists*
Chem. Rev.—*Chemical Reviews*
Comptes Rendus—*Comptes rendus hebdomadaires des séances de l'Academie des Sciences, Paris*
Dict. Sci. Biog.—*Dictionary of Scientific Biography*
Ergebn. Exakt. Naturwiss.—*Ergebnisse der Exakten Naturwissenschaften*
Hist. Stud. Phys. Sci.—*Historical Studies in the Physical Sciences*
J. Amer. Chem. Soc.—*Journal of the American Chemical Society*
J. Chem. Educ.—*Journal of Chemical Education*
J. Phys. Rad.—*Le Journal de Physique et le Radium*
K. Dansk. Vid. Selsk. Mat.-fys. Medd.—*Det Kongelige Danske Videnskabernes Selskab: Mathematisk-fysiske Meddelser*
Mitteilg. Österr. Ges. Gesch. Naturwiss.—*Mitteilungen der Österreichischen Gesellschaft für Geschichte der Naturwissenschaften*
Naturwiss.—*Die Naturwissenschaften*
Naturwiss. Rdsch.—*Naturwissenschaftliche Rundschau*
Phil. Mag.—*Philosophical Magazine*
Phys. Bl.—*Physikalische Blätter*
Phys. Bull.—*Physics Bulletin*
Phys. Rev.—*Physical Review*
Phys. Z.—*Physikalische Zeitschrift*
Proc. Amer. Phil. Soc.—*Proceedings of the American Philosophical Society*
Proc. Camb. Phil. Soc.—*Proceedings of the Cambridge Philosophical Society*
Proc. Roy. Soc. Lond.—*Proceedings of the Royal Society of London*
Radiochim. Act.—*Radiochimica Acta*
Rev. Mod. Phys.—*Reviews of Modern Physics*
Ric. Sci.—*La Ricerca Scientifica*
Roy. Soc. Lond. Obit. Not.—*Royal Society, London, Obituary Notices of Fellows*
S. Ber. Akad. Wiss. Wien—*Sitzungsberichte der Akademie der Wissenschaften in Wien*
Sci. Am.—*Scientific American*
Verh. Dt. Phys. Ges.—*Verhandlungen der Deutschen Physikalischen Gesellschaft*
Wiss. Z. Humboldt-Universität—*Wissenschaftliche Zeitschrift der Humboldt-Universität zu Berlin*

Z. Angew. Chem.—*Zeitschrift Angewandte Chemie* (later *Angew. Chem.*)
Z. Naturforschg.—*Zeitschrift für Naturforschung*
Z. Phys.—*Zeitschrift für Physik*
Z. Phys. Chem.—*Zeitschrift für Physikalische Chemie*
Z. Ver. Dt. Ing.—*Zeitschrift des Vereines Deutscher Ingenieure*

NOTES

CHAPTER 1. Girlhood in Vienna

1. Berta Karlik, "Lise Meitner: Nachruf," *Almanach der Österreichischen Akademie der Wissenschaften* 119. Jahrgang 64 (1969): 345–354; Berta Karlik, "Lise Meitner, 1878–1968," in *Neue Österreichische Biographie*, Band XX (Wien: Amalthea Verlag, 1979), 51–56; Charlotte Kerner, *Lise, Atomphysikerin: Die Lebensgeschichte der Lise Meitner* (Weinheim: Beltz, 1986), 117. "Na ja, Narrheiten kosten Geld."

2. *Geburts-Buch für die israelitische Cultusgemeinde in Wien, 1878*, entry 1820.

3. One indication: The *Geburts-Buch* entry for Elise Meitner gives no indication of a naming ceremony, which for other newborn girls took place during synagogue services soon after birth.

4. Gisela Lion-Meitner to Lise Meitner, n.d. but after 1950 (MC).

5. George E. Berkley, *Vienna and Its Jews: The Tragedy of Success, 1880s–1980s* (Cambridge: Abt Books/Lanham, Md.: Madison Books, 1988), 30–31; Arieh Tartakower, "Jewish Migratory Movements in Austria in Recent Generations," in Joseph Fraenkel, ed., *The Jews of Austria: Essays on Their Life, History and Destruction* (London: Vallentine, Mitchell, 1967), 285–310, on 286.

6. Stefan Zweig, *The World of Yesterday* (New York: Viking Press, 1943; reprint Lincoln: University of Nebraska Press, 1964), 6. "[In Moravia] the Jewish communities lived in small country villages on friendly terms with the peasants and petty bourgeoisie."

7. Ilsa Barea, *Vienna* (New York: Alfred A. Knopf, 1966), 244–248; Franz Kobler, "The Contribution of Austrian Jews to Jurisprudence," in Fraenkel, *Jews of Austria*, 25–40; P. G. J. Pulzer, *The Rise of Political Anti-Semitism in Germany and Austria* (New York: Wiley, 1964), 21–24.

8. William M. Johnston, *The Austrian Mind: An Intellectual and Social History* (Berkeley, Los Angeles, and London: University of California Press, 1972); Arthur J. May, *Vienna in the Age of Franz Joseph* (Norman: University of Oklahoma Press, 1966), 30; Carl E. Schorske, *Fin de Siècle Vienna: Politics and Culture* (New York: Vintage, 1981), 116–180; Zweig, *World of Yesterday*, 12–15.

9. Schorske, *Fin de Siècle Vienna*, 25; Kobler, "Contribution of Austrian Jews."

10. Lise Meitner to Ilse Weitsch, 9 November 1955 (MC).

11. For a description of Leopoldstadt, see Stephen S. Kalmar, *Goodbye, Vienna!* (San Francisco: Strawberry Hill Press, 1987), chap. 1; also Berkley, *Vienna and Its Jews*, 43.

For population statistics, see Tartakower, "Jewish Migratory Movements": Between 1869 and 1890, the number of Jews in Vienna increased very rapidly, then more slowly. Altogether, between 1857 and 1900, the number of Jews increased from about 7,000 to almost 150,000, nearly 10% of the city's population. Twenty-five percent of these, the largest single group, were *Ostjuden*, "eastern Jews," from Galicia.

12. The street was renamed Heinestrasse in 1919; Auguste Dick, pers. comm. 1985. A plaque commemorating Meitner's birthplace was mounted on the building in 1988; Dietmar Grieser, "Im Schatten der Bombe," in *Köpfe* (Vienna: Österreichischer Bundesverlag, 1991), 134.

13. The information comes from interviews of Otto Frisch, 1975, and Ulla Frisch, 1982, from Auguste Dick, Vienna, pers. comm. 1985, and from the records of the *israelitische Cultusgemeinde*, Vienna.

14. Berta Karlik, *Sondermarke: "100. Geburtstag von Lise Meitner,"* Österreichische Staatsdruckerei, 1978; see also Karlik, "Lise Meitner, 1878–1968."

15. Sallie A. Watkins, "Lise Meitner, 1878–1968," in Louise S. Grinstein, Rose K. Rose, and Miriam H. Rafailovich, eds., *Women in Chemistry and Physics: A Biobibliographic Sourcebook* (Westport, Conn.: Greenwood Press, 1993), 393–402.

16. Ibid.; Otto Robert Frisch, "Lise Meitner," in Charles Gillespie, ed., *Dict. Sci. Biog.* (New York: Scribners, 1974), 9:260–263.

17. Lilli Eppstein, pers. comm., Stocksund, 12 September 1987.

18. O. R. Frisch, "Lise Meitner, 1878–1968," *Biog. Mem. Fell. Roy. Soc. Lond.* 16 (1970): 405–420.

19. Gisela on 22 March 1908: *Taufmatrikel der Dompfarre St. Stefan*, Wien. Lise on 29 September 1908: *Taufschein (evangelisch)* (Baptismal certificate, Protestant) (MC), and entry 1820 in *israelitische Cultusgemeinde*, "ausgetreten 29 September 1908."

20. Kobler, "Contribution of Austrian Jews," 29, describes the severe discrimination against Jews in public office and the resultant "influx of neo-Christians into the legal profession, which probably represented a higher percentage of converts" than any other profession.

For possible insight into the Meitners' nonconversion, see Richard Willstätter, in *From My Life: The Memoirs of Richard Willstätter*, trans. Lilli S. Hornig (New York: W. A. Benjamin, 1965), 420: "Conversion to Christianity was always out of the question for me because it entailed significant advantages, while remaining a Jew carried with it only civil disadvantages."

21. Berkley, *Vienna and Its Jews*, 46–47.

22. Zweig, *World of Yesterday*, 3: "this faith in an uninterrupted and irresistible 'progress' truly had the force of a religion for that generation." Zweig also notes (p. 6) that, in contrast to the more observant Jews of eastern Europe, Jews from Moravia were "early emancipated from their orthodox religion, [becoming] passionate followers of the religion of the time, 'progress.'"

23. Elise Meitner, *Jahres-Zeugnis*, 1892 (MC).

24. Lise Meitner to Frl. Hitzenberger, 29 March/10 April 1951 (MC).

25. Lise Meitner, "Looking Back," *Bull. Atom. Sci.* 20 (November 1964): 2–7.

26. Meitner to Frl. Hitzenberger, 29 March/10 April 1951 (MC).

27. Martha Forkl und Elisabeth Koffmahn, Hrsg., *Frauenstudium und Akademische Frauenarbeit in Österreich* (Wien: Wilhelm Braumüller, 1968).

Mileva Marić, Albert Einstein's first wife, was born in 1875 in southern Hungary and received special permission to attend Gymnasium to take classes in physics. To continue, she attended secondary school in Zurich for two years before enrolling in the Swiss Federal Institute of Technology. See Gerald Holton, "Of Love, Physics and Other Passions: The Letters of Albert and Mileva, Part 1," *Physics Today* 47, no. 8 (1994): 23–29.

28. Women were admitted to legal faculties in 1910 but excluded from technical (engineering) schools until after World War I.

29. Meitner to Frl. Hitzenberger, 10 April 1951; to Hilde Hess, 20 January 1952 (MC). Frisch, *Biographical Memoirs*, 405.

30. B. Karlik and E. Schmid, *Franz S. Exner und sein Kreis* (Wien: Verlag der Österreichen Akademie der Wissenschaften, 1982), 149–150: Arthur Szarvassy (1873–1919) completed an experimental thesis under Franz Exner in 1898, was a lecturer in Vienna for several years, then a theoretical physicist in Brunn, now Brno (Czechoslovakia).

31. Meitner, "Looking Back," 2.

32. Auguste Dick, pers. comm., Vienna, 1985: The institutional records of the Akademisches Gymnasium, Beethovenplatz 1, Wien I, show that Elise Meitner, age 22 and 8/12, was awarded her *Maturitätzeugnis* on 11 July 1901 (records of the University, Rigorosen-Protokoll). Other graduates of this Gymnasium include Ludwig Boltzmann, Stefan Zweig, and Erwin Schrödinger. For Schrödinger, see Walter Moore, *Schrödinger: Life and Thought* (Cambridge: Cambridge University Press, 1989), 20–24; also Gabriele Kerber, Auguste Dick, and Wolfgang Kerber, *Dokumente, Materialien und Bilder zur 100. Wiederkehr des Geburtstages von Erwin Schrödinger* (Wien: Fassbaender, 1987), 20–22.

33. Meitner, "Looking Back," 2.

34. Auguste Dick, pers. comm., Vienna, 1985.

35. Meitner to Frl. Hitzenberger, 29 March/10 April 1951 (MC).

36. Ibid.

37. Ibid.

38. Lise Meitner, University of Vienna *Meldungsbuch* (MC).

39. Meitner, "Looking Back," 2. Gegenbauer's name is misspelled; the correct spelling is in her *Meldungsbuch*.

40. Karl Przibam, "Erinnerungen an ein altes physikalisches Institut," 2–6 in O. R. Frisch, F. A. Paneth, F. Laves, and P. Rosbaud, eds., *Trends in Atomic Physics: Essays Dedicated to Lise Meitner, Otto Hahn, and Max von Laue on the Occasion of Their 80th Birthday* (New York: Interscience, 1959).

41. Andreas Kleinert, *Anton Lampa, 1868–1938* (Mannheim: Bionomica-Verlag, 1985).

42. Meitner, "Looking Back," 3.

43. Ibid.

44. See Przibam, "Erinnerungen," 2; Karlik and Schmid, *Franz S. Exner*, 84; Hans Benndorf, "Zur Erinnerung F. S. Exner," *Physikalische Zeitschrift* 28 (1927): 397–409, quoted in Moore, *Schrödinger*, 51.

45. Przibam, "Erinnerungen," 1.

46. Ibid., 3. See also Karlik and Schmid, *Franz S. Exner*, 65–66.

47. For a discussion of contemporary physics, see E. N. daC. Andrade, *Rutherford and the Nature of the Atom* (New York: Doubleday, 1964), 4–6; J. L. Heilbron, "Lectures on the History of Atomic Physics, 1900–1922," in *History of Twentieth-Century Physics* (New York: Academic Press, 1977), 48–52: Heilbron estimates that all told there were about 700 academic physicists in Europe and the United States. Also see Jeffrey A. Johnson's review, *Science* 254 (1991): 1529–1530, of Kathryn M. Olesko, *Physics as a Calling: Discipline and Practice in the Königsberg Seminar for Physics* (Ithaca: Cornell University Press, 1991).

48. Engelbert Broda, *Ludwig Boltzmann: Mensch, Physiker, Philosoph* (Wien: Franz Deuticke, 1955), 9–10.

49. Ibid., 15.

50. Ibid., 11–12.

51. Ibid., 11.

52. Meitner, "Looking Back," 3.

53. Meitner to E. Broda, n.d. but ca. spring 1954 (MC); in Broda, *Ludwig Boltzmann*, 19.

54. Boltzmann-Aigentler correspondence, 1872–1876, in Dieter Flamm, "Aus dem Leben Ludwig Boltzmanns," in Roman Sexl and John Blackmore, eds., *Ludwig Boltzmann Gesamtausgabe: Band 8 Internationale Tagung, 5.–8. September 1981 Ausgewählte Abhandlungen* (Graz: Akademische Druck u. Verlagsanstalt, 1982), 21–56.

55. Meitner to E. Broda, spring 1954 and 3 May 1954 (MC): Meitner writes of Boltzmann's "ausgezeichneten Verhältnisse zu Frau und Kindern" and refers to their family life as "sehr harmonisch."

56. Meitner to Broda, n.d. but ca. spring 1954 (MC); in Broda, *Ludwig Boltzmann*, 19. Meitner uses the term "Reinheit seiner Seele" in a letter to Broda, 14 April 1955 (MC).

57. Flamm, "Aus dem Leben Ludwig Boltzmanns," 25; Meitner to Broda, spring 1954 and 3 May 1954 (MC).

58. Broda, *Ludwig Boltzmann*, 68; Walter Höflechner, pers. comm., 27 August 1994.

59. Meitner to Broda, 3 May 1954 (MC).

60. Using kinetic theory, gas viscosities, and condensation coefficients, Loschmidt in 1866 calculated molecular diameters and the number of molecules in a given volume of gas. Similar to Avogadro's number, which can be defined as the number of molecules in 22.4 L of gas at standard conditions of temperature and pressure, the *Loschmidt number* is the number of molecules in 1 cm^3 of gas at standard conditions.

61. The Stefan-Boltzmann Law, $E \propto T^4$, was formulated empirically by Stefan in 1879 and given a theoretical basis by Boltzmann in 1884.

62. Broda, *Ludwig Boltzmann*, 4–5.

63. Meitner to Broda, n.d., spring 1954 (MC); quoted in Broda, *Ludwig Boltzmann*, 6–7.

64. Herbert Hörz, "Helmholtz und Boltzmann," in Sexl and Blackmore, *Ludwig Boltzmann Gesamtausgabe*, 191–205, on 194.

65. Broda, *Ludwig Boltzmann*, 6, recounts the often-heard story that after the Berlin negotiations were completed and Boltzmann accepted, a dinner was arranged, where, it is said, Frau Helmholtz tartly remarked on his lack of formality: "Herr Professor, I fear you will not be comfortable in Berlin!" and Boltzmann thereupon withdrew. Without completely discounting this, Walter Höflechner attributes Boltzmann's restlessness and at times erratic decision making to what was at the time diagnosed as "neurasthenia" and is now more recognizable as manic-depressive illness. Walter Höflechner, "Ludwig Boltzmann: Sein Akademischer Werdegang in Österreich," *Mitteilg. Österr. Ges. Gesch. Naturwiss.* 2 (1982): 43–62.

66. "Hab'n S'ein's g'sehn?" Broda, *Ludwig Boltzmann*, 84.

67. "Feindselig." Ibid., 37.

68. See Gerald Holton, "The Millikan-Ehrenhaft Dispute," in *Hist. Stud. Phys. Sci.*, ed. Russell McCormmach and Lewis Pyenson (Baltimore: Johns Hopkins University Press, 1978), 9:172.

69. Ludwig Boltzmann, "A German Professor's Trip to El Dorado," trans. Bertram Schwarzschild, *Physics Today* 45, no. 1 (January 1992): 44–51.

70. Broda, *Ludwig Boltzmann*, 114–115. George Berkeley (1685–1753) was Irish, although of English descent.

71. Broda, *Ludwig Boltzmann*, 88–89. Also see Engelbert Broda, *Ludwig Boltzmann*, trans. Larry Gay and E. Broda (Woodbridge, Conn.: Oxbow Press, 1983), 98–100, and Hörz, "Helmholtz und Boltzmann."

72. Broda, *Ludwig Boltzmann*, 40.

73. Abraham Pais, '*Subtle Is the Lord* . . . ': *The Science and the Life of Albert Einstein* (Oxford: Oxford University Press, 1982), 100, 103–104. Stefan Meyer relates that Mach was won over to atomism late in life by a radioactive source producing light flashes on a scintillation screen, but other sources regard the story as implausible. See Paul K. Feyerabend, "Philosophy of Science: A Subject with a Great Past," in Roger H. Stuewer, ed., *Historical and Philosophical Perspectives of Science* (Minneapolis: University of Minnesota Press, 1970), 178 n. 15.

74. Frisch, *Biographical Memoirs*, 406.

75. Walter Höflechner, "Ludwig Boltzmann, sein akademischer Werdegang in Österreich," 59.

76. Bericht über die Dissertation von Frl. L. Meitner: *Prüfung einer Formel Maxwells*; signed by Exner and Boltzmann, 28 November 1905 (University Archives, Vienna).

77. Meitner's dissertation, "Prüfung einer Formel Maxwells," was submitted on 20 November, refereed by Exner and Boltzmann, approved on 28 November 1905 (Rigorosen Protokoll 1896–2268), and published on 23 February 1906 as "Warmeleitung in inhomogenen Korpern," *S. Ber. Akad. Wiss. Wien* IIa, Bd. 115 (February 1906): 125–137.

Meitner's *Fachprüfung* (orals) on 19 December 1905 were conducted by Exner (excellent, "*ausgezeichnet*"), Boltzmann ("*ausgezeichnet*"), and the mathematician von Escherich (sufficient, "*genügend*"). Her *Nebenrigorosum* on 22 December 1905 were held by Professors (Thirring?) and Stohr (both "*ausgezeichnet*").

78. Martin J. Klein, *Paul Ehrenfest* (New York: American Elsevier, 1970), 49. Klein suggests that their joint study took place in 1904; in "Looking Back," Meitner places Ehrenfest's influence in 1907 (p. 25). Both appear to be incorrect: Meitner had not completed her courses with Boltzmann by the earlier date, and her optics paper is dated June 1906.

79. Lise Meitner, "Über einige Folgerungen, die sich aus den Fresnel'schen Reflexionsformeln ergeben," *S. Ber. Akad. Wiss. Wien* IIa, Bd. 115 (June 1906): 259–286.

80. Meitner, "Looking Back," 3.

81. Lise Meitner, "Über die Absorption der α- und β-Strahlen," *Phys. Z.* 7 (1906): 588–590; submitted June 1906.

82. The first, Olga Steindler, received her doctorate in 1903, married the physicist Felix Ehrenhaft, and founded Vienna's first commercial school for women. A third woman, Selma Freud, who worked with Lise in the same laboratory and earned her doctorate a little later in 1906, also did not go on in physics. See Przibam, "Erinnerungen"; Karlik, *Sondermarke*.

83. Margaret Rossiter, *Women Scientists in America: Struggles and Strategies to 1940* (Baltimore: Johns Hopkins University Press, 1982).

84. According to Kerner, *Lise, Atomphysikerin*, 23, Meitner later regarded it as fortunate that she did not go to Paris; she believed she had learned more from Stefan Meyer. M. F. Rayner-Canham and G. W. Rayner-Canham, in "Pioneer

Women in Nuclear Science," *Amer. J. Phys.* 58 (1990): 1036–1043, note that men such as Ernest Rutherford may have been more supportive of women in science than was Marie Curie.

85. Meitner, "Looking Back," 3.

86. Klein, *Paul Ehrenfest,* 77.

87. Meitner to Broda, 3 May 1954 (MC).

88. Meitner, "Looking Back," 3.

89. Przibam, "Erinnerungen," 5.

90. M. Malley, "The Discovery of the Beta Particle," *Amer. J. Phys.* 39 (1971): 1454–1460.

91. Lawrence Badash, *Radioactivity in America: Growth and Decay of a Science* (Baltimore: Johns Hopkins University Press, 1979), chap. 1.

92. Przibam, "Erinnerungen," 5.

93. J. L. Heilbron, "Scattering of α and β Particles and Rutherford's Atom," *Archive for History of the Exact Sciences* 4 (Berlin: Springer Verlag, 1967–1968): 247–307.

94. Stefan Meyer, "Zur Erinnerung an die Jugendzeit der Radioaktivität," *Naturwiss.* 35 (1948): 161–163.

95. Lise Meitner, "Über die Zerstreuung der α-Strahlen," *Phys. Z.* 8 (1907): 489–491.

96. Many young Austrian scientists left for positions abroad; according to Erwin Schrödinger, the lack of opportunity caused Austria to lose its initial leadership in atmospheric electricity and radioactivity. See Moore, *Schrödinger,* 58.

97. Meitner, "Looking Back," 3.

CHAPTER 2. Beginnings in Berlin

1. Lise Meitner, "Looking Back," *Bull. Atom. Sci.* 20 (November 1964): 2–7.

2. For the first impressions of a Viennese in Berlin at the same time as Meitner, see Stefan Zweig, *The World of Yesterday* (New York: Viking Press, 1943; reprint Lincoln: University of Nebraska Press, 1964), 110–113.

3. Elisabeth Boedeker, *Marksteine der Deutscher Frauenbewegung* (Hannover: Selbstverlag, 1969), 5. Dates for admission of women to universities: Baden, 1900; Bavaria, 1903; Württemberg, 1904; Sachsen, 1906; Thüringen, 1907; Hessen, May 1908; Prussia, August 1908; Alsace-Lorraine, 1908–1909; Mecklenberg, 1909.

4. Lise Meitner to Gerta von Ubisch, 1 July 1947 (MC).

5. Gerta von Ubisch to Meitner, 28 February 1947 (MC).

6. Meitner, "Looking Back," 4.

7. Arthur Kirchhoff, Hrsg., *Die Akademische Frau* (Berlin: Hugo Steinik Verlag, 1897).

8. For more on American women in Göttingen and eventually elsewhere, see Margaret Rossiter, *Women Scientists in America: Struggles and Strategies to 1940* (Baltimore: Johns Hopkins University Press, 1982), 40–43.

9. Kirchhoff, *Die Akademische Frau:* Prof. Dr. Hubert Ludwig, zoology, Bonn, p. 279; Dr. Ludwig Fulda, München, p. 320; Prof. Dr. phil. Rudolf Sturm, mathematics, Breslau, p. 242; Felix Klein, p. 241; G. Weyer, pp. 243-255; Hugo Münsterberg, pp. 344-354.

10. Kirchhoff, *Die Akademische Frau,* 256-257.

11. Lise Meitner, "Max Planck als Mensch," *Naturwiss.* 45 (1958): 406-408.

12. Meitner, "Looking Back," 4-5.

13. Otto Hahn, *A Scientific Autobiography,* trans. and ed. Willy Ley (London: MacGibbon & Kee, 1967), 51. If, as is likely, the two were introduced at the physics colloquium, the date must be incorrect: 28 September 1907 was a Saturday.

14. Ibid., 52.

15. Ibid., 50; Otto Hahn, *My Life,* trans. Ernst Kaiser and Eithne Wilkins (New York: Herder and Herder, 1968), 86-87.

16. For a general outline of the German academic hierarchy, see Fritz K. Ringer, *The Decline of the German Mandarins: The German Academic Community, 1890-1933* (Cambridge: Harvard University Press, 1969), 33-35, 38. For a detailed study of theoretical physics professorships in Prussia and elsewhere about 1900, see Christa Jungnickel and Russell McCormmach, *Intellectual Mastery of Nature: Theoretical Physics from Ohm to Einstein.* Vol. 2: *The Now Mighty Theoretical Physics, 1870-1925* (Chicago: University of Chicago Press, 1986), esp. chap. 15, 33ff.

17. Hahn attests to his fondness for women in his biographies; his orientation toward radioactivity was based on his earlier collaboration with physicists. From his biographies (e.g., *A Scientific Biography,* 51) it is evident that Hahn was impressed with Meitner's prior accomplishments in radioactivity in Vienna. See Fritz Krafft, "Otto Hahn 1879-1968," in Lothar Gall, ed., *Die Grossen Deutschen unserer Epoche* (Berlin: Propyläen Verlag, 1985), 173-185.

For Brooks in Montreal, see Marelyne F. Rayner-Canham and Geoffrey W. Rayner-Canham, *Harriet Brooks, Pioneer Nuclear Scientist* (Montreal and Kingston: McGill-Queen's University Press, 1992), chap. 6.

18. Nikolaus Riehl, "Erinnerungen an Otto Hahn und Lise Meitner," Vortrag zur Eröffnung der Ausstellung "50 Jahre Kernspaltung" im Willi-Graf-Gymnasium, 30 November 1988, Fridolin Haugg, ed. (München, 1989), 9. Riehl, a student of Meitner's in the 1920s, describes Hahn's pre–World War I mustache as "a la Wilhelmzwo" (à la Wilhelm II).

"Anglicized Berliner": Hahn, *A Scientific Biography,* 68.

"*standesgemäß*": Interview, Ulla Frisch, 16 January 1982, and Nancy Arms, *A Prophet in Two Countries: The Life of F. E. Simon* (Oxford: Pergamon Press, 1966), 27.

19. Meitner, "Looking Back," 5.

20. In his autobiography, written in 1918, Fischer lists a number of women students in his institute (presumably after 1908) and notes that several were as good as any of the men. Fischer expresses a conservative view, however, on admitting

women to the study of chemistry: most will marry, and the effort and expense of their education will be for nothing. Emil Fischer, *Aus Meinem Leben* (Berlin: Julius Springer, 1922), 191–192.

21. Hahn noted, "Despite many years of successful work, in the Chemical Institute she was, so to speak, 'nonexistent.'" In Otto Hahn, "Lise Meitner 80 Jahre," *Naturwiss.* 45 (1958): 501–502.

The conditions in Fischer's institute are described by Meitner in "Looking Back," by Hahn in his autobiographies, and by others, including Friedrich Herneck, "Zum wissenschaftlichen Wirken von Otto Hahn und Lise Meitner in Chemischen Institut der Berliner Universität," *Wiss. Z. Humboldt-Universität* 16 (1967): 833–836; Friedrich Herneck, *Bahnbrecher des Atomzeitalters: Grosse Naturforscher vom Maxwell bis Heisenberg* (Berlin: Buchverlag der Morgen, 1965), 365–400; Fritz Krafft, *Im Schatten der Sensation: Leben und Wirken von Fritz Strassmann* (Weinheim: Verlag Chemie, 1981), chap. 2.2, 166–168; Fritz Krafft, "Lise Meitner und ihre Zeit—Zum hundertsten Geburtstag der bedeutende Naturwissenschaftlerin," *Angew. Chem.* 90 (1978): 876–892, and *Angew. Chem. Intl. Ed. Engl.* 17 (1978): 826–842.

22. Meitner, "Looking Back," 5.

23. Ibid.

24. John L. Heilbron, "The Scattering of Alpha and Beta Particles and Rutherford's Atom," *Archive for History of Exact Sciences* (Berlin: Springer-Verlag, 1967–1968), 4:256; Lawrence Badash, *Radioactivity in America: Growth and Decay of a Science* (Baltimore: Johns Hopkins University Press, 1979), 43–45.

25. Hahn, *A Scientific Autobiography*, 26–27.

26. Heilbron, "Scattering of Alpha and Beta Particles," 266–269.

27. Hahn, *A Scientific Autobiography*, 54; Otto Hahn and Lise Meitner, "Über die Absorption der β-Strahlen einiger Radioelemente," *Phys. Z.* 9 (1908): 321–333.

28. Otto Hahn and Lise Meitner, "Über die β-Strahlen des Aktiniums," *Phys. Z.* 9 (1908): 697–702.

29. Otto Hahn and Lise Meitner, "Aktinium C, ein neues kurzlebiges Produkt des Aktiniums," *Phys. Z.* 9 (1908): 649–655.

30. Lise Meitner, *Taufschein, evangelisch* (baptismal certificate, Protestant), 29 September 1908, and entry 1920 in *israelitische Cultusgemeinde*, Wien: "ausgetreten [withdrew] 29 September 1908."

For views on religion, see Meitner to Max von Laue, 15 August 1941 (MC): Meitner's friend Eva von Bahr-Bergius (a Catholic) "always says that I am so thoroughly oriented towards protestantism."

31. Max Born, "Max Karl Ernst Ludwig Planck, 1858–1947," *Roy. Soc. Lond. Obit. Not.* 6, no. 17 (1948): 161–188.

32. Otto Hahn, "Über eine neue Erscheinung bei der Aktivierung mit Aktinium," *Phys. Z.* 10 (1909): 81–88. Hahn, *A Scientific Autobiography*, 61; Dietrich Hahn, ed., *Otto Hahn, Erlebnisse und Erkenntnisse* (Düsseldorf: Econ Verlag, 1975), 28.

33. Hahn, *A Scientific Autobiography*, 61–62. Otto Hahn and Lise Meitner, "Eine neue Methode zur Herstellung radioaktiver Zerfallsprodukte: Thorium D, ein kurzlebiges Produkt des Thoriums," *Verh. Dt. Phys. Ges.* 11 (1909): 55–62.

34. Meitner, "Looking Back," 5.

35. E. N. daC. Andrade, *Rutherford and the Nature of the Atom* (Garden City, N.Y.: Doubleday, 1964), 108; Lawrence Badash, ed., *Rutherford and Boltwood: Letters on Radioactivity* (New Haven: Yale University Press, 1969), 205.

36. Hahn, *My Life*, 89.

37. Badash, *Rutherford and Boltwood*, 205–206.

38. See description in Thaddeus J. Trenn, "Rutherford and Recoil Atoms: The Metamorphosis and Success of a Once Stillborn Theory," *Hist. Stud. Phys. Sci.* 6 (1975): 530–531. For Brooks, see Rayner-Canham and Rayner-Canham, *Harriet Brooks*, 40–41. For the radioactivity context of recoil ca. 1903, see Thaddeus J. Trenn, *The Self-Splitting Atom: A History of the Rutherford-Soddy Collaboration* (London: Taylor & Francis, 1977), 91.

39. Hahn, *A Scientific Autobiography*, 63.

40. It took several months, however, for the repercussions to die down. Not long after Rutherford's visit, his Manchester co-workers Sidney Russ and Walter Makower stated that Hahn and Meitner had merely "repeated the experiments of Miss Brooks," obtaining "indications of similar processes for actinium and thorium" (*Phys. Z.* 10 [1909]: 361, a translation from *Proc. Roy. Soc. Lond.* A82 [6 May 1909]). In a rather sharp response, Hahn and Meitner ("Die Ausstößung radioaktiver Materie bei den Umwandlungen des Radiums," *Phys. Z.* 10 [1909]: 422) insisted they had proposed recoil without knowing of Brooks's experiments and objected to the term "indications" (*Anzeichen*) for the recoil they had established with "absolute certainty" (*mit absoluter Sicherheit*).

41. Badash, *Rutherford and Boltwood*, 206.

42. Hahn, *My Life*, 88.

43. Meitner to Elisabeth Schiemann, 27 March 1913 (MC).

44. Otto Robert Frisch, interview, 1975.

45. Lise Meitner, "Einige Erinnerungen an das Kaiser-Wilhelm-Institut für Chemie in Berlin-Dahlem," *Naturwiss.* 41 (1954): 97–99; Lise Meitner, "Otto Hahn zum 80. Geburtstag am 8. März 1959," *Naturwiss.* 46 (1959): 157–158.

46. Ringer, *The Decline of the German Mandarins*.

47. Sallie A. Watkins, "Lise Meitner, 1878–1968," in Marelene F. Rayner-Canham and Geoffrey W. Rayner-Canham (forthcoming). In 1911 the payment for translating an article was $10.

48. Lise Meitner, "The Status of Women in the Professions," *Physics Today* 13, no. 8 (August 1960): 16–21; also Hahn, *A Scientific Autobiography*, 65.

49. Lise Meitner, diaries, 1907, 1908 (MC).

50. Elisabeth Schiemann, "Freundschaft mit Lise Meitner," *Neue Evangelische Frauenzeitung* 3, no. 1 (Januar/Februar 1959).

51. Ibid.

52. Berta Karlik and Erich Schmid, *Franz S. Exner und sein Kreis* (Wien: Verlag der Österreichischen Akademie der Wissenschaften, 1982), 88; Walter Moore, *Schrödinger: Life and Thought* (Cambridge: Cambridge University Press, 1989), 70.

53. Elisabeth Schiemann to O. R. Frisch, 14 July 1969 (MC).

54. Meitner to James Franck, zum 29 August 1952 (Franck's 70th birthday) (MC); Heinrich G. Kuhn, "James Franck, 1882–1964," *Biog. Mem. Fell. Roy. Soc. Lond.* 11 (1965): 52–74.

55. O. R. Frisch, pers. comm., 22 March 1975, Cambridge.

56. Lise Meitner, "Max Planck als Mensch," 406, 407.

57. Ibid., 406.

58. J. L. Heilbron, *The Dilemmas of an Upright Man: Max Planck as Spokesman for German Science* (Berkeley, Los Angeles, and London: University of California Press, 1986), 37–39.

59. Lise Meitner, "Max Planck als Mensch," 406.

60. Max von Laue, "Zu Max Plancks 100. Geburtstage," *Naturwiss.* 45 (1958): 221–226.

61. Wilhelm Westphal, "Max Planck als Mensch," *Naturwiss.* 45 (1958): 234–236; Lise Meitner, "Otto Hahn zum 80."

62. Meitner to Elisabeth Schiemann, 31 December 1913 (MC).

63. Lise Meitner, "Strahlen und Zerfallsprodukte des Radiums," *Verh. Dt. Phys. Ges.* 11 (1909): 648–653. This talk, part of a study of beta emitters in the radium series, was given in Salzburg on 22 September 1909.

64. Meitner, "Looking Back," 4. Andreas Kleinert, *Anton Lampa, 1968–1938* (Mannheim: Bionomica Verlag, 1985), 10–11. Lampa also met Einstein for the first time in Salzburg.

65. Meitner, "Looking Back," 4. Einstein's Salzburg talk was titled "On the Development of Our Views on the Nature and Constitution of Radiation." See Albert Einstein, *Collected Papers*, ed. John Stachel (Princeton: Princeton University Press, 1989), 2:564–587; Ronald W. Clark, *Einstein: The Life and Times* (New York: Avon, 1971), 163–164; Jungnickel and McCormmach, *Intellectual Mastery of Nature*, 304–309; Roger H. Stuewer, *The Compton Effect: Turning Point in Physics* (New York: Science History Publications, 1975), 29–31.

66. Meitner, "Looking Back," 2.

67. Fritz Krafft, "Otto Hahn, 1879–1968," 173–185.

68. Lise Meitner to Jenny [last name unknown], 6 February 1911 (MC). Reproduced in Jost Lemmerich, ed., *Die Geschichte der Entdeckung der Kernspaltung: Ausstellungskatalog* (Berlin: Technische Universität Berlin, Universitätsbibliothek, 1988), 58.

69. Otto Hahn and Lise Meitner, "Nachweis der komplexen Natur von Radium C," *Phys. Z.* 10 (1909): 697–703.

70. Ibid.

71. Meitner, "Strahlen und Zerfallsprodukte des Radiums," 651.

72. Otto Hahn and Lise Meitner, "Über eine typische β-Strahlung des eigentlichen Radiums," *Phys. Z.* 10 (1909): 741–745.

73. Otto Hahn and Lise Meitner, "Eine neue β-Strahlung beim Thorium X; Analogien in der Uran- und Thoriumreihe," *Phys. Z.* 11 (1910): 493–497.

74. Hahn and Meitner, "Über der β-Strahlen des Aktiniums."

75. Hahn and Meitner, "Eine neue β-Strahlung beim Thorium X."

76. William Wilson, "On the Absorption of Homogeneous β Rays by Matter, and on the Variation of the Absorption of the Rays with Velocity," *Proc. Roy. Soc. Lond.* A82 (1909): 612–628.

77. Otto Hahn and Lise Meitner, "Über das Absorptionsgesetz der β-Strahlen," *Phys. Z.* 10 (1909): 948–950.

78. William Wilson, *Phys. Z.* 11 (1910): 101–104.

79. Meitner, "Looking Back," 6.

80. Dietrich Hahn, *Otto Hahn,* 29; also G. Hettner and Otto Hahn, "Zur Erinnerung an Otto v. Baeyer," *Naturwiss.* 34 (1947): 193–194.

81. Otto v. Baeyer and Otto Hahn, "Magnetische Linienspektren von β-Strahlen," *Phys. Z.* 11 (1910): 488–493. It is not clear why Meitner was not a co-author on this first deflection experiment. The paper was submitted in May 1910; possibly her father was ill and she was in Vienna. He died in December 1910 (Meitner to Hahn, 12 December 1924 [MC]).

82. Hahn and Meitner, "Eine neue β-Strahlung beim Thorium X."

83. O. v. Baeyer, O. Hahn, and L. Meitner, "Nachweis von β-Strahlen bei Radium D," *Phys. Z.* 12 (1911): 378–379.

84. Otto v. Baeyer, Otto Hahn, and Lise Meitner, "Über die β-Strahlen des aktiven Niederschlags des Thoriums," *Phys. Z.* 12 (1911): 273–279.

85. Otto v. Baeyer, Otto Hahn, and Lise Meitner, "Magnetische Spektren der β-Strahlen des Radiums," *Phys. Z.* 12 (1911): 1099–1101.

86. Baeyer, Hahn, and Meitner, "Nachweis von β-Strahlen bei Radium D."

87. Otto von Baeyer, Otto Hahn, and Lise Meitner: "Das magnetische Spektrum der β-Strahlen des Thoriums," *Phys. Z.* 13 (1912): 264–266; "Das magnetische Spektrum der β-Strahlen des Radioaktiniums und seiner Zerfallsprodukte," *Phys. Z.* 14 (1913): 321–323; "Das magnetische Spektrum der β-Strahlen des Uran X, *Phys. Z.* 15 (1914): 649–650; "Das Magnetische Spektrum der β-Strahlen von Radiothor und Thorium X," *Phys. Z.* 16 (1915): 6–7.

88. Otto Hahn and Lise Meitner, "Über die Verteilung der β-Strahlen auf die einzelnen Produkte des aktiven Niederschlags des Thoriums," *Phys. Z.* 13 (1912): 390–393.

89. J. Franck and Lise Meitner, "Über radioaktive Ionen," *Verh. Dt. Phys. Ges.* 18 (1911): 671–675.

90. Lise Meitner, "Über einige einfache Herstellungsmethoden radioaktiver Zerfallsprodukte," *Phys. Z.* 12 (1911): 1094–1099.

91. Lise Meitner, "Über das Zerfallsschema des aktiven Niederschlags des Thoriums," *Phys. Z.* 13 (1912): 623–626.

92. In Germany and elsewhere, support for radioactivity research generally came from physics, not chemistry, laboratories. See Badash, *Radioactivity in America*, 261–262.

93. Lothar Burchardt, *Wissenschaftspolitik im Wilhelminischen Deutschland: Vorgeschichte, Gründung und Aufbau der Kaiser-Wilhelm-Gesellschaft zur Förderung der Wissenschaften* (Göttingen: Vandenhoeck & Ruprecht, 1975), 33; Krafft, *Im Schatten der Sensation*, 165–166; Otto Hahn, *Der Kaiser-Wilhelm-Institut für Chemie*, Jahrbuch 1951 der MPG, Generalverwaltung der MPG (Göttingen: Hubert, 1951), 175–198; Lemmerich, *Geschichte der Entdeckung der Kernspaltung*, 62–68; Jeffrey Allan Johnson, *The Kaiser's Chemists: Science and Modernization in Imperial Germany* (Chapel Hill: University of North Carolina Press, 1990).

94. Burchardt, *Wissenschaftspolitik*, 98–99. Dietrich Stoltzenberg, *Fritz Haber: Chemiker, Nobelpreisträger, Deutscher, Jude* (Weinheim: VCH Verlagsgesellschaft, 1994), 199–201.

95. Richard Willstätter, *From My Life: The Memoirs of Richard Willstätter*, trans. Lilli S. Hornig (New York: W. A. Benjamin, 1965), 210–212.

96. Johnson, *The Kaiser's Chemists*, 125–128. In a featured address to the Kaiser Wilhelm Society in January 1911, Fischer delivered a "scientific manifesto" outlining the major achievements and direction of German chemical research. He included radioactivity, referring to Hahn's work with mesothorium.

97. Lemmerich, *Geschichte der Entdeckung der Kernspaltung*, 66–67: Hahn's contract, dated 14 June 1912, lists his salary as 5,000 marks.

98. Meitner to Max von Laue, 6 September 1955 (MC).

99. Meitner was officially informed on 18 November 1912; see Lemmerich, *Geschichte der Entdeckung der Kernspaltung*, 58, and Heilbron, *Dilemmas of an Upright Man*, 39.

100. Personnel files, MPG, Berlin-Dahlem.

101. Meitner to Elisabeth Schiemann, 22 December 1915 (MC), printed in Lemmerich, *Geschichte der Entdeckung der Kernspaltung*, 75.

CHAPTER 3. The First World War

1. Lothar Burchardt, *Wissenschaftspolitik im Wilhelminischen Deutschland: Vorgeschichte, Gründung und Aufbau der Kaiser-Wilhelm-Gesellschaft zur Förderung der Wissenschaften* (Göttingen: Vanderhoeck & Ruprecht, 1975), 146–149; Richard Willstätter, *From My Life: The Memoirs of Richard Willstätter*, trans. Lilli S. Hornig (New York: W. A. Benjamin, 1965), 217–220; Otto Hahn, *My Life*, trans. Ernst Kaiser and Eithne Wilkins (New York: Herder and Herder, 1970), 102; Dietrich Stoltzenberg, *Fritz Haber: Chemiker, Nobelpreisträger, Deutscher, Jude* (Weinheim: VCH Verlagsgesellschaft, 1994), 212ff.

2. Lise Meitner, "Einige Erinnerungen an das Kaiser-Wilhelm-Institut für Chemie in Berlin-Dahlem," *Naturwiss.* 41 (1954): 97–99. Willstätter, *From My Life*, 216–217, 222. See also Jost Lemmerich, "Dahlem—ein Deutsches Oxford," *Ausstellung*, Verein der Freunde der Domäne Dahlem, March 1992.

3. Meitner, "Einige Erinnerungen"; Willstätter, *From My Life*, 216–217; Otto Hahn, *Der Kaiser-Wilhelm-Institut für Chemie*, Jahrbuch 1951 der MPG, Generalverwaltung der MPG (Göttingen: Hubert, 1951), 177–179.

4. Otto Hahn, *A Scientific Autobiography*, trans. and ed. Willy Ley (London: MacGibbon & Kee, 1967), 81–82; Dietrich Hahn, ed., *Otto Hahn, Erlebnisse und Erkenntnisse* (Düsseldorf: Econ Verlag, 1975), 35; Otto Hahn, *My Life*, 104.

5. Telegram from Lise and Fritz Meitner for "Wedding Junghans Hahn Stettin. Breitestrasse 14," 22 March 1913 (MPG).

6. Meitner's 1913 salary was 1,500 marks, much below Hahn's 5,000 marks, presumably because her academic rank was lower than his and her assistantship with Planck, which continued until 1915, provided additional income. In 1914 her salary was raised to 3,000 marks and in 1917, when she was given her own physics section, to 4,000 marks, essentially equivalent to Hahn's 5,000 marks, which included a marriage supplement. (Marion Kazemi; personnel files of the MPG, Berlin-Dahlem.)

In *KWI für Chemie, Bericht des Direktors Okt. 1913 bis Okt. 1914* (MPG), Hahn and Meitner are both listed as *derzeit Mitglieder* (provisional as opposed to permanent [*ständige*] associates); the section is listed as Laboratorium Hahn-Meitner, but Otto Hahn refers to it as Abteilung (section) Hahn-Meitner in "Mit Lise Meitner von der Holzwerkstatt in Berlin zum Kaiser-Wilhelm-Institut in Dahlem," Auszug aus Vortrag, Akademikerinnenbund, Berlin 1958 (MPG).

7. Printed invitation (unclear if given by or for Meitner); 29 November 1913 for dinner at the Hotel Adlon (MC). For an admiring history of a luxury hotel, see Hedda Adlon, *Hotel Adlon: The Life and Death of a Great Hotel*, trans. Norman Denny (London: Barrie Books, 1958; New York: Horizon Press, 1960).

8. Lawrence Badash, *Radioactivity in America: Growth and Decay of a Science* (Baltimore: Johns Hopkins University Press, 1979), 185; Walther Gerlach, *Otto Hahn: Ein Forscherleben unserer Zeit*, ed. Dietrich Hahn (Stuttgart: Wissenschaftliche Verlagsgesellschaft, 1984, 35–36 (Grosse Naturforscher Band 45); Otto Hahn, "Otto Hahn in Berlin," *Phys. Bl.* 23 (1967): 388–389. D. Hahn, *Erlebnisse*, 25, 31–33.

9. Kaiser-Wilhelm-Gesellschaft report for 1928 (MPG); Hahn, *Kaiser-Wilhelm-Institut für Chemie*, 183; Hahn, *A Scientific Autobiography*, 83; Willstätter, *From My Life*, 218.

10. Tikvah Alper, a student of Meitner's around 1930, mentions the training students received (taped interview with Steven Weininger, Sarisbury Green, Southampton, Hampshire, 21 October 1989). The precautionary measures and Meitner's enforcement are described in Fritz Krafft, *Im Schatten der Sensation: Leben und Wirken von Fritz Strassmann* (Weinheim: Verlag Chemie, 1981), 108–

109. Meitner's strictness is also emphasized by O. R. Frisch, "Lise Meitner, 1878–1968," *Biog. Mem. Fell. Roy. Soc. Lond.* 16 (1970): 405–420.

11. The accepted value for the half-life of actinium is now 21.8 years.

12. Badash, *Radioactivity in America,* 171–175.

13. Otto Hahn and Lise Meitner, "Die Muttersubstanz des Actiniums," *Phys. Z.* 19 (1918): 208–218.

14. Otto Hahn and Lise Meitner, "Zur Frage nach der komplexen Natur des Radioactiniums und der Stellung des Actiniums im Periodischen System," *Phys. Z.* 14 (1913): 742–758.

15. Badash, *Radioactivity in America,* 208; F. Soddy, "The Origin of Actinium," *Nature* 91 (1913): 634–635.

16. Badash, *Radioactivity in America,* 202–203; Otto Hahn and Lise Meitner, "Über das Uran X$_2$," *Phys. Z.* 14 (1913): 758–759.

17. Badash, *Radioactivity in America,* 174; O. Hahn and L. Meitner, "Über das Uran Y," *Phys. Z.* 15 (1914): 236–240.

18. Hahn and Meitner, "Die Muttersubstanz des Actiniums."

19. Hahn, *A Scientific Autobiography,* 83–87.

20. Lise Meitner, "Über die α-Strahlung des Wismuts aus Pechblende," *Phys. Z.* 16 (1915): 4–6.

21. Roger H. Stuewer, "The Nuclear Electron Hypothesis," in William R. Shea, ed., *Otto Hahn and the Rise of Nuclear Physics* (Dordrecht: D. Reidel, 1983), 19–22; Abraham Pais, *Niels Bohr's Times, in Physics, Philosophy, and Polity* (Oxford: Clarendon Press, 1991), 17, 150–151.

22. Lise Meitner and Otto Hahn, "Über die Verteilung der γ-Strahlen auf einzelnen Produkte der Thoriumreihe," *Phys. Z.* 14 (1913): 873–877; O. v. Baeyer, O. Hahn, and L. Meitner, "Das magnetische Spektrum der β-Strahlen des Uran X," *Phys. Z.* 15 (1914): 649–650; James Chadwick, "Intensitätsverteilung im magnetischen Spektrum der β-Strahlen von Radium B + C," *Verh. Dt. Phys. Ges.* 16 (1914): 383–391.

23. The Prague offer came via Anton Lampa, Meitner's first-year laboratory instructor in Vienna, recently appointed professor at the Deutsche Universität in Prague. The offer came in the spring. After some delay, Planck reminded Fischer to hurry, and Meitner received Fischer's offer in July. See J. L. Heilbron, *The Dilemmas of an Upright Man: Max Planck as Spokesman for German Science* (Berkeley, Los Angeles, and London: University of California Press, 1986), 39; Max Planck to Emil Fischer, 17 May 1914; Meitner to Emil Fischer, 2 August 1914, Bancroft Library, UC Berkeley.

The negotiations to raise Meitner's salary in Dahlem were not entirely easy, but the radioactivity section was relatively inexpensive and Fischer juggled various funding sources. See Jeffrey Allan Johnson, *The Kaiser's Chemists: Science and Modernization in Imperial Germany* (Chapel Hill: University of North Carolina Press, 1990), 173–175.

24. Barbara Tuchman, *The Guns of August* (New York: Dell, 1962), 90–94.

25. Meitner to Elisabeth Schiemann, 11 August 1914 (MC). Lise's mother, Hedwig, lived in an apartment on the second floor of Hauptstrasse 1 (third district) next to Vienna's central station (Bahnhof Wien Mitte); see Meitner's 1915 *Meldzetteln* (police registration required of visitors) in Dietmar Grieser, "Im Schatten der Bombe: Lise Meitner 1878–1968," in *Köpfe* (Wien: Österreichischer Bundesverlag, 1991), 117.

26. Meitner to Schiemann, 26 August 1914 (MC).

27. Meitner to Schiemann, 9 September 1914 (MC).

28. Hahn, *My Life*, 112–115; Gerlach, *Otto Hahn*, 54.

29. Tuchman, *Guns of August*, 21.

30. P. P. Ewald, "Max von Laue," *Biog. Mem. Fell. Roy. Soc. Lond.* 6 (1960): 135–156.

31. Stoltzenberg, *Fritz Haber*, 30, 137ff.

32. Morris Goran, *The Story of Fritz Haber* (Norman: University of Oklahoma Press, 1967), 75ff. Friedrich Glum, *Zwischen Wissenschaft, Wirtschaft und Politik* (Bonn: H. Bouvier Verlag, 1964), 287–288. Stoltzenberg, *Fritz Haber*, 230ff.; Johnson, *The Kaiser's Chemists*, 184–189.

33. Sabine Ernst, ed., *Lise Meitner an Otto Hahn: Briefe aus den Jahren 1912 bis 1924: Edition und Kommentierung* (Stuttgart: Wissenschaftliche Verlagsgesellschaft, 1992), 132–133.

34. Otto v. Baeyer, Otto Hahn, and Lise Meitner, "Das magnetische Spektrum der β-Strahlen von Radiothor und Thorium X," *Phys. Z.* 16 (1915): 6–7.

35. Meitner, "Über die α-Strahlung des Wismuts aus Pechblende."

36. Lawrence Badash, ed., *Rutherford and Boltwood: Letters on Radioactivity* (New Haven: Yale University Press, 1969), 305–309. J. J. Thomson, president of the Royal Society, supported Schuster; see Lord Rayleigh, *The Life of Sir J. J. Thomson, O.M.* (London: Dawsons, 1969), 194–196.

37. The German internment may in part have been a response to the British internment of German nationals. The unique society formed by the British in Ruhleben is described in J. Davidson Ketchum, *Ruhleben: A Prison Camp Society* (Toronto: University of Toronto Press, 1965).

38. Sir Harrie Massey and N. Feather, "James Chadwick, 1891–1974," *Biog. Mem. Fell. Roy. Soc. Lond.* 22 (1976): 11–70; Sir Kenneth Hutchison, J. A. Gray, and Sir Harrie Massey, "Charles Drummond Ellis, 1895–1980," *Biog. Mem. Fell. Roy. Soc. Lond.* 27 (1981): 199–233; E. N. daC. Andrade, *Rutherford and the Nature of the Atom* (Garden City, N.Y.: Doubleday, 1964), 161; Abraham Pais, *Inward Bound: Of Matter and Forces in the Physical World* (Oxford: Clarendon Press, 1986), 160.

39. J. L. Heilbron, *H. G. J. Moseley: The Life and Letters of an English Physicist, 1887–1915* (Berkeley, Los Angeles, and London: University of California Press, 1974), 79, 116ff.

40. Walter Moore, *Schrödinger: Life and Thought* (Cambridge: Cambridge University Press, 1989), 93.

41. Stefan Meyer to Meitner, 24 September 1914 (MC).

42. Meyer to Meitner, 16 October 1914 (MC). In his obituary notice for Meyer, Lawson wrote of Meyer's "innate kindliness"; had Meyer not vouched for him and supplied him with money during the war, Lawson also would have "stagnate[d] in an internment camp." R. W. Lawson, "Stefan Meyer," *Nature* 165 (1950): 549.

43. Meyer to Meitner, 2 December 1914 (MC).

44. Fritz K. Ringer, *The Decline of the German Mandarins: The German Academic Community, 1890–1933* (Cambridge: Harvard University Press, 1969), 180–183; Glum, *Zwischen Wissenschaft, Wirtschaft und Politik*, 124–125; Otto Nathan and Heinz Norden, eds., *Einstein on Peace* (New York: Avenel Books, 1981), 3–5; Heilbron, *Dilemmas of an Upright Man*, 69–81; Willstätter, *From My Life*, 241–243.

Max Born, in *My Life: Recollections of a Nobel Laureate* (New York: Charles Scribner's Sons, 1978), 162, recounts the "Asiatic hordes" propaganda and Paul Ehrenfest's response.

Stefan Meyer thought the "Appeal" very sharp; it reinforced his opinion that "as long as weapons speak, scholars and artists should keep still" (Meyer to Meitner, 8 March 1915 [MC]).

45. Meitner to Hahn, 16 December 1914 (OHN). All letters cited from this period are from the Meitner to Hahn correspondence, OHN, and are reprinted in Ernst, *Lise Meitner an Otto Hahn*. (Hahn's letters to Meitner are not extant.)

46. L. F. Haber, *The Poisonous Cloud: Chemical Warfare in the First World War* (Oxford: Clarendon Press, 1986), 22ff.; Stoltzenberg, *Fritz Haber*, 232, 580; Johnson, *The Kaiser's Chemists*, 190ff.; Willstätter, *From My Life*, 250–254. Sackur's death was treated as a battlefield death and his widow compensated accordingly; for his work with poison gases, Haber obtained the rank of captain.

47. Meitner to Hahn, 5 January 1915 (OHN).

48. Meitner to Hahn, 14 March 1915 (OHN).

49. Hahn, *My Life*, 118–128; Gerlach, *Otto Hahn*, 54–59.

50. For a different opinion, see Lawrence Badash, "Otto Hahn, Science, and Social Responsibility," in William R. Shea, ed., *Otto Hahn and the Rise of Nuclear Physics* (Dordrecht: D. Reidel, 1983), 167–180.

51. Meitner to Hahn, 5 February 1915 (OHN).

52. Meitner to Hahn, 27 February 1915 (OHN).

53. Hahn describes the birthday "cake," a log with 2 candles, in his *My Life*, p. 119; James Franck arranged the "party" at Meitner's suggestion; see Meitner to Hahn, 16 March 1915 (OHN).

54. Meitner to Hahn, 16 December 1914; 5, 9, 24, 28 January; 5, 27 February; 14 March; 25 April; 14 May; 6 June 1915 (OHN).

55. Emilio Segrè, *From X-Rays to Quarks: Modern Physicists and Their Discoveries* (San Francisco: W. H. Freeman, 1980), 41, 43; Pais, *Inward Bound*, 236.

56. Meitner to Schiemann, 4 and 5 August 1915 (cards), 9–10 August 1915 (letter) (MC).

57. Meitner to Schiemann, 13, 22, 27 August and 24–26 September 1915 (MC).

58. Meitner to Hahn, 10 September, 14 October, 28 November 1915, 9 January 1916 (OHN).

59. Dr. Prof. Hatjidakis, Piraeus, to Meitner, [?] 1915 (MC).

60. Meitner to Schiemann, 2 December 1915 (MC).

61. Meitner to Schiemann, 15 March, 1 May, 22 May, 4 June 1916 (MC).

62. Meitner to Schiemann, 4 June, 26 June, 1 July 1916 (MC).

63. Meitner to Schiemann, 27 August 1916 (MC).

64. Meitner to Schiemann, 26 September 1916 (MC). See also Ernst, *Lise Meitner an Otto Hahn*, 133–136.

65. Meitner to Schiemann, 11 October 1916 (MC); Meitner to Hahn, 11 October 1916 (OHN). In "Looking Back," *Bull. Atom. Sci.* 20 (November 1964): 2–7, Meitner incorrectly gives the year of her return as 1917, and the mistake has been repeated elsewhere.

66. Meitner to Hahn, 25 October 1916 (OHN). Johnson, *The Kaiser's Chemists*, 191, notes that by 1917 Haber had 1,500 people on his staff, including 150 scientific workers.

67. Meitner to Hahn, 16 November 1916 (OHN).

68. Nathan and Norden, *Einstein on Peace*, 15.

69. For pay, see Marion Kazemi, MPG personnel files, Berlin-Dahlem. Fischer managed the pay increase through funding from the I.G. Farbenindustrie. See Fritz Krafft, "An der Schwelle zum Atomzeitalter: Die Vorgeschichte der Entdeckung der Kernspaltung im Dezember 1938," *Ber. Wissenschaftsgesch.* 11 (1988): 227–251, on 233; also Krafft, "Lise Meitner 7. XI. 1878–27 X. 1968," in Willi Schmidt and Christoph J. Scriba, eds., *Frauen in den exakten Naturwissenschaften: Festkolloquium zum 100. Geburtstag von Frau Dr. Margarethe Schimank (1890–1983)* (Stuttgart: Franz Steiner Verlag, 1990), 33–70, on 39. For Fischer's attitude toward the end of the war, see Johnson, *The Kaiser's Chemists*, 194–196.

70. Meitner to Hahn, 22 February 1917 (OHN).

71. Hahn and Meitner, "Die Muttersubstanz des Actiniums," 211.

72. Meitner to Hahn, 7 May 1917 (OHN).

73. Meitner to Hahn, 15 May 1917 (OHN).

74. Ibid.

75. Meitner to Hahn, 19 June 1917 (OHN).

76. Meitner to Hahn, 7 May 1917 (OHN); Meyer to Meitner, card, 22 June 1917 (MC).

77. Meitner to Hahn, 27 July–6 August 1917 (OHN).

78. Meitner to Hahn, 24 August 1917 (OHN).

79. Meitner to Hahn, 16 November 1917 (OHN).

80. Ibid.

81. This is the Geiger-Nuttall rule, a relationship between λ, the decay constant for a given species (inversely proportional to half-life), and R, the mea-

sured range of its alpha particles. *A* and *B* are empirically determined constants for a given decay series; the half-life of an unknown member of the series can thus be estimated from a measurement of its range.

82. Current value for the protactinium half-life is 33,000 years.

83. Meitner to Hahn, 17 January 1918 (OHN).

84. Hahn and Meitner, "Die Muttersubstanz des Actiniums."

85. Meitner to Hahn, 17 January 1918 (OHN). The article, submitted 25 March 1918, was published by Meitner alone; it represented nearly 10 years of measurements on relatively long-lived species: Lise Meitner, "Die Lebensdauer von Radiothor, Mesothor und Thorium," *Phys. Z.* 19 (1918): 257–263.

The incident is cited as an example of Meitner's loyalty and teamwork by Krafft, "Lise Meitner," 38–39. Krafft cites another study, also done by Meitner alone, which Hahn co-authored in 1909, simply because the laboratory was "his"; he was reprimanded by [Otto?] von Baeyer and admitted he was wrong. Meitner believed it was not "bad intentions" on Hahn's part, only "thoughtlessness." The latter incident is cited without references in Charlotte Kerner, *Lise, Atomphysikerin: Die Lebensgeschichte der Lise Meitner* (Weinheim: Beltz, 1986), 31.

86. Meyer to Meitner, 23 and 26 March 1918 (MC).

87. Meyer to Meitner, 5 June 1918 (MC).

88. Meyer to Meitner, card, 14 November 1914 (MC).

89. Meyer to Meitner, 23 November 1917 (MC). For the controversy between Fajans and others, in particular, George de Hevesy and Fritz Paneth, see Ernst, *Lise Meitner an Otto Hahn*, 145–147.

90. Hahn, *A Scientific Autobiography*, 92–93; Badash, *Radioactivity in America*, 208; F. Soddy and J. Cranston, "The Parent of Actinium," *Proc. Roy. Soc. Lond.* A94 (1918): 384–405. See also R. L. Sime, "The Discovery of Protactinium," *J. Chem. Educ.* 63 (1986): 653–657.

91. In every known case of branching, a single species underwent alpha and beta decay to form two different daughters, which would then decay by beta and alpha emission, respectively, to a single granddaughter. For example, ^{212}Bi emits beta radiation to form ^{212}Po and also alpha to form ^{208}Tl; ^{212}Po then undergoes alpha decay and ^{208}Tl undergoes beta decay to form ^{208}Pb.

92. Meitner to Hahn, 23 June 1918 (OHN).

93. Meitner, "Lebensdauer von Radiothor."

94. Meitner to Hahn, 23 June 1918 (OHN).

95. Meyer to Meitner, 30 June 1918 (MC).

96. Meyer to Meitner, 29 November 1918 (MC).

97. Meitner to Schiemann, 12 November 1919 (MC): for Meitner it was "very painful" when the movement for German-Austrian union failed. For Meitner-Schiemann correspondence from this period, see Ernst, *Lise Meitner an Otto Hahn*, 151–156.

98. Meitner to Schiemann, 29 November 1918 (MC). In full in Ernst, *Lise Meitner an Otto Hahn*, 153–155.

99. Meitner to Schiemann, 29 November 1918 (MC). Meitner notes that her difference of opinion with Hahn did not damage their friendship.

100. Meitner to Schiemann, 15 October 1918 (MC); quoted in Ernst, *Lise Meitner an Otto Hahn*, 130, 144. In 1918, Einstein wrote to Meitner, "Although in our eyes the result [proving the light quantum hypothesis] cannot be in doubt, nearly all other physicists think differently about it."

101. The "our Marie Curie" quote comes from Philipp Frank, *Einstein: His Life and Times* (London: Jonathan Cape, 1948), 139. According to Frank, in private Einstein would sometimes express the opinion that Meitner "was a more talented physicist than Marie Curie herself." Krafft, *Frauen in den exakten Naturwissenschaften*, 33, and Ernst, *Lise Meitner an Otto Hahn*, 130, wonder if Einstein's use of "our" meant "German" or "Jewish." Given Einstein's orientation and Meitner's nonaffiliation with Judaism, it is inconceivable that Einstein would have meant "Jewish"—and he may not have meant "German" either, certainly not in a heavily nationalistic way; more likely his use of "our" was a friendly acknowledgment that they both inhabited the same scientific community.

102. Max Born, *My Life*, 184-185.

103. Ibid., 187-188.

CHAPTER 4. Professor in the Kaiser-Wilhelm-Institut

1. Stefan Meyer to Meitner, 16 October 1919 (MC).

2. Letter from Einstein to Paul Ehrenfest, 22 March 1919; cited in Otto Nathan and Heinz Norden, eds., *Einstein on Peace* (New York: Avenel Books, 1960), 29; Ronald W. Clark, *Einstein: The Life and Times* (New York: Avon, 1971), 272.

3. Peter Gay, *Weimar Culture: The Outsider as Insider* (New York: Harper & Row, 1968).

4. Otto Hahn, *My Life*, trans. Ernst Kaiser and Eithne Wilkins (New York: Herder and Herder, 1968), 133; Meitner to Eva von Bahr-Bergius, 3 April 1920 (MC).

5. Meitner to Hedwig Meitner, Easter Sunday 1920 (MC).

6. J. L. Heilbron, *The Dilemmas of an Upright Man: Max Planck as Spokesman for German Science* (Berkeley, Los Angeles, and London: University of California Press, 1986), chap. 3.

7. Meitner to Hahn, 13 August 1919; letter, 19 August 1919 (OHN). Meitner's letters to Hahn through 1924 are reprinted in Sabine Ernst, *Lise Meitner an Otto Hahn: Briefe aus den Jahren 1912 bis 1924: Edition und Kommentierung* (Stuttgart: Wissenschaftliche Verlagsgesellschaft, 1992).

8. Ministerium für Wissenschaft, Kunst und Volksbildung (Science, Art, and Education) to Lise Meitner, 31 July 1919 (MC). Meitner to Hedwig Meitner, 27 September 1919 (MC).

9. Otto Hahn and Lise Meitner, "Über das Protactinium und die Lebensdauer des Actiniums," *Phys. Z.* 20 (1919): 127–130.

10. Otto Hahn and Lise Meitner, "Der Ursprung des Actiniums," *Phys. Z.* 20 (1919): 529–533.

11. Otto Hahn and Lise Meitner, "Über das Protactinium und die Frage nach der Möglichkeit seiner Herstellung als chemisches Element," *Naturwiss.* 7 (1919): 611–612; Otto Hahn and Lise Meitner, "Über die chemischen Eigenschaften des Protactiniums," *Ber. Dt. Chem. Ges.* 52 (1919): 1812–1838.

12. Dietrich Hahn, ed., *Otto Hahn, Bergründer des Atomzeitalters: Eine Biographie in Bilden und Dokumenten* (München: List Verlag, 1979), 100.

13. Meitner to Hahn, Fiskebäckskil, Sweden, 19 August 1919 (OHN).

14. Ernst, *Lise Meitner an Otto Hahn*, 99 n. 1. The original designation of the Emil Fischer Medal was "for a German chemist . . . in the field of organic chemistry . . . especially dye- or pharmaceutical chemistry . . . for the synthesis of . . . products that have been of particular service to the German chemical industry." This was certainly far afield from Meitner, but Hahn did not fit the description either.

15. Lise Meitner, "Looking Back," *Bull. Atom. Sci.* 20 (November 1964): 2–7.

16. In "Einige Erinnerungen an das Kaiser-Wilhelm-Institut für Chemie in Berlin-Dahlem," *Naturwiss.* 41 (1954): 97–99, Meitner cites the same incorrect date as in "Looking Back" but gives more details. However, her private correspondence and the official reports of the KWI für Chemie for that period (§1136, MPG Berlin-Dahlem) verify that she returned to the institute permanently in October 1916 and worked mostly alone through the publication of protactinium in March 1918. See also chap. 3.

17. Max von Laue, "Zu Max Plancks 100. Geburtstage," *Naturwiss.* 45 (1958): 221–226. Lawrence Badash, "The Completeness of Nineteenth-Century Science," *Isis* 63 (1972): 48–58.

18. Sources: Lawrence Badash, *Radioactivity in America: Growth and Decay of a Science* (Baltimore: Johns Hopkins University Press, 1979); Henry A. Boorse and Lloyd Motz, eds., *The World of the Atom* (New York: Basic Books, 1966); Abraham Pais, *Niels Bohr's Times, in Physics, Philosophy, and Polity* (Oxford: Clarendon Press, 1991); Pais, *Inward Bound: Of Matter and Physics in the Physical World* (Oxford: Clarendon Press, 1986); Emilio Segrè, *From X-Rays to Quarks: Modern Physicists and Their Discoveries* (San Francisco: W. H. Freeman, 1980); J. L. Heilbron, "Lectures on the History of Atomic Physics 1900–1922," in C. Weiner, ed., *Proceedings of the International School of Physics "Enrico Fermi," Course LVII* (New York: Academic Press, 1977), 40–108, on 58–78; Roger H. Stuewer, "The Nuclear Electron Hypothesis," in William R. Shea, ed., *Otto Hahn and the Rise of Nuclear Physics* (Dordrecht: D. Reidel, 1983), 19–21.

19. J. L. Heilbron, *H. G. J. Moseley: The Life and Letters of an English Physicist, 1887–1915* (Berkeley, Los Angeles, and London: University of California Press, 1974), 88ff.

20. For Meitner and Hahn's early work, see chap. 2; for a brief overview of beta spectra, see Sallie A. Watkins, "Lise Meitner and the Beta-Ray Energy Controversy: An Historical Perspective," *Am. J. Phys.* 51 (1983): 551–553; for beta spectra in the context of atomic physics, see Pais, *Inward Bound,* chaps. 8, 14; for an exceedingly detailed and thorough treatment, see Carsten Jensen, "A History of the Beta Spectrum and Its Interpretation, 1911–1934," Ph.D. dissertation, Niels Bohr Institute, University of Copenhagen, 1990.

21. Meitner thought other effects, such as gamma radiation, might have obscured the fast beta lines in Chadwick's spectrum, which she described as only "apparently continuous." In October 1916, just back from her war service, she was eager to repeat Chadwick's experiment, convinced that she could disprove it. Meitner to Hahn, 25 October 1916 (OHN).

22. Otto Hahn and Lise Meitner, "Über eine typische β-Strahlung des eigentlichen Radiums," *Phys. Z.* 10 (1909): 741–745; Otto v. Baeyer, Otto Hahn, and Lise Meitner, "Magnetische Spektren der β-Strahlen des Radiums," *Phys. Z.* 12 (1911): 1099–1101; O. v. Baeyer, O. Hahn, and L. Meitner, "Das magnetische Spektrum der β-Strahlen des Radioaktiniums und seiner Zerfallsprodukte," *Phys. Z.* 14 (1913): 321–323; Otto v. Baeyer, Otto Hahn, and Lise Meitner, "Das magnetische Spekktrum der β-Strahlen von Radiothor und Thorium X," *Phys. Z.* 16 (1915): 6–7.

23. Hahn and Meitner had searched unsuccessfully for beta decay products before the war; in 1917 Meitner noted (Meitner to Hahn, 24 August 1917 [OHN]) that in their previous beta studies of radium, done in 1909 and 1911, "we were fumbling in the dark regarding the chemical properties [of the possible beta product] and above all we did not think of branching." Ra β-radiation was fairly strong, and Meitner thought branching should have been evident if present but suggested a new investigation to be sure.

24. Otto Hahn and Lise Meitner, "Über die Anwendung der Verschiebungs-regel auf gleichzeitig α- und β-Strahlen aussendende Substanzen," *Z. Phys.* 2 (1920): 60–70.

25. Lise Meitner, "Über die Entstehung der β-Strahl-Spektren radioaktiver Substanzen," *Z. Phys.* 9 (1922): 131–144, on 132.

26. Hahn and Meitner, "Anwendung der Verschiebungsregel," 69–70.

27. Meitner, "Entstehung der β-Strahl-Spektren," 131–132. In 1920 (n. 24, p. 70) Meitner was "without a satisfactory explanation" (*ohne eine befriedigende Erklärung*) for the secondary beta spectra of alpha emitters, but by 1922 she had made the connection to gamma radiation, citing Rutherford's papers of 1914 (see n. 28 below) and 1917 (see n. 30 below). Jensen, in "History of the Beta Spectrum," 106, suggests Meitner may not have had access to Rutherford's 1914–1918 papers until well after the war.

28. Sir Ernest Rutherford, "Spectrum of the β Rays Excited by the γ Rays," *Phil. Mag.* 28 (1914): 281–286; "The Connexion between the β and γ Ray Spectra," *Phil. Mag.* 28 (1914): 305–319.

29. Sir Ernest Rutherford and E. N. daC. Andrade, "The Wavelength of the Soft γ Rays from Radium B," *Phil. Mag.* 27 (1914): 854–868.

30. Sir Ernest Rutherford, "Penetrating Power of the X-Radiation from a Coolidge Tube," *Phil. Mag.* 34 (1917): 153–162.

31. Meitner to Hahn, 24 April, 4 May, 18 May 1921 (OHN).

32. Robert Marc Friedman, "Karl Manne Georg Siegbahn," in Frederic L. Holmes, ed. *Dict. Sci. Biog.*, Supplement 2 (New York: Charles Scribner, 1990), 18:821–826.

33. Dr. Ada Klokke-Coster, pers. comm., 11 February 1986; Meitner to Hahn, 11/12 May 1921 (OHN), in Ernst, *Lise Meitner an Otto Hahn*; Meitner to Dirk Coster, zum 9. Dezember 1949 (MC).

34. Meitner, "Entstehung der β-Strahl-Spektren," 133–137. The logic of the experiment is readily seen from the data.

The two electron lines had energy:

$$E_{\beta_1} = 3.560 \times 10^{-7} \text{ erg} \qquad E_{\beta_2} = 2.322 \times 10^{-7} \text{ erg}$$

From x-ray data, several ionization energies for lead were known:

$$E_K = 1.417 \times 10^{-7} \text{ erg} \qquad E_{L_1} = 0.2074 \times 10^{-7} \text{ erg}$$
$$E_{L_2} = 0.242 \times 10^{-7} \text{ erg} \qquad E_{L_3} = 0.250 \times 10^{-7} \text{ erg}$$

The energy difference between the two secondary electrons corresponded most closely to the difference between the K and L_1 ionization energies:

$$E_{\beta_1} - E_{\beta_2} = 1.238 \times 10^{-7} \text{ erg}$$
$$E_K - E_{L_1} = 1.21 \times 10^{-7} \text{ erg}$$

This meant that the β_1 electrons originated in the L_1 level, and the β_2 electrons originated in the K level. Two independent calculations for gamma energy could then be made, with agreement within 1%:

$$E_\gamma = E_{L_1} + E_{\beta_1} = 3.767 \times 10^{-7} \text{ erg}$$
$$E_\gamma = E_K + E_{\beta_2} = 3.739 \times 10^{-7} \text{ erg}$$

35. Meitner, "Entstehung der β-Strahl-Spektren," 132, 143; Lise Meitner, "Über die Zusammenhang zwischen β- und γ-Strahlen," *Z. Phys.* 9 (1922): 145–152.

36. Meitner, "Entstehung der β-Strahl-Spektren," 132–133.

37. Ibid., 140–141; J. Danysz, "Sur les rayons β des radiums BCDE," *Le Radium* 10 (1913): 4–6; Jensen, "History of the Beta Spectrum," 59–61.

38. Meitner, "Entstehung der β-Strahl-Spektren," 143.

39. C. D. Ellis, "The Magnetic Spectrum of the β-Rays Excited by γ-Rays," *Proc. Roy. Soc. Lond.* 99 (1921): 261–271.

40. C. D. Ellis, "β-Ray Spectra and Their Meaning," *Proc. Roy. Soc. Lond.* A101 (1922): 1–17, on 1: "the line β-ray spectrum is, by itself, of considerable interest, but of far greater importance is the fact that these experiments give a method of finding the wave-lengths of γ-rays." For Ellis and β-spectra, see Pais,

Inward Bound, 303–309; and Sir Kenneth Hutchison, J. A. Gray, and Sir Harrie Massey, "Charles Drummond Ellis, 11 August 1895–10 January 1980," *Biog. Mem. Fell. Roy. Soc. Lond.* 27 (1981): 199–231.

41. Meitner, "Zusammenhang zwischen β- und γ-Strahlen."

42. Rutherford and Andrade, "Soft γ Rays from Radium B."

43. Ellis, "β-Ray Spectra," 13–16.

44. J. Chadwick, "Intensitätsverteilung im magnetischen Spektren der β-Strahlen von Radium B + C," *Verh. Dt. Phys. Ges.* 16 (1914): 383–391.

45. Ellis, "β-Ray Spectra."

46. C. D. Ellis, "The Interpretation of β-Ray and γ-Ray Spectra," *Proc. Camb. Phil. Soc.* 21 (1922): 121–128, on 122.

47. C. D. Ellis, "Über die Deutung der β-Strahlspektren radioaktiver Substanzen," *Z. Phys.* 10 (1922): 303–307.

48. Ibid., 307.

49. Ellis, "Interpretation of β-Ray and γ-Ray Spectra," 128.

50. Lise Meitner, "Über die β-Strahl-Spektra und ihren Zusammenhang mit der γ-Strahlung," *Z. Phys.* 11 (1922): 35–54.

51. Ibid., 50.

52. At the time, the nonlinearity of exposure and development times on photographic emulsions was also not understood. See Pais, *Inward Bound,* 157–158.

53. Meitner, "β-Strahl-Spektra und ihren Zusammenhang mit der γ-Strahlung," 52.

54. Meitner to Hahn, 17 April 1922 (OHN).

55. For a detailed discussion, see Jensen, "History of the Beta Spectrum," chap. 3.

56. James Chadwick and C. D. Ellis, "A Preliminary Investigation of the Intensity Distribution in the β-Ray Spectra of Radium B and C," *Proc. Camb. Phil. Soc.* 21 (1922): 274–280.

57. Lise Meitner, "Das β-Strahlenspektrum von UX_1 und seine Deutung," *Z. Phys.* 17 (1923): 54–66. K_α x-rays are those emitted when an electron drops from an L shell to a vacancy in a K shell.

58. Meitner to Walter Meitner, 4 May 1923 (MC).

59. Otto Hahn and Lise Meitner, "Die γ-Strahlen von Uran X und ihre Zuordnung zu Uran X_1 und Uran X_2," *Z. Phys.* 17 (1923): 157–167.

60. Meitner, "β-Strahlenspektrum von UX_1," 62.

61. See Jensen, "History of the Beta Spectrum," chaps. 3 and 4, for the theoretical contributions of Svein Rosseland and Adolf Smekal.

62. In "β-Strahlenspektrum von UX_1," pp. 61 and 64, Meitner described the processes now known as "internal conversion" (the decay electron directly ejecting a K electron) and the radiationless transition now known as the "Auger effect." See Richard Sietmann, "False Attribution: A Female Physicist's Fate," *Phys. Bull.* 39 (1988): 316–317.

63. C. D. Ellis and H. W. B. Skinner, "The Interpretation of β-Ray Spectra," *Proc. Roy. Soc. Lond.* A105 (1924): 185–198, on 196–197. It turned out that Ellis was correct: the soft gamma radiation of UX_1 is of nuclear origin.

64. Meitner, "β-Strahlenspektrum von UX_1."

65. Ellis and Skinner, "Interpretation of β-Ray Spectra," 197.

66. Meitner, "β-Strahl-Spektra und ihren Zusammenhang mit der γ-Strahlung," 51–52; "β-Strahlenspektrum von UX_1," 62–63.

67. Roger H. Stuewer, *The Compton Effect: Turning Point in Physics* (New York: Science History Publications, 1975), chap. 6.

68. See chap. 3, page 74; also Ernst, *Lise Meitner an Otto Hahn*, 130. In 1918 Einstein wrote to Meitner, "Although in our eyes the result [proving the light quantum hypothesis] cannot be in doubt, nearly all other physicists think differently about it."

69. Lise Meitner, "Über eine mögliche Deutung des kontinuierlichen β-Strahlenspektrums," *Z. Phys.* 19 (1923): 307–312, on 311.

70. Lise Meitner, "Über eine notwendige Folgerung aus dem Comptoneffekt und ihre Bestätigung," *Z. Phys.* 22 (1924): 334–342, on 341–342. For an overview of Meitner's position in 1924, see Lise Meitner, "Der Zusammenhang zwischen β- und γ-Strahlen," *Ergebn. Exakt. Naturwiss.* 3 (1924): 160–181; also Meitner, "Über die Energieentwicklung bei radioaktiven Zerfallsprozessen," *Naturwiss.* 12 (1924): 1146–1150.

71. Otto Hahn and Lise Meitner, "Das β-Strahlenspektrum von Radium und seine Deutung," *Z. Phys.* 26 (1924): 161–168.

72. Lise Meitner, "Über die Rolle der γ-Strahlen beim Atomzerfall," *Z. Phys.* 26 (1924): 169–177, on 169.

73. Meitner, "Rolle der γ-Strahlen," 169–170. Here (p. 170) she conceded, "This interpretation differs in principle from the view I expressed in my first publication [see n. 25, above], in which the energy of a γ-ray was closely linked to the energy of the primary beta particle. Such a link is entirely unnecessary and apparently does not exist."

74. C. D. Ellis and H. W. B. Skinner, "The Absolute Energies of the Groups in Magnetic β-Ray Spectra," *Proc. Roy. Soc. Lond.* A105 (1924): 60–69.

75. Ellis and Skinner, "Interpretation of β-Ray Spectra," 192.

76. Meitner, "Rolle der γ-Strahlen," 172.

77. Otto Hahn and Lise Meitner, "Die β-Strahlenspektren von Radioactinium und seinen Zerfallsprodukten," *Z. Phys.* 34 (1925): 795–806.

78. Lise Meitner, "Die γ-Strahlung der Actiniumreihe und der Nachweis, dass die γ-Strahlen erst nach erfolgtem Atomzerfall emittiert werden," *Z. Phys.* 34 (1925): 807–818.

79. C. D. Ellis to Meitner, 8 December 1925 (MC).

80. C. D. Ellis and W. A. Wooster, "The Atomic Number of a Radioactive Element at the Moment of Emission of the γ-Rays," *Proc. Camb. Phil. Soc.* 22

(1925): 844–848; and Ellis and Wooster, "The β-Ray Type of Disintegration," ibid., 849–860. See also Pais, *Inward Bound*, 307–309.

81. C. D. Ellis to Meitner, 8 December 1925 (MC).

82. Meitner, "γ-Strahlung der Actiniumreihe," 816.

83. Dietrich Hahn, ed., *Otto Hahn, Erlebnisse und Erkenntnisse* (Düsseldorf: Econ Verlag, 1975), 43. Hahn wrote in 1945, "One can certainly say that in the years after about 1920, a large part of the institute's reputation, especially abroad, was due to the work in the *Meitner* section."

84. Ibid. Otto Hahn quotes Kasimir Fajans in the mid-1920s: "Hahn, there's no point in working in radioactivity any more; there's nothing worthwhile to get out of it anymore."

Badash, *Radioactivity in America*, 213, similarly refers to radiochemistry's "suicidal success"; by the early 1920s, the field "effectively ceased to exist."

85. Meitner to Hahn, 12 and 17 April 1922 (OHN).

86. As late as August 1921, Meitner and Hahn were still using the formal "Sie." "Fachbruder" appears in a letter from Meitner to Hahn, 23 October 1922 (OHN). For the nickname "Hanno," see D. Hahn, *Otto Hahn*, 190.

87. For the lives and loves of Annemarie and Erwin Schrödinger, see Walter Moore, *Schrödinger: Life and Thought* (Cambridge: Cambridge University Press, 1989).

88. Clark, *Einstein*, 393; Otto Frisch, *What Little I Remember* (Cambridge: Cambridge University Press, 1979), 34–39.

89. For Einstein, see Pais, *Niels Bohr*, 227ff.; for Hevesy, see Hilde Levi, *George de Hevesy: Life and Work* (Bristol and Boston: Adam Hilger, 1985), 47.

90. Wilhelm Westphal, in K. E. Boeters and J. Lemmerich, eds., *Gedächtnisausstellung zum 100. Geburtstag von Albert Einstein, Otto Hahn, Max von Laue, Lise Meitner* (Bad Honnef: Physik Kongreß-Ausstellungs- und Verwaltungs, 1979), 105.

91. Meitner, "Looking Back," 7.

92. Ibid.; Finn Aaserud, *Redirecting Science: Niels Bohr, Philanthropy, and the Rise of Nuclear Physics* (New York: Cambridge University Press, 1990), 102; Meitner to Margrethe Bohr, 31 May 1921 (BPC).

93. Meitner to Hedwig Meitner, 19 October 1922 (MC).

94. Meitner to Hedwig Meitner, n.d. 1921 (MC). In 1920, for example, Meitner was renting a single room in the apartment of a family with 5 children in Lichterfelde: Otfried Bronisch to Meitner, 18 September 1955 (MC).

95. Clark, *Einstein*, 381.

96. For a general description of the effect of inflation on the KWG, see Friedrich Glum, *Zwischen Wissenschaft, Wirtschaft und Politik* (Bonn: H. Bouvier Verlag, 1964), 293ff.

For details, see annual reports (*Bericht des Direktors über den Betrieb u. die wissenschaftliche Tätigkeit*) of KWI für Chemie to the KWG Generalverwaltung (MPG). With inflation Meitner's salary rose from 4,000 marks in 1919 to 9,800 in 1920 to 78,200 in March 1922. In November 1922, it was noted that inflation had made "frightening progress" (*erschreckende Fortschritte*), prices of materials increasing by factors of 17 (for gas) and 95 (for brass) in 1922 alone. In March 1923, the institute's total budget of 222 million marks, although 100 times greater than the previous year's, was considered only half its prewar value. See Fritz K. Ringer, ed., *The German Inflation of 1923* (New York: Oxford University Press, 1969); includes Georg Schreiber, "The Distress of German Learning." See also n. 104, below.

97. Meitner to Walter Meitner, 12 March 1923 (MC).

98. Meitner to Walter Meitner, 4 May 1923 (MC).

99. Meitner, "Über eine mögliche Deutung des kontinuierlichen β-Strahlenspektrums"; Meitner, "Über eine notwendige Folgerung aus dem Compton-effekt."

100. Meitner to Walter Meitner, 26 August 1923 (MC).

101. Meitner to Hedwig Meitner, 19 September 1923 (MC).

102. Meitner to Hedwig Meitner, 3 October 1923 (MC).

103. Glum, *Zwischen Wissenschaft*, 293.

104. Meitner, "Einige Erinnerungen," 98. In November 1923 Meitner's weekly salary was 2 billion (10^9) marks on 7 November, 12 billion marks on 16 November, 116 billion marks on 17 December, with retroactive pay added on each week for loss of value of the previous salary. In November the mark went from 10^9/\$1 to 10^{12}/\$1. That month a new currency, the Rentenmark, was introduced and the process of stabilization began (Meitner, budget 1923 [MC]). For economic data, see Carl-Ludwig Holfrerich, *The German Inflation, 1914–1923: Causes and Effects in International Perspective*, trans. Theo Balderston (Berlin: Walter de Gruyter, 1986).

105. Konrad Heiden, "The Inflation and Hitler's Putsch of 1923," in Ringer, *German Inflation of 1923*.

106. Meitner to Dirk Coster, zum 9. Dezember 1949 (MC).

107. Meitner to Max von Laue, 26 December 1923 (MC).

108. A. D. Fokker to Meitner, 10 November 1923 (MC).

109. Meitner to Coster, zum 9. Dezember 1949 (MC).

110. Meitner to Hahn, 12 December 1924 (OHN).

111. Frisch, *What Little I Remember*, 13–14, 33–34. For music in Berlin in the late 1920s, see Gay, *Weimar Culture*, 130–131.

112. Heilbron, "Lectures," 90–95. It took several years for Coster and Hevesy to establish undisputed priority for element 72: Helge Kragh, "Anatomy of a Priority Conflict: The Case of Element 72," *Centaurus* 23 (1980): 275–301; Levi, *George de Hevesy*, 51–56.

113. Barbara Lovett Cline, *Men Who Made a New Physics* (Chicago: University of Chicago Press, 1965, 1987), 172. See also Victor Weisskopf, *The Joy of Insight: Passions of a Physicist* (New York: Basic Books, 1991).

114. For general references, see n. 18, above. Also see Oskar Klein, "Glimpses of Niels Bohr as a Scientist and Thinker," in S. Rozental, ed., *Niels Bohr: His Life*

and Work as Seen by His Friends and Colleagues (Amsterdam: North Holland/New York: John Wiley & Sons, 1967); George E. Uhlenbeck, "Personal Reminiscences: Fifty Years of Spin," *Physics Today* 29, no. 6 (1976): 43–48; Cline, *Men Who Made a New Physics,* chaps. 10, 11; Max Jammer, *The Conceptual Development of Quantum Mechanics,* 2d ed., vol. 12 in *History of Modern Physics, 1800–1950* (n.p.: Tomash Publishers, American Institute of Physics, 1989).

115. Joan Bromberg, "The Impact of the Neutron: Bohr and Heisenberg," *Hist. Stud. Phys. Sci.* 3 (1971): 308–309: "the need to put electrons in the nucleus was the chief reason for connecting nuclear physics with relativistic quantum theory."

116. Lise Meitner, "Experimentelle Bestimmung der Reichweite homogener β-Strahlen," *Naturwiss.* 14 (1926): 1199–1203.

117. Ellis and Wooster, "The β-Ray Type of Disintegration"; K. G. Emeleus, "The Number of β-Particles from Radium E," *Proc. Camb. Phil. Soc.* 2 (1924): 400–405. For the resolution of the controversy in this period, see also Jensen, "History of the Beta Spectrum," chap. 5.

118. As related by Meitner to Pauli, 14 October 1958 (MC): "Even before my calorimetry measurement with Orthmann (1929) I was quite uncertain about my hypothesis regarding primary β-rays, and assigned two co-workers to see if recoil nuclei (after β decay) showed an inhomogeneous energy distribution."

By then, other physicists agreed with Ellis. Writing to Meitner in 1956, Pauli remembered the "old polemic between yourself and Ellis which was finally resolved with the calorimetry experiment of yours and Orthmann's. And I remember too, that once, trying to be diplomatic, I greeted you with the words, 'It really is very nice to see you here.' That was far beyond the limited boundaries of my usual pleasantries and so you reacted quite logically, 'You don't usually pay such compliments; there must be something behind it!' That was so, since I was of the opinion Ellis was right." (Pauli to Meitner, 3 November 1956 [MC].)

Jensen, in "History of the Beta Spectrum," 237–238, notes that Pauli subsequently (and temporarily) changed his opinion, probably in early 1929, in the hope of retaining conservation of energy.

119. N. Riehl, "Die Brauchbarkeit des Geigerschen Spitzenzählers für β-Strahlen verschiedener Geschwindigkeiten und die Zahl der β-Strahlen von RaE und RaD," *Z. Phys.* 46 (1927): 478–505; K. Donat and K. Philipp, "Die Ausbeute beim β-Rückstoss von Thorium B," *Z. Phys.* 45 (1927): 512–521, and *Naturwiss.* 16 (1928): 513.

120. C. D. Ellis and W. A. Wooster, "The Average Energy of Disintegration of Radium E," *Proc. Roy. Soc. Lond.* A117 (1927): 109–123.

121. O. R. Frisch, "Lise Meitner, 1878–1968," *Biog. Mem. Fell. Roy. Soc. Lond.* 16 (1970): 408, says that Meitner was "shocked" by the result, but her beta recoil attempt (see n. 116) and her statement to Pauli in 1958 (see n. 118) makes it seem unlikely that she was very surprised. Jensen, "History of the Beta Spectrum," 230–236, 243–244, contrasts the unconcern shown by Ellis and the Rutherford

group with the dismay felt by physicists on the Continent and attributes it to a genuine difference between the two scientific communities: "Without hesitation [Ellis] was ready to abandon the fundamental accuracy of the quantum postulate," whereas Meitner "fully realized the serious consequences" of the continuous spectrum.

122. Meitner to C. D. Ellis, 14 February 1928 (MC). The Cambridge conference took place on 23–27 July 1928. See Jensen, "History of the Beta Spectrum," 235; also Kaiser-Wilhelm-Institut für Chemie 1928 budget report, §1150 KWG Generalverwaltung, MPG, Berlin-Dahlem.

123. Lise Meitner and Wilhelm Orthmann, "Über eine absolute Bestimmung der Energie der primären β-Strahlen von Radium E," *Z. Phys.* 60 (1930): 143–155.

124. Meitner to C. D. Ellis, 20 July 1929 (MC).

125. Roger H. Stuewer, "Gamow's Theory of Alpha-Decay," in Edna Ullmann-Margalit, ed., *The Kaleidoscope of Science* (Dordrecht: D. Reidel, 1986), 147–186.

126. Meitner and Orthmann, "Bestimmung der Energie," 153.

127. Ibid., 154.

128. Meitner to C. D. Ellis, 14 February 1928 (MC).

129. Stuewer, "Nuclear Electron Hypothesis," 32–42.

130. Ibid.; Laurie M. Brown, "The Idea of the Neutrino," *Physics Today* 31, no. 9 (September 1978): 23–28.

131. Wolfgang Pauli to Meitner and Hans Geiger, 4 December 1930 (MC); also in Brown, "Idea of the Neutrino," and in R. Kronig and V. F. Weisskopf, eds., "Zur alteren und neueren Geschichte des Neutrino," *W. Pauli Collected Scientific Papers* (New York: Interscience, 1964), 1313–1337, on 1316–1317.

132. As stated by Meitner to W. Pauli, 14 October 1958 (MC). Also in Kronig and Weisskopf, *Pauli Scientific Papers*, 1317.

133. George Gamow, *Constitution of Atomic Nuclei and Radioactivity* (Oxford: Oxford University Press, 1931); Pais, *Inward Bound*, 297.

CHAPTER 5. Experimental Nuclear Physics

1. Haushaltsplan (budget) in KWG Generalverwaltung records, MPG, Berlin-Dahlem; §1148 to §1151 (1926–1933).

2. Fritz Krafft, *Im Schatten der Sensation: Leben und Wirken von Fritz Strassmann* (Weinheim: Verlag Chemie, 1981), 169; Elisabeth Boedeker, *Marksteine der Deutschen Frauenbewegung* (Hannover: Selbstverlag, 1969); Elisabeth Boedeker and Maria Meyer-Plath, *50 Jahre Habilitation von Frauen in Deutschland: Eine Dokumentation über den Zeitraum von 1920–1970* (Göttingen: O. Schwarz, 1974). The first woman Privatdozent in physics was Hedwig Kohn in Breslau in 1919.

3. Armin Herrmann, *Die Neue Physik* (Müchen: Heinz Moos, 1979), 64; Herrmann, *The New Physics: The Route into the Atomic Age. In Memory of Albert Einstein, Max von Laue, Otto Hahn, Lise Meitner*, trans. David C. Cassidy (Bad

Godesberg: Inter Nationes, 1979). The documents pertaining to Meitner's Habilitation at the Friedrich-Wilhelm-Universität are given in Fritz Krafft, "Lise Meitner (7.XI.1878–27.X.1968)," in Willi Schmidt and Christoph J. Scriba, eds., *Frauen in den exakten Naturwissenschaften* (Stuttgart: Franz Steiner Verlag, 1990), 32–70, Appendix, 57ff.

4. Philosophische Fakultät der Friedrich-Wilhelms-Universität to Lise Meitner, 7 August 1922 (MC).

5. Akademische Verlagsgesellschaft in Leipzig to Meitner, 24 October 1922 (MC); quoted in Lise Meitner, "The Status of Women in the Professions," *Physics Today* 13, no. 8 (August 1960): 16–21.

6. Krafft, *Im Schatten der Sensation*, 170.

7. Announcement, Preussische Minister für Wissenschaft, Kunst und Volksbildung (Berlin) to Meitner, 1 March 1926 (MC). "Appointment of the former *Privatdozentin* Professor Dr. Lise Meitner to *nichtbeamteten außerordentlichen Professor* in the philosophy faculty of the University of Berlin."

8. O. R. Frisch, "Lise Meitner, 1878–1968," *Biog. Mem. Fell. Roy. Soc.* 16 (1970): 405–420; Prussian Academy of Sciences (Berlin) to Lise Meitner, 19 June 1924 (MC); Austrian Academy of Sciences (Vienna) to Lise Meitner, 28 May 1925 (MC); American Association to Aid Scientific Research by Women, Ellen Richards Prize, 1928 (MC).

9. Roger H. Stuewer, "The Nuclear Electron Hypothesis," in William R. Shea, ed., *Otto Hahn and the Rise of Nuclear Physics* (Dordrecht: D. Reidel, 1983), 42.

10. Sir Ernest Rutherford, James Chadwick, and C. D. Ellis, *Radiations from Radioactive Substances* (Cambridge: Cambridge University Press, 1930; reissued 1951).

11. Roger H. Stuewer, "Rutherford's Satellite Model of the Nucleus," *Hist. Stud. Phys. Sci.* 16 (1986): 321–352.

12. Sir Harrie Massey and N. Feather, "James Chadwick, 1891–1974," *Biog. Mem. Fell. Roy. Soc. Lond.* 22 (1976): 11–70, on 51; also Rutherford, Chadwick, and Ellis, *Radiations*, 202.

13. Henry A. Boorse and Lloyd Motz, eds., *The World of the Atom* (New York: Basic Books, 1966), 815; E. N. daC. Andrade, *Rutherford and the Nature of the Atom* (Garden City, N.Y.: Doubleday, 1964), 164ff.

14. Lise Meitner, "Über die verschiedenen Arten des radioaktiven Zerfalls und die Möglichkeit ihrer Deutung aus der Kernstruktur," *Z. Phys.* 4 (1921): 146–156; Lise Meitner, "Radioaktivität und Atomkonstitution," *Festschrift, Kaiser-Wilhelm-Gesellschaft 10-jährigen Jubiläum* (Berlin: Julius Springer, 1921), 154–161.

15. Rutherford, Chadwick, and Ellis, *Radiations*, 532–533; Stuewer, "Nuclear Electron Hypothesis," 24–29; Andrade, *Rutherford*, 168–170.

16. Rutherford, Chadwick, and Ellis, *Radiations*, 304–312.

17. Lise Meitner, "Neue Arbeiten über die Streuung der α-Strahlen und den Aufbau der Atomkerne," *Naturwiss.* 14 (1926): 863–869.

18. Rutherford, Chadwick, and Ellis, *Radiations*, 265ff.

19. Frisch, "Lise Meitner, 1878-1968," 409.

20. Lise Meitner and Kurt Freitag, "Photographischer Nachweis von α-Strahlen langer Reichweite nach der Wilsonchen Nebelmethode," *Naturwiss.* 12 (1924): 634-635; Lise Meitner and Kurt Freitag, "Über die α-Strahlen des ThC + C' und ihr Verhalten beim Durchgang durch verschiedene Gase," *Z. Phys.* 37 (1926): 481-517; and M. v. Laue and L. Meitner, "Die Berechnung der Reichweitestreuung aus Wilson-Aufnahmen," *Z. Phys.* 41 (1927): 397-406; also Rutherford, Chadwick, and Ellis, *Radiations*, 134ff.

21. Lise Meitner, "Über geeignete Dampf-Gasgemische für verschiedene Versuche nach der Wilsonschen Nebelmethode," *Z. Phys. Chem.* 139 (1928): 717-721.

22. Rutherford, Chadwick, and Ellis, *Radiations*, 87-88.

23. Lise Meitner, "Über die Zusammenhang zwischen β- und γ-Strahlen," *Z. Phys.* 9 (1922): 145-152, on 152.

24. Lise Meitner and Kurt Philipp, "Das γ-Spektrum von ThC″ und die Gamowsche Theorie der α-Feinstruktur," *Naturwiss.* 19 (1931): 1007; Lise Meitner and Kurt Philipp, "Die γ-Strahlen von ThC und ThC″ und die Feinstruktur der α-Strahlen," *Z. Phys.* 80 (1933): 277-284. Since gamma emission follows alpha decay, the gamma emission associated with the alpha decay of ThC → ThC″ originates in the excited ThC″ nucleus (prior to its own beta decay); Meitner assumed (p. 279 above) that excited ThC″ emits gamma radiation *before* recoil so that the gamma radiation associated with the ThC → ThC″ process would not appear in the gamma spectrum of pure ThC″. See also Carsten Jensen, "A History of the Beta Spectrum and Its Interpretation, 1911-1934," Ph.D. dissertation, Niels Bohr Institute, University of Copenhagen, 1990, 335-348.

25. Lise Meitner, "Einige Bemerkungen zur Isotopie der Elemente," *Naturwiss.* 14 (1926): 719-720. Meitner noted that the suggestion of a third uranium isotope had been made by A. Piccard in 1917 and by A. S. Russell in 1923, because they thought it unlikely that a single isotope, ^{234}U, would branch to form two long, independent series.

26. Rutherford, Chadwick, and Ellis, in *Radiations*, 24, 32-33, discuss ^{235}U with some skepticism but nevertheless list the atomic weight of Pa as 231. See also Lawrence Badash, *Radioactivity in America: Growth and Decay of a Science* (Baltimore: Johns Hopkins University Press, 1979), 209.

27. Lise Meitner, "Das γ-Strahlenspektrum des Protactiniums und die Energie der γ-Strahlen bei α- und β-Strahlenumwandlungen," *Z. Phys.* 50 (1928): 15-23.

28. Lise Meitner, "Das β-Strahlenspektrum des Radiothors als Absorptionsspektrum seiner γ-Strahlen," *Z. Phys.* 52 (1928): 637-644; and Lise Meitner, "Das γ-Strahlenspektrum des Radiothors in Emission," *Z. Phys.* 52 (1928): 645-649.

29. Lise Meitner, "Das β-Strahlenspektrum von UX_1 und seine Deutung," *Z. Phys.* 17 (1923): 54-66.

30. There is no question that Meitner was aware of Smekal's work, having had substantial correspondence with him in 1922; his approach was somewhat dogmatic and few experimentalists took him seriously at the time. See Jensen, "History of the Beta Spectrum," 13off., 349.

31. Stuewer, "Nuclear Electron Hypothesis," 41-42; Rutherford, Chadwick, and Ellis, *Radiations*, 509ff.; Jensen, "History of the Beta Spectrum," 272-273, 35off.

32. Meitner described these experiments, which were carried out by her student Gerhard Schmidt, in Lise Meitner, "Über die Ionisierungswahrscheinlichkeit innerer Niveaus durch schnelle Korpuskularstrahlen und eine Methode zu ihrem Nachweis," *Naturwiss.* 19 (1931): 497-499; Schmidt's paper is "Über die Ionisierungsprozesse der α-Strahlen und ihre Messung im Millikan-Kondensator," *Z. Phys.* 72 (1931): 275-292.

33. Boorse and Motz, *World of the Atom*, 817-824; Rutherford, Chadwick, and Ellis, *Radiations*, 519-521.

34. George de Hevesy showed in 1935 that the natural radioactivity of potassium is due to ^{40}K. See Hilde Levi, *George de Hevesy: Life and Work* (Bristol: Adam Hilger, 1985), 65, 78.

35. Meitner, "Isotopie der Elemente"; Rutherford, Chadwick, and Ellis, *Radiations*, 524-527, 543; Aaron J. Ihde, *The Development of Modern Chemistry* (New York: Harper & Row, 1964), 528-529.

Heisenberg considered the question of nuclear stability in successive beta decays as a major part of his first theoretical paper following the discovery of the neutron: Joan Bromberg, "The Impact of the Neutron: Bohr and Heisenberg," *Hist. Stud. Phys. Sci.* 3 (1971): 307-341, on 336.

36. Karen E. Johnson, "Independent-Particle Models of the Nucleus in the 1930s," *Amer. J. Phys.* 60 (1992): 164-172.

37. Otto Hahn and Lise Meitner, "Zur Enstehungsgeschichte der Bleiarten," *Naturwiss.* 21 (1933): 237-238.

38. Lise Meitner, "Über die Wechselbeziehung zwischen Masse und Energie," *Z. Ver. Dt. Ing.* 75 (1931): 977-980.

39. Robert H. Kargon, "The Evolution of Matter," in Shea, *Otto Hahn*, 69-89, on 74-75.

40. Meitner, "Wechselbeziehung zwischen Masse und Energie."

41. Andrade, *Rutherford*, 72-73, 153; Kargon, "Evolution of Matter," 74.

42. Kargon, "Evolution of Matter," 78, 82.

43. Lise Meitner, "Die Höhenstrahlung und ihre Beziehung zu physikalischen und kosmischen Vorgängen," *Z. Angew. Chem.* 42 (1929): 345-351, on 348.

44. Ibid., 348.

45. Boorse and Motz, *World of the Atom*, 1166-1198; Stuewer, "Nuclear Electron Hypothesis," 39-40; Abraham Pais, *Inward Bound: Of Matter and Forces in the Physical World* (Oxford: Clarendon Press, 1986), 312, 347-350; Bromberg, "Impact of the Neutron," 314.

46. P. A. M. Dirac, "Recollections of an Exciting Era," in C. Weiner, ed., *Proceedings of the International School of Physics "Enrico Fermi," Course LVII: History of Twentieth-Century Physics* (New York: Academic Press, 1977), 109–146, on 145; Bromberg, "Impact of the Neutron," 317; Emilio Segrè, *From X-Rays to Quarks: Modern Physicists and Their Discoveries* (San Francisco: W. H. Freeman, 1980), 172.

47. Rutherford, Chadwick, and Ellis, *Radiations*, 459–466.

48. Lise Meitner and H. H. Hupfeld, "Über das Absorptiongesetz für kurzwellige γ-Strahlung," *Z. Phys.* 67 (1931): 147–168, on 149.

49. Frisch, "Lise Meitner, 1878–1968," 410.

50. Lise Meitner and H. H. Hupfeld, "Prüfung der Streuungsformel von Klein und Nishina an kurzwelliger γ-Strahlung," *Phys. Z.* 31 (1930): 947–948.

51. L. Meitner and H. H. Hupfeld, "Über die Prüfung der Streuungsformel von Klein und Nishina an kurzwelliger γ-Strahlung, *Naturwiss.* 18 (1930): 534.

52. Meitner and Hupfeld, "Über das Absorptionsgesetz."

53. L. Meitner and H. H. Hupfeld, "Über das Streugesetz kurzwelliger γ-Strahlen," *Naturwiss.* 19 (1931): 775–776.

54. L. Meitner and H. H. Hupfeld, "Über die Streuung kurzwelliger γ-Strahlung an schweren elementen," *Z. Phys.* 75 (1932): 705–715; L. Meitner and H. Kösters, "Über die Streuung kurzwelliger γ-Strahlen," *Z. Phys.* 84 (1933): 137–144.

55. For a general discussion, see Laurie M. Brown and Donald F. Moyer, "Lady or Tiger? The Meitner-Hupfeld Effect and Heisenberg's Neutron Theory," *Amer. J. Phys.* 52 (1984): 130–136.

56. Pais, *Inward Bound*, 352.

57. Massie and Feather, "James Chadwick," 54.

58. Rutherford, Chadwick, and Ellis, *Radiations*, 326–327; Stuewer, "Neutron Electron Hypothesis," 28–29.

59. Rutherford, Chadwick, and Ellis, *Radiations*, 523–524.

60. Segrè, *From X-Rays to Quarks*, 179–184; Boorse and Motz, *World of the Atom*, 1288ff. Joliot later said that if he had known of Rutherford's prediction of the neutron, he and Irène Curie might have discovered it before Chadwick; see Pais, *Inward Bound*, 399.

61. L. Meitner and K. Philipp, "Über die Wechselwirkung zwischen Neutronen und Atomkernen," *Naturwiss.* 20 (1932): 929–932; published 23 December 1932. L. Meitner and K. Philipp, "Weitere Versuche mit Neutronen," *Z. Phys.* 87 (1934): 484–497.

62. James Chadwick to Meitner, 5 June 1932 (MC).

63. F. Rasetti, "Über die Anregung von Neutronen in Beryllium," *Z. Phys.* 78 (1932): 165–168.

64. Boorse and Motz, *World of the Atom*, 1303ff.; Stuewer, "Nuclear Electron Hypothesis," 44–45.

65. I owe most of this discussion to Roger H. Stuewer, "Mass-Energy and the Neutron in the Early Thirties," *Science in Context* 6 (1993): 195–238; and Stuewer,

"Nuclear Electron Hypothesis," 49–53. See also J. L. Heilbron and Robert W. Seidel, *Lawrence and His Laboratory: A History of the Lawrence Berkeley Laboratory* (Berkeley, Los Angeles, and Oxford: University of California Press, 1989), 153–175.

66. Meitner and Philipp, "Weitere Versuche mit Neutronen."

67. Ibid., 496–497; Heilbron and Seidel, *Lawrence and His Laboratory*, 170–172.

68. Lise Meitner, "Atomkern und periodisches System der Elemente," *Naturwiss.* 22 (1934): 733–739. In this review article Meitner assumes (p. 737) that protons and neutrons are elementary and interchangeable within the nucleus, with the exchange of either electron or positron; she points to the possibility of proton decay to neutron and positron, which was not observed for another 15 years. See Stuewer, "Mass-Energy," 227.

69. See Stuewer, "Nuclear Electron Hypothesis," 55–56; Bromberg, "Impact of the Neutron"; Segrè, *From X-Rays to Quarks*, 196–197, 203–204; Pais, *Inward Bound*, chap. 17.

70. L. Meitner and K. Philipp, "Die bei Neutronenanregung auftretenden Elektronenbahnen," *Naturwiss.* 21 (1933): 286–287; submitted 25 March 1933. The earlier observation of straight line electron tracks was in connection with neutron studies described by Meitner and Philipp in "Wechselwirkung zwischen Neutronen und Atomkernen" (see n. 61, above).

71. Rasetti notes that in addition to Meitner and her group, Auger, Curie and Joliot, and Bothe and Becker had observed the high-energy electrons (see n. 63, above).

72. Segrè, *From X-Rays to Quarks*, 190–192; Boorse and Motz, *World of the Atom*, 1261ff. Dirac, "Recollections of an Exciting Era," 146, notes that Blackett actually was the first to observe positrons but failed to publish before Anderson.

73. Meitner and Philipp, "Die bei Neutronenanregung auftretenden Elektronenbahnen," 287.

74. Ibid.

75. Meitner and Philipp's report of positron-electron pair formation (see n. 70, above) was submitted to *Naturwissenschaften* on 25 March 1933, published on 14 April; Blackett and Occhialini's report with essentially the same contents was submitted to *Nature* on 27 March, published only 5 days later, on 1 April. In a letter to Hahn of 2 April 1933 (OHN), Meitner hoped that "it won't matter. *Naturwissenschaften* simply has a much slower publication time."

76. L. Meitner and K. Philipp, "Die Anregung positiver Elektronen durch γ-Strahlen von ThC″," *Naturwiss.* 21 (1933): 468; submitted 18 May, published 16 June 1933.

77. Meitner and Kösters, "Streuung kurzwelliger γ-Strahlen," 144.

78. Frisch, "Lise Meitner, 1878–1968," 410.

CHAPTER 6. Under the Third Reich

1. Lise Meitner, diary, January 1933 (MC).

2. J. L. Heilbron, *The Dilemmas of an Upright Man: Max Planck as Spokesman for German Science* (Berkeley, Los Angeles, and London: University of California Press, 1986), 141. See also Peter Gay, *Weimar Culture: The Outsider as Insider* (New York: Harper & Row, 1968); Gordon A. Craig, *Germany, 1866–1945* (New York: Oxford University Press, 1978).

3. Walter Moore, *Schrödinger: Life and Thought* (Cambridge: Cambridge University Press, 1989), 233ff.; Gabriele Kerber, Auguste Dick, and Wolfgang Kerber, *Dokumente, Materialien und Bilder zur 100. Wiederkehr des Geburtstages von Erwin Schrödinger* (Wien: Fassbaender, 1987), 77–79. The party was on 11 February 1933.

4. Meitner to Annemarie Schrödinger, 30 January 1953 (MC).

5. Meitner to Hahn, 8 March 1933 (OHN).

6. Meitner to Hahn, 21 March 1933 (OHN).

7. Ibid.

8. Friedrich Glum, *Zwischen Wissenschaft, Wirtschaft und Politik* (Bonn: H. Bouvier Verlag, 1964), 437–438; Alan Bullock, *Hitler: A Study in Tyranny* (New York: Harper & Row, 1964), 267–268; Craig, *Germany, 1866–1945*, 569ff.

9. Lucy S. Dawidowicz, *The War Against the Jews, 1933–1945* (New York: Bantam, 1975), 67.

10. Einstein interview with the *New York World Telegram*, in Ronald W. Clark, *Einstein: The Life and Times* (New York: Avon, 1971), 557. Also see Otto Nathan and Heinz Norden, eds., *Einstein on Peace* (New York: Avenel Books, 1981), 211, 213.

11. Clark, *Einstein*, 567–570; Philipp Frank, *Einstein: His Life and Times* (London: Jonathan Cape, 1948), 281–284; Heilbron, *Dilemmas of an Upright Man*, 156–157.

12. Heilbron, *Dilemmas of an Upright Man*, 158–159.

13. Einstein to Otto Hahn, 28 January 1949, in K. E. Boeters and J. Lemmerich, eds., *Gedächtnisausstellung zum 100. Geburtstag von Albert Einstein, Otto Hahn, Max von Laue, Lise Meitner* (Bad Honnef: Physik Kongeß-Ausstellungs- und Verwaltungs, 1979), 132.

14. Meitner to Hahn, 2 April 1933 (OHN).

15. Herbert Steiner, "Lise Meitners Entlassung," in *Österreich in Geschichte und Literatur* 9 (1965): 462–466. Meitner filled out her "Fragebogen zur Feststellung der Auswirken des Beamtengesetzes vom 7 April 1933 für die Hochschulen" on 28 April 1933, one day prior to the submission deadline.

16. Carl Bosch to Meitner, 26 April 1933 (MC).

17. For general dismissal statistics, see Edward Yarnall Hartshorne, Jr., *The German Universities and National Socialism* (London: George Allen & Unwin,

1937), chap. 3, 72ff.; for physicists, see Alan D. Beyerchen, *Scientists under Hitler: Politics and the Physics Community in the Third Reich* (New Haven: Yale University Press, 1977), 14, 44; for biologists, see Ute Deichmann, *Biologen Unter Hitler: Vertreibung, Karrieren, Forschung* (Frankfurt: Campus Verlag, 1992), chap. 1 (trans. forthcoming, Harvard University Press). The statistics are difficult to obtain. Deichmann estimates the fraction of dismissed biologists at 13% and notes (p. 47) that Beyerchen's 25% for dismissed physicists is now considered too high. For contemporary documents, see also Jost Lemmerich, ed., *Max Born, James Franck: Physiker in ihren Zeit, Der Luxus des Gewissens* (Berlin: Staatsbibliothek Preussicher Kulturbesitz, Ausstellungskataloge 17, 1982), 111ff.

18. Glum, *Zwischen Wissenschaft*, 410–413, 441; Beyerchen, *Scientists under Hitler*, 144–145.

19. James Franck, in *Göttinger Zeitung*, 18 April 1933; reprinted in Lemmerich, *Max Born, James Franck*, 114.

20. Edith Hahn to James and Ingrid Franck, 22 April 1933, in Lemmerich, *Max Born, James Franck*, 115.

21. Beyerchen, *Scientists under Hitler*, 15–19.

22. Max Born, *My Life: Recollections of a Nobel Laureate* (New York: Charles Scribner's Sons, 1978), 251.

23. Albert Einstein to Max Born, 30 May 1933, in Max Born, *The Born-Einstein Letters*, trans. Irene Born (New York: Walker, 1971), 114.

24. Alan D. Beyerchen, "Anti-Intellectualism and the Cultural Decapitation of Germany under the Nazis," in Jarrell C. Jackman and Carla M. Borden, eds., *The Muses Flee Hitler: Cultural Transfer and Adaptation, 1930–1945* (Washington, D.C.: Smithsonian Institution Press, 1983), 29–44; William L. Shirer, *The Rise and Fall of the Third Reich* (New York: Simon and Schuster, 1960), 241.

25. Dietrich Stoltzenberg, *Fritz Haber: Chemiker, Nobelpreisträger, Deutscher, Jude* (Weinheim: VCH Verlagsgesellschaft, 1994).

26. Stoltzenberg, *Fritz Haber*, 575, 576; Max von Laue to Meitner, 17 October 1949 (MC). In a letter to James Franck, 19 November 1954 (Franck Papers, University of Chicago Library [hereafter FP]), Meitner recalled that only in 1933 did she become "humanly closer" to Haber.

27. Glum, *Zwischen Wissenschaft*, 443; Stoltzenberg, *Fritz Haber*, 573–591.

28. Born, *My Life*, 263; Meitner to Born, 11 January 1957 (MC).

29. Max Planck, "Mein Besuch bei Adolf Hitler," *Physikalische Blätter* 3 (1947): 143; Helmuth Albrecht, "'Max Planck: Mein Besuch bei Adolf Hitler'—Anmerkungen zum Wert einer historischen Quelle," in Helmuth Albrecht, ed., *Naturwissenschaft und Technik in der Geschichte. 25 Jahre Lehrstuhl für Geschichte der Naturwissenschaft und Technik am Historischen Institut der Universität Stuttgart* (Stuttgart: Verlag für Geschichte der Naturwissenschaften und der Technik, 1993), 41–63.

30. David C. Cassidy, *Uncertainty: The Life and Science of Werner Heisenberg* (New York: W. H. Freeman, 1992), 306–308; Heilbron, *Dilemmas of an Upright*

Man, 153–155. For the optimistic outlook of Heisenberg and others in 1933, see Finn Aaserud, *Redirecting Science: Niels Bohr, Philanthropy, and the Rise of Nuclear Physics* (New York: Cambridge University Press, 1990), 113–114. According to Albrecht, "'Max Planck: Mein Besuch bei Adolf Hitler,'" 47–49, Planck must have drawn some assurance from his meeting with Hitler that, except for the civil service dismissals, the Kaiser-Wilhelm-Gesellschaft and German physics would not be adversely affected under the new regime. Albrecht suggests (pp. 52–53) that by attempting to reverse the dismissals only of those scientists he regarded as especially valuable without publicly protesting the dismissals of all, Planck was, in effect, collaborating with the Nazi regime.

 31. Aaserud, *Redirecting Science*. For Bohr's assistance to refugees, see chap. 3, 105ff.; for Rockefeller Foundation requirements, pp. 111–112. See also Charles Weiner, "A New Site for the Seminar: The Refugees and American Physics in the Thirties," in Donald Fleming and Bernard Bailyn, eds., *The Intellectual Migration: Europe and America, 1930–1960* (Cambridge: Belknap Harvard University Press, 1969), 190–228; for aid to refugee scholars, pp. 192ff.; for Frisch and the Academic Research Council, pp. 210–211. In June 1933, Franck reported to Meitner that the Dutch government was planning to award modest two-year stipends for displaced young Germans and that Dirk Coster was interested in having Frisch work in Groningen (Franck to Meitner, 27 June 1933 [FP]).

 32. Otto Frisch, *What Little I Remember* (Cambridge: Cambridge University Press, 1979), 51–56; Sir Rudolf Peierls, "Otto Robert Frisch, 1 October 1904–22 September 1979," *Biog. Mem. Fell. Roy. Soc. Lond.* 27 (1981): 283–306.

 33. Meitner to Hahn, 23 February 1947 (MC): Meitner recalled that Hupfeld recruited von Droste into the SA prior to 1933; with the exception of his SA uniform, however, she knew nothing about von Droste's political activities outside the institute.

 Tikvah Alper, a South African student of Meitner's from 1929 to 1932, relates that Meitner got along well with her Mitarbeiter despite their party affiliation, with the exception of Herbert Hupfeld, an early and extreme Nazi. (Interview of Tikvah Alper by Steven Weininger, 21 October 1989, Sarisbury Green, Southampton, Hampshire, England.)

 For Hess, see Fritz Krafft, *Im Schatten der Sensation: Leben und Wirken von Fritz Strassmann* (Weinheim: Verlag Chemie, 1981), 43. For Erbacher and Philipp, see Dietrich Hahn, ed., *Otto Hahn, Erlebnisse und Erkenntnisse* (Düsseldorf: Econ Verlag, 1975), 54.

 34. For the Nazi reorganization of student and faculty roles, see Hartshorne, *German Universities*, 49ff.

 35. Meitner to Hahn, 5 December 1948 (MC). Also see Otto Hahn, "Kurt Philipp 70 Jahre," *Phys. Bl.* 19 (1963): 474–475.

 36. Klaus Hoffmann, *Otto Hahn* (Dresden: Radebeul, 1979), 149, describes Hahn in that period as "politisch sorglos und ahnungslos" (politically unconcerned and naive).

37. *Toronto Star Weekly*, interview with R. E. Knowles, 8 April 1933, in D. Hahn, ed., *Otto Hahn, Begründer des Atomzeitalters* (München: List Verlag, 1979), 129–130. Hahn declared that he was not a Nazi but nevertheless surmised that Germany's youth regarded Hitler as a "hero, leader, saint."

38. Glum, *Zwischen Wissenschaft*, 436, describes Planck's demeanor in 1933 as "merkwürdig zurückhaltend" (remarkably reserved) toward the National Socialists. Albrecht, "'Max Planck: Mein Besuch bei Adolf Hitler,'" 53–54, notes that Planck was sympathetic at first with certain aspects of National Socialism; according to Heilbron, *Dilemmas of an Upright Man*, 149–150, these included the "call to national cultural renewal, unity, and glory."

39. Meitner to Hahn, 3 May 1933 (OHN).

40. Meitner to Hahn, 16 May 1933 (OHN).

41. Meitner to Hahn, 8 June 1933 (OHN).

42. D. Hahn, *Otto Hahn*, 50.

43. Wilhelm Schlenk (1879–1943), a close associate of Haber's and onetime president of the German Chemical Society, tried unsuccessfully to emigrate; he retired to Tübingen. See Stoltzenberg, *Fritz Haber*, 565, 597, 603.

44. Glum, *Zwischen Wissenschaft*, 471; Heilbron, *Dilemmas of an Upright Man*, 164–165; Albrecht, "'Max Planck: Mein Besuch bei Adolf Hitler,'" 48–49; Deichmann, *Biologen Unter Hitler*, 32–33. In 1933 about half the "non-Aryan" employees of the KWG were dismissed.

45. Stoltzenberg, *Fritz Haber*, 590–594.

46. Fritz Krafft, "Otto Hahn, 1879–1968," in Lothar Gall, ed., *Die Grossen Deutschen unserer Epoche* (Berlin: Propyläen Verlag; Frankfurt: Verlag Ullstein, 1985), 173–185, on 177; Beyerchen, *Scientists under Hitler*, 62; Glum, *Zwischen Wissenschaft*, 443–450; D. Hahn, *Otto Hahn*, 50–51.

47. Spencer R. Weart and Gertrude Szilard, eds., *Leo Szilard: His Version of the Facts* (Cambridge: MIT Press, 1978), 13–14.

48. Beyerchen, *Scientists under Hitler*, 63.

49. After the war this was Heisenberg's self-justification, as reflected in the title of a biography of him written by his wife: Elisabeth Heisenberg, *Inner Exile: Recollections of a Life with Werner Heisenberg*, trans. S. Cappellari and C. Morris (Boston: Birkhäuser, 1984). See Cassidy, *Uncertainty*, chaps. 15 and 16, 299ff.

50. See Deichmann, *Biologen Unter Hitler*, 19–20 (foreword by Benno Müller-Hill), 27–28, 187–198.

51. Heilbron, *Dilemmas of an Upright Man*, 150–151; Moore, *Schrödinger*, 267ff., 320ff.

52. Frank, *Einstein*, 285.

53. Max Born, "Max Karl Ernst Ludwig Planck, 1858–1947," *Biog. Mem. Fell. Roy. Soc. Lond.* 6 (1948): 160–188, on 180.

54. Meitner to Gerta von Ubisch, 1 July 1947 (MC).

55. Aaserud, *Redirecting Science*, 129 and 311 n. 49, and Aaserud's notes, Niels Bohr Archive, Copenhagen. In November 1933, Bohr asked Lauder Jones, the

Rockefeller representative in Paris, to grant Meitner DKr6,000 (about $1,500) for a year in Bohr's institute. The formal application for Meitner, submitted by Bohr on 18 November, indicated that Planck agreed that the "best way for Meitner to overcome her present difficulties" would be to take a year's leave from the KWI without cutting her connections to Berlin; Bohr added that it was "not certain she will find satisfactory conditions in Berlin on her return." The grant, effective 1 January 1934, was approved on 23 November 1933, Jones noting to Bohr that since Meitner would have resigned had Planck not urged her not to, it was "probable" that she might decide not to return to Berlin after a year in Copenhagen. By 15 January 1934, however, Bohr informed Jones that Meitner had not given up hope of retaining her position at the KWI, Planck having "kindly insisted" that she "should not ask him for a leave of absence . . . only permission to come [to Copenhagen] . . . for a time, retaining her salary at the KWI. . . . The very uncertainty which is attached to so many cases of this kind is just characteristic of the tragical situation and is responsible for a large part of the strain to which the person suffering under it is subject." On 29 January 1934, Jones told Bohr of a letter from the director-general of the KWG, Friedrich Glum, stating that Meitner would stay in Berlin, this being Planck's "*earnest* request." As Rockefeller did not award grants for periods of less than a year, the offer to Meitner was rescinded.

56. Meitner to Frank Aydelotte, president of Swarthmore College, 22 December 1933 (MC); Albert W. Fowler, pers. comm., 25 February 1982. Hermann Weyl, an émigré mathematician, told Meitner of the opening; the position would have been associated with the Bartol Foundation, a laboratory administered by the Franklin Institute on the Swarthmore campus. There is no record of a reply.

57. For Laue also, the shortage of positions abroad was a deterrent to emigration. Max von Laue, "Mein Physikalische Werdegang," in *Gesammelte Schriften und Vorträge* (Braunschweig: Friedrich Vieweg Verlag, 1961), 3:xxx.

58. Lise Meitner, "Looking Back," *Bull. Atom. Sci.* 20 (November 1964): 7.

59. T. deVries-Kruyt, pers. comm., 28 October 1986; Meitner to Hahn, 23 February 1947, 24 June 1947 (MC). With respect to "denazification" hearings for Gottfried von Droste, Meitner recalled that although he wore a brown shirt, "I know nothing about his activities outside the institute and in the institute. . . . I never spoke to Droste about politics." Nevertheless, she regarded his contrition after the war as "dishonest from beginning to end."

60. Meitner to Gerta von Ubisch, 1 July 1947 (MC).

61. Otto Hahn to Ministerialrat Achelis, Preussischen Ministerium für Wissenschaft, Kunst und Volksbildung, 27 August 1933; Planck to Ministry, 30 August 1933; in Steiner, "Lise Meitners Entlassung."

62. Wilhelm Stuckart to Meitner, 6 September 1933 (MC). Internal memo of 11 September 1933, in Steiner, "Lise Meitners Entlassung."

63. D. Hahn, *Otto Hahn*, 54.

64. Beyerchen, *Scientists under Hitler*, 71–73; Hartshorne, *German Universities*, 87–102.

65. Fritz Krafft, "An der Schwelle zum Atomzeitalter: Die Vorgeschichte der Entdeckung der Kernspaltung im Dezember 1938," *Ber. Wissenschaftsgesch.* 11 (1988): 232; Krafft, "Lise Meitner (7.XI.1878.–27.X.1968)," in Willi Schmidt and Christoph J. Scriba, eds., *Frauen in den exakten Naturwissenschaften: Festkolloquium zum 100. Geburtstag von Frau Dr. Margarethe Schimank (1890–1983)* (Stuttgart: Franz Steiner Verlag, 1990), 47–49; letter from Hahn to R. Pummerer, 13 October 1936 (quoted in Krafft, *Im Schatten der Sensation*, 171–172).

66. Hermann Fahlenbrach to Hahn, 2 June 1947; Hahn to Fahlenbrach, [n.d.] June 1947; Rudolf Jaeckel to Hahn, 4 June 1947; Fahlenbrach to Meitner, 9 June 1947; Meitner to Fahlenbrach, 19 June 1947; Meitner to Hahn, 24 June 1947 (MC).

67. Lise Meitner and Max Delbrück, *Der Aufbau der Atomkerne: Natürliche und Künstliche Kernumwandlungen* (Berlin: Julius Springer Verlag, 1935).

68. Arnold Sommerfeld to Arnold Berliner, 31 July 1935 (DM); review by F. Kirchner (attached to above).

69. Berliner to Sommerfeld, 31 October 1933; Berliner to Sommerfeld, 8 December 1933 (DM).

70. Hugo Dingler to *Naturwissenschaften* Redaktion, 20 November 1933 (attached to letter below). To this Sommerfeld responded, in the margin of a letter from Berliner to Sommerfeld, 8 December 1933 (DM): "What do you say to our friend Dingler? All I can say is *Lout!*"

71. Ernst Gehrcke and Johannes Stark to Prussian Minister for Art and Science, 27 May 1934; copy sent by Arnold Berliner to Planck, 28 July 1934 (DM).

72. Berliner to Sommerfeld, 28 August 1935 (DM). See also Max von Laue, "Arnold Berliner, 26.12.1862–22.2.1942," *Naturwiss.* 23 (1946): 257–258.

73. Berliner to the mathematician Paul Epstein, as quoted in Beyerchen, *Scientists under Hitler*, 230 n. 84. Berliner received several invitations from friends abroad, but he would not leave. See Nancy Arms, *A Prophet in Two Countries: The Life of F. E. Simon* (Oxford: Pergamon Press, 1966), 87.

74. Meitner to Max von Laue, 13 July 1947 (MC). "Lenard and Stark . . . found it unbearable that they could no longer follow the modern theoretical and technical developments of physics."

75. Beyerchen, *Scientists under Hitler*, chaps. 5–7; Cassidy, *Uncertainty*, chap. 18.

76. Heilbron, *Dilemmas of an Upright Man*, 166–167.

77. Lise Meitner, "Max Planck als Mensch," *Naturwiss.* 45 (1958): 406–408.

78. Glum, *Zwischen Wissenschaft*, 442–443; Beyerchen, *Scientists under Hitler*, 60–61; Heilbron, *Dilemmas of an Upright Man*, 64.

79. D. Hahn, *Otto Hahn*, 54. Also, Meitner indicates in a 1933 letter to Hahn (see n. 40, above) that she would not take in Rudolf Ladenburg because "neither Philipp nor Erbacher are very enthusiastic about it."

80. Martin Nordmeyer to Max Planck, 28 May 1935, in Krafft, *Im Schatten der Sensation*, 45.

81. Meitner to Ubisch, 1 July 1947 (MC).

82. Cassidy, *Uncertainty*, chaps. 20, 21. Albrecht, "'Max Planck: Mein Besuch bei Adolf Hitler,'" 54, suggests that Planck's compromise and collaboration gave the impression abroad that Germany was not fully controlled by the Nazis, thus helping to obscure the true character of the regime. Deichmann, in *Biologen Unter Hitler*, 33, notes that in his capacity as president of the KWG, Planck would regularly include in his reports the positions taken elsewhere by dismissed scientists, without mentioning that the dismissals had been forced by the KWG, thus creating an impression of normalcy for the official record. The failure to mention Nazi persecution has evolved into a historical style that is still occasionally practiced. For example, one can read brief biographies of Meitner's scientific contemporaries in Sabine Ernst, ed., *Lise Meitner an Otto Hahn: Briefe aus den Jahren 1912 bis 1924, Edition und Kommentierung* (Stuttgart: Wissenschaftliche Verlagsgesellschaft, 1992), 230–244, without ever learning that Jews were persecuted or that scientists were dismissed (James Franck, for instance, "resigned from his position and emigrated voluntarily to America").

83. D. Hahn, *Otto Hahn*, 53.

84. Otto Hahn, *My Life*, trans. Ernst Kaiser and Eithne Wilkins (New York: Herder and Herder, 1968), 146.

85. M. von Laue, "Fritz Haber gestorben," *Naturwiss.* 22 (1934): 97; Stoltzenberg, *Fritz Haber*, 637–639.

86. Heilbron, *Dilemmas of an Upright Man*, 142.

87. Otto Hahn, *A Scientific Autobiography*, trans. and ed. Willy Ley (London: MacGibbon & Kee, 1967), 109, 133.

88. Meitner, "Max Planck als Mensch," 407.

89. Max von Laue to Meitner, 15 June 1948 (MC).

90. Glum, *Zwischen Wissenschaft*, 470–471; Stoltzenberg, *Fritz Haber*, 636–643.

91. Meitner, "Max Planck als Mensch," 407.

92. Hahn, *A Scientific Autobiography*, 112.

93. D. Hahn, *Otto Hahn*, 53.

94. Krafft, *Im Schatten der Sensation*, 35ff.

95. Ibid., 41–42.

96. Ibid., 20, 59ff. Strassmann's pay was subsequently raised but remained low.

97. Ibid., 20.

98. Ibid., 40–47. In 1986, Strassmann was honored posthumously by the Israeli government with a tree planted in his name along the "Avenue of the Righteous" in the Israeli Holocaust Memorial, Yad Vashem, Jerusalem.

99. Ibid., 466–467.

100. Max von Laue to Albert Einstein, 14 May 1933 and 30 May 1933 (DM). In Einstein's handwriting at the top of 30 May 1933: "nicht beantworten" (don't answer).

101. Heilbron, *Dilemmas of an Upright Man*, 160.
102. Max von Laue, "Ansprache bei Eröffnung der Physikertagung in Würzburg am 18. September 1933," *Phys. Z.* 34 (1933): 889–890.
103. Laue, "Mein physikalische Werdegang," 3:xxviii.
104. Einstein to Born, 7 September 1944, in *Born-Einstein Letters*, 81.
105. Laue, "Mein Physikalische Werdegang," 3:xxvii.
106. Heilbron, *Dilemmas of an Upright Man*, 168.
107. Laue, "Mein Physikalische Werdegang," 3:xxvi–xxvii.
108. James Franck, "Max von Laue (1879–1960)," *Yearbook, Amer. Phil. Soc. Biog. Mem.* (1960): 155–159.
109. Laue to Meitner, 15 June 1948 (MC).
110. Laue, "Mein Physikalische Werdegang," 3:xxvi, xxx.
111. Clark, *Einstein*, 639.
112. Laue to Meitner, November 1958 (80th birthday) (MC). For Franck's visit to Berlin, see Meitner to Hahn, 8 June 1933 (OHN), n. 41.
113. Meitner, "Max Planck als Mensch," 407.

CHAPTER 7. Toward the Discovery of Nuclear Fission

1. Emilio Segrè, *From X-Rays to Quarks: Modern Physicists and Their Discoveries* (San Francisco: W. H. Freeman, 1980), photograph, 194–195; Roger H. Stuewer, "The Nuclear Electron Hypothesis," in William R. Shea, ed., *Otto Hahn and the Rise of Nuclear Physics* (Dordrecht: D. Reidel, 1983), 51.
2. I. Curie and F. Joliot, "Un nouveau type de radioactivité," *Comptes Rendus* 198 (1934): 254–256.
3. Lise Meitner, "Über die von I. Curie und F. Joliot entdeckte künstliche Radioaktivität," *Naturwiss.* 22 (1934): 172–174, submitted 23 February, published 16 March 1934. On 1 June 1934 (see n. 11, below), Meitner reported artificial radioactivities from the alpha particle bombardment of Li, F, and Zn.
4. Lise Meitner, "Das Energiespektrum der positiven Elektronen aus Aluminium," *Naturwiss.* 22 (1934): 388–390, published 1 June 1934.
5. Franco Rasetti, telephone interview, 15 June 1985; Emilio Segrè, *A Mind Always in Motion: The Autobiography of Emilio Segrè* (Berkeley, Los Angeles, and Oxford: University of California Press, 1993), 86–88; Segrè, *Enrico Fermi, Physicist* (Chicago: University of Chicago Press, 1962), 58, 68, 72–73; E. Amaldi, "Personal Notes on Neutron Work in Rome in the '30s and Post-War European Collaboration on High-Energy Physics," in C. Weiner, ed., *Proceedings of the International School of Physics "Enrico Fermi," Course LVII, History of Twentieth-Century Physics* (New York: Academic Press, 1977), 297.
6. For an extremely comprehensive overview, see Edoardo Amaldi, "From the Discovery of the Neutron to the Discovery of Nuclear Fission, *Physics Reports* 111 (1984): 1–332. Also, Segrè, *Enrico Fermi, Physicist*, 73; Amaldi, "Personal Notes," 298.

7. E. Fermi, "Radioattività provocata da bombardamento di neutroni I," *Ric. Sci.* 5, no. 1 (1934): 283; English translation in E. Segrè, ed., *Enrico Fermi: Collected Papers (Note e Memorie)* (Chicago: University of Chicago Press, 1962), 1:674–675 (Accademia Nazionale dei Lincei, Roma).

8. E. Fermi, "Radioattività provocata da bombardamento di neutroni II," *Ric. Sci.* 5, no. 1 (1934): 330–331; English translation in Segrè, *Enrico Fermi: Collected Papers*, 676. An article in *Nature* included the contents of the previous *Ricerca Scientifica* notes: E. Fermi, "Radioactivity Induced by Neutron Bombardment," *Nature* 133 (1934): 757; submitted 10 April, appeared 19 May 1934.

9. Segrè, *Enrico Fermi, Physicist*, 74; Amaldi, "Personal Notes," 304.

10. Meitner to Enrico Fermi, 16 May 1934 (MC).

11. Lise Meitner, "Über die Erregung künstlicher Radioaktivität in verschiedenen Elementen," *Naturwiss.* 22 (1934): 420; published 1 June 1934.

12. Segrè, *Enrico Fermi: Collected Papers*, 641.

13. Otto Frisch, *What Little I Remember* (Cambridge: Cambridge University Press, 1979), 88; Frisch, "The Interest Is Focussing on the Atomic Nucleus," in S. Rozental, ed., *Niels Bohr: His Life and Work as Seen by His Friends and Colleagues* (Amsterdam: North-Holland; New York: John Wiley & Sons, 1967), 140.

14. E. Fermi, E. Amaldi, O. D'Agostino, F. Rasetti, and E. Segrè, "Radioattività provocata da bombardamento di neutroni III," *Ric. Sci.* 5, no. 1 (1934): 452–453; English translation in Segrè, *Enrico Fermi: Collected Papers*, 677–678.

15. Laura Fermi, *Atoms in the Family: My Life with Enrico Fermi* (Chicago: University of Chicago Press, 1954), 91.

16. E. Fermi, "Possible Production of Elements of Atomic Number Higher than 92," *Nature* 133 (1934): 898–899; published 16 June 1934.

17. Lise Meitner, "Wege und Irrwege zur Kernenergie," *Naturwiss. Rdsch.* 16 (1963): 167–169. The *Nature* reference is undoubtedly that cited in n. 16, above, published 16 June 1934. Meitner may have meant *Ricerca Scientifica* (nn. 7, 8, 14) rather than the *Nuovo Cimento* review articles (*Nuovo Cimento* 11 [1934]: 429–441, 442–451) published by the Rome group in July, which include no results not reported previously in *Ricerca Scientifica* and *Nature*. In any event, Meitner knew of Fermi's uranium results by June.

18. Meitner to Max von Laue, 4 September 1944 (MC).

19. Ibid.

20. Dietrich Hahn, ed., *Otto Hahn, Erlebnisse und Erkenntnisse* (Düsseldorf: Econ Verlag, 1975), 40–41; O. Hahn and L. Meitner, "Notiz über die Entdeckung des Protactiniums," *Naturwiss.* 19 (1931): 738; A. v. Grosse, "Zur Entdeckung und Isolierung des Elements 91," response by O. Hahn and L. Meitner, *Naturwiss.* 20 (1932): 362–363.

21. A. von Grosse and H. Agruss, "The Chemistry of Element 93 and Fermi's Discovery," *Phys. Rev.* 46 (1934): 241; published 1 August 1934, it probably arrived in Germany toward the end of August.

22. D. Hahn, *Otto Hahn*, 47.

23. Among many others, Otto Hahn in *Nobel Lectures Chemistry, 1942–1962* (Amsterdam: Elsevier, 1964), 172; Otto Hahn, *New Atoms* (New York: Elsevier, 1950), 17; Hahn, *Naturwiss. Rdsch.* 2 (1953): 45–49; Hahn, *Naturwiss.* 46 (1959): 158–163; Hahn, *A Scientific Autobiography*, trans. and ed. Willy Ley (London: MacGibbon & Kee, 1967), 141; Hahn, *My Life*, trans. Ernst Kaiser and Eithne Wilkins (New York: Herder and Herder, 1968), 147–148.

24. Meitner, "Über die Erregung künstlicher Radioaktivität"; Meitner, "Über die Umwandlung der Elemente durch Neutronen," *Naturwiss.* 22 (1934): 759.

25. Otto Hahn and Lise Meitner, "Über die künstliche Umwandlung des Urans durch Neutronen," *Naturwiss.* 23 (1935): 37–38; submitted 22 December 1934, appeared 11 January 1935.

26. Lise Meitner, "Atomkern und periodisches System der Elemente," *Naturwiss.* 22 (1934): 733–739 (talk given in Leningrad 11 September 1934). Contrast the content of Meitner's Leningrad talk with Hahn's "recollection" (in *My Life*, p. 147) that he and Meitner did not learn of Fermi's experiments until they returned from the Soviet Union and were prodded by Meitner's assistant Max Delbrück (!) to repeat them.

27. Meitner, "Über die Umwandlung der Elemente durch Neutronen"; submitted October 1934, appeared 9 November 1934.

28. E. Fermi, E. Amaldi, B. Pontecorvo, F. Rasetti, and E. Segrè, "Azione di sostanze idrogenate sulla radioattività provocata da neutroni I," *Ric. Sci.* 5, no. 2 (1934): 282–283; translation in Segrè, *Enrico Fermi: Collected Papers*, 761–762. Also see Segrè, *Enrico Fermi, Physicist*, 80–82; Amaldi, "Personal Notes," 312–313.

29. Meitner to Fermi, 26 October 1934 (MC).

30. Hahn and Meitner, "Über die künstliche Umwandlung des Urans," 37–38.

31. Lise Meitner and Max Delbrück, *Der Aufbau der Atomkerne: Natürliche und künstliche Kernumwandlungen* (Berlin: Julius Springer Verlag, 1935), 48ff.; Spencer R. Weart, "The Discovery of Fission and a Nuclear Physics Paradigm," in Shea, *Otto Hahn*, 91–133, on 102–104. For a discussion of Gamow's alpha decay theory, his liquid-drop model, and its effect on the thinking of Meitner and others, see Roger H. Stuewer, "The Origin of the Liquid-Drop Model and the Interpretation of Nuclear Fission," *Perspectives on Science* 2 (1994): 76–129.

32. Ida Noddack, "Über das Element 93," *Z. Angew. Chem.* 47 (1934): 653–655. English translation in H. G. Graetzer and D. L. Anderson, *The Discovery of Nuclear Fission: A Documentary History* (New York: Van Nostrand Reinhold, 1971), 16–20.

33. Segrè, *Enrico Fermi, Physicist*, 76: "The reason for our blindness is not clear. Fermi said, many years later, that the available data on mass defect at that time were misleading and seemed to preclude the possibility of fission." See also Amaldi, "From the Discovery of the Neutron," 277.

It appears that the Berlin team was biased against Noddack, in part because she and her husband had claimed but never conclusively verified the discovery of "masurium" (now technetium, element 43) in 1925. See Fritz Krafft, *Im Schatten der Sensation: Leben und Wirken von Fritz Strassmann* (Weinheim: Verlag Chemie, 1981), 314–317.

The issue of Noddack's scientific reliability has generated considerable debate: Pieter Van Assche, "Ignored Priorities: First Fission Fragment (1925) and First Mention of Fission (1934)," *Nuclear Europe* 6–7 (1988): 24–25; Günter Herrmann, "Technetium or Masurium: A Comment on the History of Element 43," *Nuclear Physics* A505 (1989): 352–360; Teri Hopper, " 'She was Ignored': Ida Noddack and the Discovery of Fission," Master's thesis, Stanford University, 1990.

34. Glenn T. Seaborg, "Origin of the Actinide Concept," chapter 118 in K. A. Gschneider, Jr., L. Eyring, G. R. Choppin, G. H. Lander, eds., *Handbook on the Physics and Chemistry of Rare Earths*, vol. 18 (New York: Elsevier Science B.V. 1994), 1–27. A. von Grosse, "The Chemical Properties of Elements 93 and 94," *J. Amer. Chem. Soc.* 57 (1935): 440–441.

35. Otto Hahn and Lise Meitner, "Über die künstliche Umwandlung des Urans durch Neutronen (II. Mitteil.)," *Naturwiss.* 23 (1935): 230–231.

36. Krafft, *Im Schatten der Sensation*, 40ff., 48–51.

37. Otto Hahn, Lise Meitner, and Fritz Strassmann, "Einige weitere Bemerkungen über die künstlichen Umwandlungsprodukte beim Uran," *Naturwiss.* 23 (1935): 544–545; submitted 15 July, published 2 August 1935.

38. For a detailed contemporary review of the investigation, see Lawrence L. Quill, "The Transuranium Elements," *Chem. Rev.* 23 (1938): 87–155; for a recent treatment, see Günter Herrmann, "Five Decades Ago: From the 'Transuranics' to Nuclear Fission," *Angew. Chem. Intl. Ed. Engl.* 29 (1990): 481–508 (*Angew. Chem.* 102 [1990]: 469–496).

39. Hahn, Meitner, and Strassmann, "Einige weitere Bemerkungen"; and Lise Meitner, "Über die β- und γ-Strahlen der Transurane," *Ann. Phys.* 29 (1937): 246–250, submitted 15 April 1937.

40. E. Amaldi, O. D'Agostino, B. Pontecorvo, E. Fermi, F. Rasetti, and E. Segrè, "Artificial Radioactivity Produced by Neutron Bombardment, Part II," *Proc. Roy. Soc. Lond.* A149 (1935): 522–558 (15 February 1935).

41. Fritz Krafft, "Internal and External Conditions for the Discovery of Nuclear Fission by the Berlin Team," in Shea, *Otto Hahn*, 135–165, on 144–147. Also Krafft, *Im Schatten der Sensation*, 213–215, 222–227; Herrmann, "Five Decades Ago," 484.

42. Otto Hahn, Lise Meitner, and Fritz Strassmann, "Neue Umwandlungsprozesse bei Bestrahlung des Urans; Elemente jenseits Uran," *Ber. Dt. Chem Ges.* 69 (1936): 905–919, on 918.

43. Lise Meitner and Otto Hahn, "Neue Umwandlungsprozesse bei Bestrahlung des Urans mit Neutronen," *Naturwiss.* 24 (1936): 158–159; submitted 10 February, published 6 March 1936.

44. L. Meitner, O. Hahn, and F. Strassmann, "Über die Umwandlungsreihen des Urans, die durch Neutronenbestrahlung erzeugt werden," *Z. Phys.* 106 (1937): 249–270; submitted 14 May 1937.

45. O. Hahn, L. Meitner, and F. Strassmann, "Über die Trans-Urane und ihr chemisches Verhalten," *Ber. Dt. Chem. Ges.* 70 (1937): 1374–1392; submitted 15 May 1937.

46. Ibid., 1391.

47. Meitner, Hahn, and Strassmann, "Über die Umwandlungsreihen des Urans," 255.

48. Ibid., 267.

49. Ibid., 268.

50. Ernst Berninger, "Discovery of Uranium Z by Otto Hahn," in Shea, *Otto Hahn*, 213–220, on 219. Berninger contends that by providing the first theoretical understanding of Hahn's 1922 discovery of UZ, Weizsäcker's interpretation of nuclear isomerism was particularly welcome in Berlin and thus served to prolong Hahn and Meitner's belief in the false transuranes. See also Weart, "Discovery of Fission," 99–100; Carl Friedrich von Weizsäcker, "Vorwort," in Dietrich Hahn, ed., *Otto Hahn: Leben und Werk in Texten und Bildern* (Frankfurt: Insel, 1988); Weizsäcker, recorded talk, Deutsches Museum, Munich, July 1991. For Meitner's association with Weizsäcker, see Stuewer, "Origin of the Liquid-Drop Model," 109.

51. Meitner, Hahn, and Strassmann, "Über die Umwandlungsreihen des Urans," 269.

52. Weart, "Discovery of Fission,", 101, 108; Quill, "Transuranium Elements," 137–138, 141, 146; J. L. Heilbron and Robert W. Seidel, *Lawrence and His Laboratory: A History of the Lawrence Berkeley Laboratory* (Berkeley, Los Angeles, and Oxford: University of California Press, 1989), 446.

53. Meitner, "Wege und Irrwege," 168.

54. Armin Hermann, *Die Neue Physik, Der Weg in das Atomzeitalter* (München: Heinz Moos Verlag, 1979), 98. In English: *The New Physics: The Route into the Atomic Age. In Memory of Albert Einstein, Max von Laue, Otto Hahn, Lise Meitner*, trans. David C. Cassidy (Bad Godesberg: Inter Nationes, 1979).

55. Krafft, *Im Schatten der Sensation*, 220–221.

56. Lise Meitner, interview by Thomas Kuhn in Cambridge, England (with O. R. Frisch), 12 May 1963; American Institute of Physics (AIP), New York, Oral History Project, Tape 65a, transcript pp. 19–20.

In his *Scientific Biography*, 146–147, Hahn admits that "suspicion should have been aroused" by the differing irradiation conditions that formed the 23-minute U, "but the situation was so complicated that even this strange fact was accepted."

57. AIP, Oral History Project, Tape 65a, transcript p. 18.

58. Krafft, *Im Schatten der Sensation*, 55.

59. Otto Hahn and Lise Meitner, "Die künstliche Umwandlung des Thoriums durch Neutronen: Bildung der bisher fehlenden radioaktiven 4n+1-Reihe," *Naturwiss.* 23 (1935): 320; submitted 2 May, published 17 May 1935.

60. Weart, "Discovery of Fission," 104–105.

61. Krafft, *Im Schatten der Sensation*, 54–55.

62. Hahn, *A Scientific Autobiography*, 147–148.

63. Lise Meitner, Fritz Strassmann, and Otto Hahn, "Künstliche Umwandlungsprozesse bei Bestrahlung des Thoriums mit Neutronen; Auftreten isomerer Reihen durch Abspaltung von α-Strahlen," *Z. Phys.* 109 (1938): 538–552. According to Krafft, *Im Schatten der Sensation*, 56, the fact that Hahn's name was last indicates that Strassmann was chiefly responsible for the chemistry and Hahn is listed primarily for his role as institute director.

64. Meitner, Strassmann, and Hahn, "Künstliche Umwandlungsprozesse bei Bestrahlung des Thoriums," 550.

65. A. Braun, P. Preiswerk, and P. Scherrer, "Detection of α-Particles in the Disintegration of Thorium," *Nature* 140 (1937): 682.

66. Meitner, Strassmann, and Hahn, "Künstliche Umwandlungsprozesse bei Bestrahlung des Thoriums," 550–551.

67. Meitner referred to fn. 1, p. 15, of N. Bohr and F. Kalckar, "On the Transmutation of Atomic Nuclei by Impact of Material Particles. I. General Theoretical Remarks," *K. Dansk. Vid. Selsk. Mat.-fys. Medd.* 14, no. 10 (1937): 1–40; in *Niels Bohr, Collected Works*, 9:225–264. See also Pais, *Niels Bohr's Times*, 339–340. The footnote refers to Weizsäcker's isomeric theory and suggests that other effects, suggested in the body of the article, might also account for the persistence of metastable states. The article as a whole is a detailed discussion of the model described as a "liquid-drop," which would later underly Meitner and Frisch's theoretical interpretation of nuclear fission. See Stuewer, "Origin of the Liquid-Drop Model," 99–102.

68. I. Curie and P. Savitch, "Sur les radioéléments formés dans l'uranium irradié par les neutrons," *J. Phys. Rad.* 8 (1937): 385–387.

69. Meitner, "Wege und Irrwege," 168.

70. Meitner's 20 January 1938 letter to Curie is in the Joliot-Curie papers, Radium Institute, Paris; Meitner refers to it in "Wege und Irrwege," 168, and in a letter to Hahn of 24 January 1957 (MC). See also Krafft, *Im Schatten der Sensation*, 79, 206.

71. Irène Curie and Paul Savitch, "Sur le radioélément de période 3.5 heures formé dans l'uranium irradié par les neutrons," *Comptes Rendus* 206 (1938): 906–908.

72. Krafft, *Im Schatten der Sensation*, 207 (reproduced from Fritz Strassmann, *Kernspaltung: Berlin Dezember 1938* [Mainz: Privatdruck, 1978]).

73. Irène Curie and Paul Savitch, "Sur la nature du radioélément de période 3.5 heures formé dans l'uranium irradié par les neutrons," *Comptes Rendus* 206

(1938): 1643–1644. Translation, "Concerning the Nature of the Radioactive Element with 3.5-Hour Half-Life, Formed from Uranium Irradiated by Neutrons," in Graetzer and Anderson, *Discovery of Nuclear Fission*, 37–38.

CHAPTER 8. Escape

1. Gordon Brook-Shepherd, *Anschluss: The Rape of Austria* (London: Macmillan/Philadelphia: Lippincott, 1963); Ulla Frisch, interview, January 1982; William L. Shirer, *The Nightmare Years, 1930–1940* (Boston: Little, Brown, 1984), 314; George E. Berkley, *Vienna and Its Jews: The Tragedy of Success, 1880s–1980s* (Cambridge: Abt Books, 1988/Lanham, Md.: Madison Books, 1988), 259ff., 301ff.

Not all Austrians supported the Nazi takeover; Vienna in particular had a strong socialist movement ("red Vienna") that opposed the Nazis, especially, at first, in the labor unions. But repressive measures were immediate and strong, and opposition was not openly expressed.

2. Dietrich Hahn, ed., *Otto Hahn, Erlebnisse und Erkenntnisse* (Düsseldorf: Econ Verlag, 1975), 54.

3. Fritz Krafft, *Im Schatten der Sensation: Leben und Wirken von Fritz Strassmann* (Weinheim: Verlag Chemie, 1981), 44.

4. Lise Meitner, pocket diary, 14 March 1938 (MC).

5. For more on Mentzel and the various education and scientific ministries, see Alan D. Beyerchen, *Scientists under Hitler: Politics and the Physics Community in the Third Reich* (New Haven: Yale University Press, 1977), 155–156; Friedrich Glum, *Zwischen Wissenschaft, Wirtschaft und Politik* (Bonn: H. Bouvier Verlag, 1964), 450ff.

6. D. Hahn, *Otto Hahn*, 54.

7. Otto Hahn, Taschenkalender, 17 March and 18 March 1938 (OHN).

8. Meitner, pocket diary and calendar, 20, 21 March 1938 (MC).

9. Meitner, diary, 22 March 1938 (MC).

10. Meitner, calendar, 22 March 1938 (MC).

11. D. Hahn, *Otto Hahn*, 54.

12. Meitner, calendar, 23 March 1938 (MC).

13. Krafft, *Im Schatten der Sensation*, 89; Arnold Kramish, *The Griffin: The Greatest Untold Espionage Story of World War II* (Boston: Houghton Mifflin, 1986), 14–17.

14. Meitner, diary and calendar, 26–27 March 1938 (MC).

15. D. Hahn, *Otto Hahn*, 53; Meitner to Hahn, 20 October 1946 and 23 March 1948 (MC); Dietrich Stoltzenberg, *Fritz Haber: Chemiker, Nobelpreisträger, Deutscher, Jude* (Weinheim: VCH Verlagsgesellschaft, 1994), 578, 590–591.

16. Meitner, calendar, 31 March 1938 (MC). Also D. Hahn, *Otto Hahn*, 54.

17. Paul Scherrer to Meitner, 14 March 1938 (MC).

18. Bohr to Meitner, 21 April 1938 (MC). Although Bohr and Meitner had long addressed each other with the familiar "Du," this letter is written with the formal "Sie."

19. Wolfgang L. Reiter, "Österreichische Wissenschaftsemigration am Beispiel des Instituts für Radiumforschung der Österreichischen Akademie der Wissenschaften," in Friedrich Stadler, ed., *Vertriebene Vernunft II: Emigration und Exil österreichischer Wissenschaft* (Wien: Jugend und Volk, 1988), 708–729; Reiter, "Das Jahr 1938 und seine Folgen für die Naturwissenschaften an Österreichs Universitäten," in Stadler, *Vertriebene Vernunft II*, 664–680. Also see Ute Deichmann, *Biologen Unter Hitler: Vertreibung, Karrieren, Forschung* (Frankfurt: Campus Verlag, 1992), 38.

20. Inscription by Meitner in Max von Laue's *Autogästebuch*, Easter 1938, Deutsches Museum, Munich; Theodore Von Laue, Worcester, Mass.

21. Krafft, *Im Schatten der Sensation*, 173; Meitner, diary, 22 April 1938 (MC).

22. Meitner, diary and calendar, 23 April to 7 May 1938 (MC).

23. Auguste Dick, pers. comm., June 1985, Vienna.

24. James Franck, Affidavit for Meitner, 2 June 1938 (MC).

25. Meitner, diary, 9 May 1938 (MC).

26. Hahn, pocket calendar, 10 May 1938 (OHN); Meitner, diary, 10 May 1938 (MC).

27. Meitner, diary, 11 May 1938 (MC).

28. Meitner diary, 12–16 May 1938 (MC).

29. Wilhelm Frick, an associate of Hitler's from the earliest days in Munich, was one of his highest-ranking government officials, promulgator of the April 1933 civil service law and cosigner of the Austrian Anschluss of 13 March 1938. Bosch's letter to Frick is in Krafft, *Im Schatten der Sensation*, 173.

30. Meitner, diary, 22 April–4 June 1938 (MC).

31. Ada Klokke-Coster, pers. comm., 11 February 1986.

32. Meitner to Dirk Coster, 6 June 1938 (collection of Ada Klokke-Coster). The letter could not be entrusted to the mail; it was taken out of Germany and mailed from outside, probably by Niels Bohr.

33. Scherrer to Meitner, 9 June 1938 (from M.-L. Rehder, OHN).

34. Meitner, diary, 6 June 1938 (MC).

35. Peter Debye (1884–1966) stayed in Germany until 1940 when he emigrated to the United States. His ability to survive, indeed thrive, in Nazi Germany has provoked some discussion. In J. L. Heilbron, *The Dilemmas of an Upright Man: Max Planck as Spokesman for German Science* (Berkeley, Los Angeles, and London: University of California Press, 1986), 177, 179, Debye is characterized as an able administrator with the toughness to stand up to Nazi bureaucrats, the "only undepressed person" in German science, who insisted that the former KWI for Physics be named for Max Planck despite opposition from the proponents of *Deutsche Physik*. The assessment is not inconsistent with that of Hendrik B. G. Casimir, *Haphazard Reality: Half a Century of Science* (New York: Harper & Row,

1983), 197, but a number of Debye's contemporaries, including Dutch colleague H. A. Kramers, regarded him as an opportunist who tolerated the excesses of Nazi Germany too long. See Max Dresden, *H. A. Kramers: Between Tradition and Revolution* (New York: Springer Verlag, 1987), 515–516. In his position Debye seems to have willingly and actively participated in the expulsion of Jews from the Deutsche Physikalische Gesellschaft in 1938 (Debye correspondence, MPG, 1938–1939).

36. Hendrik Antonie Kramers (1894–1952), a Dutch theoretical physicist who had been an early and close collaborator of Bohr's, came to Leiden in 1934 as the successor to Paul Ehrenfest. See Dresden, *H. A. Kramers*.

37. Dirk Coster to A. D. Fokker, 11 June 1938 (MB). Unless otherwise specified, the language of the Boerhaave documents is Dutch, and translations are by the author.

38. Fokker to Coster, 14 June 1938 (MB).

39. Coster to Fokker, 11 June 1938 (MB).

40. Coster to Fokker, 16 June 1938 (MB).

41. Coster to A. Bouwers, 13 June 1938; from Prof. L. K. ter Veld (UG). The language of all UG documents cited is Dutch, and all translations are by the author. For more on the N. V. Philips Company and A. Bouwers, see Casimir, *Hapahazard Reality*, 288–289.

42. Coster to G. J. Sizoo, 20 June 1938 (UG).

43. Johanna Westerdijk to Fokker, 12 May 1938 (MB).

44. Fokker to Westerdijk, 11 June 1938 (MB).

45. Westerdijk to Fokker, 14 June 1938 (MB).

46. W. H. van Leeuwen to Fokker, 16 June 1938; Jhr. H. Loudon to Fokker, 16 June 1938; Fokker to Loudon, 17 June 1938 (MB).

47. Fokker to Westerdijk, 17 June 1938 (MB).

48. Fokker to Coster, 14 June 1938 (MB).

49. Coster to Fokker, 16 June 1938 (MB).

50. Fokker to Coster, 17 June 1938 (MB).

51. Fokker to Coster, 21 June 1938 (MB).

52. Meitner, diary, 14 and 15 June 1938 (MC).

53. Reichsministerium (Wilhelm Frick) to Carl Bosch, 16 June 1938. In Ernst Berninger, *Otto Hahn: Eine Bilddokumentation* (München: Heinz Moos Verlag, 1969), the first page of the stenogram is shown and a transcription given on pp. 42–43. The transcription in Krafft, *Im Schatten der Sensation*, p. 174, is similar but probably somewhat more reliable.

54. Krafft, *Im Schatten der Sensation*, 175.

55. Telegram, Scherrer to Meitner, 17 June 1938 (MC).

56. Debye to Bohr (in German), 16 June 1938 (MB).

57. Bohr to Fokker (in English), 21 June 1938 (MB).

58. Bohr to Fokker (in English), 18 June 1938 (MB).

59. Coster to Fokker, 20 June 1938 (MB).

60. Fokker to Kramers, 13 July 1938 (MB).

61. Fokker to Coster, 21 June 1938 (MB).

62. Prof. D. Cohen to Fokker, 20 June 1938 (MB). Emigration and placement statistics as of 1935 (which includes the bulk of the 1933 dismissals) show that Holland took in approximately 30 displaced German scholars (4 in permanent positions, 15 temporary, 10 no position, 1 uncertain). Compared to the United States, which took in 167 (96 permanent, 61 temporary, 8 no position, 2 uncertain), Holland's per capita record is higher. See Edward Yarnall Hartshorne, Jr., *The German Universities and National Socialism* (London: George Allen & Unwin, 1937), 95–98.

63. Fokker to Coster, 21 June 1938 (MB).

64. Fokker/Coster to Minister van Justitie, 28 June 1938 (MB).

65. Fokker to Minister van Onderwijs, Kunsten en Wetenschappen, 28 June 1938; Fokker to A. J. L. van Baeck Calkoen (Onderwijs) and Pannenborg (Justitie) (MB).

66. W. J. de Haas to Fokker, 24 June 1938 (MB); translation by Eleonore Watrous.

67. Coster to Fokker, 27 June 1938 (MB).

68. Fokker to Coster, 29 June 1938; A. E. van Arkel to Fokker, 1 July 1938 (MB).

69. Coster to A. J. L. van Baeck Calkoen, 29 June 1938 (UG).

70. Coster to Calkoen, 29 June 1938 (MB).

71. Paul Jaffe to Fokker, 24 June 1938 (MB).

72. Van Arkel to Fokker, 1 July 1938 (MB).

73. Fokker to Coster, 29 June 1938 (MB).

74. Fokker to A. F. Philips, 22 June 1938 (MB).

75. For efforts to place and support refugee physicists in America, see Charles Weiner, "A New Site for the Seminar: The Refugees and American Physics in the Thirties," in Donald Fleming and Bernard Bailyn, eds., *The Intellectual Migration: Europe and America, 1930–1960* (Cambridge: Belknap/Harvard University Press, 1969), 190–232. Between 1933 and 1941, about 100 refugee physicists came to the United States. Many had come via England, which afforded temporary positions but had fewer permanent positions to offer. For displaced German scholars in general, see Hartshorne, *German Universities*, 97.

76. W. van Beuningen to Fokker, 28 June 1938 (MB).

77. Robert May to Fokker, 24 June 1938 (MB).

78. The next day he reconsidered and pledged f.1,000, providing that the entire 5-year goal could be reached: J. L. Pierson to Fokker, 29 and 30 June 1938 (MB).

79. P. F. S. Otten to Fokker, 28 June 1938 (MB). The smug reply angered Fokker; he responded (Fokker to P. F. S. Otten, 1 July 1938 [MB]) that private grants to Netherlands universities were "nothing compared to those to English universities or the university in Copenhagen." Fokker also noted that he found

"another remark"—undoubtedly the one about political considerations—"not very clear" and intended to pursue it in person. Casimir, *Haphazard Reality*, 225–228, traces the background of the Philips family to Jewish antecedents.

80. Fokker to Coster, 29 June 1938 (MB).

81. Sizoo to Coster, 22 June 1938 (UG).

82. H. R. Kruyt to Fokker, 24 June 1938 (MB).

83. Fokker to Kramers, 13 July 1938 (MB).

84. Coster to Fokker, 27 June 1938 (MB).

85. Fokker to Coster, 27 June 1938 (MB).

86. Meitner diary, 27 June 1938 (MC). Ebbe Rasmussen had functioned for Bohr as an emissary in Germany since the beginning of the Hitler regime; see Finn Aaserud, *Redirecting Science: Niels Bohr, Philanthropy, and the Rise of Nuclear Physics* (New York: Cambridge University Press, 1990), 108ff.

87. Bohr wrote, "Lise Meitner and her German friends have thought it most advisable to accept the invitation to work in Siegbahn's institute in Stockholm, even if it is formally only for a year, because the conditions in Sweden as regards permissions [*sic*] from the authorities to stay and work seem easier there, where the invitation will come directly from the [Royal] Swedish Academy [of Sciences] to which Siegbahn's new institute belongs." Bohr to Fokker (in English), 30 June 1938 (MB).

88. Debye to Coster, 29 June 1938 (MB).

89. Fokker to Kramers, 13 July 1938 (MB).

90. K. C. Honig to Fokker, Paul Jaffé to Fokker, 2 July 1938 (MB).

91. Meitner, calendar, 30 June 1938 (MC); Meitner, diary, 1 July 1938 (MC).

92. Bohr to Fokker (in English), 30 June 1938 (MB).

93. Meitner, diary, 4 July 1938 (MC). Debye's 22-year-old son, an active Nazi, was present during their discussions. Meitner found it "unpleasant that everything [was said] in front of Debye's son."

94. Debye to Coster, 6 July 1938 (MB); for assistance with translation, I am grateful to Eleonore Watrous. The letter took three days to reach Coster, an unusually long time; possibly it was intercepted and read by the German authorities.

95. C. F. von Weizsäcker remembers that his father in some way helped her to escape (transcript of talk, *Deutsches Museum Ehrensaal*, June 1991; pers. comm., 9 April 1992), but that is not evident from Meitner's diary, 7 and 9 July 1938 (MC), or any other source. Baron Ernst von Weizsäcker's directives to German embassies were dated 8 July 1938; see Raul Hilberg, *The Destruction of the European Jews* (New York: Octagon Books, 1978), 93. The concern that the attention of two ministries had been drawn to Meitner was expressed by Fokker to Bohr (in English), 16 July 1938 (MB): "Owing to the fact that the [G]erman passport for her had been denied officially when Director Bosch applied for it [and that] . . . Lise Meitner had been told that the *auswärtiges Amt* [foreign office] would not back her application for her passport . . . the attention of authorities had already been drawn to her case."

96. Meitner to Hahn, 13 May 1966 (MC).

97. Fokker to Kramers, 13 July 1938 (MB).

98. Telegram, Debye to Fokker, 12:32 P.M., 11 July 1938 (MB).

99. The telegram from Fokker to Debye is cited in Fokker to Kramers, 13 July 1938 (MB); Fokker added, "Het is toch rottig, dat je niet eene telegraferen kunt, dat er een officiele Einreisebewilligung was."

100. Meitner diary, 11 July 1938 (MC).

101. Hans P. Coster (Belleaire, Texas), pers. comm., 1 February 1986; Ada Klokke-Coster (Epse, Netherlands), pers. comm., 11 February 1986.

102. Meitner, diary, 12 July 1938 (MC); Fokker to Bohr, 16 July 1938 (MB).

103. Meitner to Gerta von Ubisch, 1 July 1947 (MC).

104. Meitner, diary, 12 July 1938 (MC).

105. Meitner to Hahn, 13 May 1966 (MC).

106. Otto Hahn, *My Life*, trans. Ernst Kaiser and Eithne Wilkins (New York: Herder and Herder, 1970), 149.

107. D. Hahn, *Otto Hahn*, 55.

108. Meitner to Ubisch, 1 July 1947 (MC).

109. Otto Hahn, *My Life*, 149.

110. Kramish, *The Griffin*, 49.

111. Hans Coster, pers. comm., 1 February 1986; Ada Klokke-Coster, pers. comm., 11 February 1986. Miep Coster thought the German border guards may have let Meitner through because they assumed the "Frau Professor" was a professor's wife.

112. Meitner, diary, 13 July 1938 (MC).

113. George Graue, "Erinnerungen an meinen Doktorvater," *Chemie in Labor und Betrieb* 30 (1979): 92–94, quoted in Krafft, *Im Schatten der Sensation*, 171–174 n. 29. According to Krafft, Hess's denunciation was sent not to the SD but to the Dahlem regional office of the NSDAP, whose leader passed it on to Graue.

114. Max von Laue to Meitner, November 1958 (MC).

115. Hahn to Coster, card, 15 July 1938 (MC).

116. As quoted by Ada Klokke-Coster, pers. comm., 11 February 1986.

117. Fokker to Miep Coster, 16 July 1938 (MB).

118. Miep Coster to Fokker, n.d., probably 15 July 1938 (MB).

119. Ibid.

120. Hahn to Coster [Meitner], card, 15 July 1938 (MC).

121. Paul Scherrer to Meitner, 15 July 1938 (MC).

122. Meitner to Scherrer, 20 July 1938 (MC).

123. Fokker to Manne Siegbahn (in English), 16 July 1938 (MB).

124. Meitner, diary, 14 to 19 July 1938 (MC).

125. Fokker to Siegbahn (in English), 22 July 1938 (MB). Siegbahn responded with a telegram (in English) on 25 July 1938 (MB): "Swedish legacy get new instructions today."

126. Meitner, diary, 25 July 1938 (MC).

127. Fokker to Meitner, 27 July 1938 (MC).

128. Meitner to Ubisch, 1 July 1947 (MC).

129. Meitner diary, 29 July to 1 August 1938 (MC).

130. E. Barrett (Pasadena) to Meitner, 2 August 1938; E. O. Lawrence (Berkeley) to Meitner, 25 July 1938 (MC).

131. C. D. Ellis to W. B. Brander, 9 September 1938; Brander to Master of Balliol College, Oxford, 21 September 1938; A. D. Lindsey to F. A. Lindemann, 22 September 1938; Lindemann to Lindsey, 23 September 1938 (Cherwell Papers, The Library, Nuffield College, Oxford). I am grateful to Arnold Kramish for calling these to my attention. Also see Kramish, *The Griffin*, 186–187.

132. W. L. Bragg to Manne Siegbahn, 23 August 1938 (MC).

133. O. R. Frisch to J. D. Cockcroft, 28 August 1938 (MC).

134. Casimir, *Haphazard Reality*, 184–185.

135. J. D. Cockcroft to Frisch, 30 August 1938 (MC).

136. O. R. Frisch, pers. comm., February 1975.

137. Meitner to Coster, 9 August 1938 (collection of Ada Klokke-Coster, Epse, Netherlands).

CHAPTER 9. Exile in Stockholm

1. Otto Hahn, Siemens-Taschenkalender, 13 July 1938 (OHN).

2. Reichsministerium für Wissenschaft, Erziehung und Volksbildung (Dames) to Kaiser-Wilhelm-Gesellschaft (Telschow), n.d., but August 1938 (MC). It is unclear why Meitner's Jewish ancestry is given as 25 percent; in her 1933 *Fragebogen* (see chap. 6) she listed all four grandparents as "nichtarisch."

3. Ernst Telschow to Otto Hahn, 18 August 1938 (MC).

4. Hahn to Eva von Bahr-Bergius (i.e., Meitner), card, 2 August 1938 (MC).

5. Meitner to Hahn, 24 August 1938 (MC).

6. Hahn to Meitner, 27 August 1938 (MC).

7. Hahn to Meitner, 1 September 1938 (MC).

8. Meitner to Hahn, 6 September 1938 (MC).

9. Hahn to Meitner, 8 September 1938 (MC).

10. Meitner to Hahn, 10 September 1938 (MC).

11. Hahn to Meitner, 27 August 1938 (MC). For Mattauch, see Fritz Krafft, *Im Schatten der Sensation: Leben und Wirken von Fritz Strassmann* (Weinheim: Verlag Chemie, 1981), 202–203, 365–366.

12. Hahn to Meitner, 29 August 1938; Meitner to Hahn, 2 September 1938 (MC).

13. Hahn to Meitner, 1 September 1938 (MC).

14. Meitner to Hahn, 14 September 1938 (MC).

15. Meitner to Hahn, 2–3 September 1938 (MC). In 1951, Meitner wrote (Meitner to Hahn, 4 March 1951 [MC]) that for nine years her salary in Siegbahn's

institute had "not even been that of a first assistant." As the pay of a first assistant would be higher than others, one can estimate Meitner's stipend as roughly that of an ordinary assistant.

16. Coster to Meitner, 9 September 1938 (MC). Beginning in summer 1938, the Reich froze the bank accounts and other assets of emigrating Jews for pragmatic reasons (currency regulation and eventual confiscation), under the rationale that Jews could not have acquired their assets honestly. Ernst von Weizsäcker, the *Staatssekretär* to whom Meitner had appealed in early July, was at that time sending instructions to German embassies to that effect. See Raul Hilberg, *The Destruction of the European Jews* (New York: Octagon Books, 1978), 93–94, 302–303.

17. Meitner to Hahn, 6 September 1938 (MC).

18. Meitner to Hahn, 9 September 1938 (MC).

19. Meitner to Hahn, 14 September 1938 (MC).

20. Meitner to Hahn, 10 September 1938 (MC).

21. Hahn to Meitner, 12, 15, 17 September 1938 (MC).

22. Inventory of Meitner's apartment, Thielallee 63 (MC).

23. Meitner to Hahn, 18 September 1938 (MC). Max junior and Max senior were Laue and Planck, respectively. In a response to Meitner of 27 September 1938 (MC), Hahn wrote that Laue was "quite serious. Today he again said, that you are to be envied that you are in Sweden."

24. Hahn to Meitner, 20 September 1938 (MC).

25. Meitner to Hahn, 21 September 1938 (MC).

26. Lucy S. Dawidowicz, *The War Against the Jews, 1933–1945* (New York: Bantam, 1975), 139–140; George E. Berkley, *Vienna and Its Jews: The Tragedy of Success, 1880s–1980s* (Cambridge: Abt Books, 1988/Lanham, Md.: Madison Books, 1988), 266ff. For the July 1938 Evian meeting on the Jewish refugee problem and the restrictive national policies worldwide, see Alfred A. Häsler, *The Lifeboat Is Full: Switzerland and the Refugees, 1933–1945*, trans. Charles Lam Markmann (New York: Funk & Wagnalls, 1969), 25–29.

27. Built as a neutron generator, Meitner and Reddemann's machine was designed to accelerate deuterium nuclei, which, on striking a target of heavy ice (D_2O, ^2H_2O), produced helium-3 and neutrons: $^2H + ^2H \rightarrow {}^3He + {}^1n$. It is likely that the paper Meitner was correcting the night before she left Berlin was Reddemann's manuscript for *Zeitschrift für Physik*; in its printed form, Meitner's name is completely absent. See Burghard Weiss, "Lise Meitners Maschine," *Kultur & Technik* (March 1992): 22–27.

28. Meitner to Hahn, 25 September 1938 (MC).

29. Dietrich Hahn, ed., *Otto Hahn, Erlebnisse und Erkenntnisse* (Düsseldorf: Econ Verlag, 1975), 55–56.

30. Hahn to Meitner, 24 September 1938 (MC).

31. Hahn to Meitner, 27 and 28 September 1938 (MC).

32. Meitner to Hahn, 29 September 1938 (MC).

33. Hahn to Meitner, 1 October 1938 (MC).

34. Walter Moore, *Schrödinger: Life and Thought* (Cambridge: Cambridge University Press, 1989), 268 ff., 337–344.

35. Annemarie Schrödinger to Meitner, 18 October 1938 (MC).

36. Meitner was granted emeritus status as of 1 October 1938. For a short time her pension was paid, but into a frozen account, which she could, however, transfer to her relatives in Vienna. See Krafft, *Im Schatten der Sensation*, 179.

37. Hahn to Meitner, 1 October 1938 (MC).

38. Hilberg, *The Destruction of the European Jews*, 57, 59, 302. See also n. 16, above.

39. Meitner to Hahn, 6 October 1938 (MC). Hahn's suggestion that Meitner congratulate Philipp was in his letter of 1 October 1938.

40. Meitner to Hahn, 23 October 1938 (MC).

41. Elisabeth Schiemann to Otto Robert Frisch, 14 July 1969 (MC).

42. Meitner to Hahn, 6 October 1938 (MC).

43. Lilli Eppstein, Stocksund, Sweden, pers. comm., 3 May 1987.

44. Institute directors' reports in *Kungliga Svenska Vetenskapsakadamiens Årsbok* (Stockholm: Almqvist & Wiksells Boktryckeri, 1940–1946). I am grateful to Urban Wråkberg of the KVA for assistance.

45. Robert Marc Friedman, "Karl Manne Georg Siegbahn," in Frederick L. Holmes, ed., *Dict. Sci. Biog.*, Supplement 2 (New York: Charles Scribner, 1990), 18:821–826; Elisabeth Crawford, "The Benefits of the Nobel Prizes," in Tore Frängsmyr, ed., *Science in Sweden: The Royal Swedish Academy of Sciences, 1739–1989* (Canton, Mass: Science History Publications, 1989), 227–248, on 242; Sigvard Eklund, "Forskningsinstitutet för Atomfysik 1937–1987," in Per Carlson, ed., *Fysik i Frescati: Föredrag från Jubileumskonferens den 23 Oktober 1987* (Stockholm: Manne Siegbahn Institute, 1989); H. Atterling, "Karl Manne Georg Siegbahn, 1886–1978," *Biog. Mem. Fell. Roy. Soc. Lond.* 37 (1991): 428–444.

46. Meitner to Hahn, 15 October 1938 (MC).

47. Meitner to Hahn, card, 16 October 1938 (MC).

48. Meitner to Hahn, 23 October 1938 (MC).

49. Meitner to Hahn, 1 November 1938 (MC).

50. Meitner to Hahn, 23 October 1938 (MC).

51. Ibid.

52. I. Curie and P. Savitch, "Sur les radioéléments formés dans l'uranium irradié par les neutrons," *J. Phys. Rad.* 8 (1937): 385–387; "Sur le radioélement de période 3,5 heures formé dans l'uranium irradié per les neutrons," *Comptes Rendus* 206 (1938): 906–908; "Sur la nature du radioélement de période 3,5 heures formé dans l'uranium irradié par les neutrons," *Comptes Rendus* 206 (1938): 1643–1644. See also chap. 7.

53. I. Curie and P. Savitch, "Sur les radioéléments formés dans l'uranium irradié par les neutrons II," *J. Phys. Rad.* 9 (1938): 355–359. The article was submitted 12 July 1938 and appeared in October.

54. Fritz Strassmann, *Kernspaltung: Berlin Dezember 1938* (Mainz: Privat-druck, 1978), 16; reprinted in Krafft, *Im Schatten der Sensation*, 203–211, on 207.

55. Hahn to Meitner, 25 October 1938 (MC).

56. Meitner to Hahn, card, 28 October 1938 (MC).

57. Strassmann, *Kernspaltung*, 17; Krafft, *Im Schatten der Sensation*, 80–81, 208, 233–234.

58. Hahn to Meitner, 30 October 1938 (MC).

59. Meitner to Hahn, 1 November 1938 (MC).

60. Hahn to Meitner, 2 November 1938 (MC).

61. Ibid.

62. Meitner to Hahn, 4 November (incorrectly dated 4 October) 1938 (MC).

63. Hahn to Meitner, 5 November 1938 (MC).

64. O. Hahn and F. Strassmann, "Über die Entstehung von Radiumisotopen aus Uran beim Bestrahlen mit schnellen und verlangsamten Neutronen," *Natur-wiss.* 26 (1938): 755–756; submitted 8 November, appeared 18 November 1938. Hahn wrote the articles by himself: Strassmann, *Kernspaltung*, 19; Krafft, *Im Schatten der Sensation*, 209.

65. Gottlieb Klein, Oskar's father, was a Stockholm rabbi whose liberal congregation held services on Saturdays *and* Sundays; he focused on the ethics of Judaism and was a scholar of intellectual movements in Israel at the time of Jesus. (Lilli Eppstein, pers. comm., 20 March 1994.) For Oskar Klein's career, see Abraham Pais, *Niels Bohr's Times, in Physics, Philosophy, and Polity* (Oxford: Clarendon Press, 1991), 360.

Around 1930, Sweden was known for its experimental physics, but the state of theoretical physics was considered (in the words of Wolfgang Pauli) to be "deplorably backward." See Suzanne Gieser, "Philosophy and Modern Physics in Sweden: C. W. Oseen, Oskar Klein, and the Intellectual Traditions of Uppsala and Lund, 1920–1940," in Svante Lindqvist, ed., *Center on the Periphery: Historical Aspects of 20th-Century Swedish Physics* (Canton, Mass.: Science History Publications, 1993), 25.

66. Lilli Eppstein, pers. comm., 18 February 1987.

67. Hahn to Meitner, 5 November 1938 (MC).

68. Laue to Meitner, 60th birthday, November 1938 (MC).

69. Albert Einstein to Meitner, 31 October 1938 (MC).

70. Sixtieth birthday greetings, Meitner Collection 7/1 (MC).

71. Werner Heisenberg to Meitner, 60th birthday, 7 November 1938 (MC).

72. Arnold Berliner to Meitner, 1 November 1938 (MC).

73. For the "*J*" stamp, see Häsler, *The Lifeboat Is Full*, 30–53; also Hilberg, *Destruction of the European Jews*, 56ff., 83–84, 90–92, 118–119. For refugees in Switzerland, see Häsler, above, and Helmut F. Pfanner, "The Role of Switzerland for the Refugees," in Jarrell C. Jackman and Carla M. Borden, eds., *The Muses Flee Hitler: Cultural Transfer and Adaptation, 1930–1945* (Washington, D.C.: Smithsonian Institution Press, 1983), 235–248. For the Evian conference of July 1938,

see Anthony Read and David Fisher, *Kristallnacht: The Nazi Night of Terror* (New York: Random House, 1989), 198–210.

74. Rita Thalmann and Emmanuel Feinermann, *Crystal Night 9–10 November 1938*, trans. Gilles Cremonesi (London: Thames and Hudson, 1974); Read and Fisher, *Kristallnacht*.

75. Otto Hahn, Taschenkalender, 9 and 10 November 1938 (OHN). It is possible, although unlikely, that Hahn's "schöne Abend" comment for 9 November was sardonic. For the Frisch couple's situation in Vienna, see Otto Frisch, "A Walk in the Snow," *New Scientist* 60 (1973): 833.

76. Meitner stayed in Copenhagen from 10 to 17 November 1938; Guest book, Institute for Theoretical Physics (NBA). I am grateful to Dr. Hilde Levi for the information.

77. Meitner alluded to their meeting only once, when Otto Robert Frisch was making the arrangements (Meitner to Hahn, 4 November 1938 [MC]): "You will, or perhaps already have received a letter from O.R. I dare not assume that you would say yes, but you can scarcely imagine what a yes would mean to me."

78. For the Carlsberg Residence and the Bohrs' hospitality, see Pais, *Niels Bohr's Times*, 332–335.

79. Hahn, *Taschenkalender*, 13 and 14 November 1938 (OHN).

80. D. Hahn, *Otto Hahn*, 58. Hahn cites Bohr as "etwas unglücklich" and recalls seeing Frisch (but not Meitner) in Copenhagen. This account was written without notes in 1945 while Hahn was interned in England, but Hahn did not correct it in subsequent memoirs, when he did have personal calendars and other records available. See, e.g., Otto Hahn, *My Life*, trans. Ernst Kaiser and Eithne Wilkins (New York: Herder and Herder, 1970), 150; and Hahn, "Die 'falschen' Trans-Urane: Zur Geschichte eines wissenschaftlichen Irrtums," *Naturwiss. Rdsch.* 15 (1962): 43–47.

81. There were theoretical objections to the double alpha emission as well. The possibilities were an $(n,2\alpha)$ process:

$$^{238}U + {}^1n \rightarrow 2\,{}^4He + {}^{231}Ra$$

or an (n,α) process followed immediately by the α-decay of Th:

$$^{238}U + {}^1n \rightarrow {}^4He + {}^{235}Th \rightarrow {}^{231}Ra + {}^4He$$

Of these, the first was considered virtually impossible, the second, very improbable. In "Die 'falschen' Trans-Urane," Hahn later recalled that "Bohr" objected strongly to the radium isomers (to Bohr's name one must add, or perhaps substitute, Meitner's): "Bohr was quite unhappy. For him the splitting-off of two α-particles from uranium was unimaginable. He did not consider [the radium isomers] to be possible."

82. L. Meitner, F. Strassmann, and O. Hahn, "Künstliche Umwandlungsprozesse bei Bestrahlung des Thoriums mit Neutronen: Auftreten isomerer Reihen durch Abspaltung von α-Strahlen," *Z. Phys.* 109 (1938): 538–552.

83. Strassmann, *Kernspaltung*, 18, 20; Krafft, *Im Schatten der Sensation*, 208, 210. See also Strassmann, p. 20, and Krafft, p. 210, where Strassmann writes, "To this day I remain convinced that it was L. Meitner's critical demand that motivated us to test our findings once again, after which the result came to us. The relevant letter has not yet been found." (Such a letter probably does not exist; Meitner's criticism was communicated verbally by Hahn on his return from Copenhagen.)

84. A direct link between the physicists' (Meitner's and Bohr's) objections and Hahn's and Strassmann's ensuing experiments was clearly stated in their next publication. O. Hahn and F. Strassmann, "Über den Nachweis und das Verhalten der bei der Bestrahlung des Urans mittels Neutronen entstehenden Erdalkalimetalle," *Naturwiss.* 27 (1939): 11–15: "As the formation of radium isotopes under bombardment with slow neutrons was energetically not easy to understand, a particularly thorough determination of the chemical character of the newly produced artificial radioelements was essential."

85. Meitner to Hahn, 5 December 1938 (MC).

86. Meitner to Hahn, 26 November 1938 (MC).

87. Meitner to Max von Laue, 20 November 1938 (MC).

88. Hahn to Meitner, 19 November 1938; reprinted in Jost Lemmerich, ed., *Die Geschichte der Entdeckung der Kernspaltung* (Berlin: Technische Universität Berlin, Universitätsbibliothek, 1988), 184.

89. Meitner to Hahn, 5 December 1938 (MC).

90. Meitner to Hahn, 26 November and 5 December 1938 (MC).

91. Emilio Segrè, *Enrico Fermi, Physicist* (Chicago: University of Chicago Press, 1970), 98–99, 214ff.

92. Laura Fermi, *Atoms in the Family: My Life with Enrico Fermi* (Chicago: University of Chicago Press, 1954), 156.

CHAPTER 10. The Discovery of Nuclear Fission

1. Fritz Strassmann, *Kernspaltung: Berlin Dezember 1938* (Mainz: Privatdruck, 1978); reprinted in Fritz Krafft, *Im Schatten der Sensation: Leben und Wirken von Fritz Strassmann* (Weinheim: Verlag Chemie, 1981), 203–211. Strassmann stresses (Strassmann, p. 20; Krafft, p. 210) that the impetus came from Meitner: "To this day I remain convinced that it was L. Meitner's critical demand that motivated us to test our findings once again, after which the result came to us." Hahn also stated this plainly six months later (Hahn to Norman Feather [in English], 2 June 1939 [MC]): "Only after several physicists had expressed their astonishment that slow neutrons should initiate two successive α-processes in uranium, did Strassmann and I, in order to dispel the doubts of the physicists, investigate still more carefully the properties of our radium-isotopes." For the experiments, see Krafft, pp. 84–85, 244ff.

2. Meitner to Hahn, 26 November and 5 December 1938 (MC). Meitner asked about thorium, either *before* radium (^{239} U $\overset{\alpha}{\rightarrow}$ ^{235} Th $\overset{\alpha}{\rightarrow}$ ^{231} Ra) or *after* radium

(^{231} Ra $\xrightarrow{\beta}$ ^{231} Ac $\xrightarrow{\beta}$ ^{231} Th). If present, ^{231}Th was of some interest because it would be identical or isomeric with the known UY and thus amenable to test. On 4 December, Hahn noted the possibility of a fourth Ra isomer and that formation of all Ra activities was enhanced by the use of slow neutrons; on 10 December, he told Meitner they had found two thoriums *after* radium, none before (Hahn to Meitner, 4 and 10 December 1938 [MC]). Thus there was no direct evidence for the process ^{238}U (n,2α) ^{231}Ra. For relevant passages from the Meitner-Hahn correspondence in the context of the experimental record, see Krafft, *Im Schatten der Sensation*, 250–253.

3. Meitner to Hahn, 21 December 1938 (MC).

4. Richard Willstätter, *From My Life: The Memoirs of Richard Willstätter*, trans. Lilli S. Hornig (New York: W. A. Benjamin, 1965), 424–431. Willstätter left the KWI for Chemistry for Munich in 1915, the same year he was awarded the Nobel Prize in chemistry; in 1924, he resigned his Munich position in protest of the university's anti-Semitic hiring practices. Unlike Meitner, Willstätter could not bring himself to emigrate illegally. Between November 1938, when he was nearly arrested, and February 1939, when he finally got out, everything he had was systematically taken from him as a precondition for "legal" emigration. He died in Switzerland in 1942 at age 70.

5. Hahn to Meitner, 10 December 1938 (MC).

6. J. L. Heilbron, *The Dilemmas of an Upright Man: Max Planck as Spokesman for German Science* (Berkeley, Los Angeles, and London: University of California Press, 1986), 168; Alan D. Beyerchen, *Scientists under Hitler: Politics and the Physics Community in the Third Reich* (New Haven: Yale University Press, 1977), 158; Peter Gay, *Weimar Culture: The Outsider as Insider* (New York: Harper & Row, 1968), 17, 137.

7. Dietrich Hahn, ed., *Otto Hahn, Erlebnisse und Erkenntnisse* (Düsseldorf: Econ Verlag, 1975), 52: "In Frankfurt there were Jewish bankers named Hahn, a very distinguished family; in Dahlem there was the Haber Institute."

8. Hahn to Meitner, 19 December 1938 (MC). In 1945 Hahn recalled, "'The Eternal Jew' was a traveling exhibit containing all the defamations and accusations that had been made against Jews since 1933. . . . Among the Jewish professors dismissed in 1933, my name was also listed. . . . One day I received an excited phone call from the KWG . . . asking what I had done to remove [my name]. I said, 'Nothing.' They were quite upset about that, and I had to prove once again that I was 'Aryan.' It was all due to the fact that I resigned from the faculty of the University of Berlin in 1933." Hahn resigned to avoid the party functions required of professors (see D. Hahn, *Otto Hahn, Erlebnisse und Erkenntnisse*, 50, 52.) His official letter of resignation, shown in Dietrich Hahn, ed., *Otto Hahn: Begründer des Atomzeitalters* (München: List Verlag, 1979), 135, is dated 31 January 1934.

9. In D. Hahn, *Otto Hahn: Begründer des Atomzeitalters*, 135, Dietrich Hahn, the grandson of Otto Hahn, quotes without comment a passage from Bernt

Engelmann, *Germany Without Jews*, trans. D. J. Beer (New York: Bantam Books, 1984), 183, which notes that Hahn had Jewish ancestors on his mother's side, "an anxiously guarded secret [*ängstlich gehütetes Geheimnis*] that was hushed up because German science could not afford the loss of this scientist of highest international repute." Caution is in order: Engelmann makes numerous factual errors. Possibly it was a rumor planted by Hahn's enemies during the Third Reich, maintained by his supporters after. For a similar situation with Max Planck, see Heilbron, *Dilemmas of an Upright Man*, 191.

10. Hahn to Meitner, 4 December 1938 (MC). Hahn cautiously refers to the "communication from [Dr.] Leist [lawyer] about the remittance to Vienna." See also Krafft, *Im Schatten der Sensation*, 179.

11. Hahn to Meitner, 19 December 1938 (MC).

12. Willstätter, *From My Life*, 424-431.

13. Hahn to Meitner, 19 December 1938 (MC).

14. In 1945 (D. Hahn, *Otto Hahn, Erlebnisse und Erkenntnisse*, 58), Hahn gave no reason for the fractionation: "For some reason or another we wanted to enrich our Ra-isotopes somewhat." Later he said the fractionations were needed to concentrate the weak radiation of a long-lived Ra isomer (Nobel Lecture, 13 December 1946, in *Les Prix Nobel en 1946* [Stockholm, 1948], 167-183, on 174, reprinted in Otto Hahn, *Mein Leben* [München: Bruckmann, 1968], 255-256; Hahn, "Die 'falschen' Trans-Urane: Zur Geschichte eines wissenschaftlichen Irrtums," *Naturwiss. Rdsch.* 15 [1962]: 43-47; Hahn, *New Atoms* [New York: Elsevier, 1950], 20), a reason contemporary nuclear chemist Günter Herrmann regards as "rather trivial." For the experiments, see Krafft, *Im Schatten der Sensation*, chap. 3; for a detailed discussion, see Günter Herrmann, "Discovery and Confirmation of Fission," *Nuclear Physics* A502 (1989): 141c-158c; Herrmann, "Five Decades Ago: From the 'Transuranics' to Nuclear Fission," *Angew. Chem. Intl. Ed. Engl.* 29 (1990): 481-508 (*Angew. Chem.* 102 [1990]: 469-496). Hahn and Strassmann performed the first fractionation on 25 November 1938 using barium chloride, the second on 28 November with barium bromide. Given the timing, Herrmann considers it obvious that the fractionation experiments were a response to the doubts expressed by Meitner and other physicists in Copenhagen on 13 November. See also Hahn to Norman Feather, 2 June 1939 (MC), in n. 1, above.

15. Hahn to Meitner, 19 December 1938 (MC).

16. Although the Deutsches Museum in Munich displays the apparatus on a single table (the *Hahn-Meitner-Strassmann Arbeitstisch* display), the physical measurements were carried out in two separate rooms—room 29 for neutron irradiations, room 23 for radioactivity measurements—and the chemical separations in a third, room 20, all in Meitner's former section on the ground floor. See Herrmann, "Five Decades Ago," 483-485; Krafft, *Im Schatten der Sensation*, 219-220, 222-226.

17. Walther Gerlach, *Otto Hahn: Ein Forscherleben unserer Zeit* (München: Oldenbourg, 1969), 53. Reprinted (edited and with additions by Dietrich Hahn)

in *Grosse Naturforscher*, Band 45 (Stuttgart: Wissenschaftliche Verlagsgesellschaft, 1984), 90.

18. Krafft, *Im Schatten der Sensation*, 104. Strassmann always believed that the team had been so close that Meitner's "thought processes were present even when she was not, kept alive by the criticisms, questions and suggestions in her letters." For the persistence of the team, even after Meitner's departure from Berlin, see pp. 103ff., and Krafft, "An der Schwelle zum Atomzeitalter: Die Vorgeschichte der Entdeckung der Kernspaltung im Dezember 1938," *Ber. Wissenschaftsgesch.* 11 (1988): 227–252, on 230–233.

19. Hahn to Meitner, 21 December 1938 (MC).

20. Strassmann, *Kernspaltung*, 19; reprinted in Krafft, *Im Schatten der Sensation*, 209.

21. See Otto Hahn, *A Scientific Autobiography*, trans. and ed. Willy Ley (London: MacGibbon & Kee, 1967), 157; also in Krafft, *Im Schatten der Sensation*, 266 n. 82.

22. O. Hahn and F. Strassmann, "Über den Nachweis und das Verhalten der bei der Bestrahlung des Urans mittels Neutronen entstehenden Erdalkalimetalle," *Naturwiss.* 27 (1939): 11–15.

23. Ibid.

24. Meitner to Hahn, 21 December 1938 (MC). Reproduced in Jost Lemmerich, ed., *Die Geschichte der Entdeckung der Kernspaltung: Austellungskatalog* (Berlin: Technische Universität Berlin, Universitätsbibliothek, 1988), 176.

25. Hahn to Meitner (addressed to Eva von Bahr-Bergius), 23 December 1938, in Krafft, *Im Schatten der Sensation*, 267.

26. According to Erich Bagge (R. Fleischmann to P. Van Assche, pers. comm., 22 November 1982), Hahn told him in 1945, "If Frl. Meitner had still been in the institute in December 1938, she would have talked us out of barium." Similarly, Werner Heisenberg, in "Gedenkworte für Otto Hahn und Lise Meitner," *Orden pour le mérite für Wissenschaft und Künste, Reden und Gedenkworte* 9 (1968–1969): 111–119, noted that over a glass of wine, Hahn "might let the statement slip: 'I don't know; I'm afraid Lischen would have forbidden me to split uranium.'"

27. Meitner to Otto Robert Frisch, 31 October 1954 (MC): ". . . unser Weihnachtsessen bei Bergius's, wo wir beide an dem traditionell-schwedischen Lutfisch gewürgt haben." An alternative translation—less delicate but perhaps more accurate—of the word *gewürgt* could be "gagged." Lutfisk, a Scandinavian favorite (Norwegian: lutefisk), is lye-soaked cod that is rinsed, boiled, and eaten with butter.

28. About dates: On 21 December, Meitner told Hahn she would travel to Kungälv on Friday, the 23d; Hahn did not learn of it until he received her letter on 23 December. Meanwhile on 22 December, he mailed a carbon copy of the *Naturwissenschaften* manuscript to her Stockholm address; Meitner would not see it until 30 December.

There is no question that both Meitner and Frisch were in Kungälv for New

Year's Eve: they sent a joint letter to Hahn before leaving the next day. Frisch's date of arrival in Kungälv, however, varies somewhat in different accounts. In "The Discovery of Fission: How It All Began," *Physics Today* 20, no. 11 (November 1967): 43–48, Frisch says he and Meitner spent only 2 or 3 days together. But Meitner later refers to the Christmas lutfisk, and Frisch remembers their "habit of celebrating Christmas" together ("The Interest Is Focussing on the Atomic Nucleus," in S. Rozental, ed., *Niels Bohr: His Life and Work as Seen by His Friends and Colleagues* [Amsterdam: North Holland; New York: John Wiley & Sons, 1967], 137–148; *What Little I Remember* [Cambridge: Cambridge University Press, 1979], 114–116), that he "joined her in Sweden for Christmas" ("Atomic Energy: How It All Began," *Brit. J. Appl. Phys.* 5 [March 1954]: 81–84), and, most specifically, that Meitner showed him Hahn's 19 December letter when they met in Kungälv for Christmas ("Lise Meitner, Nuclear Pioneer," *New Scientist* [9 November 1978]: 426–428). Considering the importance of the holiday and the homesickness of the two immigrants, it seems very likely that Meitner and Frisch were together for Christmas and the entire week in Kungälv.

Estimates also vary for the time they took to develop the fission interpretation: "an hour or so" ("Lise Meitner"), "very gradually" ("Atomic Energy"), sitting on a tree trunk in the snow ("Interest Is Focussing" and *What Little I Remember*), and so on. (In a letter to Hahn of 29 December [see n. 36, below], after several days with Frisch, Meitner writes, "Otto R. and I have really racked our brains," an indication that it was not instantaneous.) Frisch may occasionally condense time for dramatic effect, but the scientific and personal content of his accounts is consistent.

29. Frisch, "Interest Is Focussing," 143–148, and *What Little I Remember*, 115–116; the part between . . . and . . . is taken from Frisch, "Discovery of Fission."

30. For this discussion I have relied primarily on Roger H. Stuewer, "The Origin of the Liquid-Drop Model and the Interpretation of Nuclear Fission," *Perspectives on Science* 2 (1994): 76–129. (The article includes a discussion of the dates of Meitner's and Frisch's stay in Kungälv, 113–114 n. 58.)

31. C. F. von Weizsäcker was Meitner's Assistent in 1936 for a few months as a temporary replacement for Max Delbrück; the term "Haustheoretiker" applied to a theoretician who worked in a predominantly experimental institute. After Weizsäcker moved to the nearby KWI for Physics, he and Meitner maintained contact. Carl Friedrich von Weizsäcker, pers. comm., 9 April 1992; Weizsäcker, foreword to Dietrich Hahn, ed., *Otto Hahn, Leben und Werk in Texten und Bildern* (Frankfurt: Insel, 1988); Weizsäcker, recorded speech on the induction of Heisenberg and Meitner into the *Ehrensaal*, Deutsches Museum, Munich, July 1991.

32. The compound nucleus would have interested Meitner in any case: its quantized vibrations were expected to account for gamma emission spectra, a field of interest to Meitner since the early 1920s. In a 1938 publication on the products of the neutron irradiation of thorium (L. Meitner, F. Strassmann, and O. Hahn,

"Künstliche Umwandlungsprozesse bei Bestrahlung des Thoriums mit Neu-tronen: Auftreten isomerer Reihen durch Abspaltung von α-Strahlen," *Z. Phys.* 109 [1938]: 538–552), Meitner concluded that no satisfactory theory existed for multiple inherited isomerism. With the comment, "Perhaps one must search for a more general basis for the existence of isomeric nuclei," she referred to Bohr and Kalckar's 1937 article (p. 14 in N. Bohr and F. Kalckar, "On the Transmutation of Atomic Nuclei by Impact of Material Particles. I. General Theoretical Remarks," *K. Dansk. Vid. Selsk. Mat.-fys. Medd.* 14, no. 10 [1937]: 1–40; in Niels Bohr, *Collected Works* 9:225–264).

33. See Krafft, *Im Schatten der Sensation,* 266–267. Hahn's telephoned addition to the page proofs (on 27 December 1939) reads: "Concerning the 'trans-uranes,' these elements are chemically related, but not identical to their lower homologues rhenium, osmium, iridium, platinum. Whether they are chemically identical to their still lower homologues masurium, ruthenium, rhodium, palla-dium has not yet been tested. One could not, of course, think of that before. The sum of the atomic masses of Ba + Ma, e.g., 138 + 101, gives 239!"

34. Hahn to Meitner, 28 December 1938 (MC). In Krafft, *Im Schatten der Sensation,* 267–268.

35. Friedrich Herneck, *Bahnbrecher des Atomzeitalters* (Berlin: Buchverlag der Morgen, 1969), 454–455.

In Otto Hahn, "The Discovery of Fission," *Sci. Am.* 198 (1958): 76–84, Hahn notes (p. 82): "Not being physicists, we thought of uranium's atomic weight (238) rather than the number of its protons (92)." As Hahn wrote the article without input from Strassmann, the "we" should be "I."

It is possible, even likely, that if Meitner had been in Berlin, the correct interpretation would have been part of the first barium publication. See Fritz Krafft, "Lise Meitner (7. XI.1878.–27.X.1968)," in Willi Schmidt and Christoph J. Scriba, eds., *Frauen in den Exakten Naturwissenschaften: Festkolloquium zum 100. Geburtstag von Frau Dr. Margarethe Schimank (1890–1983)* (Stuttgart: Franz Steiner Verlag, 1990), 33–70, on 40.

36. Meitner to Hahn, 29 December 1938 (MC).

37. Hahn and Meitner's first neutron-uranium publication was published in January 1935.

38. Meitner to Hahn (with note from Frisch), 1 January 1939 (MC). In Krafft, *Im Schatten der Sensation,* 268–269.

39. In their 1918 protactinium publication, Hahn was even the first author: O. Hahn and L. Meitner, *Phys. Z.* 19 (1918): 208–218. According to Fritz Krafft, "She might certainly have expected a corresponding loyalty in 1938." See Fritz Krafft, *Lise Meitner,* Hahn-Meitner-Institut HMI-B448, January 1988 (talk given 2 December 1987), and Krafft, "Lise Meitner (7.XI.1978–27.X.1968)," 39.

40. Strassmann, *Kernspaltung,* 23, and Krafft, *Im Schatten der Sensation,* 211.

41. Meitner to Hahn, 3 January 1939 (MC). In Krafft, *Im Schatten der Sensation,* 271.

42. Hahn to Meitner, 2 January 1939 (MC). In Krafft, *Im Schatten der Sensation*, 269–271.

43. Meitner to Hahn, card, 4 January 1939 (MC).

44. Frisch to Hahn, 4 January 1939. In Krafft, *Im Schatten der Sensation*, 271–272.

45. Frisch to Meitner, 3 January 1939 (MC). In Lemmerich, *Geschichte der Entdeckung*, 177.

46. Frisch, "Interest Is Focussing," 145, and *What Little I Remember*, 116–117.

47. Meitner to Frisch, n.d. but must be 4 January 1939 (MC). In Lemmerich, *Geschichte der Entdeckung*, 179.

48. Hahn to Meitner, 7 January 1939; in Krafft, *Im Schatten der Sensation*, 275–276.

49. Meitner to Frisch, draft: undated but immediately following receipt of Hahn's letter of 7 January (MC).

50. As Frisch's English was more fluent than Meitner's, he formulated the note.

51. Frisch to Meitner, 8 January 1939 (MC); in Lemmerich, *Geschichte der Entdeckung*, 179–182. According to Frisch, "Discovery of Fission: How It All Began," 47–48, and *What Little I Remember*, 117, Placzek's skepticism prompted him to find physical evidence for fission.

52. As described by Meitner in a letter to Hahn of 12 February 1939 (MC). Also in Meitner to Frisch, undated draft or possibly notes for telephone discussion, ca. 9 or 10 January 1939 (MC), she warned Frisch that his initial recoil proposal was not possible, due to the natural α-decay of uranium. In Dahlem, her assistant Gottfried von Droste had used a uranium-lined ionization chamber to search (unsuccessfully) for energetic alpha particles from neutron-irradiated uranium, but he covered the source with aluminum foil to screen out uranium's natural α-particles and thus screened out the large fission fragments as well. Frisch solved the problem electronically.

53. Hahn to Meitner, 10 January 1939 (MC); in Krafft, *Im Schatten der Sensation*, 276–277.

54. Frisch, "Interest Is Focussing," 146, and *What Little I Remember*, 117; Lemmerich, *Geschichte der Entdeckung*, 180–185.

55. O. R. Frisch, "Physical Evidence for the Division of Heavy Nuclei under Neutron Bombardment," *Nature* 143 (1939): 276.

56. Frisch to Meitner, 17 January 1939 (MC). Meitner had been chafing somewhat at the delay: "Our Nature-note will finally be sent off tomorrow." (Meitner to Hahn, 14 January 1939 [MC]; in Krafft, *Im Schatten der Sensation*, 278.)

57. As described by Frisch in "Discovery of Fission: How It All Began." See also Richard Rhodes, *The Making of the Atomic Bomb* (New York: Simon & Schuster, 1988), 263.

58. Meitner to Hahn, 18 January 1939 (MC); in Krafft, *Im Schatten der Sensation*, 281–282.

59. Lise Meitner and O. R. Frisch, "Disintegration of Uranium by Neutrons: A New Type of Nuclear Reaction," *Nature* 143 (1939): 239–240.

60. Meitner to Hahn, 14 January 1939 (MC).

61. Ibid.

62. Meitner's suggestion, as cited by Frisch in "Physical Evidence for the Division of Heavy Nuclei."

63. Frisch to Meitner, 24 January 1939 (MC).

64. For the American reaction, see Roger H. Stuewer, "Bringing the News of Fission to America," *Physics Today* 38, no. 10 (October 1985): 48–56. See also Lawrence Badash, Elizabeth Hodes, and Adolph Tiddens, "Nuclear Fission: Reaction to the Discovery, 1939," *Proc. Amer. Phil. Soc.* 130 (1986): 196–231; John A. Wheeler, "The Mechanism of Fission," *Physics Today* 20, no. 11 (November 1967): 49–52; Frisch, "Discovery of Fission: How It All Began"; Rhodes, *Making of the Atomic Bomb*, 264–271.

65. Frisch, "Interest Is Focussing," 146–147.

66. Ibid., 147.

67. Ibid., 146.

68. Frisch to Meitner, 6 February 1938 (MC). Telegram, Bohr (from Princeton) to Meitner, 4 February 1939 (MC): "Just received copies of your and Frisch Nature notes. Heartiest congratulations on most important discovery. Best wishes for continuation of work on wonderful new phenomena. Bohr."

69. Frisch to Meitner, 6 February 1939 (MC).

70. Meitner to Hahn, 25 January 1939 (MC).

71. Frisch to Meitner, 24 January 1939; Meitner to Eva von Bahr-Bergius, 6 February 1939 (MC).

72. Meitner to Hahn, 18 January 1939 (MC).

73. Elizabeth Rona, *How It Came About: Radioactivity, Nuclear Physics, Atomic Energy* (Oak Ridge: Oak Ridge Associated Universities, 1978), 45. Rona reports that during a visit to Dahlem in 1939, Hahn "complained bitterly" that he was a prisoner of his collaborators.

74. Hahn to Meitner, 16 January 1939 (MC); in Krafft, *Im Schatten der Sensation*, 278–279.

75. Meitner to Hahn, 18 January 1939 (MC); in Krafft, *Im Schatten der Sensation*, 281–282. According to Krafft, p. 282 n. 103, Meitner's letter reached Dahlem on the 20th; on that day, in response to Meitner's question about "ekaPt" being silver, Strassmann tested the "transurane" mixture for the presence of silver.

76. Krafft, *Im Schatten der Sensation*, 92–93, suggests that Strassmann may have tested for strontium rather than krypton because he had considerable prior experience with its analysis, and perhaps because he and Hahn thought they might have mistaken strontium for barium in the indicator trials.

77. Hahn to Meitner, 24 January 1939 (MC). In Krafft, *Im Schatten der Sensation*, 282–283. One sees how firmly atomic mass was embedded in Hahn's

thinking: it seems he abandoned the mass argument only because it "didn't work," not because he appreciated the necessity for considering atomic charge.

78. According to Krafft, *Im Schatten der Sensation*, 93, 282 n. 105, 283, Strassmann began planning the krypton experiment on the evening of 24 January, after copies of the Meitner-Frisch manuscripts had been received and studied. No data on the krypton experiment were entered into the laboratory record prior to 25 January.

79. As expressed by Meitner in a letter to Frisch, who responded (Frisch to Meitner, 6 February 1939 [MC]) that Hahn may have been a "bit envious" and "somewhat annoyed that the proof for [fission] is so much simpler and easier with physical methods" than with chemical separations.

80. Meitner to Hahn, 25 January 1939 (MC); in Krafft, *Im Schatten der Sensation*, 284.

81. Krafft, *Im Schatten der Sensation*, 282 n. 103.

82. Hahn to Meitner, 25 January 1938 (MC); in Krafft, *Im Schatten der Sensation*, 285.

83. Meitner to Hahn, 26 January 1939 (MC); in Krafft, *Im Schatten der Sensation*, 286.

84. Otto Hahn and Fritz Strassmann, "Nachweis der Entstehung aktiver Bariumisotope aus Uran und Thorium durch Neutronenbestrahlung; Nachweis weiterer aktiver Bruchstücke bei der Uranspaltung," *Naturwiss.* 27 (1939): 89–95.

85. This sentence appears in the draft but not in the final letter.

86. The last phrase is omitted from the final letter.

87. This sentence is omitted from the final letter.

88. Draft and letter, Meitner to Hahn, 4 and 5 February 1939 (MC).

89. Meitner to Walter Meitner, 6 February 1939 (MC).

90. Hahn to Meitner, 7 February 1939 (MC); in Krafft, *Im Schatten der Sensation*, 295, 298. Probably "Dr. K." was a certain Professor Krauch who was taking advantage of Hahn's "politically tainted [*politisch belastet*]" status by trying to install one of his associates in Meitner's former apartment. See D. Hahn, *Otto Hahn, Erlebnisse und Erkenntnisse*, 64–65.

91. Meitner to Hahn, 12 February 1939 (MC); in Krafft, *Im Schatten der Sensation*, 298.

92. Meitner to von Bahr-Bergius, 1 February 1939 (incorrectly dated 1 January) (MC).

93. Ulla Frisch, pers. comm., 10 June 1990.

94. Sir Rudolf Peierls, "Otto Robert Frisch: 1 October 1904–22 September 1979," *Biog. Mem. Fell. Roy. Soc. Lond.* 27 (1981): 289–290. Peierls regards Meitner and Frisch's theoretical note as more significant than Frisch's experimental determination of the existence of fission fragments, which was done independently by others soon after: "There is no question about the importance of the note by Meitner and Frisch, with its clear discussion of the physics of the fission process in simple terms."

95. Gustav Korlén, "Politik und Wissenschaft im schwedischen Exil," *Bei Wissenschaftsgesch.* 7 (1984): 11–21; Svante Lindqvist, ed., *Center on the Periphery: Historical Aspects of 20th-Century Swedish Physics* (Canton, Mass.: Science History Publications, 1993), xxi; Helmut Müssener, "Österreichische Wissenschaftler im schwedischem Exil," in Friedrich Stadler, ed., *Vertriebene Vernunft II: Emigration und Exil österreichischer Wissenschaft* (Wien: Jugend und Volk, 1988), 965–975.

96. Ingmar Bergström, Stockholm, pers. comm, May 1992.

CHAPTER 11. Priorities

1. F. Joliot, "Preuve expérimentale de la rupture explosive des noyaux d'uranium et de thorium sous l'action des neutrons," *Comptes Rendus* 208 (1939): 341–343. The experiment, performed on 26 January, was published 30 January, well before Frisch's note (*Nature* 143 [1939]: 276), which was submitted 16 January and appeared 18 February 1939.

2. Louis A. Turner, "Nuclear Fission," *Rev. Mod. Phys.* 12 (1940): 7.

3. Edwin McMillan, "Radioactive Recoils from Uranium Activated by Neutrons," *Phys. Rev.* 55 (1939): 510 (17 February 1939).

4. Turner, "Nuclear Fission," 8.

5. F. Joliot, "Observation par la méthode de Wilson des trajectoiries de brouillard des produits de l'explosion des noyaux d'uranium," *Comptes Rendus* 208 (1939): 647–649 (27 February 1939).

6. D. R. Corson and R. L. Thornton, "Disintegration of Uranium," *Phys. Rev.* 55 (1939): 509 (15 February 1939).

7. Philip Abelson, "Cleavage of the Uranium Nucleus," *Phys. Rev.* 55 (1939): 418 (3 February 1939). According to Strassmann, Abelson had come close to an independent discovery of fission in late 1938. See Fritz Krafft, *Im Schatten der Sensation: Leben und Wirken von Fritz Strassmann* (Weinheim: Verlag Chemie, 1981), 97 n. 71.

8. N. Feather and E. Bretscher, "Atomic Numbers of the So-called Transuranic Elements," *Nature* 143 (1939): 516.

9. J. L. Heilbron and Robert W. Seidel, *Lawrence and His Laboratory: A History of the Lawrence Berkeley Laboratory*, vol. 1 (Berkeley, Los Angeles, and Oxford: University of California Press, 1989), 446–447.

10. Günter Herrmann, "Five Decades Ago: From the 'Transuranics' to Nuclear Fission," *Angew. Chem. Intl. Ed. Engl.* 29 (1990) 481–508. In June 1939, Hahn and Strassmann resolved the iodine question: the co-precipitation of "ekaPt" with ammonium hexachloroplatinate had been taken as evidence of its resemblance to platinum, but in fact "ekaPt" was radioactive iodine that was co-precipitating with the chlorine in the compound. See Otto Hahn and Fritz Strassmann, "Zur Frage nach der Existenz der 'Trans-Urane.' I. Endgültige Streichung von Eka-Platin und Eka-Iridium," *Naturwiss.* 27 (1939): 451–453, submitted 19 June 1939; reprinted in Krafft, *Im Schatten der Sensation*, 311–313.

Subsequent attempts to identify the former "transuranes" are described in Krafft, pp. 232–233; Otto Hahn, *A Scientific Autobiography*, trans. and ed. Willy Ley (London: MacGibbon & Kee, 1967), 172–175; H. Menke and G. Herrmann, "Was waren die 'Transurane' der dreißiger Jahre in Wirklichkeit?" *Radiochim. Act.* 16 (1971): 119–123.

11. L. Meitner, O. Hahn, and F. Strassmann, "Über die Umwandlungsreihen des Urans, die durch Neutronenbestrahlung erzeugt werden," *Z. Phys.* 106 (1937): 249–270.

12. N. Bohr, "Disintegration of Heavy Nuclei," *Nature* 143 (1939): 330 (submitted 20 January 1939); N. Bohr, "Resonance in Uranium and Thorium Disintegrations and the Phenomenon of Nuclear Fission," *Phys. Rev.* 55 (1939): 418–419 (7 February 1939).

13. N. Bohr and J. A. Wheeler, "The Mechanism of Nuclear Fission," *Phys. Rev.* 56 (1939): 426–450.

14. For the Bohr-Wheeler theory and its relationship to the subsequent development of fission, see Abraham Pais, *Niels Bohr's Times, in Physics, Philosophy, and Polity* (Oxford: Clarendon Press, 1991), 455–459; John A. Wheeler, "The Mechanism of Fission," *Physics Today* 20, no. 11 (November 1967): 49–52; Wheeler, "Some Men and Moments in the History of Nuclear Physics: The Interplay of Colleagues and Motivations," in Roger H. Stuewer, ed., *Nuclear Physics in Retrospect: Proceedings of a Symposium on the 1930s* (Minneapolis: University of Minnesota Press, 1979), 273ff.; Turner, "Nuclear Fission," 17–20; Richard Rhodes, *The Making of the Atomic Bomb* (New York: Simon & Schuster, 1988), 282–288.

A personal description of the experimental verification of the fission of ^{235}U is given by Alfred O. Nier in Stuewer, *Nuclear Physics in Retrospect*, 310.

15. H. von Halban, F. Joliot, and L. Kowarski, "Liberation of Neutrons in the Nuclear Explosion of Uranium," *Nature* 143 (1939): 470–471; H. L. Anderson, E. Fermi, and H. B. Hanstein, "Production of Neutrons in Uranium Bombarded by Neutrons," *Phys. Rev.* 55 (1939): 797–798.

16. Otto Robert Frisch, "The Interest Is Focussing on the Atomic Nucleus," in S. Rozental, ed., *Niels Bohr: His Life and Work as Seen by His Friends and Colleagues* (Amsterdam: North Holland/New York: John Wiley & Sons, 1967), 147–148. In *What Little I Remember* (Cambridge: Cambridge University Press, 1979), Frisch remembers that he first heard mention of a chain reaction from neutron multiplication from the Copenhagen physicist Christian Møller (p. 118).

17. For an overview of the research and reaction from the physics community and the press, see Lawrence Badash, Elizabeth Hodes, and Adolph Tiddens, "Nuclear Fission: Reaction to the Discovery in 1939," *Proc. Amer. Phil. Soc.* 130 (1986): 196–231.

18. After the war, when German physicists learned that Meitner and Frisch knew of the barium finding two weeks before they did, some were furious. The same group subsequently adopted the view that Meitner's part in the uranium

investigation had impeded the discovery of fission. For example, R. Fleischmann to P. Van Assche, pers. comm., 22 November 1982: "[Frisch] did the same [i.e., detecting the fission fragments] as [Viennese physicists] W. Jentschke and F. Prankl, also G. von Droste, only Frisch was informed about the Hahn-Strassmann findings 14 days before v. Droste and 17 days before Jentschke. . . . L. Meitner emphasized over and over again, 'only protons and alpha-particles can leave the nucleus, anything else is impossible.'" See also Krafft, *Im Schatten der Sensation*, 94–95, and chap. 16, page 374.

19. The institute's high-tension apparatus, constructed by Reddemann under Meitner's direction and operational since the summer of 1938, was not functioning in January 1939, a crucial month for fission experiments. See Krafft, *Im Schatten der Sensation*, 276.

20. Krafft, *Im Schatten der Sensation*, 95 n. 68.

21. In Hahn to Meitner, 7 February 1939 (MC), fission was "a heaven-sent gift" that would protect him and his institute.

22. Hahn to Meitner, 3 March 1939 (MC); in Krafft, *Im Schatten der Sensation*, 324–326.

23. Ibid.

24. Ibid.

25. Friedrich Herneck, *Bahnbrecher des Atomzeitalters* (Berlin: Buchverlag der Morgen, 1969), 454–455.

26. Hahn to Meitner, 3 March 1939 (MC).

27. See chap. 10, pp. 252.

28. In 1935, Meitner and Hahn believed that Irène Curie had not properly acknowledged their priority for some of the early results of the neutron irradiation of thorium (see chap. 7). The comment about radioactive recoil goes back to 1908 when Hahn discovered it in the process ^{227}Th → ^{223}Ra and Rutherford insisted he had proposed and described it for ^{218}Po → ^{214}Pb some years before (see chap. 2).

29. Hahn to Meitner, 3 March 1939 (MC).

30. Mark Walker, *German National Socialism and the Quest for Nuclear Power, 1939–1949* (Cambridge: Cambridge University Press, 1989), 17–20, 42. Rhodes, in *Making of the Atomic Bomb*, 296, indicates that Hahn was invited to the secret first fission meeting on 29 April 1939 but chose not to attend. In Dietrich Hahn, ed., *Otto Hahn, Erlebnisse und Erkenntnisse* (Düsseldorf: Econ Verlag, 1975), 62–63, Hahn himself remembers that he was at first not invited to the April meeting (ostensibly because he was not a physicist, but more likely because he was politically suspect) and after a late invitation sent Josef Mattauch in his place; he sent Siegfried Flügge to the next meeting and attended subsequent meetings himself.

31. D. Hahn, *Otto Hahn, Erlebnisse und Erkenntnisse*, 54.

32. Meitner to Hahn, 6 March 1939 (MC); in Krafft, *Im Schatten der Sensation*, 326–327.

33. Meitner to Hahn, 7 March 1939 (MC).

34. Pais, *Niels Bohr's Times*, 400–401; Frisch, "Interest Is Focussing," 142. The high-tension apparatus in the Bohr institute was begun in 1935; Frisch was involved in its construction off and on until its completion in February 1939, when it first accelerated deuterons to produce neutrons of energy in the 1 million electron volt range. Meitner and Frisch's "transurane" experiment was one of the first successful applications of the new equipment.

35. L. Meitner and O. R. Frisch, "Products of the Fission of the Uranium Nucleus," *Nature* 143 (1939): 471–472; submitted 6 March, appeared 18 March 1939.

36. Meitner to Hahn, 10 March 1939 (MC); in Krafft, *Im Schatten der Sensation*, 302–303. According to Krafft (p. 302), the experiments were done using the Bohr institute's cyclotron, but in their article, "Products of the Fission of the Uranium Nucleus," Meitner and Frisch state that they used the institute's high-tension equipment.

37. Hahn to Meitner, 13 March 1939 (MC); in Krafft, *Im Schatten der Sensation*, 304–305.

38. Lise Meitner, "New Products of the Fission of the Thorium Nucleus," *Nature* 143 (1939): 637; submitted 26 March, published 15 April 1939.

39. Meitner to Bohr, 24 March 1939 (MC).

40. Meitner to Hahn, 28 March 1939 (MC); 31 March 1939, in Krafft, *Im Schatten der Sensation*, 178. Unlike Meitner, the Frisch couple emigrated legally and thus had less trouble sending for their belongings than she did. See Krafft, p. 179.

41. Hahn to Meitner, 20 March 1939; in D. Hahn, *Otto Hahn, Erlebnisse und Erkenntnisse*, 115.

42. In D. Hahn, *Otto Hahn, Erlebnisse und Erkenntnisse*, 117–118.

43. Hahn to Meitner, 24 March 1939; in D. Hahn, *Otto Hahn, Erlebnisse und Erkennntnisse*, 116–117.

44. Correspondence with Dr. F. W. Arnold (MC).

45. Meitner to Hahn, 10 April 1939 (MC).

46. Hahn to Meitner, 14 April 1939 (MC).

47. Paul Rosbaud to Meitner, 18 April 1939 (MC).

48. Meitner to Hahn, 13 May 1939 (MC).

49. Meitner to Hahn, 18 May 1939 (MC).

50. Meitner to Hahn, 2 June 1939 (MC). Edith begged Lise to gain weight (Edith Hahn to Meitner, 30 June 1939 [MC]): "With willpower and tireless eating one can get back to one's former weight."

51. Max von Laue to Meitner, 14 January 1939 (MC).

52. Ibid.

53. Laue to Meitner, 15 February 1939 (MC).

54. Laue to Meitner, 3 March 1939 (MC).

55. Laue to Meitner, 7 April 1939 (MC).

56. Max von Laue, "Meine physikalischer Werdegang," in *Gesammelte Schrifte und Vorträge* (Braunschweig: Friedrich Vieweg Verlag, 1961), 3:xxvi.

57. Laue to Meitner, 24 April 1939 (MC).

58. Hahn to Meitner, 20 March 1939 (MC).

59. Meitner to Hahn, 28 March 1939 (MC).

60. Ida Noddack, "Über das Element 93," *Z. Angew. Chem.* 47 (1934): 653–654.

61. Krafft, *Im Schatten der Sensation*, 316. Pieter Van Assche, pers. comm., 5 June 1990, states that although the Noddacks lacked expertise in radioactivity and therefore were unable to pursue Ida Noddack's suggestion, they did check Fermi's chemical separations in 1934, and they frequently spoke to Hahn about the radioactive consequences of a possible nuclear disintegration.

62. Ida Noddack, "Bemerkung zu den Untersuchungen von O. Hahn, L. Meitner und F. Straßmann über die Produkte, die bei der Bestrahlung von Uran mit Neutronen entstehen," *Naturwiss.* 27 (1939): 212–213; submitted 10 March, appeared 31 March 1939. For the text of Noddack's letter and Strassmann and Hahn's unpublished response, see Krafft, *Im Schatten der Sensation*, 317–322.

63. Paul Rosbaud to Meitner, 18 April 1939 (MC). Meitner to Rosbaud, draft, April 1939 (MC).

64. Krafft, *Im Schatten der Sensation*, 316.

65. Elisabeth Crawford, J. L. Heilbron, and Rebecca Ullrich, *The Nobel Population 1901–1937: A Census of the Nominators and Nominees for the Prizes in Physics and Chemistry* (Berkeley: University of California, Office for History of Science and Technology/Uppsala: Uppsala University, Office for History of Science, 1987). The Noddacks were nominated for the Nobel Prize in chemistry in 1932, 1933, 1934, 1935, and 1937.

66. According to Pieter Van Assche, pers. comm., 5 June 1990, during the period 1935–1938, whenever Walter Noddack tried to remind Hahn of Ida Noddack's suggestion of nuclear disintegration, Hahn would always dismiss it, saying, "Ein Fehler reicht [One mistake is enough]." Recently there has been a call for reevaluation of the Noddacks' claim to element 43. Van Assche believes the Noddack-Berg x-ray apparatus was sufficiently sensitive to detect element 43 in the ores they analyzed; he notes that long-lived Tc isotopes are indeed present in uranium ores, presumably the products of spontaneous natural fission. See Pieter H. M. Van Assche, "Ignored Priorities: First Fission Fragment (1925) and First Mention of Fission (1934)," *Nuclear Europe* 6–7 (1988): 24–25; Van Assche, "The Ignored Discovery of the Element Z=43," *Nuclear Physics* A480 (1988): 205–214. This raises the intriguing possibility that Ida Noddack not only suggested nuclear disintegration in 1934 but may have unknowingly identified the first fission fragment in 1925, as stated by Van Assche and by Cornelius Keller, "Verpasste Chancen: Warum wurde die Kernspaltung nicht schon früher entdeckt?" *Bild der Wiss.* 2 (1988): 102–111. The Noddacks did not, however, note any correlation between their presumed Ma and uranium ores; on this basis and

also with regard to spectral sensitivity, Günter Herrmann disagrees with Van Assche and concludes the opposite: Günter Hermann, "Technetium or Masurium: A Comment on the History of Element 43," *Nuclear Physics* A505 (1989): 352–360. See also Teri Hopper, *"She was Ignored"*: *Ida Noddack and the Discovery of Nuclear Fission*, Master's thesis, Stanford University, 1990, 15–18, 21; Fathi Habashi, "Ida Noddack 1896–1978;" *CIM Bulletin* (May 1985): 90–93; B. Voland, "Zur Geschichte der Entdeckung des Elementes Technetium," *Isotopenpraxis* 24 (1988): 445–488.

67. Interview, Emilio Segrè, 21 May 1985, Lafayette, Calif. Following Segrè and Perrier's finding of element 43 in 1937, Walter Noddack visited Segrè in Palermo, insisting that he and his wife had isolated 0.5 mg of nonradioactive element 43; for his visit Noddack wore an irregular military uniform complete with swastikas. According to Segrè, the Noddacks were ultranationalists who had chosen the name *rhenium* for a World War I German victory on the Rhine, and *masurium* for a lake in East Prussia, site of another German victory. See also Segrè, *A Mind Always in Motion: The Autobiography of Emilio Segrè* (Berkeley, Los Angeles, and Oxford: University of California Press, 1993), 115–118. Other sources, including Van Assche, Voland, and Hopper (see n. 66), give a more prosaic version, namely, that the elements were named for Ida's Rhine birthplace and Walter's home region of Masuren in the former East Prussia, now Poland. Van Assche, in his 1988 article in *Nuclear Europe* (see n. 66), notes that the Noddacks' alleged Nazi leanings have been strongly disputed.

68. Hahn to Meitner, 7 April 1939 (MC); in Krafft, *Im Schatten der Sensation*, 319. Paul Rosbaud believed the Noddacks were sympathetic to national socialism although probably not party members; in a statement written 5 August 1945 (Samuel Goudsmit correspondence, AIP), Rosbaud claims that after 1933 the Noddacks failed to cite their former collaborator, Otto Berg, and also V. M. Goldschmidt, both Jewish. In 1946 (Rosbaud to Goudsmit, 9 August 1946, Goudsmit correspondence, AIP), Rosbaud listed the Noddacks among rehabilitated pro-Nazis who were "in high spirits" in Germany.

69. Hopper, *"She Was Ignored,"* 23–24.

70. Hahn to Major Rittner, Farm Hall, 8 August 1945 (MC). Reprinted in Ruth Lewin Sime, "Lise Meitner and Fission: Fallout from the Discovery," *Angew. Chem. Intl. Ed. Engl.* 30 (1991): 942–953; also *Angew. Chem.* 103 (1991): 956–967.

71. W. Heisenberg, *Orden pour le mérite für Wissenschaft und Künste, Reden und Gedenkworte* 9 (1968–1969): 111–119.

72. Hahn to Meitner, 20 March 1939 (MC); in Krafft, *Im Schatten der Sensation*, 327.

73. In the 18 March 1939 issue of *Nature*, the article by H. von Halban, F. Joliot, and L. Kowarski (*Nature* 143 [1939]: 470–471) is followed immediately by Meitner and Frisch's (*Nature* 143 [1939]: 471–472).

74. Meitner to Hahn, 28 March 1939 (MC); in Krafft, *Im Schatten der Sensation*, 327.

75. N. Feather, "Fission of Heavy Nuclei: A New Type of Nuclear Disintegration," *Nature* 143 (1939): 877–879.

76. Hahn to Norman Feather (in English), 2 June 1939 (MC); reprinted in Krafft, *Im Schatten der Sensation*, 330–331.

77. Meitner to Hahn, 2 and 3 June 1939, draft and final copy (MC); also Krafft, *Im Schatten der Sensation*, 329–333.

78. Hahn and Strassmann, "Existenz der 'Trans-Urane.'"

79. Meitner to Hahn, 12 July 1939 (MC). Hahn agreed that his citation was incorrect, and changed it (Hahn to Meitner, 20 July 1939; in Krafft, *Im Schatten der Sensation*, 123).

80. Meitner to Hahn, 15 July 1939 (MC); in Krafft, *Im Schatten der Sensation*, 122. In Hahn to Meitner, 20 July 1939 (Krafft, p. 123), Hahn admitted he had forgotten to cite Feather and Bretscher's work.

81. Hahn to Meitner (in English), card, 21 June 1939 (MC). Rosbaud to Goudsmit, 9 August 1946 (Goudsmit correspondence, AIP). According to Rosbaud, Wever reported to the Nazi authorities on Hahn and his impressions of England.

82. Meitner to Walter Meitner, 18 June 1939 (MC).

83. Meitner-Hahn correspondence, June-July 1939; in Krafft, *Im Schatten der Sensation*, 120–123.

84. Hahn to Meitner, 1 July 1939; in Krafft, *Im Schatten der Sensation*, 120.

85. Meitner to Hahn, 12 July 1939 (MC).

86. See n. 12, above. Bohr refers to the earlier uranium investigations and the "striking peculiarities" in the products of uranium fission, which, he notes, "could not be disentangled on the ordinary ideas of nuclear disintegrations." This alludes to the fact that the radiochemical properties of the "transurane" mixture were independent of neutron energy, so that all investigators, including Meitner, were misled into believing that the slow- and fast-neutron processes were the same and that the nucleus of origin was, therefore, ^{238}U. Once it was understood that the "transuranes" were fission fragments, the puzzle cleared up. Fission, Bohr pointed out, occurs in "a large number of different ways, in which a wide range of mass and charge numbers of the fragments may occur," so that the slow-neutron fission of ^{235}U and the fast-neutron fission of ^{238}U will have some fission products in common.

87. Meitner to Hahn, 15 July 1939 (MC); in Krafft, *Im Schatten der Sensation*, 122.

88. Meitner to Hahn, 12 July 1939 (MC); in Krafft, *Im Schatten der Sensation*, 121.

89. Siegfried Flügge, "Kann der Energieinhalt der Atomkerne technisch nutzbar gemacht werden?" *Naturwiss.* 27 (1939): 402–410.

90. Hahn to Meitner, 13 July 1939; in Krafft, *Im Schatten der Sensation*, 121.

91. Meitner to Hahn, 15 July 1939 (MC).

92. Walker, *German National Socialism*.

93. Albert Einstein to President Roosevelt, 2 August 1939, in Otto Nathan and Heinz Norden, eds., *Einstein on Peace* (New York: Avenel Books, 1981), 294–295; and Badash, Hodes, and Tiddens, "Nuclear Fission."

94. Meitner to Hahn, 16 June 1939 (MC).

95. Meitner, diary, 18 July 1939 (MC).

96. Meitner, diary, 3 August 1939 (MC).

97. Meitner to Hahn, 26 February 1942 (MC).

98. Hahn to Meitner, 12 August 1939 (MC).

99. Hahn to Meitner, 24 August 1939 (MC).

100. Meitner to Hahn, 25 August 1939 (MC).

CHAPTER 12. Again, World War

1. Edith Hahn to Lise Meitner, 4 September 1939 (MC).

2. Otto Hahn to Meitner, 25 October 1939 (MC).

3. Hahn to Meitner, 13 October 1939 (MC).

4. Max von Laue to Meitner, 6 September 1939; Meitner to Hahn, 2 September 1939 (MC).

5. Laue to Meitner, 11 September 1939 and 16 September 1939 (MC).

6. Meitner to James Chadwick, 14 September 1939; J. D. Cockcroft to Meitner, 16 October 1939; Meitner to Cockcroft, 3 November 1939; W. L. Bragg to Meitner, 25 October 1939; Meitner to Cockcroft, 18 March 1940 (MC). See also Guy Hartcup and T. E. Allibone, *Cockcroft and the Atom* (Bristol: Adam Hilger, 1984), 87; Margaret Gowing, *Britain and Atomic Energy, 1939–1945* (London: Macmillan/New York: St. Martin's Press, 1964), 37.

7. Meitner to Hahn, 12 November 1939 (MC).

8. Meitner to Laue, 12 November 1946 (MC).

9. L. Meitner and O. R. Frisch, "Products of the Fission of Uranium and Thorium under Neutron Bombardment," *K. Dansk. Vid. Selsk. Mat.-fys. Medd.* 17 (1939): 5.

10. Meitner to Hahn, 17 December 1939 (MC).

11. Meitner to Hahn, 30 January 1940 (MC).

12. Lise Meitner, "Capture Cross-Sections for Thermal Neutrons in Thorium, Lead and Uranium 238," *Nature* 145 (1940): 422–423.

13. Lise Meitner, "Über das Verhalten einiger seltenen Erden bei Neutronenbestrahlung," *Ark. Mat. Astr. Fys.* 27A, no. 17 (1940).

14. Hilde Levi, *George de Hevesy: Life and Work* (Bristol: Adam Hilger, 1985), 78–79.

15. In 1936, Meitner and Hahn chemically identified the 23-minute activity as uranium and showed that it was a beta emitter by photographing its beta particles in a cloud chamber, which proved that element 93 must exist. Meitner's 1937 determination of neutron capture by ^{238}U to form ^{239}U established the conditions for producing element 93. See Lise Meitner and Otto Hahn, "Neue

Umwandlungsprozesse bei Bestrahlung des Urans mit Neutronen," *Naturwiss.* 24 (1936): 158–159 (submitted 10 February, published 6 March 1936); Lise Meitner, Otto Hahn, and Fritz Strassmann, "Über die Umwandlungsreihen des Urans, die durch Neutronenbestrahlung erzeugt werden," *Z. Phys.* 106 (1937): 1374–1392.

16. E. McMillan, "Radioactive Recoils from Uranium Activated by Neutrons," *Phys. Rev.* 55 (1939): 510 (17 February 1939).

17. Emilio Segrè, *A Mind Always in Motion: The Autobiography of Emilio Segrè* (Berkeley, Los Angeles, and Oxford: University of California Press, 1993), chap. 6; Segrè, "Fifty Years Up and Down a Strenuous and Scenic Trail," *Ann. Rev. Nucl. Part. Sci.* 31 (1981): 1–18.

18. Emilio Segrè, "An Unsuccessful Search for Transuranic Elements," *Phys. Rev.* 55 (1939): 1104. See also Segrè, *Autobiography*, 152–153; J. L. Heilbron and Robert W. Seidel, *Lawrence and His Laboratory: A History of the Lawrence Berkeley Laboratory* (Berkeley, Los Angeles, and Oxford: University of California Press, 1989), 1:456–457; Richard Rhodes, *The Making of the Atomic Bomb* (New York: Simon & Schuster, 1986), 348–349.

19. Abraham Pais, *Niels Bohr's Times, in Physics, Philosophy, and Polity* (Oxford: Clarendon Press, 1991), 204–207. Bohr first proposed that a second rare-earth series might begin beyond uranium in June 1922, in a series of lectures in Göttingen. See also Fritz Krafft, *Im Schatten der Sensation: Leben und Wirken von Fritz Strassmann* (Weinheim: Verlag Chemie, 1981), 96–97, and Edoardo Amaldi, "From the Discovery of the Neutron to the Discovery of Nuclear Fission," *Physics Reports* 111 (1984): 276, 295. In 1934, Aristide von Grosse pointed out that transuranics might be higher homologues of the lanthanides, but this was disregarded (even by Bohr) during studies of the false transuranes, probably because the chemistry of elements Ac to U did appear more like that of transition elements than lanthanides.

20. Meitner to Hahn, 3 October 1940 (MC). Meitner retained her interest in recoil, which remained a valuable experimental tool. As late as 1952, she surveyed the theoretical and experimental applications of α, β, and γ recoil in a review article, "Die Anwendung des Rückstoßes bei Atomkernprozessen," *Z. Phys.* 133 (1952): 140–152.

21. Otto Frisch, *What Little I Remember* (Cambridge: Cambridge University Press, 1979), 108.

22. Stefan Rozental, "The Forties and Fifties," in S. Rozental, ed., *Niels Bohr: His Life and Work as Seen by His Friends and Colleagues* (Amsterdam: North Holland/ New York: John Wiley, 1967), 155.

23. George de Hevesy to Max von Laue, 6 January 1957 (MC). Hevesy also dissolved James Franck's medals. See Pais, *Niels Bohr's Times,* 480; Ruth Moore, *Niels Bohr: The Man, His Science, and the World They Changed* (New York: Knopf, 1966), 302–303.

24. Institute for Theoretical Physics, guest book (NBA): Meitner's visit was from 8 April to 29 April 1940. For Jews in occupied Denmark, see Pais, *Niels Bohr's Times,* 476–479.

25. Meitner to 23 May 1940 (MC). According to Pais, *Niels Bohr's Times*, 479, the attack on Denmark had been foreseen in Norway. For Frisch in Birmingham, see Sir Rudolf Peierls, "Otto Robert Frisch, 1 October 1904–22 September 1979" *Biog. Mem. Fell. Roy. Soc. Lond.* 27 (1981): 290.

26. Hartcup and Allibone, *Cockcroft and the Atom*, 121–122; Rudolf Peierls, *Bird of Passage: Recollections of a Physicist* (Princeton: Princeton University Press, 1985), 152–156; Frisch, *What Little I Remember*, 131.

27. Edwin McMillan and Philip H. Abelson, "Radioactive Element 93," *Phys. Rev.* 57 (1940): 1185–1186 (submitted 27 May 1940). See also Gowing, *Britain and Atomic Energy*, 44ff.; Heilbron and Seidel, *Lawrence and His Laboratory*, 457–458; Rhodes, *Making of the Atomic Bomb*, 349–350.

28. Meitner to Hahn, 3 October 1940 (MC). According to McMillan's 1951 Nobel lecture, extracted in Rhodes, *Making of the Atomic Bomb*, 349–350, McMillan also realized that the activity within the uranium had to be transuranic, as a fission product would have been expelled by recoil.

29. According to Frisch (interview, 22 March 1975, Cambridge), Meitner felt she really had discovered the first true transuranium element.

30. Systematic deportations of German Jews to Poland started from towns and cities (like Breslau) that were near the German-Polish border. See Raul Hilberg, *The Destruction of the European Jews* (New York: Octagon Books, 1978), 137ff.

31. Max Born to Meitner, 22 December 1939 (MC). Kohn could not be admitted to Britain unless she had been in a neutral country before 3 September 1939.

32. The £50 was given by the International Federation of University Women in London; Dr. Karl Herzfeld of the Catholic University of America contributed $300.

33. Meitner to Dr. Hollitscher, 10 March 1940 (MC).

34. Hedwig Kohn to Meitner, 9 April 1940 (MC).

35. Hedwig Kohn to Meitner, card, 18 October 1940; Kohn to Meitner, letter, 23 July 1941 (MC).

Hedwig Kohn, born in 1887 in Breslau, specialized in optics and radiation measurements. She received her doctorate from Breslau in 1913, was habilitiert in 1918, and worked there as an Assistent from 1914 to 1930 and as a Privatdozent until she was dismissed in 1934. In the United States, she was an instructor in the Women's College of the University of North Carolina from 1940 to 1942; then at Wellesley College, lecturer (1942–1945), associate professor (1945–1948), and professor until she retired in 1952 and returned to North Carolina as a research associate at Duke University. She died in 1965. Her brother did not survive the Holocaust. Elisabeth Boedeker and Maria Meyer-Plath, *50 Jahre Habilitation von Frauen in Deutschland: Eine Dokumentation über den Zeitraum von 1920–1970* (Göttingen: Verlag Otto Schwartz, 1974), 174.

36. Hilberg, *Destruction of the European Jews*, 83–90, 98–122.

37. Meitner to Hahn, 1 May 1941 (MC).

38. For Meyer, see Wolfgang L. Reiter, "Österreichische Wissenschaftsemigration am Beispiel des Instituts für Radiumforschung der Österreichischen Akademie der Wissenschaften," in Friedrich Stadler, ed., *Vertriebene Vernunft II: Emigration und Exil österreichischer Wissenschaft* (Wien: Jugend und Volk, 1988), 709–729. In 1939, the Joliot-Curies in Paris and George de Hevesy in Copenhagen tried to find a position for Meyer, without success. See also Levi, *George de Hevesy*, 98.

For the situation in Austrian universities, see Reiter, "Das Jahr 1938 und seine Folgen für die Naturwissenschaften an Österreichs Universitäten," in Stadler, *Vertriebene Vernunft II*, 664–680. For "Mischlinge," see Hilberg, *Destruction of the European Jews*, 268ff.

39. Meitner to Hans Pettersson, 6 November 1941 (MC).

40. Meitner to Pettersson, 8 December 1941 (MC). Europeans were aware of American anti-Semitism; prior to World War II, relatively few Jews occupied teaching positions in American colleges and universities, Jewish student enrollments were restricted, and discrimination in accommodations ("restricted clientele") and employment ("no Jews need apply") was obvious. James Franck regarded it as worse than in Germany before 1933. See Victor F. Weisskopf, "Vertriebene Vernunft aus Österreich," in Stadler, *Vertriebene Vernunft II*, 698–701; Weisskopf, *The Joy of Insight: Passions of a Physicist* (New York: Basic Books, 1991), 113; Segrè, *Autobiography*, 102, 168; Paul Hoch, "Flight into Self-Absorption and Xenophobia," *Physics World* (January 1990): 23–26.

41. Meitner to Pettersson, 5 March 1942 (MC).

42. Meitner to Pettersson, 24 April 1942 (MC).

43. Hilberg, *Destruction of the European Jews*, 264–266; Gordon A. Craig, *Germany, 1866–1945* (New York: Oxford University Press, 1978), 748–749.

44. Meitner to Pettersson, 18 December 1942 (MC). The Meitner-Pettersson correspondence, most of it concerning Stefan Meyer, numbers close to 300 letters.

45. Pettersson to Meitner, 8 July 1944 (MC).

46. Lise Meitner, "Resonance Energy of the Th Capture Process," *Phys. Rev.* 60 (1941): 58; Lise Meitner, "Radiations Emitted by Scandium 46," *Ark. Mat. Astr. Fys.* 28B 4, no. 14 (1942); Lise Meitner, "Disintegration Scheme of Scandium 46," *Ark. Mat. Astr. Fys.* 32A, no. 6 (1945); L. Meitner, "A Simple Method for the Investigation of Secondary Electrons Excited by γ-Rays and the Interference of These Electrons with Measurements of Primary β-Ray Spectra," *Phys. Rev.* 63 (1943): 73 and 384 (erratum), and *Ark. Mat. Astr. Fys.* 29A, no. 17 (1943).

47. Meitner to Eva von Bahr-Bergius, 27 May 1941 (MC).

48. Meitner to O. R. Frisch, 9 March 1941 (MC).

49. Meitner to Hahn, 4 March 1944 (MC).

50. Meitner to Laue, 15 August 1941 (MC). The cyclotron was tested in the fall of 1939, underwent a major redesign, and first produced deuterons in December 1941, approximately 2 years behind schedule. See Hugo Atterling, "Karl

Manne Georg Siegbahn, 1886–1978," *Biog. Mem. Fell. Roy. Soc. Lond.* 37 (1991): 439.

51. Meitner to Hahn, 26 February 1942 (MC).

52. Meitner to Hahn, 4 March 1944 and 21 June 1944 (MC).

53. Frisch, *What Little I Remember*, 13–14. In 1951, Kai Siegbahn was appointed "without much discussion" to a professorship at the Royal Institute of Technology. See Ulf Larsson, "Physics in a Stronghold of Engineering: Professorial Appointments at the Royal Institute of Technology, 1922–1985," in Svante Lindqvist, ed., *Center on the Periphery: Historical Aspects of 20th-Century Swedish Physics* (Canton, Mass.: Science History Publications, 1993), 58–75.

54. In 1941, Siegbahn asked Meitner to help his son prepare some ThB (^{212}Pb) to test a beta spectrograph, "whose existence I thereby heard of for the first time." She did help but was never told the outcome of the test. (Meitner to Hahn, 3 September 1941 [MC]; reprinted in Fritz Krafft, "Lise Meitner und ihre Zeit: Zum hundertsten Geburtstag der bedeutenden Naturwissenschaftlerin," *Angew. Chem.* 90 [1978]: 876–892, and *Angew. Chem. Intl. Ed. Engl.* 17 [1978]: 826–842.)

55. Meitner to Hahn, 30 December 1942, in Krafft, "Lise Meitner und ihre Zeit," 889.

56. Robert Marc Friedman, "Text, Context, and Quicksand: Method and Understanding in Studying the Nobel Science Prizes," *Hist. Stud. Phys. Sci.* 20, no. 1 (1989): 74.

57. Pettersson to Meitner, 8 February 1942 (MC).

58. Robert Vestergaard to Meitner, n.d. (MC).

59. Meitner to Pettersson, 15 June 1943 (MC). In this letter Meitner wanted to refute certain "allegations" by Siegbahn, which, she said, led to "unjustified, malicious rumors being spread about me."

60. Meitner to Bahr-Bergius, 25 January 1943 (MC).

61. Ingmar Bergström, Sigvard Eklund, interviews, May 1992; Sigvard Eklund, "Forskningsinstitutet för Atomfysik, 1937–1987," in Per Carlson, ed., *Fysik i Frescati: Föredrag från jubileumskonferens den 23 oktober 1987* (Stockholm: Manne Siegbahn Institute, 1989); Ingmar Bergström and Wilhelm Forsling, *I Demokritos Fotspår: En vandring genom urämnesbegreppets historia från antiken till Nobelprisen* (Stockholm: Natur och Kultur, 1992), 335.

62. Robert Marc Friedman, "Karl Manne Georg Siegbahn," *Dict. Sci. Biog.*, Supplement 2, ed. Frederic L. Holmes (New York: Charles Scribners, 1990), 18:821–826; Friedman, "Text, Context, and Quicksand."

63. In 1922, Siegbahn had a preferred candidate for professor of physics at the Royal Institute of Technology, but the appointment went to Gudmund Borelius, another of Meitner's friends in Sweden. See Larsson, "Physics in a Stronghold of Engineering," 62–64.

64. Meitner to Hahn, 26 February 1942 (MC).

65. Meitner to Hahn, 5 December 1938 (MC).

66. Meitner to Hahn, 9 September 1940 (MC).

67. In Meitner to Hahn, 20 August 1940 (MC), Meitner was pleased that Louis A. Turner, in his January 1940 review article "Nuclear Fission," *Rev. Mod. Phys.* 12 (1940): 1–29, gave appropriate recognition to Meitner and Frisch, "because in Germany it has been gradually forgotten that the basis for the explanation originated with Otto Robert and me."

68. Karl Friedrich Frhr. von Weizsäcker, "Die theoretische Deutung der Spaltung von Atomkernen," *Forschungen und Fortschritte* 17 (1941): 10–11. (In this and other wartime publications, Weizsäcker's first name, Carl, was Germanicized). See also Fritz Krafft, "Lise Meitner und die Entdeckung der Kernspaltung," *Mitteilg. Österr. Ges. Gesch. Naturwiss.* 4 (1984): 1–6 n. 15.

69. Meitner to Hahn, 15 April 1943 (MC).

70. Primo Levi, *The Drowned and the Saved*, trans. Raymond Rosenthal (New York: Vintage International, 1989), 31.

71. Meitner to Bahr-Bergius, 27 May 1941 (MC).

72. Meitner to Laue, 4 September 1942 (MC).

73. Meitner to Bahr-Bergius, 27 May 1941 (MC).

74. Meitner to Bahr-Bergius, 25 January 1943 (MC).

75. Meitner to Bahr-Bergius, 27 May 1941 (MC).

76. Erwin Schrödinger to Meitner, 22 October 1939 (MC). See Walter Thirring, "Die Emigration Erwin Schrödingers," in Stadler, *Vertriebene Vernunft II*, 730–732; Walter Moore, *Schrödinger: Life and Thought* (Cambridge: Cambridge University Press, 1989).

77. Meitner to Max Born, 22 October 1944 (MC); Meitner to Theodor von Laue, 9 December 1944 (MC); Max von Laue to Theodor von Laue, 21 June 1945 (LP).

78. Meitner to Hahn, 4 March 1944 (MC).

79. Hilde Levi, pers. comm., May 1992. Levi, *George de Hevesy*, 99–100.

80. For the Bohr family's escape from Denmark, see Pais, *Niels Bohr's Times*, 487–488.

81. *Kungliga Svenska Vetenskapsakademiens Årsbok*, 1944, 1945 (Stockholm: Almqvist & Wiksells Boktryckeri, 1944, 1945). I am grateful to Urban Wråkberg for assistance.

82. Meitner to Annemarie Schrödinger, 30 January 1953 (MC).

83. Meitner-Schiemann correspondence; Meitner to Hahn, 14 June 1940 and 22 July 1940; Meitner to Hahn, 26 February 1942 (MC).

Between 1940 and 1943, Schiemann received some grants and worked at the botanical museum in Dahlem; in 1943, she was made section head at the KWI für Kulturpflanzenforschung (Cultivated Plant Research), a position she kept even after she was reinstated as professor in Berlin in 1945. See Elisabeth Schiemann, "Autobiographie," in *Nova Acta Leopoldina, N.F. (Leipzig)* Bd. 21, Nr. 143 (1959): 291–292; Ute Deichmann, *Biologen Unter Hitler: Vertreibung, Karrieren, Forschung* (Frankfurt: Campus Verlag, 1992), 45.

84. Meitner-Rosbaud correspondence (MC). On a visit to the United States just before the war, Laue expressed the same to Einstein: "I hate them so much I must be close to them. I have to go back." Ronald W. Clark, *Einstein: The Life and Times* (New York: Avon, 1971), 639; Alan D. Beyerchen, *Scientists under Hitler: Physics and the Physics Community in the Third Reich* (New Haven: Yale University Press, 1977), 65–66.

85. Report by H. Sarkowski for Springer internal publication *Zentralblatt*, 1987; Arnold Kramish, *The Griffin: The Greatest Untold Espionage Story of World War II* (Boston: Houghton Mifflin, 1986); review of *The Griffin* by R. V. Jones, "A Merchant of Light," *Nature* 325 (1987): 203–204.

86. Meitner to Paul Rosbaud, 4 August 1946 (MC).

87. Rosbaud to Meitner, 5 April 1944 (MC).

88. As quoted by Elisabeth Schiemann in a letter to O. R. Frisch, 14 July 1969 (MC).

89. Laue to Meitner, 3 March 1939, 24 April 1939; Magda von Laue to Meitner, 20 April 1941 (MC).

90. Laue to Meitner, 26 October 1938 (letter written and mailed from Copenhagen) and 14 March 1942 (MC); Stefan Meyer was dismissed from the Prussian Academy 24 November 1938, as were corresponding members Victor Hess, Erwin Schrödinger, and Viktor Goldschmidt.

91. Max von Laue to Meitner, 3 March 1939 (MC). It appears that Wilhelm Lenz's article was not published after all; see Beyerchen, *Scientists under Hitler*, 170, 256 n. 9.

92. Correspondence Max von Laue–C. F. von Weizsäcker, 1943 (LP, IMC 29 1051).

93. Laue to Meitner, 2 March 1940 (MC); Laue to Theodor von Laue, 9 December 1945, p. 4 (LP). In a memorandum written in February 1946, Paul Rosbaud noted, "Food rations for Dr. Berliner, as for all his fellow sufferers, were curtailed; they received neither vegetables nor fruit nor meat nor coffee." (I am grateful to Arnold Kramish for a copy of the Rosbaud document.) For anti-Jewish measures, see Hilberg, *Destruction of the European Jews*, 102, 116–119.

94. Max von Laue, "Arnold Berliner, 26.12.1862–22.3.1942," *Naturwiss.* 33 (1946): 17–18. This issue is dated 15 November 1946, but appeared only in April 1947. Also see Laue to Theodor von Laue, 9 December 1945, p. 5 (LP).

95. Arnold Berliner to Meitner, 27 February 1940 (MC).

96. Berliner to Meitner, 21 November 1940 (MC).

97. Laue to Meitner, card, 22 March 1942 (MC). According to Rosbaud (see n. 93, above), Berliner feared he would be sent to a *Massenquartier* (barracks), a first step to deportation.

98. Laue to Meitner, 24 March 1942; Meitner to Laue, 8 April 1942; Laue to Meitner, 14 April 1942 (MC). For the poison, Rosbaud (see n. 93, above) states that Fritz Haber gave Berliner the "rapid-acting" poison some years before; Laue

(to Theodor von Laue, n. 94) says that Berliner prepared by having a chemist friend give him a vial of pure hydrocyanic acid several years before.

99. According to Rosbaud, Berliner was buried next to his sister in the cemetery on Heerstrasse; Ludwig Ruge gave the eulogy (see n. 93, above). Laue (see n. 98) notes that Berliner's apartment was "cleaned out by the police."

100. Laue, "Arnold Berliner."

101. Laue to Meitner, 5 June 1942 (MC). The deported colleague was A. Byk; Laue added that Göttingen professor Adolf Windaus "strongly interceded on his behalf, but nothing helped."

102. Laue to Meitner, 13 September 1942 and 25 October 1942 (MC).

103. Laue to Meitner, 18 December 1943 (MC).

104. Laue to Meitner, 26 January 1943, 13 March 1943, and 20 February 1944 (MC). Two years later, Laue vividly described the fire that destroyed the KWI for Chemistry in a letter to his son (Laue to Theodor von Laue, 27 May 1945 [LP]; in Beyerchen, *Scientists under Hitler,* 194–195).

105. For example, Laue to Meitner, 3 April 1941 (MC).

106. David C. Cassidy, *Uncertainty: The Life and Science of Werner Heisenberg* (New York: W. H. Freeman, 1992), 417ff.; Mark Walker, *German National Socialism and the Quest for Nuclear Power, 1939–1949* (Cambridge: Cambridge University Press, 1989), chap. 1.

107. Spencer Weart and Gertrude Weiss Szilard, *Leo Szilard: His Version of the Facts* (Cambridge: MIT Press, 1978), 53; Otto Nathan and Heinz Norden, eds., *Einstein on Peace* (New York: Avenel, 1981), 289–297; Rhodes, *Making of the Atomic Bomb,* 303–308. Cassidy, *Uncertainty,* 420, suggests that Heisenberg's behavior on his visit to the United States earlier in 1939 may have added urgency to the warning to Roosevelt.

108. Frisch, *What Little I Remember,* 126–127; Peierls, *Bird of Passage,* 153–155; Rhodes, *Making of the Atomic Bomb,* 321–325. For the Frisch-Peierls memorandum, see Gowing, *Britain and Atomic Energy,* 40–42, 389–393. Attempts to separate ^{235}U by gaseous diffusion, centrifugation, and mass spectroscopy were under way in numerous laboratories: Heilbron and Seidel, *Lawrence and His Laboratory,* 449–454.

109. Alfred O. Nier, E. Booth, J. Dunning, and A. von Grosse, "Nuclear Fission of Separated Uranium Isotopes," *Phys. Rev.* 57 (1940): 546; and "Further Experiments on Fission of Separated Uranium Isotopes," *Phys. Rev.* 57 (1940): 748. See Heilbron and Seidel, *Lawrence and His Laboratory,* 450–451; Roger H. Stuewer, ed., *Nuclear Physics in Retrospect: Proceedings of a Symposium on the 1930s* (Minneapolis: University of Minnesota Press, 1979), 310.

110. Meitner to Laue, 29 June 1940 (MC).

111. McMillan and Abelson, "Radioactive Element 93."

112. Turner, "Nuclear Fission," 18.

113. The 1941 work on the chemical, nuclear, and fissile properties of Pu was classified until 1946: G. T. Seaborg, E. M. McMillan, J. W. Kennedy, A. C. Wahl,

Phys. Rev. 69 (1946): 366–367; G. T. Seaborg, A. C. Wahl, J. W. Kennedy, *Phys. Rev.* 69 (1946); 367; J. W. Kennedy, G. T. Seaborg, E. Segrè, A. C. Wahl, *Phys. Rev.* 70 (1946): 555–556. Also see Rhodes, *Making of the Atomic Bomb*, 353–355.
    114. Louis A. Turner, "Atomic Energy from U-238," *Phys. Rev.* 69 (1946): 366; Turner's article was published in 1946 along with other declassified reports. See Rhodes, *Making of the Atomic Bomb*, 346–347, 350; Weart and Szilard, *Leo Szilard*; Spencer R. Weart, "Scientists with a Secret," *Physics Today* 29, no. 2 (February 1976): 23–30; Henry D. Smyth, *Atomic Energy for Military Purposes* (Princeton: Princeton University Press, 1945), 28–29.
    115. David Irving, *The Virus House* (London: W. Kimber, 1967), 66–67; Cassidy, *Uncertainty*, 424.
    116. Cassidy, *Uncertainty*, 380–381.
    117. Max Born, *My Life: Recollections of a Nobel Laureate* (New York: Charles Scribner's Sons, 1978), 269–270.
    118. See chap. 9.
    119. Cassidy, *Uncertainty*, 411–415; Werner Heisenberg, *Physics and Beyond: Encounters and Conversations* (New York: Harper & Row, 1971), 170–172; Max Dresden, letter to the editor, *Physics Today* 44, no. 5 (May 1991): 92, 94.
    120. Cassidy, *Uncertainty*, 436, notes that the German cultural institutes in Copenhagen and elsewhere were propaganda centers, run by a subdivision of the German Foreign Office administered by Baron Ernst von Weizsäcker, Carl Friedrich's father; he suggests (p. 441) that Heisenberg's Copenhagen lecture in the German Culture Institute, which he must have known was odious to the Danes, was probably an effort to prove his reliability with the Nazi regime. Walker, *German National Socialism*, 105–118, documents Heisenberg's extensive wartime travels under the auspices of similar propaganda institutes as evidence (p. 117) of his full support for the German war effort.
    121. Meitner to Hahn, 24 January 1954; in Krafft, *Im Schatten der Sensation*, 182 n. 48: "Heisenberg and Weizsäcker tried to set up a German congress in occupied Copenhagen—surely a very unpsychological idea. I know that the entire Bohr institute (except for Bohr himself) was invited to it and everyone turned down the invitation." Meitner's secondhand recollection was mostly correct, but neither Cassidy (in *Uncertainty*) nor Walker (in *German National Socialism*) indicate that Bohr was not invited.
    122. Cassidy, *Uncertainty*, 442.
    123. Meitner to Paul Scherrer, 26 June 1945 (MC). Meitner's information is consistent with the conclusions of Cassidy and Walker.
    124. Stefan Rozental, quoted in Pais, *Niels Bohr's Times*, 483.
    125. Aage Bohr, "The War Years and the Atomic Weapons," in Rozental, *Niels Bohr*, 193; Moore, *Niels Bohr*, 290–293; Cassidy, *Uncertainty*, 436ff.; Walker, *German National Socialism*, 225ff.; Pais, *Niels Bohr's Times*, 481–485.
    126. The debate continues to the present, primarily because, as Rudolf Peierls has noted, "The question of Heisenberg's integrity is somewhat at issue on the

476   NOTES TO PAGES 301-303

topic of whether the German physicists *could not* or *would not* work on the bomb."
See Rudolf Peierls, "Atomic Germans," *New York Review of Books* 16 (1 July 1971):
23–24, a review of Heisenberg's *Physics and Beyond*.

127. Meitner to Laue, 20 April 1942 (MC).

128. Laue to Meitner, 26 April 1942 (MC). The confusion Laue described is
evident in Heisenberg's memoir, *Physics and Beyond*, a collection of remembered
"conversations" (p. 175). Heisenberg has Weizsäcker say, "It is quite possible that
I have unconsciously fallen victim to wishful thinking. For while no one in his
senses can hope for Hitler's victory, no German can wish for the complete defeat
of his country with all the terrible consequences that would entail."

129. Elisabeth Heisenberg, *Inner Exile: Recollections of a Life with Werner Heisenberg*, trans. S. Cappellari and C. Morris (Boston: Birkhäuser, 1984), 77–81; Heisenberg, *Physics and Beyond*, 181–182; Thomas Powers, *Heisenberg's War* (New York:
Alfred A. Knopf, 1993), 114–116. Heisenberg's claim that he wanted to discuss
with Bohr the morality of nuclear weapons work is disputed by Cassidy, *Uncertainty*, 437–438, 442, and Walker, *German National Socialism*, 225–226.

130. Cassidy, *Uncertainty*, 474. For "inner emigration," see Ernst Glaser,
"Zum Problem der 'Inneren Emigration' am beispiel von Hans Thirring," in
Stadler, *Vertriebene Vernunft II*, 1065–1074. Cassidy notes that while Heisenberg
"invoked the rationale of using warfare for physics," he was in fact "making physics
useful for warfare and himself acceptable to the rulers of the Reich" (p. 427).

131. See Pais, *Niels Bohr's Times*, 483–484; Walker, *German National Socialism*,
225.

132. Meitner to Scherrer, 26 June 1945 (MC). Shortly after the war, Meitner
read Weizsäcker's *Zum Weltbild der Physik* and commented, "Reading it, I had the
recurring feeling that he was fleeing from an unbearable reality. That is to a certain
extent intellectual dishonesty. I think that Heisenberg, in other ways, also suffers
from intellectual dishonesty."

133. Paul Rosbaud to Samuel Goudsmit, 25 October 1950 (Goudsmit papers,
AIP). Rosbaud added, "[Heisenberg] has not changed a bit and has not learned
anything." In an earlier report to Goudsmit (Rosbaud to Goudsmit, August 1945
[AIP]), Rosbaud described a frank 1940 discussion with Heisenberg in which they
completely disagreed on political matters: Heisenberg was ready to work with the
regime to advance his science and "never failed to mention" that he was a school-
and *Du*-friend of Hans Frank, the governor-general of occupied Poland. Powers,
in *Heisenberg's War*, 110–111, regards the frankness of their discussion as evidence
that Rosbaud "did not think Heisenberg was a Nazi who would betray him to the
Gestapo." That is true, but in 1945 Rosbaud also stated that he never had a
personal talk with Heisenberg again, and, since Heisenberg was "in complete
accord" with C. F. von Weizsäcker, Rosbaud avoided further contact with Weizsäcker as well.

134. Hendrik B. G. Casimir, *Haphazard Reality: Half a Century of Science* (New
York: Harper & Row, 1983), 207–208; Walker, *German National Socialism*, 105.

135. Heisenberg to Dirk Coster, 16 February 1943 (DM, IMC 29 1051). Powers, in *Heisenberg's War*, 431-432, cites Weizsäcker's interpretation of Heisenberg's letter as an "urgent but despairing appeal which Heisenberg had necessarily hidden between the lines"; similarly, Walker, in *German National Socialism*, 108-109, and also in "Heisenberg, Goudsmit and the German Atomic Bomb," *Physics Today* 43, no. 1 (January 1990): 52-60, suggests that Heisenberg was courageous in writing this "public statement" to help Goudsmit's parents. But Jonothan L. Logan (letter to the editor, *Physics Today* 44, no. 5 [May 1991]: 13, 15, 90-91) notes that Heisenberg's letter was written not to German authorities but privately to Coster, and in any case was too late: according to records later obtained by Goudsmit, his parents were gassed in Auschwitz on 11 February 1943, his father's seventieth birthday (Samuel A. Goudsmit, *Alsos* [New York: Henry Schuman, 1947], 46-49; Cassidy, *Uncertainty*, 485 n. 48). See also A. van der Ziel, letter to the editor, *Physics Today* 44, no. 5 (May 1991): 94, and Cassidy, *Uncertainty*, 484-485.

136. In describing her husband's "virtues," Elisabeth Heisenberg demonstrates her own (and undoubtedly her husband's) moral blindness. In a typical passage in her biography of Heisenberg, *Inner Exile*, 82, she makes the persecution of Jews seem beside the point and perhaps illusory—"In September 1943, Bohr had fled from Denmark, taking a small boat over the Sound at night *apparently fearing a threatened pogrom against the Jews*" (italics mine)—while citing the "hardships" her husband endured to secure the independence of Bohr's institute ("He had to eat and drink with [the Gestapo]. . . . [T]here was no cheaper way").

137. Peierls, *Bird of Passage*, 168-169.

138. Laue to Meitner, 5 June 1942 (MC).

139. Meitner to Max Born (in English), 20 September 1942 (MC). See Cassidy, *Uncertainty*, 443-446.

140. Discussions with Lilli Eppstein, Stocksund, July 1989; Interviews, Sigvard Eklund, Vienna, June 1985 and May 1992.

141. Kramish, *The Griffin*, chap. 34.

142. Sigvard Eklund, pers. comm., 26 July 1987.

143. Meitner to Rosbaud, 4 August 1946 (MC).

144. For the British in Los Alamos, see Gowing, *Britain and Atomic Energy*, 260ff.

145. Meitner's invitation to Los Alamos is repeated by Frisch: O. R. Frisch, *Dict. Sci. Biog.*, 9:260; Frisch, "Lise Meitner, 1878-1968," *Biog. Mem. Fell. Roy. Soc. Lond.* 16 (1970): 414; Frisch, "Lise Meitner, Nuclear Pioneer," *New Scientist* (4 November 1978): 426-428, on 428. Rudolf Peierls was unaware of the invitation to Meitner but regards Frisch as reliable (Peierls, pers. comm., 16 November 1993). Meitner's Swedish friends also knew of an invitation (Lilli Eppstein, pers. comm., 18 February 1987, 29 December 1987). Arnold Kramish, Paul Rosbaud's biographer, was told that Meitner was offered an opportunity to go to England but

refused because it was tied to work on the bomb (Kramish, pers. comm., 6 October 1985).

146. Otto Hahn, in "Lise Meitner 70 Jahre," *Z. Naturforschg.* 1, no. 3a (1948): 425–428, states that she refused an invitation to work on a weapon because it might be used against Germany.

147. Meitner to Max Born, 22 October 1944; Meitner to Schiemann, n.d. 1945; Meitner to Dirk Coster, 15 October 1945 (MC).

148. J. L. Heilbron, *The Dilemmas of an Upright Man: Max Planck as Spokesman for German Science* (Berkeley, Los Angeles, and London: University of California Press, 1986). Planck traveled to Sweden in June 1943.

149. Meitner to Bahr-Bergius, 21 June 1944 (MC).

150. Meitner to Born, 22 October 1944 (MC). According to Heilbron, *Dilemmas of an Upright Man*, 194, Erwin Planck knew the plotters and sympathized with their goal but did not participate in the assassination attempt. See also Erich Zimmermann and Hans-Adolf Jacobsen, eds., *Germans Against Hitler: July 20, 1944*, trans. Allan Yahraes and Lieselotte Yahraes (Bonn: Press and Information Office of the Federal Republic of Germany, 1964), 12, 218, 226, 333.

151. Nikolaus Riehl, "Erinnerungen an Otto Hahn und Lise Meitner," in Fridolin Haugg, ed., *Vortrag zur Eröffnung der Ausstellung "50 Jahre Kernspaltung" im Willi-Graf-Gymnasium, 30 November 1988*, 12. Riehl, a former student of Meitner's, describes a visit by colleagues to Hahn to congratulate him on his Nobel Prize in 1944, an indication that many others besides Hahn knew of it. The story may not be reliable: Riehl places the visit in Dahlem in November 1944, and Hahn was in Tailfingen, southern Germany, at the time. The official reaction to Hahn's prize and the Swedish communication are given in Walther Gerlach, *Otto Hahn, Ein Forscherleben unserer Zeit* (Grosse Naturforscher, Band 45) (Stuttgart: Wissenschaftliche Verlagsgesellschaft, 1984), 119–120.

152. Meitner to Pettersson, 26 October 1944 (MC).

153. Meitner to Bahr-Bergius, 21 June 1944 (MC).

154. Meitner to Pettersson, 8 October 1944 (MC).

155. Meitner to Bahr-Bergius, 30 March 1945 (MC).

CHAPTER 13. War Against Memory

1. Lise Meitner to Eva von Bahr-Bergius, 28 May 1945 (MC).

2. Meitner to Otto Hahn, 27 June 1945 (MC). Reprinted (in German) in Fritz Krafft, *Im Schatten der Sensation: Leben und Wirken von Fritz Strassmann* (Weinheim: Verlag Chemie, 1981), 181–182, and in Ute Deichmann, *Biologen Unter Hitler: Vertreibung, Karrieren, Forschung* (Frankfurt: Campus Verlag, 1992), 316–318. In his foreword to *Biologen Unter Hitler*, p. 20, Benno Müller-Hill agrees with Meitner's statement that German scientists lost their "standards of justice and fairness"; he believes this was the basis for the decline of science in Germany under Hitler and afterward.

3. Meitner to Dirk Coster, 15 October 1945 (MC).

4. Meitner to Hahn, 27 June 1945 (MC).

5. Thomas Powers, *Heisenberg's War: The Secret History of the German Bomb* (New York: Alfred A. Knopf, 1993), 273–276, 294–297; David C. Cassidy, *Uncertainty: The Life and Science of Werner Heisenberg* (New York: W. H. Freeman, 1992), 491–492.

6. Meitner to Paul Scherrer, 26 June 1945 (MC).

7. Thomas Powers, pers. comm., 27 May 1991; notes by Morris Berg, Stockholm, 9 January 1946.

8. Coster to Meitner (in English), card, 30 June 1945 (MC).

9. Hendrik B. G. Casimir, *Haphazard Reality: Half a Century of Science* (New York: Harper & Row, 1983), 217. Goudsmit was in Holland shortly after the liberation, seeking information about the German fission effort.

10. Coster to Meitner (in English), 4 August and 17 August 1945 (MC).

11. Stefan Meyer to Meitner, 26 August 1947 (MC). Meyer's brother Hans was a professor of organic chemistry in the German University in Prague. Meyer was made honorary professor in 1946 and retired in 1947 to Bad Ischl where he died on 29 December 1949, at the age of 78. Wolfgang L. Reiter, "Österreichische Wissenschaftsemigration am Beispiel des Instituts für Radiumforschung der Österreichischen Akademie der Wissenschaften," in Friedrich Stadler, ed., *Vertriebene Vernunft II: Emigration und Exil österreichischer Wissenschaft* (Wien: Jugend und Volk, 1988), 709–729, on 711–715.

12. Meitner, diary, 7 August 1945 (MC).

13. Leksand *Falu-Kuriren*, 8 August 1945, pp. 1, 6: "Landsflyktig kvinnlig fysiker en av atombombens banbrytare."

14. Meitner, diary, 7 August 1945 (MC).

15. Meitner to Frida Frischauer, 7 September 1945 (MC).

16. Meitner to Hilde Rosbaud, 1 September 1945 (MC).

17. Stockholm *Expressen*, 9 August 1945, 14–15.

18. Stockholm *Expressen*: "*FLYENDE JUDINNA*," 7 August 1945.

19. Meitner, diary, 9 August and 10 August 1945 (MC).

20. Meitner to Frida Frischauer, 7 September 1945 (MC).

21. William L. Laurence, "The Atom Gives Up," *Saturday Evening Post*, 7 September 1940, 12ff. Laurence's 1940 story appeared in various permutations from 1945 on (e.g., in *Time*, 20 August 1945, 31) and was quoted almost verbatim in *Current Biography 1945*, 393–395, from which extracts appeared elsewhere (e.g., Jay Walz, "Her Specialty: Atoms," *New York Times*, 10 March 1946). Based on his 1940 article and reputation at the *New York Times*, Laurence was hand-picked in 1945 by General Groves to report on the Manhattan Project. See Leslie R. Groves, *Now It Can Be Told: The Story of the Manhattan Project* (New York: Harper & Row, 1962), 325ff.

22. Hilde Rosbaud to Meitner, 27 August 1945 (MC).

23. Meitner to Max Born, 1 June 1948 (MC).

24. For a recent version of the story, see Mark E. Smith, letter to *New York Review of Books*, 3 February 1994, 45; see also response, M. F. Perutz, pp. 45–46.

25. Meitner to Frida Frischauer, 7 September 1945 (MC).

26. Ibid.

27. Hilde Rosbaud to Meitner, 27 August 1945 (MC).

28. Paul Rosbaud to Meitner, 25 October 1945 (MC).

29. Hilde Rosbaud to Meitner, 11 December 1945 (MC). According to Arnold Kramish, *The Griffin: The Greatest Untold Espionage Story of World War II* (Boston: Houghton Mifflin, 1986), 250, Rosbaud was smuggled out of Berlin in British military uniform by his "spymaster," Eric Welsh.

30. Samuel A. Goudsmit, *Alsos* (New York: Henry Schuman, 1947), 255–258; David Irving, *Virus House* (London: W. Kimber, 1967), 2, 333–340; Groves, *Now It Can Be Told*, 325ff.; Walther Gerlach, *Otto Hahn, Ein Forscherleben unserer Zeit*, ed. D. Hahn (Stuttgart: Wissenschaftliche Verlagsgesellschaft, 1984) (Grosse Naturforscher, Band 45), 114–118; Cassidy, *Uncertainty*, 497–500.

31. Meitner to Hahn, 20 September 1945 (MC).

32. Erich Zimmermann and Hans-Adolf Jacobsen, eds., *Germans Against Hitler, July 20, 1944*, trans. Allan Yahraes and Lieselotte Yahraes (Bonn: Press and Information Office of the Federal Government of Germany, 1964), 226, note that Erwin Planck, "under pretext of checking armament orders," was a "direct channel of communication" between one of the early planners of the assassination attempt, Friedrich Olbricht, and other staff. See also J. L. Heilbron, *The Dilemmas of an Upright Man: Max Planck as Spokesman for German Science* (Berkeley, Los Angeles, and London: University of California Press, 1986), 194–195. Berlin had no gallows, so that the method used for hangings was unusually gruesome; see William Shirer, *The Rise and Fall of the Third Reich* (New York: Simon and Schuster, 1960), 1069ff.

Death Notice (Meitner papers [MC]): Erwin Planck, born 12 May 1893, Berlin; arrested 23 July 1944; sentenced to death 23 October 1944; executed 23 January 1945. The January execution date, also given in Zimmermann and Jacobsen, must be incorrect; Heilbron's date, 23 February, is based on correspondence.

33. Rough notes and report by Morris Berg, 9 January 1946, of his meeting with Meitner in Stockholm. I am indebted to Thomas Powers for a copy of the document.

34. Hahn to Meitner, ? October 1945 (date obscured but on or before 6 October) (MC). In *Operation Epsilon: The Farm Hall Transcripts* (Berkeley, Los Angeles, and Oxford: University of California Press, 1993), 227, the commanding officer notes that Hahn's letter to Meitner was sent on 6 October, "translations having been made."

35. Hahn knew of the use of slave labor; he was among those who helped Rosbaud in illegally assisting prisoners of war stationed in Berlin (Kramish, *The Griffin*, 220). Also, in a 1943 letter to Meitner (12 February 1943, in Fritz Krafft,

Im Schatten der Sensation, 131), Edith Hahn casually mentions that "the Frenchmen" (i.e., slave laborers) repaired windows in the institute that were blown out in air raids.

Mark Walker, in *German National Socialism and the Quest for Nuclear Power, 1939–1949* (Cambridge: Cambridge University Press, 1989), 132–133, notes that German industry's extensive use of slave labor under appalling conditions, with exceptionally high mortality rates, included the manufacturing associated with the German fission project, which relied heavily on Polish and Russian prisoners of war and Jewish concentration camp inmates. For Germany's economic exploitation of the occupied countries and slave labor, see Gordon A. Craig, *Germany, 1866–1945* (New York: Oxford University Press, 1978), 746–748.

Meitner refers to Hahn's "bled dry" remark in Meitner to Hahn, 1 April 1946 (MC).

36. *Operation Epsilon,* 71.

37. *Operation Epsilon,* 70–91. Max von Laue recorded his own, Hahn's, and Gerlach's reaction in a letter to his son (Laue to Theodor von Laue, 7 August 1945 [LP]); Laue's wording and the corresponding wording in the Farm Hall transcripts (p. 79) are remarkably similar, in particular about Gerlach: "der sich . . . etwa wie ein geschlagener Feldherr vorkam," the same as the British "He appeared to consider himself in the position of a defeated General." Quite possibly the British, who read and translated all mail, appropriated Laue's term.

Paul Rosbaud to Samuel Goudsmit, 18 July 1958 (Goudsmit papers, AIP), notes that Weizsäcker and others thought it would take "50 years to make [a bomb]. . . . Therefore they did not believe . . . the news of Hiroshima and thought all this is American propaganda."

38. *Operation Epsilon,* 76–77.

39. Ibid., 90.

40. Goudsmit, *Alsos,* 46–49; Powers, *Heisenberg's War,* 431; Cassidy, *Uncertainty,* 485, 513ff. Max Dresden (letter to the editor, *Physics Today* 44, no. 5 [May 1991]: 92, 94) notes that Goudsmit was deeply disappointed and angry with Heisenberg, regarding his allegiance to Nazi Germany as a violation of the international physics community's standards of integrity and accountability.

41. Mark Walker, in "Heisenberg, Goudsmit and the German Atomic Bomb," *Physics Today* 43, no. 1 (January 1990): 52–60, describes the interchange as it appeared in the *Bulletin of the Atomic Scientists,* the *New York Times,* and private correspondence. See also Walker, *German National Socialism,* 211–221; on pp. 223 and 228, Walker explicitly refers to Heisenberg's interpretation as "postwar apologia."

42. Cassidy, *Uncertainty,* 517. Powers, in *Heisenberg's War,* 481–484, agrees that Heisenberg never told "the whole truth" about his wartime work but argues that Heisenberg, who "had done what he could to stop the one weapon which might have won the war," could not speak out because it would mean "years of

abuse and acrimony," possibly even accusations of treason from his former associates.

43. Dan Bar-On, *Legacy of Silence: Encounters with Children of the Third Reich* (Cambridge: Harvard University Press, 1989), demonstrates the "forgetting" and silence of the Nazi generation of parents and its profound psychological effect on their children. Ute Deichmann, in *Biologen Unter Hitler*, p. 313, notes that German biologists were isolated from the international community for several decades after the war, not only because of Germany's crimes and their individual behavior under the Third Reich, but also because their desire to "forget" estranged them from colleagues elsewhere.

44. Walker's article in *Physics Today* (see n. 41, above) and the ensuing response, particularly from Jonothan Logan and Max Dresden ("Heisenberg, Goudsmit and the German 'A-Bomb,'" letters to the editor, *Physics Today* 44, no. 5 [1991]: 13, 15, 90–96), shows that the facts of the German fission project are still being debated and in any case have not discharged the emotions surrounding the project and Heisenberg's role in it.

45. This thesis was most sensationally advanced by Robert Jungk, *Brighter than a Thousand Suns: A Personal History of the Atomic Scientists*, trans. James Cleugh (New York: Harcourt Brace, 1958). Jungk, a German Jewish journalist who emigrated to the United States, relied heavily on interviews with Heisenberg and Weizsäcker to conclude that the Germans had refused to build a bomb and were therefore morally superior to the Americans.

Recently Jungk has stated that he was too willing to believe Heisenberg and Weizsäcker, because at a time of McCarthyism in the United States he was anxious to show that scientists could resist the power of the state. See Mark Walker, "Legenden um die Deutsche Atombombe," *Vierteljahrhefte für Zeitgeschichte* 38, no. 1 (1990): 45–74, n. 39.

46. This is the thesis of Powers's *Heisenberg's War*. He concludes, as does Jungk, in *Brighter than a Thousand Suns*, that Heisenberg knowingly prevented the building of an atomic bomb. See also n. 42.

47. Based on wartime documents and scientific reports, Paul Lawrence Rose ("Did Heisenberg Misconceive A-Bomb?" letter to the editor, *Physics Today* 45, no. 2 [February 1992]: 126; and pers. comm., unpublished letter to *Physics Today*, 20 February 1992) concludes that Heisenberg never understood the essential principle of a fast-neutron chain reaction, so that his estimates for an explosive mass of ^{235}U were far too large, thereby forestalling the German bomb project. In his survey of the Farm Hall transcripts, Irving Klotz ("Germans at Farm Hall Knew Little of A-Bombs," letter to the editor, *Physics Today* 46, no. 10 [October 1993]: 11, 13, 15, 135) notes that Heisenberg's initial estimates for the explosive mass of ^{235}U were off by many orders of magnitude and that the Farm Hall scientists knew almost nothing about plutonium. Protactinium was mentioned several times, however, and Alvin M. Weinberg and Jaroslav Franta ("Was Nazi Know-how Enough for an A-Bomb?" letters to the editor, *Physics Today* 47, no.

12 [December 1994], 84) suggest that the Farm Hall scientists were aware that protactinium is a possible fast-fission material for nuclear weapons.

48. Cassidy, in *Uncertainty*, p. 510, explicitly refutes "the notion that Heisenberg hindered the project in any way" and on p. 519 refers to the heroic legends propagated by Jungk as the "old apologetic Farm Hall story carried to its ultimate conclusion," a position consistent with Walker (in "Legenden" and *German National Socialism*, chap. 7, n. 85, p. 264), who faults subsequent writers from Jungk to the present for uncritically accepting Heisenberg's story.

49. *Operation Epsilon*, 42.

50. Ibid., 83.

51. Ibid., 146.

52. Ibid., 83, 86–87.

53. Ibid., 168.

54. Ibid., 230. In the same report, dated 25 October 1945, Major T. H. Rittner drily noted that Wirtz and Weizsäcker blamed Commodore Perry's expedition for the war between the United States and Japan.

55. Ibid., 46 (translation correction, p. 63). The comment was made by Paul Harteck, 18 July 1945.

56. Ibid., 55. Comments by Heisenberg and Wirtz, 30 July 1945.

57. Ibid., 50. Comment by Bagge, 26 July 1945.

58. Otto Hahn, *My Life*, trans. Ernst Kaiser and Eithne Wilkins (New York: Herder and Herder, 1970), 122.

59. Max von Laue to Theodor von Laue, 19 August 1945 (LP). When Sir Charles Darwin visited Farm Hall in August, he told the scientists that Rudolf Peierls in Bristol and Franz Simon at Oxford had done some of the initial work; Laue did not know of Frisch's theoretical work with Peierls.

60. Conversation with Carl Friedrich von Weizsäcker, in Dieter Hoffmann, ed., *Operation Epsilon: Die Farm-Hall-Protokolle oder Die Angst der Alliierten vor der deutschen Atombombe*, trans. Wilfried Sczepan (Berlin: Rowohlt, 1993), 331–360, on 338–340; also Weizsäcker interview, "Ich gebe zu, ich war verrückt," in *Der Spiegel* 17 (1991): 227ff.

61. *Operation Epsilon*, 92. Statement by Weizsäcker, 7 August 1945.

62. For the moral dimension, see Cassidy, *Uncertainty*, 505–510. For analysis of the scientific dishonesty, see Walker, *German National Socialism*, 162–165. German scientists were aware in 1941 that reactor-bred element 94 could—in theory—be a fissile alternative to ^{235}U. They may also have considered ^{231}Pa (see n. 47, above).

63. *Operation Epsilon*, 92–94, 102–106. There is no evidence that the Farm Hall memorandum was ever released to the press. The scientists who first refused to sign were Kurt Diebner, Erich Bagge, Horst Korsching, Karl Wirtz, and Weizsäcker. On 7 August, Heisenberg said, "If the Americans had not got so far with the [reactor] as we did—that's what it looks like—then we are in luck. There is a possibility of making money."

64. See Walker, *German National Socialism*, chap. 7.

65. In Max von Laue to Theodor von Laue, 7 August 1945 (LP), Laue states his agreement with the memorandum; a copy of the memorandum is with his papers. It is unclear when Laue's letter to his son was actually mailed.

66. Max von Laue to Paul Rosbaud, 4 April 1959, reprinted (in English) in Kramish, *The Griffin*, 245–246; see also Cassidy, *Uncertainty*, 519. Laue's letter was prompted by the publication and ensuing discussion surrounding Jungk's book (see n. 45, above).

67. Max von Laue to Theodor von Laue, 7 August 1945 (LP).

68. Dietrich Hahn, ed., *Otto Hahn, Erlebnisse und Erkenntnisse* (Düsseldorf: Econ Verlag, 1975), 72. For a typical account, see *Life*, 20 August 1945, 89B: Hahn did not realize he had split the atom, but Meitner realized the implications of his work and "carried out the historic experiment of splitting the atom so that energy was released." Also see *Time*, 20 August 1945, 31: Hahn wrote a "diffident note" to *Naturwissenschaften*, but Meitner, "having fled Hitler to Copenhagen," passed the idea to Niels Bohr who brought it to America, where scientists "sprang to their atom-smashing machines," then "stood gallantly back" so that Meitner could publish first.

69. *Operation Epsilon*, 102–103.

70. Otto Hahn to Major Rittner, 8 August 1945. Copy in Meitner's papers (MC); not reproduced in the Farm Hall transcripts.

71. Hahn to Meitner, 7 February 1939 (MC).

72. So stated by Karl-Erik Zimen, "Die Saga des Urans und der Januskopf des Fortschritts," *Atomwirtschaft* (December 1988): 578–584; and Zimen, "Otto Hahn, Lise Meitner und die Kernspaltung im Ausblick auf die Zukunft," *Phys. Bl.* 35 (1979): 200–210.

73. This term and the chapter title are taken from Primo Levi, *The Drowned and the Saved*, trans. Raymond Rosenthal (New York: Vintage International, 1989), 31.

CHAPTER 14. Suppressing the Past

1. For the Farm Hall reaction to Hahn's award, see *Operation Epsilon: The Farm Hall Transcripts* (Berkeley, Los Angeles, and Oxford: University of California Press, 1993), 244ff.

2. Lilli Eppstein, Stocksund, pers. comm., 18 February 1987.

3. Hans Pettersson to Lise Meitner, 16 November 1945 (MC). The term *krokbensläggare* refers to a person who tries to trip others up: a mean person playing dirty tricks. For the explanation I am grateful to Thomas Blitz, Djursholm, pers. comm., October 1988.

4. Lilli Eppstein, pers. comm., 18 February 1987.

5. Birgit Broomé Aminoff to Meitner (in Swedish), 16 November 1945 (MC); for the translation, I am grateful to Ulla McDaniel. Aminoff's husband, a physicist and mineralogist, was a member of the Royal Academy, director of the National

Museum, and a board member of the Nobel Foundation. I am grateful to Urban Wråkberg of the Royal Academy of Sciences (KVA) for the information.

6. Meitner to Birgit Aminoff, 20 November 1945 (MC).

7. The article about Hahn's award in *Dagens Nyheter*, 16 November 1945, refers to Meitner several times as Hahn's principal *medarbetare;* it was written by L. G. Sillén, a physical-organic chemist (not a radiochemist) who could have, but evidently did not, contact Meitner prior to publication. For the translation, I am grateful to Lilli Eppstein.

8. The English and German are quite similar in their implication of rank. An assistant, for example, could be described as the co-worker or Mitarbeiter (female: *Mitarbeiterin*) of a professor, but never the reverse.

9. Meitner to Eva von Bahr-Bergius, 5 December 1945 (MC).

10. Klein was elected to the Royal Academy of Sciences on 28 February 1945. I am grateful to Urban Wråkberg of the KVA for the information.

11. Oskar Klein to Niels Bohr (in Danish), 16 September 1945 (BSC). Klein indicates that another factor in the decision to delay the award was that Hahn's "own position at the moment" was unclear, a possible reference to his internment at Farm Hall. At any rate, Klein told Bohr, "I believe I had something to do with this decision [to delay]; I did what I could." In a previous letter to Klein (Bohr to Klein, 11 September 1945 [BSC]), Bohr described Meitner's and Frisch's contributions as essential to the development of fission and urged that the chemistry prize be shared by all three. For the translations, I am grateful to Hilde Levi of the Niels Bohr Archive and Ulla McDaniel, Sacramento.

12. Klein to Bohr, 17 November 1945 (BSC supplement, not yet microfilmed). For the translation, I am grateful to Hilde Levi and Ulla McDaniel.

13. For the mismatch between the Nobel tradition of awards for individual creativity and the practice of big science, see John L. Heilbron, "Creativity and Big Science," *Physics Today* 45, no. 11 (November 1992): 42–47. Heilbron notes that the three-person rule was adopted by the academy, although the Nobel statutes do not limit the number of scientists who may share a prize. See Elisabeth Crawford, *The Beginnings of the Nobel Institution: The Science Prizes, 1901–1915* (Cambridge: Cambridge University Press, 1984), 223.

14. It is unlikely that there were jurisdictional objections to a physicist receiving a chemistry award, as there was considerable precedent, including Svante Arrhenius (1903), Rutherford (1908), and the Joliot-Curies (1935). See Crawford, *Beginnings of the Nobel Institution*, 116–123, 128–133.

15. Nobel decisions are traditionally tilted to more senior scientists, and it appears the Nobel committee incorrectly regarded Strassmann as a very junior, very young associate of Hahn's who joined the uranium investigation only at the end for the discovery of barium. (In fact, Strassmann already had his doctorate when he came to Berlin, and he was 36 years old at the time of the discovery.) See Fritz Krafft, *Im Schatten der Sensation: Leben und Wirken von Fritz Strassmann* (Weinheim: Verlag Chemie, 1981), 1–2 n. 3.

16. On 31 January 1939, Svedberg nominated Hahn, or Hahn and Meitner (Meitner and Frisch's *Nature* article had not yet appeared); on 31 March 1939, Svedberg submitted an extensive review of the already quite complex field and nominated Hahn and Meitner for a prize in chemistry for radioactivity and fission in 1939. (Nomination proposal for chemistry 1939: *Protokoll vid Kung. Vetenskaps-akademiens*, Samankomster för Behandlung af Ärenden Rörande Nobelstiftelsen, År 1939.)

Although in 1939 Svedberg seems to have understood Meitner's role quite well, Borelius was later of the opinion that the chemistry committee did not consider Meitner in 1944 because she was not in Berlin at the time of the barium finding. See Asta Ekenvall, "Lise Meitner: Kärnfysiker mot Atombomb," *Biblioteka Feminarum* 21 (1984): 172–201, on 187. I am grateful to Kerstin Klein, Södertälje, for the translation.

17. Max von Laue to Theodor von Laue, 18 November 1945 (LP). Laue indicates that a Nobel Prize for Hahn "was an open secret for us for years."

18. Paul Rosbaud to Samuel Goudsmit, 18 July 1958 (Goudsmit papers, AIP). In 1944, Rosbaud sent Arne Westgren, secretary of the KVA, a coded list of "trustworthy" German scientists, including Hahn, Strassmann, Laue, and Mattauch, some 24 names in all. (To Goudsmit, Rosbaud emphasized that Heisenberg and Weizsäcker were *not* on the list.)

Friedrich Glum, in *Zwischen Wissenschaft, Wirtschaft und Politik* (Bonn: H. Bouvier Verlag, 1964), 520, indicates that as of 1940 another member of the chemistry committee, H. von Euler-Chelpin, had a "pro-German, that is, pro-Hitler" bias. Euler-Chelpin served on the committee until 1946. For the Nobel committees, see Elisabeth Crawford, J. L. Heilbron, and Rebecca Ullrich, *The Nobel Population: 1901–1937* (Berkeley: University of California, Office for History of Science and Technology/Uppsala: Uppsala University, Office for History of Science, 1987), 6–7.

19. *Operation Epsilon*, 244–245; Krafft, *Im Schatten der Sensation*, 483–484.

20. Karl-Erik Zimen, "Otto Hahn, Lise Meitner und die Kernspaltung im Ausblick auf die Zukunft," *Phys. Bl.* 35 (1979): 200–210. "The majority of the Nobel committee [*sic:* Royal Academy] did not permit their view to be darkened by the hate-filled atmosphere of the year 1945 and thus eased the rebuilding of German science." Zimen, "Fifty Years of Nuclear Fission," *IANCAS Special Bulletin* 5, no. 1 (1989): 1–4: "The fact that the majority of the Nobel committee [*sic*] voted for Hahn in this time upset by atomic bombs and war crimes is, in my opinion, not only an honor for Otto Hahn but also for the Swedish committee."

In a letter to his son Theo at the time (see n. 17, above), Laue noted that Arne Westgren informed Hahn of the possible delay in September 1945. "From this we gathered that the atomic bombs made such a terrible impression in Sweden that they did not want to give a prize for the discovery that led to them in the same year." Evidently Westgren did not inform Hahn that the committee's decision to delay was based on scientific considerations.

21. Klein to Bohr, 17 November 1945 (BSC, supplement to microfilm).

22. Ibid.

23. Klein to Bohr, 26 April 1946 (BSC). I am grateful to Hilde Levi for the translation.

24. Bohr to Klein, 28 December 1945 (BSC). Nominating Meitner and Frisch for physics in 1946 (and for chemistry in 1947 and 1948), Bohr cited their explanation of the fission process and Meitner's earlier investigations, with Hahn, of the neutron reactions of U and Th that became the basis for theoretical understanding of the slow-neutron fission of ^{235}U and the fast-neutron fission of ^{232}Th.

25. Lilli Eppstein, pers. comm., 18 February 1987.

26. Meitner to Margrethe Bohr, 25 November 1945 (BSC, supplement to microfilm).

27. Otto Robert Frisch to Meitner (in English), 12 January 1946 (MC).

28. Time, 4 February 1946, 43; Meitner to Bahr-Bergius, 16 February 1946, and Meitner to Marga Planck, 5 May 1946 (MC). In the New York Times, 26 January 1946, 4, Meitner is the "niece [sic] of Prof. Rudolf Allers."

29. Meitner to Bahr-Bergius, 16 February 1946; Meitner to Frisch, 19 February 1946 (MC). In Time, 18 February 1946, 44, Meitner was again the "pioneer contributor to the atomic bomb"; a photograph of her and the other National Women's Press Club honorees, including Georgia O'Keeffe and Agnes de Mille, was "as notable for its variety of necklines [all very modest] as for its collection of female talent." See also the New York Times, 10 February 1946, 13.

30. Meitner to Frisch, 19 February 1946 (MC).

31. A sampling: George Axelsson, "Is the Atom Terror Exaggerated?" Saturday Evening Post, 5 January 1946, 34ff.; New York Times Magazine, 10 March 1946; Science News Letter, 16 March 1946; San Francisco News (crossword puzzle), 16 May 1946 (MC).

32. Irene Orgel, "Sonnet to Lise Meitner," American Scholar 15, no. 2 (1946): 146.

33. James Franck to Meitner, n.d. but probably February 1946 (MC, U.S. folder).

34. Meitner to Frisch, 19 February 1946 (MC).

35. Ibid.

36. On 9 September 1946, after she had left the United States, Meitner gave James Franck, T. R. Hogness, and H. C. Urey power of attorney to sue if any woman scientist was portrayed in the film, even under a fictitious name; later, she refused permission to use her name in many other proposed plays and films (Meitner to Mary Koues Sachs, 9 June 1953 [MC]).

37. O. R. Frisch, interview by author, 16 June 1975, Cambridge; Frisch, What Little I Remember (Cambridge: Cambridge University Press, 1979), 194. The quote may not be entirely accurate: Kerstin Klein, a niece of Oskar Klein's and a good friend of Meitner's in Stockholm, believes Meitner would never have used such an expression. (Kerstin Klein, interview, July 1989, Södertälje.)

38. Meitner to Frisch, 19 February 1946 (MC).

39. Meitner, memorandum, 27 February 1946 (MC).

40. Meitner to Margrethe Bohr (in English), 17 August 1946 (MC).

41. Meitner to Margrethe Bohr, 22 November 1946 (MC).

42. James Franck to Meitner, 27 June 1957 (MC).

43. Meitner to Rudolf Ladenburg, 13 July 1947 (MC).

44. Meitner to Margrethe Bohr (in English), 17 August 1946 (MC).

45. Ibid.

46. J. L. Heilbron, *The Dilemmas of an Upright Man: Max Planck as Spokesman for German Science* (Berkeley, Los Angeles, and London: University of California Press, 1986), 197.

47. Meitner to Margrethe Bohr (in English), 17 August 1946 (MC).

48. Max von Laue to Theodor von Laue, 12–17 July 1946 (LP).

49. Meitner to Ladenburg, 13 July 1947 (MC).

50. Laue to Theodor von Laue, 16 July 1946 (LP).

51. Meitner to Hahn, 1 April 1946 (MC).

52. Meitner to Walter and Lotte Meitner, 1 April 1946 (MC).

53. Samuel Goudsmit, *Alsos* (New York: Henry Schuman, 1947), 48.

54. Albert Einstein to James Franck, 30 December 1945; in Jost Lemmerich, ed., *Max Born, James Franck: Physiker in ihren Zeit, Der Luxus des Gewissens* (Berlin: Staatsbibliothek Preussicher Kulturbesitz, Ausstellungskatalog 17, 1982), 141–142, 144.

55. Meitner to Frisch, 19 February 1946 (MC).

56. Meitner to Bahr-Bergius, 22 May 1946 (MC).

57. Theodor von Laue to Meitner, 8 March 1946 (MC).

58. Meitner to Ladenburg, 13 July 1947 (MC).

59. Hahn to Meitner, 17 September 1946 (MC). For the views of Hitler among Western governments prior to 1939, see Gordon A. Craig, *Germany, 1866–1945* (New York: Oxford University Press, 1978), 673ff.

60. Meitner to Hahn, 20 October 1946 (MC).

61. Otto and Edith Hahn to Meitner, 10 November 1946 (MC).

62. Edith Hahn to Meitner, 10 October 1946; in Krafft, *Im Schatten der Sensation,* 483.

63. Meitner to Hahn, card, 15 November 1946 (MC).

64. Meitner to Margrethe Bohr, 22 November 1946 (MC).

65. Meitner to Lola Allers, 27 October 1946 (MC).

66. Meitner to Frisch, 28 November 1946 (MC).

67. *Dagens Nyheter, Stockholms Tidningen, Svenska Dagbladet,* 5 December 1946; for research and translations, I am grateful to Lilli Eppstein, pers. comm., 23 October 1987, and Svante Lindqvist, Royal Institute of Technology, Stockholm.

68. *Svenska Dagbladet*, 5 December 1946. I am grateful to Lilli Eppstein for the translation.

69. Ernst Berninger, *Otto Hahn: Eine Bilddokumentation* (München: Heinz Moos Verlag, 1969), 74.

70. Otto Hahn, *My Life*, trans. Ernst Kaiser and Eithne Wilkins (New York: Herder and Herder, 1970), 199–200. A similar description appears in Meitner to Lola Allers, 29 December 1946 (MC).

71. Meitner, diary, 4–14 December 1946 (MC).

72. Lilli Eppstein, pers. comm., 23 October 1987, 29 December 1987.

73. Otto Hahn, *Mein Leben* (München: Bruckmann, 1968), 208–210; Hahn, *My Life*, 201–202. Hahn must have been very fond of this speech; he included it in his memoirs more than twenty years later.

74. *Aftonbladet*, 11 December 1946. For the translation, I am grateful to Lilli Eppstein.

75. Otto Hahn, "Von den Natürlichen Umwandlungen des Urans zu seiner Künstlichen Zerspaltung," *Le Prix Nobel en 1946* (Stockholm, 1948), 167–183. English translation: "From the Natural Transmutations of Uranium to Its Artificial Fission," *Nobel Lectures, Chemistry, 1942–1962* (Amsterdam: Elsevier, 1964), 51–66.

76. Meitner, diary, 14 December 1946 (MC).

77. Hahn to Meitner, 21 December 1938 (MC).

78. Hahn, *Mein Leben*, 206; *My Life*, 199; Krafft, *Im Schatten der Sensation*, 484. Among the many factual inaccuracies, Hahn had the hotel wrong: he stayed at the Grand Hotel, not the Savoy.

79. In 1966, Hahn asked Frisch (Hahn to Frisch, 19 August 1966 [MC]): "Does Tante Lise still have resentments from July 1938 when she, accompanied by Herr Coster, left Germany? For a while she believed it was not at all necessary." Frisch replied (Frisch to Hahn, 23 August 1966 [MC]): "She certainly regards it as right that she left Germany in 1938; she has often told me the story, and always with gratitude for the people who helped her leave."

80. Meitner, 1950–1951 memorandum booklet, 27–28 June 1951 (MC): "OH and I, dinner. . . . [He] reproached me for leaving so late: had forgotten that I left on my own initiative, and what happened between him and Hörlein . . . [also] that he kept me [from leaving] in 1933."

81. Hahn, *My Life*, 203 ("a largish sum"); *Mein Leben*, 210 ("ein größeren Betrag"); Krafft, *Im Schatten der Sensation*, 486.

82. Hahn to Meitner, card, 16 December 1946 (MC).

83. Hahn to Meitner, 28 December 1946 (MC).

84. Meitner to Eva von Bahr-Bergius, 24 December 1946 (MC).

85. Meitner to Lola Allers, 29 December 1946 (MC).

86. Meitner to James Franck, 16 January 1947 (incorrectly dated 1946) (MC).

CHAPTER 15. No Return

1. Lise Meitner to Otto Hahn, 23 February 1947 (MC).

2. Karl-Erik Larsson, "Kärnkraftens historia i Sverige," *Kosmos* 64 (1987): 121–161, on 123ff.; Stefan Lindström, "Implementing the Welfare State: The Emergence of Swedish Atomic Energy Policy," in Svante Lindqvist, ed., *Center on the Periphery: Historical Aspects of 20th-Century Swedish Physics* (Canton, Mass.: Science History Publications, 1993), 181–187. Lindström notes (pp. 182–183) that Meitner's presence in Sweden and her contacts with Bohr contributed to the awareness among scientists and politicians of fission developments during the war, preparing for a more rapid consensus afterward.

3. For the structure of Swedish war research and an overview of military defense projects, see Hans Weinberger, "Physics in Uniform: The Swedish Institute of Military Physics 1939–1945," in Lindqvist, *Center on the Periphery*, 141–163.

4. Meitner to Margrethe Bohr, 7 March 1947 (MC).

5. Meitner to Hahn, 30 March 1947 (MC). The proposed high-voltage apparatus was not as large as the one Meitner built in her institute in Dahlem in the 1930s.

6. Meitner to Margrethe Bohr, 17 August 1946, 22 November 1946, 7 March 1947; Meitner to Hahn, 30 March 1947 (MC). In 1951, Meitner told Hahn (Meitner to Hahn, 4 March 1951 [MC]) that for nine years her salary in Siegbahn's institute had "not even been that of a first assistant." According to Lindström, "Implementing the Welfare State," 182–183, Tage Erlander had met Niels Bohr and also Lise Meitner during the war. I am grateful to Professor Inga Fischer-Hjalmars for calling Erlander's role to my attention.

7. Meitner to Eva von Bahr-Bergius, 31 August 1946 (MC).

8. Meitner to Bahr-Bergius, 24 December 1946 (MC).

9. Meitner to Bahr-Bergius, 31 August 1946 (MC).

10. Meitner to James Franck, 14 July 1957 (FP).

11. Meitner to Franck, 22 August 1952 (FP).

12. Meitner to Franck, 14 July 1957 (FP).

13. Ibid.

14. Meitner to Frau [?] Schindler, 20 February 1957 (MC): "The exclusion of women from higher positions is quite great."

15. Meitner to Franck, 14 July 1957 (FP); Meitner to Elisabeth Schiemann, 13 December 1954 (MC).

16. Meitner to Franck, 14 July 1957 (FP); Meitner to Schiemann, 18 November 1961 (MPG).

17. Meitner to Hahn, 23 February 1947 (MC).

18. Meitner to Margrethe Bohr, 25 October 1949 (BSC, supplement). Meitner included Oskar and Gerda Klein among those who "spoke her language."

19. Meitner to Max von Laue, 12 November 1946 (MC).

20. Meitner to Lilli Eppstein, 14 December 1960 (EC). Meitner wrote that her friendships, with Eppstein and a few others, "shielded me from the fear that my life had lost all valuable meaning when I left Dahlem."

21. Meitner to Laue, 12 November 1946 (MC).

22. Meitner to Hahn, 30 March 1947 (MC).

23. Raul Hilberg, *The Destruction of the European Jews* (New York: Octagon Books, 1978), 699–702; David C. Cassidy, *Uncertainty: The Life and Science of Werner Heisenberg* (New York: W. H. Freeman, 1992), 529.

24. Gottfried von Droste to Meitner, 22 December 1946 and 28 January 1947 (MC).

25. Meitner to Hahn, 23 February 1947 (MC).

26. Hermann Fahlenbrach to Meitner, 9 June 1947; see also letters from Fahlenbrach to Hahn, 2 June 1947, and Rudolf Jaeckel to Hahn, 4 June 1947 (MC).

27. Meitner to Hahn, 24 June 1947 (MC).

28. Meitner to Margrethe Bohr, 17 June 1947 (MC).

29. Meitner to Gottfried von Droste, 28 February 1947 (MC).

30. Meitner to Hermann Fahlenbrach, 19 June 1947 (MC).

31. Meitner to Hahn, 24 June 1947 (MC).

32. Meitner to Hahn, 23 March 1948 (MC).

33. Paul Rosbaud to Samuel Goudsmit, 9 August 1946 and other letters, 1946–1947 (Goudsmit Papers, AIP). Meitner frequently discussed the German issue with Rosbaud, including the following occasions: Meitner to Rosbaud, 4 August 1946 (MC); Rosbaud to Meitner, 5 August 1946 (MC); Meitner, diary, 2 April 1947 (MC).

34. Meitner to Hahn, 20 October 1946 (MC). Telschow, an assistant to General Director Friedrich Glum, joined the Nazi party early; according to Glum, Telschow displaced him in 1937 because he "had the trust of the Nazis." See Friedrich Glum, *Zwischen Wissenschaft, Wirtschaft und Politik* (Bonn: H. Bouvier Verlag, 1964), 439, 614–615.

35. Meitner to Gerta von Ubisch, 1 July 1947 (MC); Elisabeth Schiemann, "Autobiographie," *Nova Acta Leopoldina* Bd. 21, Nr. 143 (Leipzig, 1959): 291–292; Elisabeth Schiemann, "Erinnerungen an meine Berliner Universitätsjahre," in *Studium Berolinense* (Berlin: Walter de Gruyter, 1960), 850–851.

36. Meitner, diary, 24 April 1947 (MC). Meitner to Hahn, 8 June 1947; in Fritz Krafft, *Im Schatten der Sensation: Leben und Wirken von Fritz Strassmann* (Weinheim: Verlag Chemie, 1981), 188.

37. Wolffenstein was hidden by the Schiemanns from January to March 1943, by the Strassmanns from March to May 1943, then by a succession of Protestant pastors. Andrea Wolffenstein, "Fernsehaufnahme zum 75. Geburtstag von Prof. Dr. Straßmann," 3. Fernsehprogramm des Südwestfunks, 1 November 1977, in Krafft, *Im Schatten der Sensation*, 46–47; Wolffenstein, "Kein Risiko gescheut," in *In Memoriam Fritz Strassmann* (Mainz: Privatdruck, Verlag H. Schmidt, 1980), 10.

38. Irmgard Strassmann, Mainz, pers. comm., 1986; *Attestation: Yad Vashem,* 18 September 1986; *Allgemeine Zeitung Mainz,* 26 July 1986, 3 October 1986; *Süddeutscher Zeitung,* 2/3 August 1986. Strassmann was posthumously honored with a tree on the Avenue of the Righteous, Yad Vashem, Jerusalem.

39. Schiemann's and Meitner's opposing views of German history were quite typical: on the one hand, Nazism was an "accident in an otherwise commendable development"; on the other, Nazism was "the culmination of centuries of German cultural and political misdevelopment." See Ian Kershaw, *The Nazi Dictatorship: Problems and Perspectives of Interpretation,* 2d ed. (London: Edward Arnold, 1989), 7.

Meitner to James Franck, 19 June 1957 (FP): the Schiemann sisters were sometimes "difficult in their vehement and in part extremely unrealistic political statements."

Elisabeth Schiemann to Otto Robert Frisch, 14 July 1969 (MC). Schiemann cites the Meitner letters in the 1940s as having a "wounded tone, often embittered"; Meitner "fought [brawled] with Hahn as if it had not been necessary to leave Berlin and her work." Meitner did not fight with Hahn about this (see chap. 14, page 342): Schiemann was repeating Hahn's statements, perhaps after reading his autobiography, (Otto Hahn, *My Life,* trans. Ernst Kaiser and Eithne Wilkins [New York: Herder and Herder, 1970], 199), which appeared in Germany in 1968.

40. Meitner to Laue, 4 August 1947 (MC).

41. Meitner to Franck, 16 January 1947 (misdated 1946) (MC).

42. Meitner to Bahr-Bergius, 21 October 1947; Meitner to Hedy Born, 23 November 1947 (MC).

43. J. L. Heilbron, *The Dilemmas of an Upright Man: Max Planck as Spokesman for German Science* (Berkeley, Los Angeles, and London: University of California Press, 1986), 199.

44. Fritz Strassmann to Meitner, 11 September 1947; in Krafft, *Im Schatten der Sensation,* 184. For Mattauch's reaction, see Krafft, pp. 186–187.

45. Meitner to Fritz Strassmann, 21 December 1947 (MC); in Krafft, *Im Schatten der Sensation,* 184–185.

46. Meitner to Bahr-Bergius, 10 January 1948 (MC).

47. Meitner to Margrethe Bohr, 2 October 1947 (supplement to BSC); Meitner to Bahr-Bergius, 21 October 1947 (MC). Lise Meitner, "Max Planck als Mensch," *Naturwiss.* 45 (1958): 406–408.

48. Max Born, "Max Karl Ernst Ludwig Planck, 1858–1947," *Roy. Soc. Lond. Obit. Not.* 6, no. 17 (1948): 161–188.

49. Meitner to Hahn, 23 March 1948 (MC).

50. The Allies objected to Telschow for some time, however. See Hahn, *My Life,* 210–213. Hahn always preferred Telschow (one of his first doctoral students) to his predecessor, Friedrich Glum, who had been displaced for political reasons in 1937 and tried to reacquire his former position after the war. See Glum, *Zwischen Wissenschaft,* 614–615, 624–625; also Hahn to Meitner, 17 September 1946 (MC).

Meitner believed that Hahn was unduly influenced by Telschow, also later in the choice of Hahn's successor, Adolf Butenandt, in 1959. See Meitner to Franck, 23 October 1959 (FP).

51. Max von Laue, "The Wartime Activities of German Scientists," *Bull. Atom. Sci.* 4 (1948): 103.

52. Mark Walker, "Legenden um die deutsche Atombombe," *Vierteljahrhefte für Zeitgeschichte* 38, no. 1 (1990): 45–74, on 69–73; Cassidy, *Uncertainty*, 508–514.

53. Samuel A. Goudsmit, *Alsos* (New York: Henry Schuman, 1947), 134–139.

54. Philip Morrison, "*Alsos:* The Story of German Science," *Bull. Atom. Sci.* 3 (1947): 354, 365.

55. Jonathan L. Logan (letter to *Physics Today* 44, no. 5 [May 1991]: 13, 15, 90–91) notes Laue's "insupportable implication that the word of the oppressors is intrinsically more, rather than less, reliable than the word of their victims."

56. Philip Morrison, "A reply to Dr. von Laue," *Bull. Atom. Sci.* 4 (1948): 104. In response to Laue's implication that Morrison must have "suffered personally" (i.e., that anyone holding such opinions was Jewish), Morrison replied that he had not. See also Mark Walker, *German National Socialism and the Quest for Nuclear Power, 1939-1949* (Cambridge: Cambridge University Press, 1989), 213–215.

57. Laue to Meitner, 15 June 1948 (MC).

58. Meitner to Margrethe Bohr, 10 July 1948 (MC).

59. Meitner to Hahn, 6 June 1948 (MC); reprinted in Krafft, *Im Schatten der Sensation*, 185–186.

60. Hahn to Meitner, 16 June 1948 (MC).

61. Meitner to Hahn, 23 July 1948 (MC). Expressing the same view as Meitner did, Ute Deichmann, in *Biologen Unter Hitler* (Frankfurt: Campus Verlag, 1992), 188, notes that Germans did not understand that as a group they were deeply distrusted by foreign scientists, particularly by those from the "ruined lands" of Holland, Belgium, France, and Norway. As an example, Deichmann cites the blanket exclusion of Germans from the 1948 Stockholm genetics conference (p. 191); a few individual Germans were invited who were known to be politically untainted (*politisch unbelastet*).

62. In 1946, Rosbaud also thought it "very serious" that many "decent scientists"—he almost certainly had Hahn and Laue in mind, as there were not very many others he regarded as "decent"—were talking about "preserving national honor" as if they had forgotten the extremes such nationalism could take. (Rosbaud to Goudsmit, 9 August 1946 [Goudsmit Papers, AIP]).

63. Meitner to Franck, (n.d.) summer 1948 (MC).

64. Jost Lemmerich, ed., *Max Born, James Franck: Physiker in ihrer Zeit, Der Luxus des Gewissens* (Berlin: Staatsbibliothek Preussischer Kulturbesitz, Ausstellungskatalog 17, 1982), 144.

65. Meitner to Hahn, 5 December 1948 (MC).

66. Meitner to Margrethe Bohr, 25 October 1949 (supplement to BSC). Rosbaud, who also attended the 1949 Basel meeting, reported that Rudolf Fleisch-

mann, a Strassburg physicist who worked on fission during the war, told him that he "and almost *everybody* in Germany" [original italics] regarded *Alsos* as "the most infamous book" ever written. (Rosbaud to Goudsmit, 13 November 1949 [Goudsmit Papers, AIP]).

67. Meitner to Margrethe Bohr, 10 June 1948 (MC); Margrethe Bohr to Meitner (in English), 8 February 1948 (MC). At the Planck commemoration in April 1948, Meitner wrote, "I had a strange [*merkwürdiges*] talk with Weizsäcker, and the day after this conversation he wrote me a letter that was even more strange." Weizsäcker's letter (C. F. von Weizsäcker to Meitner, 24 April 1948 [MC]) is oddly self-centered: "Last night you let me see deeply into the fate that befell you. . . . I know that people cannot reduce each other's isolation. And, the way things have gone, I see that I, especially, cannot reduce yours. I frankly ask you not to answer this letter. I simply wanted to tell you that I saw something of what has happened to you, and I am thinking of it." He ended by enclosing a story written by his best friend shortly before he fell in Russia.

68. Dietrich Hahn, ed., *Otto Hahn, Begründer des Atomzeitalters: Eine Biographie in Bildern und Dokumenten* (München: List Verlag, 1979), 207. Einstein was responding (28 January 1949) to a letter from Hahn (18 December 1948) asking if he would accept foreign membership in the Max-Planck-Gesellschaft.

69. Lemmerich, *Max Born, James Franck,* 145.

70. Meitner to Laue, 25 April 1949, in K. E. Boeters and Jost Lemmerich, eds., *Gedächtnisausstellung zum 100. Geburtstag von Albert Einstein, Otto Hahn, Max von Laue, Lise Meitner* (Bad Honnef: Physik Kongreß-Ausstellungs- und Verwaltungs, 1979), 109.

71. Meitner to Margrethe Bohr, 17 June 1947 (MC); Meitner to Rudolf Ladenburg, 13 July 1947 (MC). Tage Erlander was prime minister from late 1946 until 1969. For his influence, see Lindström, "Implementing the Welfare State," 182.

72. As stated in a contemporary Swedish encyclopedia: "med prof.'s lön" (Prof. Inga Fischer-Hjalmars, pers. comm., Stockholm, May 1992). At first Meitner had no pension rights in Sweden but eventually she received a small pension and also a small pension from Germany (Meitner to Hahn, 28 January 1952 [MC]).

73. Berta Karlik, "In memoriam Lise Meitner," *Phys. Bl.* 35 (1979): 49-52.

74. *Protokoll,* Kungl. Vetenskapsakademiens, Klasser och Kommittéer, 3d *klass* (physics). Sir William Napier Shaw nominated Meitner to foreign membership in 1945; it was approved 8 October 1945. The conversion to full membership is dated 3 December 1951.

75. O. R. Frisch, "Lise Meitner, 1878–1968," *Biog. Mem. Fell. Roy. Soc. Lond.* 16 (1970): 405–420.

76. Larsson, "Kärnkraftens historia i Sverige," 131ff.

77. Lise Meitner, "Einige Bemerkungen zu den Einfangquerschnitten langsamer und schneller Neutronen bei schweren Elementen," *Ann. Phys.* 3 (1948): 115–119.

78. Lise Meitner, "An Attempt to Single Out Some Fission Processes of Uranium by Using the Differences in Their Energy Release," *Rev. Mod. Phys.* 17 (1945): 287–291. The article was published in the April-July 1945 issue, well before the Allied data accumulated during the war were made available.

79. See chap. 5.

80. Karen E. Johnson, "Independent-Particle Models of the Nucleus in the 1930s," *Amer. J. Phys.* 60 (1992): 164–172; Joan Dash, *A Life of One's Own: Three Gifted Women and the Men They Married* (New York: Harper & Row, 1973), 310ff.; Maria Goeppert Mayer, "The Structure of the Nucleus," *Sci. Am.*, March 1951; *Les Prix Nobel en 1963* (Stockholm: Nobel Foundation), 1964.

81. Lise Meitner, "Fission and Nuclear Shell Model," *Nature* 165 (1950): 561 (letter, submitted 26 January, published 8 April 1950). Also "Spaltung und Schalenmodell des Atomkerns," *Ark. Mat. Astr. Fys.* 4 (1950): 383; "Spaltung und Schalenmodell der Atomkerne," in *Manne Siegbahn 1886 3/12 1951* (65th birthday Festschrift) (Uppsala: Almqvist & Wiksells Boktryckeri, 1951), 575–578.

82. Meitner to Hahn, 6 February 1950 (MC); Meitner to O. R. Frisch, 21 August 1954 (MC).

83. Lise Meitner, "Die Anwendung des Rückstoßes bei Atomkernprozessen," *Z. Phys.* 133 (1952): 140–152. In this 1952 issue of *Zeitschrift für Physik* dedicated to Max Born and James Franck, Meitner reviewed 40 years of nuclear recoil experiments.

84. Larsson, "Kärnkraftens historia i Sverige," 134–136.

85. Lindström, "Implementing the Welfare State," 180, notes that little is known about Swedish nuclear weapons development except that it was "close to the threshold."

86. Meitner to James Franck, 14 March 1954 (MC).

87. Eklund (1911–   ), director of the IAEA from 1961 to 1981, is currently director emeritus.

CHAPTER 16. Final Journeys

1. Otto Hahn, "Zum 75. Geburtstag von Stefan Meyer," *Z. Naturforschg.* 2a (1947): 364; Stefan Meyer, "Zur Erinnerung an die Jugendzeit der Radioaktivität: Lise Meitner (geb. 7 Nov. 1878) und Otto Hahn (geb. 8 März 1879) zu ihrem siebzigsten Geburtstag gewidmet)," *Naturwiss.* 35 (1948): 161–163; Otto Hahn, "Im Memoriam Stefan Meyer," *Z. Naturforschg.* 5a (1950): 407.

2. Meitner to Otto Hahn, 12 July 1949 (MC).

3. Richard Willstätter, *From My Life*, trans. Lilli S. Hornig (New York: W. A. Benjamin, 1965), 36off., 422.

4. Meitner to Frieda Frischauer, 7 September 1945 (MC).

5. Meitner to Dirk Coster, zum 9 December 1949 (MC).

6. H. A. Kramers, "Herdenking van Dirk Coster (5 October 1889–12 Februari 1950)," *Jaarboek der Koninklijke Nederlandse Akademie van Wetenschappen*, 1951–1952; Hans Coster, Belleaire, Texas, pers. comm., 1 February 1986.

7. Meitner to James Franck, zum 29 August 1952; 19 August 1957 (MC).

8. Laue to Meitner, 17 October 1949 (MC).

9. Max von Laue, "Arnold Berliner (26.12.1862–22.3.1942)," *Naturwiss.* 33 (1946): 257–258 (appeared April 1947).

10. P. P. Ewald, "Max von Laue, 1879–1960," *Biog. Mem. Fell. Roy. Soc. Lond.* 6 (1960): 135–156.

11. Laue to Meitner, 80th birthday greetings, 7 November 1958 (MC).

12. Laue to Paul Rosbaud, 4 April 1959. Printed in English in Arnold Kramish, *The Griffin: The Greatest Untold Espionage Story of World War II* (Boston: Houghton Mifflin, 1986), 245–247. See also chap. 13, p. 323. Laue is somewhat unclear, but he apparently means that *genuine* ethical points of view were not mentioned. See also David C. Cassidy, *Uncertainty: The Life and Science of Werner Heisenberg* (New York: W. H. Freeman, 1992), 519.

In a "conversation with C. F. Weizsäcker," appended to the German translation of the Farm Hall transcripts (Dieter Hoffmann, ed., *Operation Epsilon: Die Farm-Hall-Protokolle oder Die Angst der Alliierten vor den deutschen Atombombe*, trans. Wilfried Sczepan [Berlin: Rowohlt, 1993], 331–360), Weizsäcker shrugs off Laue's 1959 letter by saying (p. 349) that in Farm Hall he did not intend to initiate a "legend," but meant what he said.

13. Robert Jungk, *Brighter than a Thousand Suns: A Personal History of the Atomic Scientists*, trans. James Cleugh (New York: Harcourt Brace, 1958), 105. The German original, *Heller als tausend Sonnen*, was published in 1956 by Alfred Scherz, Bern.

14. Mark Walker, "Legenden um die deutsche Atombombe," *Vierteljahrhefte für Zeitgeschichte* 38, no. 1 (1990): 45–74, on 61–64 n. 39.

15. Meitner to Hahn, 24 January 1957 (MC).

16. Laue to Rosbaud, 4 April 1959 (see n. 12).

17. Rosbaud to Meitner, card, n.d. 1957 (MC). Rosbaud also thought the book was a "successful attempt to wash clean [*rein zu waschen*] the German 'Uranverein,' especially Weizsäcker."

18. Rosbaud to Samuel Goudsmit, 30 June 1958 (Goudsmit Papers, AIP).

19. Otto Hahn, "Lise Meitner 70 Jahre," *Z. Naturforschg.* 3a (1948): 425–428. Hahn gives the wrong dates for Meitner's years as Max Planck's Assistent (correct: 1912–1915).

20. Lise Meitner, "Otto Hahn zum 8 März 1949," *Z. Naturforschg.* 4a (1949): 81.

21. Otto Hahn, "Lise Meitner 75 Jahre," *Z. Naturforschg.* 8a (1953): 679–680.

22. Lise Meitner, "Einige Erinnerungen an das Kaiser-Wilhelm-Institut für Chemie in Berlin-Dahlem," *Naturwiss.* 41 (1954): 97–99. Meitner's dates for her military service and description of the discovery of protactinium are incorrect.

23. Otto Hahn, "Lise Meitner 80 Jahre," *Naturwiss.* 45 (1958): 501–502; "Lise Meitner 85 Jahre," *Naturwiss.* 50 (1963): 653–654.

Lise Meitner, "Otto Hahn zum 80. Geburtstag am 8. März 1959," *Naturwiss.* 46 (1959): 157–158; "Otto Hahn zum 85. Geburtstag," *Naturwiss.* 51 (1964): 97.

24. Meitner to Eva von Bahr-Bergius, 10 January 1948 (MC). Meitner was pleased with the 1947 award from Vienna, but, she wrote, "the sum is small and anyway I probably won't ever be able to get it transferred here." For Meitner's reaction to her "golden diploma" (from the university of Vienna in 1956), see Bertha Karlik, "Gedenkwörte für Lise Meitner," *Akademische Gedenkfeier zu Ehren von Otto Hahn und Lise Meitner am 21. Februar 1969 in Berlin* (Berlin: Max-Planck-Gesellschaft, 1969), 35–42, on 40. To Otto Stern (Meitner to Stern, 20 January 1956 [MC]), Meitner wrote, "I know very well that one needs outer recognition in one's early years, in order to develop confidence in one's chosen path." For a similar statement, see Meitner to Laue, 6 September 1955 (MC).

25. Fritz Strassmann, *Kernspaltung: Berlin Dezember 1938* (Mainz: Privat-druck, 1978), 23; reprinted in Fritz Krafft, *Im Schatten der Sensation: Leben und Wirken von Fritz Strassmann* (Weinheim: Verlag Chemie, 1981), 211.

26. Hildegard Pusch, Munich, pers. comm., 16 August 1989, 24 September 1989. Pusch's aunt, Lotte Volmer, recounted the following: At the Volmers, in the 1920s or 1930s, "Meitner and Hahn were having a heated debate, at which Max Volmer remarked, 'Let it go, Hähnchen; of the two of you, Lischen is the cleverer one.' Hahn did not like hearing this; he knew it was true."

27. Walther Gerlach, *Otto Hahn: Ein Forscherleben unserer Zeit*, ed. Dietrich Hahn (Grosse Naturforscher, Band 45) (Stuttgart: Wissenschaftliche Verlags-gesellschaft, 1984), 205–211.

28. Krafft, *Im Schatten der Sensation*; Günter Herrmann, "Ein Forscher, der Geschichte machte," in *In Memoriam Fritz Strassmann* (Mainz: Privatdruck, 1980), 5–9. Between 1946 and his retirement in 1970, Strassmann established the new MPI for Chemistry and the university's institute for inorganic chemistry, taught nearly all classes in analytical and inorganic chemistry, and oversaw the construc-tion of a neutron generator, a research reactor, and a new institute for nuclear chemistry.

29. Burghard Weiss, "Hahn und Meitner, Merton und Matthäs: Zur Na-mengebung einer deutschen Großforschungseinrichtung," *Ber. Wissenschaftsgesch.* 13 (1990): 219–231. In June 1988, the Otto-Hahn-Bau auditorium was named for Meitner and a bust placed near the door, but her contributions are not indicated.

30. Weiss, "Hahn und Meitner," 228.

31. Karl-Erik Zimen, "Otto Hahn, Lise Meitner und die Kernspaltung im Ausblick auf die Zukunft," *Phys. Bl.* 35 (1979): 200–210.

32. R. Spence, "Otto Hahn, 1879–1968," *Biog. Mem. Fell. Roy. Soc. Lond.* 16 (1970): 279–313, on 300.

33. Meitner to Hahn, 22 October 1951 (MC). Meitner's reference is to Adelbert von Chamisso's *Peter Schlemihls wundersame Geschichte*, the story of a man without a shadow, considered an allegory for the author's situation as a man without a country.

34. This was true even for Laue. In a well-intended tribute to Meitner at the Otto-Hahn-Bau dedication in 1956, he referred to her as Hahn's "engste [closest] Mitarbeiterin." See Weiss, "Hahn und Meitner," 225.

35. Meitner to Hahn, 22 June 1953 (MC); in Krafft, *Im Schatten der Sensation*, 188. The sentence, "What would you say if you were *only* characterized as the longtime Mitarbeiterin of me?" is an insert handwritten by Meitner at the bottom of the page and is difficult to read; Krafft substitutes *auch* (also) for *nur* (only) and does not indicate the intended position of the insert (italics not in original).

Fritz Krafft, "Lise Meitner (7.XI.1878–27.X.1968)," in Willi Schmidt and Christoph J. Scriba, eds., *Frauen in den exakten Naturwissenschaften: Festkolloquien zum 100. Geburtstag von Frau Dr. Margarethe Schimank (1890–1983)* (Stuttgart: Franz Steiner Verlag, 1990), 33–70; see pp. 55–56. Krafft notes that "it has not been possible to root out the term *Mitarbeiterin* among older physicists." Krafft gives the excerpt from Heisenberg's 1953 article, which not only denotes Meitner as Hahn's Mitarbeiterin but includes the often-repeated (false) allegation that she received news of barium from Hahn and immediately telegraphed it to a Washington meeting of physicists.

36. Otto Hahn, *My Life*, trans. Ernst Kaiser and Eithne Wilkins (New York: Herder and Herder, 1970), 88, 147.

37. From a sampling of writings by Otto Hahn:

Physicists considered fission "impossible": *New Atoms* (New York: Elsevier, 1950), 42; "50 Jahre Radioaktivität," *Orion* 26a (October 1950) 805–811; "Radium- und Atomkernforschung: Internationale Wissenschaft," (Reichenhall XIII: Europa Verlag, 1951), 25–32; "The Discovery of Fission," *Sci. Am.* 198 (February 1958): 76–84.

Elements of midweight from neutron irradiation of uranium were "completely ruled out": "Die Entwicklung der Radiochemie und die Spaltung des Urans," *Naturwiss. Rdsch.* 6 (1953): 45–49.

Fission was "forbidden" by physics: "25 Jahre Uranspaltung," *Urania (Berlin, DDR)* 1 (1964): 8; "Erinnerungen an einige Arbeiten: Anders geplant als verlaufen," *Naturwiss. Rdsch.* 18 (1965): 86–91.

38. That Hahn had a political message is evident in articles written after the war. In 1951, for example, in "Radium- und Atomkernforschung," p. 31, Hahn states that his institute had "never thought of military uses" for fission (untrue) and continued to publish openly throughout the war (true only for work considered unimportant to the fission project).

Mark Walker, in *German National Socialism and the Quest for Nuclear Power, 1939–1949* (Cambridge: Cambridge University Press, 1989), notes (p. 163) that research was done in Hahn's institute—in particular, studies of resonance absorption and the chemistry of transuranic elements—that would have been necessary for the manufacture of nuclear weapons had the German fission project ever gotten that far. In Otto Hahn, *A Scientific Autobiography*, trans. and ed. Willy Ley (London: MacGibbon & Kee, 1967), 170, is a table of some 100 fission fragments identified in Hahn's institute during the war; knowledge of the chemistry and neutron absorption characteristics of such fission products would have been vital for running a reactor, either for energy production or breeding plutonium. In *Operation Epsilon: The Farm Hall Transcripts* (Berkeley, Los Angeles, and Oxford: University of California Press, 1993), for example, Hahn says (p. 125), "[Element] 93 can be quantitatively separated from 92. Strassmann and I have worked out the quantitative separation"; it is evident that they were quite prepared to work on the chemistry of element 94, had they been able to produce it in quantity. In his *Autobiography*, p. 171, Hahn complains about his weak neutron sources, but in *Operation Epsilon*, p. 143, he states that Heisenberg's laboratory had a high-tension apparatus from which he obtained stronger preparations, even after his institute was moved to Tailfingen. Altogether, it is evident from the Farm Hall scientific discussions that Hahn and his co-workers in the KWI for Chemistry were knowledgeable contributors to the fission project. For comparison with the Allies' early work on the chemistry of plutonium, see Bertrand Goldschmidt, *Atomic Rivals*, trans. Georges M. Temmer (New Brunswick: Rutgers University Press, 1990).

39. Otto Hahn, *Vom Radiothor zur Uranspaltung: Eine wissenschaftliche Selbstbiographie* (Braunschweig: H. Vieweg, 1962).

40. Lise Meitner, "Wege und Irrwege zur Kernenergie," *Naturwiss. Rdsch.* 16 (1963): 167–169; Lise Meitner, "Right and Wrong Roads to the Discovery of Nuclear Energy," *Advancement of Science* 19 (1963): 363–365, reprinted from the *International Atomic Energy Agency Bulletin.*

41. Oral History Interview, Lise Meitner with Thomas S. Kuhn (Otto Robert Frisch present), 12 May 1963, Archive for the History of Quantum Physics, American Institute of Physics, 18–20. See chap. 7.

42. The process that physicists *did* consider impossible was the chipping off of sizable nuclear chunks of mass 10 or 20. Meitner emphasized this point in her interview with Thomas Kuhn (see n. 41, above) and again in "25 Jahre Uranspaltung," *Urania (Berlin, DDR)* 1 (1964): 9.

43. Krafft, *Im Schatten der Sensation*, 222–227. The plain wooden table and apparatus are authentic but probably not original. Although the apparatus is displayed on one table, irradiation and measurement were done separately, in rooms 29 and 23, respectively. The display also includes a suction flask to symbolize the chemical separations, done in room 20, all in Meitner's section on the ground floor of the institute.

Krafft notes (p. 224) that photographs of the table generally only showed the sign *Arbeitstisch von Otto Hahn*, another instance of Strassmann's name disappearing from view.

44. In 1990, after considerable input from women's groups and historians of science, the museum expanded and greatly improved the accuracy of the materials accompanying the Arbeitstisch. A sign behind the table now reads, in German and English: "The Experimental Apparatus with which the Team of Otto Hahn, Lise Meitner and Fritz Strassmann Discovered Nuclear Fission in 1938." Text and

pictures in a panel to the side include the three scientists and also Otto Robert Frisch. In 1991, the Deutsches Museum added a bust of Lise Meitner to their Ehrensaal (Hall of Honor), the first woman in a room of German scientific greats. (Of course, she was not German, strictly speaking, but neither was Copernicus [a Pole], who is also represented. The notion of a science museum devoted to *German* science now seems outdated and may have run its course.)

In the museum in Haigerloch, where Heisenberg's group worked on a uranium "machine" (reactor), a full-scale reproduction of the old Deutsches Museum display is shown. As of 1994, it has not been updated.

45. Kurt Starke, "The Detours Leading to the Discovery of Nuclear Fission," *J. Chem. Educ.* 56 (1979): 771–775.

46. Karl-Erik Zimen, "Otto Hahn, Lise Meitner"; Zimen, "50 Jahre Kernspaltung," *Atomwirtschaft* (December 1988): 577–584; Zimen, "Some Recollections on Radioactivity and Fission," *IANCAS Special Bulletin* 5, no. 1 (January 1989): 1–4.

47. Zimen indicates, in "50 Jahre Kernspaltung" and "Some Recollections" (see n. 46, above), that he "witnessed" the controversy surrounding the 1945 decision to give the award to Hahn alone; he could not have been ignorant about the poor relationship between Siegbahn and Meitner.

48. Siegfried Flügge, "Zur Entdeckung der Uranspaltung vor zehn Jahren," *Z. Naturforschg.* 4a (1949): 82–84. The article, written to counter the almost exclusively chemical perception of the discovery, emphasizes the importance of theoretical physics. Flügge was not aware of Meitner's continued collaboration in the fall of 1938 or her meeting with Hahn in Copenhagen; he states that institute physicists, including himself, objected to the radium isomers on theoretical grounds, but he did not know to what extent their objections influenced the final experiments. Unlike others who avoided the topic of Meitner's forced emigration, Flügge describes it as "deplorable." According to Paul Rosbaud, Flügge was among the scientists who always had been decent under the Third Reich (Rosbaud to Samuel Goudsmit, 3 October 1948, Goudsmit Papers, AIP) (see also n. 52, below).

49. Carl Friedrich von Weizsäcker, foreword to Dietrich Hahn, ed., *Otto Hahn: Leben und Werk in Texten und Bildern* (Frankfurt: Insel Taschenbuch, 1988); Weizsäcker, recorded talk, Deutsches Museum Ehrensaal, June 1991; Weizsäcker to Vincent C. Frank-Steiner, 25 March 1992; Weizsäcker, pers. comm., 9 April and 27 May 1992. For the absence of help from the father, Baron von Weizsäcker, see chap. 8.

50. Rudolf Fleischmann to Pieter Van Assche, 22 November 1982. (I am grateful to Pieter Van Assche for permission to quote from this letter.) Goudsmit (in *Alsos* [New York: Henry Schuman, 1947], 70) notes that although Fleischmann was extremely uncooperative with the *Alsos* scientists, his papers were useful because he was a "great collector of gossip and wrote it all down." See also Mark Walker, *German National Socialism*, 154–156. For Bagge in Farm Hall, see chap. 13.

51. Werner Heisenberg, "Gedenkworte für Otto Hahn und Lise Meitner," *Orden pour le mérite für Wissenschaft und Künste, Reden und Gedenkworte* 9 (1968–1969): 111–119.

52. Fleischmann complained that Frisch knew of barium 14 days earlier than Siegfried Flügge and Gottfried von Droste, and 17 days earlier than Willibald Jentschke and Friedrich Prankl in Vienna (see n. 50, above). Flügge and von Droste's interpretation of the fission process was submitted on 22 January 1939, shortly after Meitner and Frisch's, but published considerably later; Jentschke and Prankl physically confirmed fission shortly after Frisch and submitted their note on 14 February 1939; by then Joliot in France and many physicists in the United States had done similar experiments. See also chap. 11, n. 18.

Zimen (in "50 Jahre Kernspaltung," 579, and "Some Recollections," 3) claims that "in the English-speaking world" Meitner and Frisch are "regrettably" given credit for the theoretical explanation alone, although Siegfried Flügge and Gottfried von Droste's explanation was given "at the same time independently." Flügge himself, in "Zur Entdeckung der Uranspaltung," makes no such claim; he accords Meitner and Frisch the "uncontestable priority" (*unbestreitbare Priorität*) for the theoretical interpretation and the physical confirmation (see also n. 48, above). Starke, in "Detours," believes Hahn was trying "to be fair" in telling Meitner first.

53. Ingmar Bergström and Wilhelm Forsling, *I Demokritos fotspår: En vandring genom urämnesbegreppets historia från antiken till Nobelpreisen* (Stockholm: Natur och Kultur, 1992), 332ff. Bergström, who knew Jentschke personally, also relates (p. 339) Jentschke's claim that Hahn indicated the possibility of nuclear fission already in 1938(!) at a lecture in Vienna, an example of the pitfalls of unconfirmed oral history.

54. For Meitner's 80th birthday, Oskar Klein composed "a conversation with Goethe" in which he seeks advice and solace for his friend Lise who "cannot really have any pleasure" in her part of the famous discovery. Oskar Klein, "Lise Meitner zum achtzigsten Geburtstag 7/11 1958" (pers. comm., Lilli Eppstein). Frisch states this explicitly in O. R. Frisch to Hahn, 22 August 1966 (OHN).

55. Meitner to James Franck, 8 September 1955 (FP).

56. Meitner to James Franck, ? March 1958 (FP). Meitner did not sign a petition drafted by Linus Pauling that called for an international ban on nuclear testing; the petition, signed by over 11,000 scientists, was presented to the UN in 1958. Pauling won the 1962 Nobel Peace Prize; a partial nuclear test ban went into effect on 10 October 1963.

57. Meitner to Franck, 14 July 1957 (FP).

58. Richard Rhodes, "The Complementarity of the Bomb," *J. Chem. Educ.* 66 (1989): 376–379.

59. Lise Meitner, Lecture, Austrian UNESCO Commission, 30 March 1953, in *Atomenergie und Frieden: Lise Meitner und Otto Hahn* (Vienna: Wilhelm Frick-Verlag, 1954), 11–26, on 23–24.

60. Meitner to Franck, ? March 1958 (FP).

61. Meitner to Franck, 6 October 1957 (FP).

62. Meitner to Lilli Eppstein, 27 September 1962 (EC).

63. Meitner to James Franck, 6 May 1960 (FP).

64. Meitner to Franck, 19 August 1957 (FP).

65. Meitner to Franck, 15 March 1956 (FP).

66. Meitner to Hahn, 22 June 1953 (MC).

67. Meitner to Franck, 25 July 1961; also 30 November 1963 (FP) (incorrectly given as 20 November 1963 in Jost Lemmerich, *Max Born, James Franck: Physiker in ihrer Zeit: Der Luxus der Gewissens* [Berlin: Staatsbibliothek Preussischer Kulturbesitz, Ausstellungskatalog 17, 1982], 172; the letter was written shortly after the assassination of President John F. Kennedy).

68. Meitner to Berta Karlik, 13 July 1959. Quoted in Karlik, "Lise Meitner: Nachruf," *Alman. Österr. Akad. Wiss.* 119 (1969): 345-354.

69. Meitner to Franck, 9 October 1958 (FP).

70. Lise Meitner, "The Status of Women in the Professions," *Physics Today* 13, no. 8 (August 1960): 17-21.

71. Meitner to Franck, 19 June 1959 (FP).

72. Meitner to Franck, 9 October 1958; 29 March 1959; 20 August 1959 (FP).

73. Meitner to Berta Karlik, 3 April 1953. Quoted in Karlik, "Lise Meitner," 353. Also see Berta Karlik, "In Memoriam Lise Meitner," *Phys. Bl.* 35 (1979): 49-52.

74. Meitner to Berta Karlik, 24 August 1950 (MC).

75. Meitner to Eppstein, 4 October 1963 (EC).

76. Lise Meitner, "Looking Back," *Bull. Atom. Sci.* 20, no. 11 (November 1964): 2-7.

77. Otto Robert Frisch, interview, 22 March 1975, Cambridge.

78. Meitner to Hahn, 15 November 1953 (MC); Meitner to Hertha Sponer, 16 December 1953 (FP).

79. Meitner to Eppstein, 1 November 1960 (EC).

80. Meitner to Eppstein, 4 October 1963 (EC).

81. Meitner to Eppstein, 12 December 1963 (EC).

82. Meitner to Franck, 25 June 1961 (FP).

83. Meitner-Franck correspondence, 1960 (FP); Meitner to Eppstein, 16 January 1964 (EC).

84. Meitner to Eppstein, 1 November 1960 (EC).

85. Meitner to Margrethe Bohr, 17 June 1947 (MC).

86. Meitner to Elisabeth Schiemann, 18 November 1961 (MPG).

87. Meitner to Eppstein, 16 January 1964 (EC).

88. Karlik, "Lise Meitner," in *Neue Österreichische Biographie*, Band XX (Wien: Amalthea Verlag, 1979), 51-56.

89. Meitner to Franck, 20 March 1961; 25 July 1961 (FP); Meitner to Elisabeth Schiemann, 18 November 1961 (MPG).

90. Meitner to Franck, 30 November 1963 (FP).

91. A sample taken from Meitner's notebook ca. 1960 (MC):

O.R. Fragen:

Why are e^- and e^+ particle and antip[article] but not π^- and π^+?

Non-singular force (velocity dependent)

effective nuclear moments arising from the question of the free nuclear moment inside nuclear matter

Pauling's resonating valence bond theory

The spectral density of charged particles

highfalutin?

Fermi surface in metals? anomalous skin effect?

juke box?

Stonehenge (archaeology)?

Spin-spin relaxation effects?

What is the meaning of "nuclear form factor"?

92. Meitner to Franck, 20 March 1961 (FP). In 1961, Mössbauer was being considered for the Nobel Prize in physics; having heard that others had contributed substantially to his work, Meitner thought the prize should wait (very likely she was reminded of the 1945 controversy over the award to Hahn). Mössbauer was awarded the prize that year, however.

93. Meitner to Franck, 21 December 1954 (FP).

94. Meitner to Max Born, 6 April 1956 (MC). For a similar description of Irène Curie, see Elizabeth Rona, *How It Came About: Radioactivity, Nuclear Physics, Atomic Energy* (Oak Ridge: Oak Ridge Associated Universities, 1978), 27.

95. Meitner to Max von Laue, 22 February 1956 (MC).

96. Meitner to Hertha and James Franck, 20 August 1959 (FP); Walter Moore, *Schrödinger: Life and Thought* (Cambridge: Cambridge University Press, 1989), 472ff.

97. Meitner to Franck, 25 July 1961 (FP).

98. Ibid.

99. Meitner to Schiemann, 3 July 1961 (MPG).

100. Meitner to Eppstein, 19 November 1962 (EC).

101. Meitner to Franck, n.d. but ca. July 1963 (FP).

102. Meitner to Hertha Sponer and James Franck, 30 November 1963 (FP).

103. Meitner to Franck, 14 May 1964 (FP).

104. Meitner to Hertha Franck-Sponer, 22 May 1964 (FP).

105. Meitner to Lilli Eppstein, 7 December 1965 and ? February 1966 (the last letter from Meitner to Eppstein) (EC).

106. According to Marie-Luise Rehder, Hahn's secretary for many years, Hahn repeatedly sought more recognition for Strassmann and proposed him alone for the Fermi Prize. As to Meitner, Hahn believed "she already had so many prizes." (M.-L. Rehder, Göttingen, pers. comm., 30 June 1985.)

107. The designation of the award differed for each person; see Krafft, *Im Schatten der Sensation*, 475–476. Meitner's citation was for "pioneering research in the naturally occurring radioactivities and extensive experimental studies leading to the discovery of fission." In his presentation, Seaborg also noted her research in the 1920s, her initiative for the uranium investigation, and her theoretical interpretation with Frisch. See Glenn T. Seaborg, "Lise Meitner," in *Nuclear Milestones* (San Francisco: W. H. Freeman, 1972), 120–123.

Krafft, p. 474, indicates that the Fermi award was an intentional correction to the decision of the Nobel committee, but Glenn Seaborg (pers. comm., 20 October 1988) does not believe he and other members of the AEC were motivated by a desire "'to correct the decision of the Nobel committee.'"

108. Meitner to Hahn, 8 August 1966 (MPG).

109. Hahn to O. R. Frisch, 19 August 1966 (MPG). Meitner most certainly did not believe she had left Germany too soon, but Hahn suggests that she did, in this letter and also in Hahn, *My Life*, 199. See also chap. 14.

110. O. R. Frisch to Hahn, 23 August 1966 (MPG).

111. Max F. Perutz (1914–   ) completed his studies in Vienna and emigrated to England in 1936; for his x-ray crystallographic studies of proteins, he was awarded the Nobel Prize in chemistry in 1962. Perutz thought the Nobel Committee had been "narrow-minded" in not awarding the Nobel Prize jointly to Hahn and Meitner and when he had the chance, proposed her for the Fermi Prize in 1964 (Max F. Perutz, Cambridge, pers. comm., 15 August 1994; M. F. Perutz, "Lise Meitner's Genius," reply to letter, *New York Review of Books*, 3 February 1994, 46).

112. Glenn T. Seaborg, pers. comm., September 1988.

113. Frisch to Marie-Luise Rehder, 15 May 1967 (OHN).

114. Frisch to Rehder, 19 October 1968; 27 October 1968 (OHN).

115. Ulla Frisch, Cambridge, pers. comm., 24 June 1994; Anne Meitner, Sparsholt, Winchester, pers. comm., 28 June 1994.

SELECTED BIBLIOGRAPHY

Aaserud, Finn. *Redirecting Science: Niels Bohr, Philanthropy, and the Rise of Nuclear Physics.* New York: Cambridge University Press, 1990.

Albrecht, Helmuth. "'Max Planck: Mein Besuch bei Adolf Hitler': Anmerkungen zum Wert einer historischen Quelle." In Helmuth Albrecht, ed., *Naturwissenschaft und Technik in der Geschichte: 25 Jahre Lehrstuhl für Geschichte der Naturwissenschaft und Technik am Historischen Institut der Universität Stuttgart.* Stuttgart: Verlag für Geschichte der Naturwissenschaften und der Technik, 1993. 41–63.

Amaldi, Edoardo. "From the Discovery of the Neutron to the Discovery of Nuclear Fission." *Physics Reports* 111 (1984): 1–332.

Andrade, E. N. daC. *Rutherford and the Nature of the Atom.* Garden City, N.Y.: Doubleday, 1964.

Arms, Nancy. *A Prophet in Two Countries: The Life of F. E. Simon.* Oxford: Pergamon Press, 1966.

Atterling, Hugo. "Karl Manne Georg Siegbahn, 1886–1978." *Biog. Mem. Fell. Roy. Soc. Lond.* 37 (1991): 428–444.

Badash, Lawrence, ed. *Rutherford and Boltwood: Letters on Radioactivity.* New Haven: Yale University Press, 1969.

Badash, Lawrence. *Radioactivity in America: Growth and Decay of a Science.* Baltimore: Johns Hopkins University Press, 1979.

Badash, Lawrence, Elizabeth Hodes, and Adolph Tiddens. "Nuclear Fission: Reaction to the Discovery, 1939." *Proc. Amer. Phil. Soc.* 130 (1986): 196–231.

Barea, Ilsa. *Vienna.* New York: Alfred A. Knopf, 1966.

Bar-On, Dan. *Legacy of Silence: Encounters with Children of the Third Reich.* Cambridge: Harvard University Press, 1989.

Berkley, George E. *Vienna and Its Jews: The Tragedy of Success, 1880s–1980s.* Cambridge: Abt Books/Lanham, Md.: Madison Books, 1988.

Berninger, Ernst. *Otto Hahn: Eine Bilddokumentation*. München: Heinz Moos Verlag, 1969.

Beyerchen, Alan D. *Scientists under Hitler: Politics and the Physics Community in the Third Reich*. New Haven: Yale University Press, 1977.

Boedeker, Elisabeth. *Marksteine der Deutschen Frauenbewegung*. Hannover: Selbstverlag, 1969.

Boedeker, Elisabeth, and Maria Meyer-Plath. *50 Jahre Habilitation von Frauen in Deutschland: Eine Dokumentation über den Zeitraum von 1920–1970*. Göttingen: O. Schwarz, 1974.

Boeters, K. E., and J. Lemmerich, eds. *Katalog, Gedächtnisausstellung zum 100. Geburtstag von Albert Einstein, Otto Hahn, Max von Laue, Lise Meitner*. Bad Honnef: Physik Kongreß-Ausstellungs- und Verwaltungs, 1979.

Boltzmann, Ludwig. "A German Professor's Trip to El Dorado." Trans. Bertram Schwarzschild. *Physics Today* 45, no. 1 (January 1992): 44–51.

Boorse, Henry A., and Lloyd Motz, eds. *The World of the Atom*. New York: Basic Books, 1966.

Born, Max. "Max Karl Ernst Ludwig Planck, 1858–1947." *Roy. Soc. Lond. Obit. Not.* 6, no. 17 (1948): 161–188.

———. *My Life: Recollections of a Nobel Laureate*. New York: Charles Scribner's Sons, 1978.

Broda, Engelbert. *Ludwig Boltzmann*. Trans. Larry Gay and Engelbert Broda. Woodbridge, Conn.: Oxbow Press, 1983.

———. *Ludwig Boltzmann: Mensch, Physiker, Philosoph*. Wien: Franz Deuticke, 1955.

Bromberg, Joan. "The Impact of the Neutron: Bohr and Heisenberg." *Hist. Stud. Phys. Sci.* 3 (1971): 307–341.

Burchardt, Lothar. *Wissenschaftspolitik im Wilhelminischen Deutschland: Vorgeschichte, Gründung und Aufbau der Kaiser-Wilhelm-Gesellschaft zur Förderung der Wissenschaften*. Göttingen: Vandenhoeck & Ruprecht, 1975.

Casimir, Hendrik B. G. *Haphazard Reality: Half a Century of Science*. New York: Harper & Row, 1983.

Cassidy, David C. *Uncertainty: The Life and Science of Werner Heisenberg*. New York: W. H. Freeman, 1992.

Clark, Ronald W. *Einstein: The Life and Times*. New York: Avon, 1971.

Craig, Gordon A. *Germany, 1866–1945*. New York: Oxford University Press, 1978.

Crawford, Elisabeth. *The Beginnings of the Nobel Institution: The Science Prizes, 1901–1915*. Cambridge: Cambridge University Press, 1984.

Crawford, Elisabeth, J. L. Heilbron, and Rebecca Ullrich. *The Nobel Population 1901–1937: A Census of the Nominators and Nominees for the Prizes in Physics and Chemistry*. Berkeley: University of California, Office for History of Science and Technology/Uppsala: Uppsala University, Office for History of Science, 1987.

Dawidowicz, Lucy S. *The War Against the Jews, 1933–1945*. New York: Bantam, 1975.

Deichmann, Ute. *Biologen Unter Hitler: Vertreibung, Karrieren, Forschung*. Introduction by Benno Müller-Hill. Frankfurt: Campus Verlag, 1992. Translation forthcoming, Harvard University Press.

Dick, Jutta, and Marina Sassenberg, eds. *Jüdische Frauen im 19. und 20. Jahrhundert: Lexikon zu Leben und Werk*. Reinbek: Rowohlt, 1993.

Dresden, Max. *H. A. Kramers: Between Tradition and Revolution*. New York: Springer Verlag, 1987.

Eklund, Sigvard. "Forskningsinstitutet för Atomfysik 1937–1987." In Per Carlson, ed., *Fysik i Frescati: Föredrag från Jubileumskonferens den 23 Oktober 1987*. Stockholm: Manne Siegbahn Institute, 1989.

Ernst, Sabine, ed. *Lise Meitner an Otto Hahn: Briefe aus den Jahren 1912 bis 1924, Edition und Kommentierung*. Stuttgart: Wissenschaftliche Verlagsgesellschaft, 1992.

Ewald, P. P. "Max von Laue." *Biog. Mem. Fell. Roy. Soc. Lond.* 6 (1960): 135–156.

Fermi, Laura. *Atoms in the Family: My Life with Enrico Fermi*. Chicago: University of Chicago Press, 1954.

Fischer, Emil. *Aus Meinem Leben*. Berlin: Julius Springer, 1922.

Flamm, Dieter. "Aus dem Leben Ludwig Boltzmanns." In Roman Sexl and John Blackmore, eds., *Ludwig Boltzmann Gesamtausgabe: Band 8 Internationale Tagung, 5.–8. September 1981 Ausgewählte Abhandlungen*. Graz: Akademische Druck u. Verlagsanstalt, 1982. 21–56.

Forkl, Martha, and Elisabeth Koffmahn, eds. *Frauenstudium und Akademische Frauenarbeit in Österreich*. Wien: Wilhelm Braumüller, 1968.

Fraenkel, Joseph, ed. *The Jews of Austria: Essays on their Life, History and Destruction*. London: Vallentine, Mitchell, 1967.

Franck, James. "Max von Laue (1879–1960)." *Yearbook, Amer. Phil. Soc. Biog. Mem.* (1960): 155–159.

Frank, Philipp. *Einstein: His Life and Times*. London: Jonathan Cape, 1948.

Friedman, Robert Marc. "Karl Manne Georg Siegbahn." *Dict. Sci. Biog.*, Supplement 2. Ed. Frederic L. Holmes. New York: Charles Scribner, 1990. 18: 821–826.

———. "Text, Context, and Quicksand: Method and Understanding in Studying the Nobel Science Prizes." *Hist. Stud. Phys. Sci.* 20, no. 1 (1989): 63–77.

Frisch, O. R. "The Discovery of Fission: How It All Began." *Physics Today* 20, no. 11 (November 1967): 43–48.

———. "The Interest Is Focussing on the Atomic Nucleus." In S. Rozental, ed., *Niels Bohr: His Life and Work as Seen by His Friends and Colleagues*. Amsterdam: North Holland/New York: John Wiley, 1967. 137–148.

———. "Lise Meitner." In Charles Gillespie, ed., *Dict. Sci. Biog.* New York: Scribners, 1974. 9:260–263.

———. "Lise Meitner, 1878–1968." *Biog. Mem. Fell. Roy. Soc. Lond.* 16 (1970): 405–420.

———. *What Little I Remember*. Cambridge: Cambridge University Press, 1979.

Gay, Peter. *Weimar Culture: The Outsider as Insider.* New York: Harper & Row, 1968.

Gerlach, Walther. *Otto Hahn: Ein Forscherleben unserer Zeit.* Ed. Dietrich Hahn. Stuttgart: Wissenschaftliche Verlagsgesellschaft, 1984. Grosse Naturforscher Band 45.

Glum, Friedrich. *Zwischen Wissenschaft, Wirtschaft und Politik.* Bonn: H. Bouvier Verlag, 1964.

Goudsmit, Samuel A. *Alsos.* New York: Henry Schuman, 1947.

Gowing, Margaret. *Britain and Atomic Energy, 1939–1945.* London: Macmillan/ New York: St. Martin's Press, 1964.

Graetzer, H. G. and D. L. Anderson. *The Discovery of Nuclear Fission: A Documentary History.* New York: Van Nostrand Reinhold, 1971.

Grieser, Dietmar. "Im Schatten der Bombe: Lise Meitner 1878–1968." In *Köpfe.* Vienna: Österreichischer Bundesverlag, 1991.

Hahn, Dietrich, ed. *Otto Hahn, Begründer des Atomzeitalters: Eine Biographie in Bilden und Dokumenten.* München: List Verlag, 1979.

——. *Otto Hahn, Erlebnisse und Erkenntnisse.* Düsseldorf: Econ Verlag, 1975.

——. *Otto Hahn: Leben und Werk in Texten und Bildern.* Frankfurt: Insel, 1988.

Hahn, Otto. *Mein Leben.* München: Bruckmann, 1968.

——. *A Scientific Autobiography.* Trans. and ed. Willy Ley. London: MacGibbon & Kee, 1967.

——. *My Life.* Trans. Ernst Kaiser and Eithne Wilkins. New York: Herder and Herder, 1968.

Hartshorne, Edward Yarnall, Jr. *The German Universities and National Socialism.* London: George Allen & Unwin, 1937.

Heilbron, J. L. *The Dilemmas of an Upright Man: Max Planck as Spokesman for German Science.* Berkeley, Los Angeles, and London: University of California Press, 1986.

——. *H. G. J. Moseley: The Life and Letters of an English Physicist, 1887–1915.* Berkeley, Los Angeles, and London: University of California Press, 1974.

——. "Lectures on the History of Atomic Physics 1900–1922." In Heilbron, *History of Twentieth-Century Physics.* New York: Academic Press, 1977.

——. "The Scattering of α and β Particles and Rutherford's Atom." *Archive for History of the Exact Sciences* 4. Berlin: Springer Verlag, 1967–1968.

Heilbron, J. L., and Robert W. Seidel. *Lawrence and His Laboratory: A History of the Lawrence Berkeley Laboratory.* Vol. 1. Berkeley, Los Angeles, and Oxford: University of California Press, 1989.

Heisenberg, Elisabeth. *Inner Exile: Recollections of a Life with Werner Heisenberg.* Trans. S. Cappellari and C. Morris. Boston: Birkhäuser, 1984.

Heisenberg, Werner. "Gedenkworte für Otto Hahn und Lise Meitner." *Orden pour le mérite für Wissenschaft und Künste, Reden und Gedenkworte* 9 (1968–1969): 111–119.

Herneck, Friedrich. *Bahnbrecher des Atomzeitalters: Grosse Naturforscher vom Maxwell bis Heisenberg.* Berlin: Buchverlag der Morgen, 1965.

———. "Zum wissenschaftlichen Wirken von Otto Hahn und Lise Meitner im Chemischen Institut der Berliner Universität." *Wissenschaftliche Zeitschrift der Humboldt-Universität zu Berlin* 16 (1967): 833–836.

Hermann, Armin. *Die Neue Physik, Der Weg in das Atomzeitalter.* München: Heinz Moos Verlag, 1979. Trans. David C. Cassidy, *The New Physics: The Route into the Atomic Age. In Memory of Albert Einstein, Max von Laue, Otto Hahn, Lise Meitner.* Bad Godesberg: Inter Nationes, 1979.

Herrmann, Günter. "Five Decades Ago: From the 'Transuranics' to Nuclear Fission." *Angew. Chem. Int. Ed. Engl.* 29 (1990): 481–508 (*Angew. Chem.* 102 [1990]: 469–496).

Hilberg, Raul. *The Destruction of the European Jews.* New York: Octagon Books, 1978.

Höflechner, Walter. "Ludwig Boltzmann: Sein Akademischer Werdegang in Österreich." *Mitteilg. Österr. Ges.. Gesch. Naturwiss.* 2 (1982): 43–62.

Hutchison, Sir Kenneth, J. A. Gray, and Sir Harrie Massey. "Charles Drummond Ellis, 1895–1980." *Biog. Mem. Fell. Roy. Soc. Lond.* 27 (1981): 199–233.

Jensen, Carsten. "A History of the Beta Spectrum and Its Interpretation, 1911–1934." Ph.D. dissertation, Niels Bohr Institute, University of Copenhagen, 1990.

Johnson, Jeffrey Allan. *The Kaiser's Chemists: Science and Modernization in Imperial Germany.* Chapel Hill: University of North Carolina Press, 1990.

Johnston, William M. *The Austrian Mind: An Intellectual and Social History.* Berkeley, Los Angeles, and London: University of California Press, 1972.

Jungnickel, Christa, and Russell McCormmach. *Intellectual Mastery of Nature: Theoretical Physics from Ohm to Einstein.* Vol. 2: *The Now Mighty Theoretical Physics 1870–1925.* Chicago: University of Chicago Press, 1986.

Karlik, B., and E. Schmid. *Franz S. Exner und sein Kreis.* Wien: Verlag der Österreichen Akademie der Wissenschaften, 1982.

Kerber, Gabriele, Auguste Dick, and Wolfgang Kerber. *Dokumente, Materialien und Bilder zur 100. Wiederkehr des Geburtstages von Erwin Schrödinger.* Wien: Fassbaender, 1987.

Kerner, Charlotte. *Lise, Atomphysikerin: Die Lebensgeschichte der Lise Meitner.* Weinheim: Beltz, 1986.

Kirchhoff, Arthur, ed. *Die Akademische Frau.* Berlin: Hugo Steinik Verlag, 1897.

Krafft, Fritz. "An der Schwelle zum Atomzeitalter: Die Vorgeschichte der Entdeckung der Kernspaltung im Dezember 1938." *Ber. Wissenschaftsgesch.* 11 (1988): 227–251.

———. *Im Schatten der Sensation: Leben und Wirken von Fritz Strassmann.* Weinheim: Verlag Chemie, 1981.

———. "Internal and External Conditions for the Discovery of Nuclear Fission by the Berlin Team." In William R. Shea, ed., *Otto Hahn and the Rise of Nuclear Physics*. Dordrecht: D. Reidel, 1983. 135–165.

———. "Lise Meitner 7.XI.1878–27X.1968." In Willi Schmidt and Christopher J. Scriba, eds., *Frauen in den exakten Naturwissenschaften: Festkolloquium zum 100. Geburtstag von Frau Dr. Margarethe Schimank (1890–1983)*. Stuttgart: Franz Steiner Verlag, 1990. 33–70.

———. "Lise Meitner und ihre Zeit: Zum hundertsten Geburtstag der bedeutende Naturwissenschaftlerin." *Angew. Chem.* 90 (1978): 876–892. *Angew. Chem. Intl. Ed. Engl.* 17 (1978): 826–842.

———. "Otto Hahn, 1879–1968." In Lothar Gall, ed., *Die Grossen Deutschen unserer Epoche*. Berlin: Propyläen Verlag, 1985. 173–185.

Kramish, Arnold. *The Griffin: The Greatest Untold Espionage Story of World War II*. Boston: Houghton Mifflin, 1986.

Kuhn, Heinrich G. "James Franck, 1882–1964." *Biog. Mem. Fell. Roy. Soc. Lond.* 11 (1965): 52–74.

Laue, Max von. "Arnold Berliner, 26.12.1862–22.3.1942." *Naturwiss.* 33 (1946): 17–18.

———. "Mein physikalische Werdegang: Eine Selbstdarstellung." In *Gesammelte Schriften und Vorträge*. Vol. III. Braunschweig: Friedrich Vieweg Verlag, 1961. v–xxxiv.

Lemmerich, Jost, ed. *Die Geschichte der Entdeckung der Kernspaltung: Ausstellungskatalog*. Berlin: Technische Universität Berlin, Universitätsbibliothek, 1988.

———. *Max Born, James Franck: Physiker in ihren Zeit, Der Luxus des Gewissens*. Berlin: Staatsbibliotek Preussicher Kulturbesitz, Ausstellungskataloge 17, 1982.

Levi, Hilde. *George de Hevesy: Life and Work*. Bristol: Adam Hilger, 1985.

Lindqvist, Svante, ed. *Center on the Periphery: Historical Aspects of 20th-Century Swedish Physics*. Canton Mass.: Science History Publications, 1993.

Massey, Sir Harrie, and N. Feather. "James Chadwick, 1891–1974." *Biog. Mem. Fell. Roy. Soc. Lond.* 22 (1976): 11–70.

Meitner, Lise. "Einige Erinnerungen an das Kaiser-Wilhelm-Institut für Chemie in Berlin-Dahlem." *Naturwiss.* 41 (1954): 97–99

———. "Looking Back." *Bull. Atom. Sci.* 20 (November 1964): 2–7.

———. "Max Planck als Mensch." *Naturwiss.* 45 (1958): 406–408.

———. "The Status of Women in the Professions." *Physics Today* 13, no. 8 (August 1960): 16–21.

———. "Wege und Irrwege zur Kernenergie." *Naturwiss. Rdsch.* 16 (1963): 167–169.

Meyer, Stefan. "Zur Erinnerung an die Jugendzeit der Radioaktivität." *Naturwiss.* 35 (1948): 161–163.

Moore, Walter. *Schrödinger: Life and Thought*. Cambridge: Cambridge University Press, 1989.

Nathan, Otto, and Heinz Norden, eds. *Einstein on Peace.* New York: Avenel Books, 1981.

Operation Epsilon: The Farm Hall Transcripts. Berkeley, Los Angeles, and Oxford: University of California Press, 1993.

Pais, Abraham. *Inward Bound: Of Matter and Forces in the Physical World.* Oxford: Clarendon Press, 1986.

———. *Niels Bohr's Times, in Physics, Philosophy, and Polity.* Oxford: Clarendon Press, 1991.

———. *"Subtle Is the Lord . . .": The Science and the Life of Albert Einstein.* Oxford: Oxford University Press, 1982.

Peierls, Rudolf. *Bird of Passage: Recollections of a Physicist.* Princeton: Princeton University Press, 1985.

———. "Otto Robert Frisch, 1 October 1904–22 September 1979." *Biog. Mem. Fell. Roy. Soc. Lond.* 27 (1981): 283–306.

Powers, Thomas. *Heisenberg's War: The Secret History of the German Bomb.* New York: Alfred A. Knopf, 1993.

Przibam, Karl. "Erinnerungen an ein altes physikalisches Institut." In O. R. Frisch, F. A. Paneth, F. Laves, and P. Rosbaud, eds., *Trends in Atomic Physics: Essays Dedicated to Lise Meitner, Otto Hahn, and Max von Laue on the Occasion of Their 80th Birthday.* New York: Interscience, 1959. 2–6.

Rayner-Canham, Marelyne F., and Geoffrey W. Rayner-Canham. *Harriet Brooks, Pioneer Nuclear Scientist.* Montreal and Kingston: McGill-Queen's University Press, 1992.

———. "Pioneer Women in Nuclear Science." *Amer. J. Phys.* 58 (1990): 1036–1043.

Rhodes, Richard. *The Making of the Atomic Bomb.* New York: Simon & Schuster, 1988.

Ringer, Fritz K. *The Decline of the German Mandarins: The German Academic Community, 1890–1933.* Cambridge: Harvard University Press, 1969.

Rossiter, Margaret. *Women Scientists in America: Struggles and Strategies to 1940.* Baltimore: Johns Hopkins University Press, 1982.

Rozental, S., ed. *Niels Bohr: His Life and Work as Seen by His Friends and Colleagues.* Amsterdam: North Holland/New York: John Wiley, 1967.

Rutherford, Sir Ernest, James Chadwick, and C. D. Ellis. *Radiations from Radioactive Substances.* Cambridge: Cambridge University Press, 1930. Reissued 1951.

Schiemann, Elisabeth. "Freundschaft mit Lise Meitner." *Neue Evangelische Frauenzeitung* 3, no. 1 (Januar/Februar 1959).

Schorske, Carl E. *Fin de Siècle Vienna: Politics and Culture.* New York: Vintage, 1981.

Segrè, Emilio. *Enrico Fermi, Physicist.* Chicago: University of Chicago Press, 1962.

———. *From X-Rays to Quarks: Modern Physicists and Their Discoveries.* San Francisco: W. H. Freeman, 1980.

———. *A Mind Always in Motion: The Autobiography of Emilio Segrè*. Berkeley, Los Angeles, and Oxford: University of California Press, 1993.

Segrè, Emilio, ed. *Enrico Fermi: Collected Papers (Note e Memorie)*. Chicago: University of Chicago Press, 1962 Accademia Nazionale dei Lincei, Roma.

Stadler, Friedrich, ed. *Vertriebene Vernunft II: Emigration und Exil österreichischer Wissenschaft*. Wien: Jugend und Volk, 1988.

Stoltzenberg, Dietrich. *Fritz Haber: Chemiker, Nobelpreisträger, Deutscher, Jude*. Weinheim: VCH Verlagsgesellschaft, 1994.

Strassmann, Fritz. *Kernspaltung: Berlin Dezember 1938*. Mainz: Privatdruck, 1978.

Stuewer, Roger H. "Bringing the News of Fission to America." *Physics Today* 38, no. 10 (October 1985): 48–56.

———. "Gamow's Theory of Alpha-Decay." In Edna Ullmann-Margalit, ed., *The Kaleidoscope of Science*. Dordrecht: D. Reidel, 1986. 147–186.

———. *The Compton Effect: Turning Point in Physics*. New York: Science History Publications, 1975.

———. "Mass-Energy and the Neutron in the Early Thirties." *Science in Context* 6 (1993): 195–238.

———. "The Nuclear Electron Hypothesis." In William R. Shea, ed., *Otto Hahn and the Rise of Nuclear Physics*. Dordrecht: D. Reidel, 1983. 19–67.

———. "The Origin of the Liquid-Drop Model and the Interpretation of Nuclear Fission." *Perspectives on Science* 2 (1994): 76–129.

Stuewer, Roger H., ed. *Nuclear Physics in Retrospect: Proceedings of a Symposium on the 1930s*. Minneapolis: University of Minnesota Press, 1979.

Trenn, Thaddeus J. "Rutherford and Recoil Atoms: The Metamorphosis and Success of a Once Stillborn Theory." *Hist. Stud. Phys. Sci.* 6 (1975) 513–547.

Turner, Louis A. "Nuclear Fission." *Rev. Mod. Phys.* 12 (1940): 1–29.

Walker, Mark. *German National Socialism and the Quest for Nuclear Power, 1939–1949*. Cambridge: Cambridge University Press, 1989.

———. "Legenden um die Deutsche Atombombe." *Vierteljahrhefte für Zeitgeschichte* 38, no. 1 (1990): 45–74.

Watkins, Sallie A. "Lise Meitner, 1878–1968." In Louise S. Grinstein, Rose K. Rose, and Miriam H. Rafailovich, eds., *Women in Chemistry and Physics: A Biobibliographic Sourcebook*. Westport, Conn.: Greenwood Press, 1993. 393–402.

Weart, Spencer R. "The Discovery of Fission and a Nuclear Physics Paradigm." In William R. Shea, ed., *Otto Hahn and the Rise of Nuclear Physics*. Dordrecht: D. Reidel, 1983. 91–133.

Weisskopf, Victor. *The Joy of Insight: Passions of a Physicist*. New York: Basic Books, 1991.

Willstätter, Richard. *From My Life: The Memoirs of Richard Willstätter*. Trans. Lilli S. Hornig, ed. in orig. German Arthur Stoll. New York: W. A. Benjamin, 1965.

Zweig, Stefan. *The World of Yesterday*. New York: Viking Press, 1943. Reprint Lincoln: University of Nebraska Press, 1964.

INDEX